INFORMATION TRANSMISSION, MODULATION, AND NOISE

McGraw-Hill Series in Electrical Engineering

Consulting Editor
Stephen W. Director, Carnegie–Mellon University

Networks and Systems
Communications and Information Theory
Control Theory
Electronics and Electronic Circuits
Power and Energy
Electromagnetics
Computer Engineering and Switching Theory
Introductory and Survey
Radio, Television, Radar, and Antennas

Communications and Information Theory

Consulting Editor
Stephen W. Director, Carnegie–Mellon University

INFORMATION TRANSMISSION, MODULATION, AND NOISE
A Unified Approach to Communication Systems

Third Edition

Mischa Schwartz

*Professor of Electrical Engineering
and Computer Science
Columbia University*

McGraw-Hill Book Company

New York   St. Louis   San Francisco   Auckland   Bogotá   Hamburg
Johannesburg   London   Madrid   Mexico   Montreal   New Delhi
Panama   Paris   São Paulo   Singapore   Sydney   Tokyo   Toronto

INFORMATION TRANSMISSION, MODULATION, AND NOISE

1234567890 DODO 89876543210

This book was set in Times Roman by Bi-Comp, Inc.
The editor was Frank J. Cerra;
The production supervisor was Donna Piligra.
The cover was designed by Robin Hessel.
R. R. Donnelley & Sons Company was printer and binder.

Library of Congress Cataloging in Publication Data

Schwartz, Mischa.
    Information transmission, modulation, and noise.

    (McGraw-Hill series in electrical engineering)
    Includes bibliographical references and index.
        1. Telecommunication. 2. Digital communications.
3. Modulation (Electronics) 4. Electronic noise.
I. Title. II. Series.
TK5101.S3 1980   621.38   80-11555
ISBN 0-07-055782-9

To my son
David

CONTENTS

Chapter 5 Performance of Communication Systems: Limitations Due to Noise   317

Chapter 6 Statistical Communication Theory and Digital Communications   453

PREFACE

This book has again undergone a complete revision for this third edition. The ever-growing emphasis in communications on digital systems has dictated the introduction of even more material on digital communications in this edition. The more classical concepts of modulation theory still receive a thorough and comprehensive treatment, however, in keeping with their widespread utilization in practice.

New sections have been added in Chapter 3, on delta modulation, duobinary and partial response signaling, and time-division multiplexing of digital signals. New material on nonuniform signal quantization has been added. New application examples have been included.

Chapter 4 now includes new material on digital signal constellations, QAM transmission, and the basic concepts of modem design. (The new material on data multiplexers and modems should make the book particularly useful to students, educators, and practicing professionals interested in data transmission and computer communications.) New sections on frequency-division multiplexing have been added in Chapter 4, with examples drawn from telephony and satellite communications. Current satellite multiplexing techniques are described in detail.

New material added to Chapter 5 on the performance evaluation of communication systems with noise as a disturbing factor includes the discussion of several current space missions, and signal-to-noise calculations for satellite communication systems. A section on thermal noise and the calculation of thermal noise spectral density in Chapter 5 strengthens the study of noise in linear systems and enables these calculations to be carried out using real numbers.

Most of Chapter 6 is new. This chapter incorporates new material on digital signaling over the gaussian white noise channel, with application to the performance of QAM and M-ary orthogonal signal constellations. A completely revised and enlarged section on error correction and detection has been added.

Chapter 2 has been substantially revised. The material on probability that previously appeared in Chapter 5 has been moved to the Appendix. This enables the discussion of Chapters 5 and 6 to proceed in a smoother, more logical fashion. Those readers lacking knowledge of probability theory will find that the material of the Appendix provides a detailed, comprehensive introduction to the subject, complete with problems. The Appendix can also be used by those readers requiring a review of the subject.

The book in its overall format provides a complete treatment of modern communication systems. It assumes no prior knowledge of communication systems, yet takes the reader, by the end of the book, to up-to-date developments in digital communications and communication theory. As was the case with the two previous editions, the material progresses gradually from the simpler ideas to the more complex. Theoretical concepts are explained in terms of concrete, real-life examples. This blend of theory and practice, and the introduction of examples of real systems where appropriate, were, in fact, the hallmarks of the previous editions. These features have been expanded for this edition. Applications drawn from the telephony, space, and satellite communications fields appear throughout the book. Examples of up-to-date commercial systems from these three application areas are included throughout.

With appropriate selection of material, the book should lend itself to various types of courses. Thus, in those schools allowing a full year for the study of modern communication systems, essentially the entire book can be studied. Alternatively, Chapters 1 to 4 could serve as the text for a one-semester comprehensive introduction to the field. Portions of Chapters 3 and 4, plus Chapters 5 and 6, could serve as a one-semester introduction to digital communications, or provide the text material for an introductory course in communication theory.

At Columbia University we offer a two-semester sequence at the senior level that covers most of the material in the book. The first semester, labeled Introduction to Communication Systems, covers Chapters 1 and 2, portions of Chapter 3, Chapter 4, and portions of Chapter 5. Students entering this course have had some Fourier analysis at the junior level, so that Chapter 2 is gone over rather quickly. They take a course on probability theory concurrently with this course, so that the material in Chapter 5 requiring probability theory comes after they have had the necessary introduction to the subject and in fact provides much-welcomed application of probability and statistical concepts. Those schools that do not require probability theory as either a prerequisite or a corequisite could cover more of the topics in Chapter 3 and defer Chapter 5 until a later course. Alternatively, the Appendix could be used to provide the necessary background in probability theory, as noted previously, with correspondingly less material covered from the main body of the text.

The second semester of our two-semester communication sequence provides an introduction to digital communication systems. In addition to those students who have completed the first-semester course, it is taken as well by first-year graduate students who have had an equivalent introductory communications systems course elsewhere. The course is therefore handled somewhat independently of the first course. It reviews some of the material in Chapters 3 and 5, and then spends most of its time on Chapter 6. Time permitting, selected topics from the current literature on digital communications are introduced at the end of the semester.

ACKNOWLEDGMENTS

The author has again profited immensely in the writing of this book from comments, criticisms, and suggestions provided by many readers, and colleagues at various universities. He would particularly like to acknowledge the support of Professor T. E. Stern, who joined in teaching the two-semester sequence mentioned above, and Professor Y. Tsividis, who participated in teaching the first introductory course. Help with the application examples was obtained from a number of friends and colleagues in industry. Particular thanks are due to M. Robert Aaron of Bell Laboratories, Stanley Butman of the Jet Propulsion Laboratory, G. David Forney of Codex Corp., and Adam Lender of GTE–Lenkurt.

Mischa Schwartz

INFORMATION TRANSMISSION, MODULATION, AND NOISE

INTRODUCTION TO INFORMATION TRANSMISSION

This book is devoted to a study of communication systems, or systems used to transmit information. In the course of the development we shall emphasize the limitations imposed on the information transmitted by the system through which it was passed and shall attempt some comparison of different systems on the basis of information-handling capabilities.

A complete system will generally include a transmitter, a transmission medium over which the information is transmitted, and a receiver which produces at its output a recognizable replica of the input information. In most communication work information transmission is closely related to the modulation or time variation of a particular sinusoidal signal called the *carrier*. A typical system diagram would thus appear as in Fig. 1-1.

The transmitter generally includes the source of the information to be transmitted—voice signals, TV signals, computer printouts, telemetry data in the case of space probes, or perhaps telemetry data transmitted from a remote automatically operated plant to a central control station.

As the signals traverse the transmission medium (or *channel* as it is most often referred to), they are distorted; noise and interfering signals are added, and it becomes a major task to correctly interpret the signals as finally received at the desired destination.

It is the purpose of this book to discuss in detail the effect of transmitting signals of various types over distorting channels—whether they be cables or wires for telephonic communication, space for radio transmission, or more exotic channels such as water (underwater communications) or the earth (seismic communications). We shall consider the limitations on information trans-

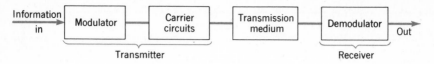

Figure 1-1 Communication system.

mission introduced by transmission distortion and additive noise, using simple models for these effects, where possible, in this first treatment. We shall compare various ways of transmitting the desired information, as well as various techniques used at the receiver, to see how one goes about improving signal transmission in the presence of perturbing influences.

Although the stress in this book is placed on the type of communication systems shown in Fig. 1-1, much of the material is also particularly appropriate to the broad area of *signal processing*. Here one is concerned less with the aspects of transmitting signals and more with the problem of *interpreting signals* once received. These signals could very well be the types of signals mentioned earlier, but might also include such types of data as biological signals (EEG and EKG data, for example), computer printouts of the results of scientific experiments, stock-market quotations, weather data, etc., etc. The basic problem here would be the analysis of the received data, that is, extracting the desired and pertinent information from the obscuring factors (or ''noise'') usually present. The techniques developed for the processing of these types of *random signals,* most commonly digital in form, draw heavily on principles developed in studies of information transmission.[1]

1-1 TYPICAL DIGITAL COMMUNICATION SYSTEM DESIGN

To develop an overview to the general problem of information transmission and to pinpoint more specifically some of the problems encountered in communications, we shall detail in this section the problem of transmitting a typical digital data message from one point to another. Many of the problems raised will in turn be developed and studied in much more detail throughout this book. We consider a digital (or *binary*) data message since this is rapidly becoming the most common form of signal transmission, whether the message is directly digital in form, as in the case of computer printouts, or whether converted to a digital format, as is true with telemetered data and much of telephonic voice communication. The binary data message we introduce in this section will then be utilized, as a particularly simple form of signal, throughout this book.

By a binary message or signal we mean a sequence of two types of pulses of known shape, occurring at regularly spaced intervals, as shown in Fig. 1-2.

---

[1] See M. Schwartz and L. Shaw, *Signal Processing,* McGraw-Hill, New York, 1975, for a thorough discussion.

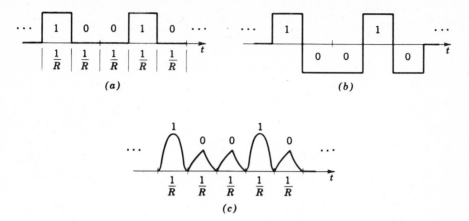

Figure 1-2 Binary signal transmission. (*a*) On-off sequence. (*b*) Polar sequence. (*c*) Arbitrary waveshapes.

Although the shape of the pulses is presumed known in advance, the occurrence of one or the other, say the 1 or 0 in Fig. 1-2, is not known beforehand, and the information carried is actually given by the particular sequence of binary 1's and 0's coming in. We shall most commonly use the rectangular pulse shapes of Fig. 1-2*a* and *b* in this book for simplicity's sake, but other types of signal shape could and are being used as well.

These pulses are shown occurring regularly every $1/R$ seconds, or at a rate of R/second. $1/R$ is commonly referred to as the *binary interval*, and the signal source is said to be putting out R binary digits or *bits* per second (bits from *bi*nary dig*its*). As an example, if $1/R = 10^{-3}$ s, R is 1,000 bits/s. If $1/R = 1$ μs, $R = 10^6$ bits/s.

We now assume that this sequence of binary symbols is to be transmitted to a distant destination. A typical system block diagram, a more detailed version of Fig. 1-1, is shown in Fig. 1-3. The two filters shown, one at the transmitter and the other at the receiver, represent the filtering of the signals either innately present in the system circuitry or purposely introduced as part of the design. The demodulator at the receiver serves to strip away the high-frequency sine-wave modulation introduced at the transmitter modulator. The modulation pro-

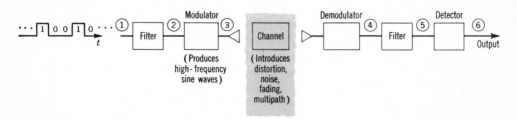

Figure 1-3 Transmission of digital message.

cess is necessary to enable the signals to be effectively radiated into space (or whatever other medium is represented by the channel shown). The purpose of the detector at the receiver is to reproduce, "as best as possible," the original signal sequence representing the digital data to be transmitted.

Some typical waveshapes, corresponding to the numbered points in Fig. 1-3, are shown sketched in Fig. 1-4. Note that the filters cause excessive symbols to overlap into adjacent time slots. If carried too far, these interfere with symbols actually transmitted in the adjacent time slots, leading to confusion in symbol interpretation and possible errors at the system output. *Intersymbol interference* is a significant problem in many data communication systems, being particularly troublesome in the transmission of data via telephone lines. We shall have occasion to discuss this problem further in later chapters, and will indicate some design techniques used to overcome it.

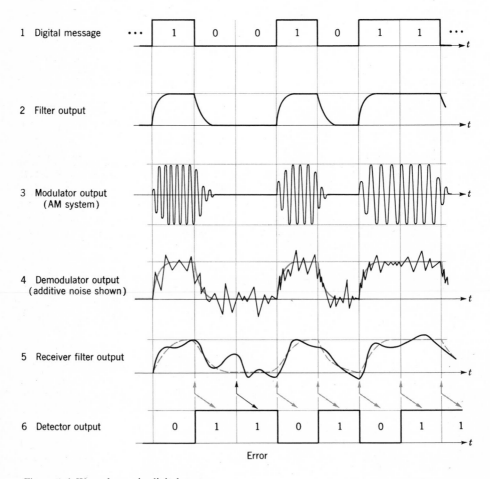

Figure 1-4 Waveshapes in digital system.

The modulator chosen in this example happens to be of the amplitude-modulation (AM) type, in which a sine-wave oscillator adjusts its amplitude to the incoming signals. We could equally well have depicted the output of a frequency-shift-keyed (FSK) modulator, in which the carrier frequency alternates between two frequencies, depending on the symbol coming in (this is the digital version of an FM signal), or a phase-shift-keyed output (PSK), in which the polarity ($\pm$) of the sine wave depends on the incoming signal.

In this example the channel is shown as having introduced noise during transmission, so that the demodulator output represents the sum of signal plus noise waveshapes. Note that the noise introduced tends to obscure the signal if the two are comparable in magnitude. Some channels (the telephone line, for example) introduce signal distortion as well (although this effect could be modeled by incorporating it in the filter following the demodulator). Others introduce signal fading, in which the received signal amplitude is found to fluctuate randomly, or so-called *multipath* effects, in which the radiated transmitter energy for one signal symbol, following several alternate paths to the receiver, appears at the receiver as a sequence of received symbols. Examples of this latter type of randomly fading channel include short-wave radio transmission via the ionosphere, as well as underwater and seismic communications, among others. For simplicity's sake we shall stress additive noise effects in this book.[2]

The receiver filter serves to eliminate some of the noise, at the expense, however, of further distorting the signal transmitted. This is shown pictorially in Fig. 1-4, in which the dashed lines of parts (4) and (5) represent the signal term, while the solid lines represent the composite sum of signal and noise.

To finally reproduce the original signal message, the detector must sample the receiver filter output once every bit interval and decide "as best as possible" whether a 1 or 0 was transmitted. It is apparent from part (5) of Fig. 1-4 that the appropriate sampling times, in this example, occur at the end of each bit interval, since it is there that the filtered signals reach their maximum amplitudes. Mistakes can be made, however, with noise obscuring the correct symbol. An example of an error occurring is shown in part (6) of Fig. 1-4.

We have purposely used the words "as best as possible" twice in the discussion above because they play a key role in the overall design of any communication system. In the context of the simple digital communication system we have been describing, these words have a rather precise meaning. It is apparent that we would like to transmit any arbitrary sequence of data symbols with as small a number of errors occurring as is possible, costwise. The fact that errors will occur is apparent. Noise is always introduced in a system (we shall have a great deal to say on this subject later in the book), physical transmitters have power limitations so that one cannot generally increase the signal strength as much as necessary in order to overcome the noise, intersym-

---

[2] See M. Schwartz, W. R. Bennett, and S. Stein, *Communication Systems and Techniques,* McGraw-Hill, New York, 1966, for a comprehensive treatment of such channels.

bol interference is generally a problem with which to reckon, signal fluctuations during transmission give rise to possible errors, etc.

A comprehensive engineering design of a data system such as this one would have to take into account all possible sources of error and try to minimize their effect. This includes appropriate design of the signals at the transmitter. That is: How does one shape the signals or design the transmitter filter? Given various means of transmitting the binary symbols at the required high frequency—the AM technique shown in Fig. 1-4, FSK, PSK, etc., which is most appropriate for the problem at hand, including the particular channel over which transmission will take place? We shall find in discussing this particular question later in the book that there is the usual engineering trade-off between the various modulation techniques. Thus PSK, although generally most effective in terms of conserving power or minimizing errors, is difficult to use over fading channels, and its use introduces severe phase control problems. FSK is generally less effective in the presence of noise and requires wider system bandwidths (a term to be defined in the next chapter), but is more effective over fading channels.

With a specific modulation scheme chosen how does one design the receiver? How does one minimize intersymbol interference and noise? How does one design the detector and the decision-making circuitry associated with it?

These are just a few of the design questions that arise in the actual development of a digital system of the type we have been considering. (We have, for example, ignored timing problems associated with the sequential transmission of 1's and 0's every $1/R$ s. This is particularly important in the design of systems with the transmitter and receiver located thousands and even millions of miles apart, where the receiver must nonetheless always maintain the same timing as the transmitter.)

It is one of the purposes of this book to approach problems such as these systematically, indicating how one does go about designing a particular system. Although we shall stress for simplicity's sake the effects of noise and filtering, since their effect is important in every communication system, some of the other problems noted will be treated as well.

Now assume the system has been "optimumly" designed. We have minimized the number of errors that will occur, on the average, by appropriate design at both transmitter and receiver. How good is it? Is the resultant design appropriate for our needs (i.e., for the particular application for which intended)? This requires a quantitative system evaluation, modeling the presumed effect of the channel, noise introduced, etc., and such an evaluation will be one of the subjects treated in this book. For this purpose we shall find it necessary to introduce statistical concepts and approach the combined problems of system design and evaluation statistically.

If the error rate of the system, the average number of errors occurring per unit time, is too high, more complex system configurations, with their attendant increased cost, may be called for. These include signal coding techniques, sophisticated error detection and correction procedures, etc.

This raises another extremely important question: For a given channel over which we desire to communicate, is it possible to keep improving the system performance, i.e., reduce its error rate, as much as we like, with appropriate increased complexity of the system design? This is obviously a basic question in all communications design, for if the answer is "no," there is no sense in even trying to design more complex systems.

To answer this question we choose to phrase it in somewhat more precise fashion. The question is now: With the rate of transmission of binary symbols, R bits/s, fixed, as is the power available at the transmitter, is it possible, for a given channel, to reduce the error rate as much as desired (with appropriate system design and complexity)? The answer, as first established by Claude Shannon in 1948 to 1949 in a monumental piece of work,[3] is "yes," with one qualification. That is, his work was restricted to the study of a channel introducing noise only (intersymbol interference and fading effects were not included). This has since been extended to a few other channels by other investigators.

Shannon found that the chance of an error occurring may ideally be reduced as low as one likes by appropriate coding of the incoming signals, providing the binary signaling rate R, in bits per second, is less than a specified number determined by the transmitter power, channel noise, and channel response time or bandwidth (this latter concept will be discussed in detail in the next chapter). If one tries to push too many bits per second over the channel, the errors begin to mount up rapidly. The maximum rate of transmission of signals over the channel is referred to as the *channel capacity*.

Since the channel capacity is obviously an important concept in systems design (one can determine from this whether it pays to develop more complex systems), we shall devote some time to exploring its significance. The remainder of this chapter is devoted to a qualitative discussion of this concept, indicating why one physically expects a channel capacity to exist. There is a semantic difficulty that must be mentioned, however. We choose to use the word *channel* here most often to denote the physical medium over which transmission takes place. Many authors include in the channel various portions of the transmitter and receiver as well. Generally the meaning is clear from the context of the discussion. Shannon's channel capacity actually refers to this more general class of channels. To avoid confusion, we shall use the term *system capacity*, and, as we explain in the next section, all portions of the transmitter and receiver, as well as the physical channel, contribute to determining the capacity.

Since the maximum rate of transmission of binary symbols over a given channel is fixed, one would like to know what the binary rate R of a given signal source is. This is particularly true in the case of sources that are initially nonbinary in nature and that must be converted to binary symbols. These

[3] C. E. Shannon, "A Mathematical Theory of Communication," *Bell System Tech. J.*, vol. 27, pp. 379–423, July 1948; pp. 623–656, October 1948. C. E. Shannon, "Communication in the Presence of Noise," *Proc. IRE*, vol. 37, pp. 10–21, January 1949.

include speech, TV or facsimile, telemetry signals, etc. The concept of information content of a given message, in terms of the bits needed to represent it, is thus also explored in this chapter.

One last word before we close this section. We have emphasized digital communications here because of its relative simplicity as well as its technological importance. The questions raised here hold as well for other types of communication systems. Thus one is always interested in "optimizing" system performance. The major difficulty arising in much of communication system design and evaluation, however, is that no simple criterion exists as to "optimum performance." How does one determine whether a particular speech signal is reproduced as effectively as possible? When does a TV picture have an "optimum" appearance? It is apparent that for continuously varying (analog) signals such as these, simple performance measures may be hard to justify, since much of the essential system evaluation can only be done subjectively. Yet the techniques of minimizing noise or of appropriate signal filtering that we shall discuss are significant in these types of systems as well as in strictly digital ones, and we will be able to establish some measure of their performance.

1-2 INFORMATION AND SYSTEM CAPACITY

As we mentioned in the previous section, the information content of a message to be transmitted must be established in order to determine whether or not the message may be transmitted over a given channel. By the information content we mean the number of binary symbols that will ultimately be necessary, on the average, to transmit it. Although we may in actuality not transmit binary symbols, but choose to transmit a more complex signal pattern, we prefer to normalize all signal messages to their binary equivalents for simplicity's sake. (This conversion to a binary equivalent will be discussed further both in this chapter and in chapters that follow.)

Since all communication systems transmit *information* in one form or another and since we desire some measure of the information content of messages to be transmitted, it is important first to establish some measure of what is actually meant by the concept *information*. Although a precise mathematical definition can be set up for this concept, we shall rely on our intuitive sense in this introductory text.

Consider a student attending a class in which the teacher spends the entire time whistling one continuous note. Obviously, attending such a class would be a waste of time. What could one possibly learn from the one note? (Even if it were important for the student to repeat the sound exactly, he would be better off staying at home listening to a recording.) Perhaps in the next class the teacher chooses to devote the entire hour to a reading, word for word, from the text: no time for questions, no pauses, no original thoughts. Again why come to class (aside from the irrelevant fact that many persons might not then take the opportunity to read the book for themselves)?

What is the point to these hypothetical and obviously made-up stories? A student comes to class to receive *information*. That is, the teacher and student are in class to discuss new material or, at least, review old material in a new way.

The words and phrases used should thus be *changing* continuously; they should, in most cases, be *unpredictable* (otherwise—why come to class?).

The key phrase here is *unpredictable* change. If information is to be conveyed, we must presumably have sounds or, more generally, signals changing unpredictably with time. A continuous trilling of one note conveys no information to you. If the note is varied in a manner that you can interpret, however, the "signal" begins to convey meaning and information.

The binary sequence of the previous section was one simple example of a signal to be transmitted that is changing unpredictably in time. There the specific sequences of 1's and 0's were unknown beforehand and corresponded to the message to be transmitted.

So the transmission of information is related to signals changing with time, and changing in an unpredictable way. (For a well-known melody or old story conveys no new information although made up of changing notes or words.)

Why is it so important to stress these points? As noted previously, if we, as engineers, are to design systems to transmit information and are interested in the best possible type of system given practical equipment and a limited budget, we must know (at least intuitively) what it is we are transmitting and the effect of the system on this quantity.

To see how these concepts fit into our work in communication, consider the voltage-time diagram of Fig. 1-5. Assume that we have an interval of time T seconds long in which to transmit information and a maximum voltage amplitude (because of power limitations) that we can use. (In Fig. 1-5, T is 10 s, and the maximum voltage is 3 V.) A natural question to ask is: How much information can we transmit in this interval? Can we put a tag on the amount, and how does it depend on our system? (Note that the system has already introduced one limitation—that of power.)

The next question is: Why a limit to the amount of information? If information transmission is related to signals changing unpredictably with time, why not just change the signal as rapidly as we like and over as many subdivisions of

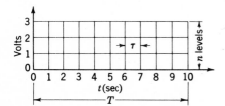

Figure 1-5 Voltage-time diagram.

the maximum amplitude as we like? This would imply increasing the information content indefinitely.

We deal with physical systems, however, and these systems do not allow us to increase indefinitely the rate of signal change and to distinguish indefinitely many voltage amplitudes, or levels:

1. All our systems have energy-storage devices present, and changing the signal implies changing the energy content. There is a limit on the rate of doing this determined by the particular system.
2. Every system provides inherent (even if small) variations or fluctuations in voltage, or whatever parameter is used to measure the signal amplitude, and we cannot subdivide amplitudes indefinitely. These unwanted fluctuations of a parameter to be varied are called noise. This noise is exactly the noise we noted as being introduced during transmission over a channel in the previous section. The channels discussed there are all examples of physical systems of the type with which we deal.

There is thus a *minimum* time τ required for energy change and a minimum detectable signal-amplitude change. As an example, τ is given as 1 s in Fig. 1-5. If the inherent voltage fluctuations of the system may be assumed as an example to vary within ± 1 V most of the time, the minimum detectable voltage change due to the signal is 1 V. With a maximum voltage amplitude of 3 V, there are thus four detectable levels of signal (0 voltage being assumed to be a possible signal value). For if the signal were to change by less than 1 V, it could not be distinguished from the undesired noise fluctuations introduced by the system.

If the "amount of information" transmitted in T seconds is related to the number of different and distinguishable signal-amplitude combinations we can transmit in that time—as we might intuitively feel to be the case—it is apparent that the information capacity of the system is limited. It is exactly these arguments, phrased in a much more quantitative manner, that were used by Shannon in developing his capacity expression referred to in the previous section. (Recall that the word *channel* was used in that section to refer to the physical medium over which signals are transmitted. Limitations on signal-time response may be produced anywhere in the systems of Figs. 1-1 or 1-3. The filters shown in Fig. 1-3, for example, definitely introduce specific time responses. Also, noise is often introduced at the receiver as well as during transmission. We shall thus refer here to *system capacity,* considering the effect of the overall system, rather than *channel capacity,* as commonly done in the literature.)

The system capacity, or maximum rate at which it can transmit information, should be measurable in terms of τ and n, the number of distinguishable amplitude levels. (Both limitations may be produced again anywhere in the system of Fig. 1-1, but we are speaking of the effect of the overall system.)

We can derive a more quantitative measure of system capacity in the following manner. We assume that the information transmitted in the 10-s interval of Fig. 1-5 is directly related to the number of different signal-amplitude combinations in that time. For example, two different signals that might be transmitted are shown in Fig. 1-6. They differ over the first two intervals and have the same amplitudes over the remaining 8 s. How many such combinations can we specify? There are four different possibilities in the first interval, and corresponding to each such possibility there are four more in the second interval, or a total of $4^2 = 16$ possibilities in two intervals. (The reader should tabulate the different combinations to check this result.) Repeating this procedure, we find that there are 4^{10} combinations of different signal amplitudes in 10 s.

If, instead of 4, we had had n levels and, instead of 1-s, τ-s intervals, the number of combinations in T seconds would have been

$$n^{T/\tau}$$

Under our basic assumption the information transmitted in T seconds is related to this number of signal combinations. We might feel intuitively, however, that information should be proportional to the length of time of transmission. Doubling T (10 s here) should double the information content of a message. The information content can be made proportional to T by taking the logarithm of $n^{T/\tau}$, giving us

$$\text{Information transmitted in } T \text{ seconds} \propto \frac{T}{\tau} \log n \qquad (1\text{-}1)$$

The proportionality factor will depend on the base of logarithm used. The most common choice is the base 2, or

$$\text{Information} = \frac{T}{\tau} \log_2 n \qquad (1\text{-}2)$$

The unit of information defined in this manner is the *bit* (mentioned earlier and to be explained in the next section). The information content of the 10-s strip of Fig. 1-5 is, for example,

$$10 \log_2 4 = 20 \text{ bits}$$

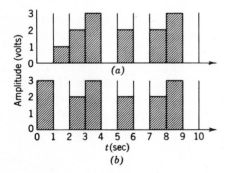

(a)

(b)

Figure 1-6 Two different signals.

A 5-s strip would have 10 bits of information. If there had been only two possible voltage levels (say 0 and 1), the information conveyed in 10 s would have been 10 bits.

The system capacity can be defined as the maximum *rate* of transmitting information. From Eq. (1-2) this is simply

$$C = \frac{\text{information}}{T} = \frac{1}{\tau} \log_2 n \tag{1-3}$$

and the units are given in bits per second.

System capacity is thus inversely proportional to the minimum interval τ over which signals can change and proportional to the logarithm of n.

We shall show in Chap. 2, in reviewing some simple concepts of networks, that there is an intimate inverse relationship between time response and frequency response. This will enable us to relate information transmitted and system capacity to the system "bandwidth."

These two parameters of system behavior, τ (or its inverse, bandwidth) and n (or, as we shall see, the signal-to-noise ratio in a system), are basic in any study of communication systems. Much of the material of this book will thus be devoted to a study of the time (or frequency) and noise characteristics of different networks and the frequency-noise characteristics of various practical communication systems.

1-3 BINARY DIGITS IN INFORMATION TRANSMISSION

The information content of a signal was defined in the previous section by Eq. (1-2),

$$\text{Information} = \frac{T}{\tau} \log_2 n \qquad \text{bits}$$

The use of the logarithm to the base 2 in defining the unit of information can be justified in an alternative and instructive way. Assume that a signal to be transmitted will vary anywhere from 0 to 7 V with any one voltage range as likely as the next. Because of the system limitations described in Sec. 1-2, the signal can be uniquely defined only at the integral voltage values and will not change appreciably over an interval τ seconds long. (The noise fluctuations are assumed to have the same magnitudes on the average as in the example of Sec. 1-2.)

The signal can thus be replaced by a signal of the type shown in Fig. 1-6; during any interval τ s long it will occupy one of eight voltage levels (0 to 7 V), each one equally likely to be occupied.

The process of replacing a continuous signal by such a discrete signal is called the *quantizing process*. A typical signal and its quantized equivalent are shown in Fig. 1-7.

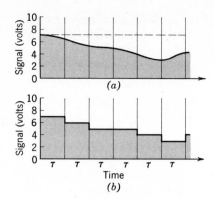

Figure 1-7 Quantization of a signal. (*a*) Original signal. (*b*) Quantized signal.

The signal can of course be transmitted by simply sending the successive integral voltage values as they appear. In any one interval any one of eight different voltages must be sent. The informational content of the signal is thus related to these eight different voltage levels (i.e., one of eight choices).

We ask ourselves, however: Is there another way of sending this information so that fewer than eight numbers are needed completely to specify the signal in any one interval? The informational content will then be assumed equal to the smallest number needed. The answer is yes: the simplest way uniquely to label a particular level and to indicate its selection is by means of a series of yes-no instructions. For this particular type of signal with eight levels three such yes-no instructions are needed.

To indicate the procedure followed, assume that the signal is at 7 V during a particular instant. We first decide whether the proper level lies among the first four or the last four levels. If "yes," we use a designating symbol 1 for each level in the group of four; if "no," a designating symbol 0. In this case, then, levels 0 to 3 are labeled 0, and levels 4 to 7 are labeled 1 (see Fig. 1-8*a*). We can thus immediately reject the 0 levels and concentrate on choosing one of the remaining four. Our area of choice has been reduced considerably.

Again we separate the remaining levels into two parts. That half which does not contain the desired level (7 in this case) is labeled no, or 0; the other yes, or

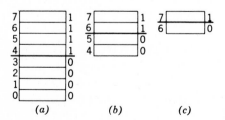

Figure 1-8 Binary selection of a signal. (*a*) First choice. (*b*) Second choice. (*c*) Third choice.

1. This is shown in Fig. 1-8*b*. Again half the levels are eliminated, leaving only 6 and 7. This time we find level 7 uniquely singled out. Note that this method required three consecutive yes-no responses.

Proceeding in a similar way, we could single out any one of the eight possible voltages. They can thus be uniquely identified by means of three 0 or 1 labels. This method of identification is called *binary coding*. A typical identification table would appear as follows:

	Binary coding
7	111
6	110
5	101
4	100
3	011
2	010
1	001
0	000

Instead of transmitting this signal as one of eight different voltage levels, we need transmit only three successive yes-no (voltage–no voltage) voltages during a particular interval. Any one yes-no label is a *bit*. Three bits are thus required to transmit the desired information as to a particular voltage level occupied for the eight-level signal under discussion.

This process of binary coding is the simplest one that can be devised for uniquely tagging a signal. With binary selection only three consecutive numbers, or 3 bits, are needed to transmit the informational content of this particular signal in any one time interval. For 16 levels 4 bits are required, for 32 levels 5 bits, and so on. For n levels $\log_2 n$ bits are required.

If the information in three successive intervals, each containing eight possible levels, is to be transmitted, then 3 bits for each interval, or 9 in all, are required for the signal transmission.

For T/τ intervals and n levels, $T/\tau \log_2 n$ bits must be transmitted. The information content of a signal is defined to be equal to the number of bits, or binary choices, needed for transmission.

1-4 RELATION BETWEEN SYSTEM CAPACITY AND INFORMATION CONTENT OF MESSAGES

In Sec. 1-2 we pointed out that the capacity (or ability) of a system to transmit information depended on the system time response and its ability to distinguish among different levels of a signal.

The capacity of a given system is defined as the maximum amount of information per second (in bits per second) that the system can transmit. The following chapters of this book will be devoted to relating the system limitations on information transmission to bandwidth and system noise properties. Here we take time to consider the relation between system capacity and the information content of signals to be transmitted.

Being able to measure the ability of a particular system to transmit information is not enough. The more basic question in information transmission should perhaps be phrased: Which system or group of systems will have sufficient capacity to transmit a specified class of signals or information-bearing messages? In order to answer this question, we must be able to measure the information content of a signal.

For example, assume that we are interested in transmitting a speech delivered in the English language. In order to choose a system for transmission, we must be able to determine the information content of the speech and the rate at which it is to be transmitted. The system chosen will then obviously have to accommodate the rate of information to be transmitted, or will have to possess a capacity greater than the rate of information transmission desired. The system must thus be "matched" to the class of signals being transmitted.

In order to choose a system or group of systems with the proper information capacity, we first determine or measure the information content of the signals that are to be transmitted. The detailed consideration of this point is in the realm of information theory, and the reader is referred to the ever-growing literature on this subject for thorough discussions.[4] We shall, however, indicate the approach used, and we shall find some of the material of value in our later work.

We have already shown, in Secs. 1-2 and 1-3, that, where different values of a signal are *equally likely* and where the signal appears at discrete voltage levels, the information content of a signal is readily evaluated. It is simply the logarithm of the number of equally likely signal combinations possible in a given interval. (In the example we considered, $n^{T/\tau}$ was the number of equally likely signal combinations in T seconds, since each signal level was as likely to be occupied as any other. The information content of the strip T seconds long was thus $\log_2 n^{T/\tau} = T/\tau \log_2 n$.)

The equally likely case is a very restricted one, however. For example, if we were transmitting a speech in the English language by transmitting the different letters in the successive words uttered, this would correspond to assuming that each letter was equally likely to occur. This assumption is obviously not true, since we know, for example, that the letter e occurs much more frequently than the letter z, or any other letter for that matter. We could guess

[4] Shannon, "A Mathematical Theory of Communication," *op. cit.* See R. G. Gallager, *Information Theory and Reliable Communication,* Wiley, New York, 1968, for a comprehensive treatment.

that a particular letter to come would be an e and be much more sure that we were right than if we had guessed $q, z,$ or u.

But this prior knowledge of the greater chance of e rather than z occurring reduces the information content of the speech being transmitted. For, as we noted in Sec. 1-2, the amount of information transmitted depends on the *uncertainty* of the message. In particular, if e were known to be the only letter occurring in this speech, no information at all would be transmitted, since all uncertainty as to the message would vanish.

Thus, although the different number of signals in any one interval is still the gamut of all letters from a to z, the fact that some occur more frequently than others reduces the information content of the message. In T seconds the different signal combinations possible occur with differing relative frequencies, and the information content of the T-second message is reduced as compared with the equally likely case.

The same considerations obviously hold true in our representation of different signals by different voltage levels in Sec. 1-2. If a 3-V signal were to be the only signal transmitted, it would carry no information and transmission might just as well cease. If the 3-V signal were to be expected more often than any other, a message T seconds long would carry information but the *information content* of the message would be less than if all voltage levels were equally likely. (Those signal combinations containing the 3-V level would be expected to occur more frequently than any of the others.)

The information content of a message thus relates not only to the number of possible signal combinations in the message but also to their *relative frequency of occurrence*. This in turn depends on the source of the message. The information content of a message in the English language depends on a knowledge of the structure of the language and its alphabet: the relative frequency of occurrence of each letter, of different combinations of letters, of word combinations, of sentence combinations, etc. All these structural properties of English affect the different possible signal combinations and their relative frequency of occurrence, and hence the information content of a particular signal.

The decrease in the information content of a message due to unequally likely signals results in the requirement of a correspondingly reduced system capacity for information transmission. Telegraph transmission of messages in the English language has long taken this into account by coding the letter e with the shortest telegraph symbol. This thus reduces the average time required for transmission of a message.

But how do we *quantitatively* measure the information content of a message in this more general case of signals with differing frequencies of occurrence? To study this case, we shall first assume that successive signals (the individual letters in the case of English) are independent of one another and then attempt some further generalization. The assumption of *independence* implies that the occurrence of any one signal does not affect in any way the occurrence of any other signal. In the case of a message in the English language this assumption implies that the occurrence of one letter does not affect the occurrence of any other. (A q coming up could then be followed by an x or a z, as well as by a u. This assumption of independence is thus an oversimplification of a much more complicated situation in the case of English, but it does serve to simplify the analysis.)

To develop a quantitative measure of the information content of a message, we shall first rewrite our result for the equally likely case in a different form. We recall that we showed that $n^{T/\tau}$ was the total possible number of signal combinations in T seconds if each signal lasted τ seconds and there were n possible levels in each interval.

If we were to look at many messages, each T seconds long, we would find that on the average each possible signal combination would occur with a relative frequency of $1/n^{T/\tau}$. For example, with $\tau = 1$ s and $n = 4$, 64 different combinations are possible in an interval 3 s long. The relative frequency of occurrence of each 3-s message would be $\frac{1}{64}$. In 10,000 such 3-s messages there would be approximately 10,000/64 messages of each of the 64 possibilities. The greater the number of 3-s messages we were to look at, the more closely any one signal combination would approach a relative frequency of $\frac{1}{64}$.

The relative frequency of occurrence of any one combination, or *event,* we define to be its probability of occurrence, or, symbolically, P. Thus

$$P \equiv \frac{\text{number of times event occurs}}{\text{total number of possibilities}} \tag{1-4}$$

where the total number of possibilities must be very large compared with the number of possible events (10,000 as compared with 64 in the example just cited) if we are to have an accurate measure of the relative frequency.

For example, if we were interested in the probability of occurrence of a letter in the English alphabet, we would pick letters at random (say, from words on successive pages of a book to ensure independence of choice) and determine the number of times a given letter showed. We would have to pick many more than 26 total letters for our results to be valid, however.

If n possible events are specified to be the n possible signal levels, at any instant, of Sec. 1-2, then $P = 1/n$ for equally likely events. The information carried by the appearance of any one event in one interval is then

$$H_1 = \log_2 n = -\log_2 P \qquad \text{bits/interval} \tag{1-5}$$

Over m intervals of time (an interval is τ sec long) we should have m times as much information, assuming that each signal or event in time is independent. Therefore,

$$H = mH_1 = -m \log_2 P \qquad \text{bits in } m \text{ intervals} \tag{1-6}$$

The information available in T sec is thus ($m = T/\tau$)

$$H = -\frac{T}{\tau} \log_2 P = \frac{T}{\tau} \log_2 n \qquad \text{bits in } T \text{ seconds} \tag{1-7}$$

as in Sec. 1-2.

Now consider the case where the different signal levels (or events) are not equally likely. For the sake of simplicity we first assume just two levels to be transmitted, 0 or 1, the first with probability p, the second with probability q. Then

$$p \equiv \frac{\text{number of times 0 occurs}}{\text{total number of possibilities}} \qquad (1\text{-}8)$$

$$q \equiv \frac{\text{number of times 1 occurs}}{\text{total number of possibilities}} \qquad (1\text{-}9)$$

Since either a 0 or a 1 must always come up, $p + q = 1$. (The number of times 0 comes up plus the number of times 1 comes up equals the total number of possibilities.)

For example, say that the message to be transmitted by this two-level signal device represents the birth of either a boy or a girl in the United States; 1 corresponds to boy, 0 to girl. After counting 1,000,000 births, we find that 480,000 boys and 520,000 girls were born. Then we estimate $p = 0.52$, $q = 0.48$, and $p + q = 1$.

What is now the information content of a particular message consisting of a group of 0's and 1's? Each time a 0 appears, we should gain $-\log_2 p$ bits of information, and each time 1 appears we gain $-\log_2 q$ bits. If p and q are approximately each 0.5, either event occurring (0 or 1) carries almost the same amount of information. The two events are nearly equally likely. This is of course the case in the births of boys and girls in the United States to which we referred. But now assume that $p \gg q$ (0 occurs more frequently, on the average). Since $-\log_2 q \gg -\log_2 p$, the occurrence of a 1, the more *rarely occurring* event, carries *more* information.

This seems to agree with our previous discussion, where we point out that, the greater the uncertainty of an event occurring, the more the information carried. Does this again agree with intuition? We use births as an example once more, but this time consider the case of a family with five sons and no daughters. The father has given up all hope of a daughter, especially since both his family and that of his wife have a long history of a preponderance of male children. The father, waiting expectantly for his wife to give birth again, receives word that his wife has given birth to—a *boy*. So? That is nothing new. A boy was expected. But had his wife given birth to—a *girl!* This news would be something tremendously different, it would carry much more information, it would be the completely unexpected!

More rarely occurring events thus carry more information than frequently occurring events in an intuitive sense, and our use of the $-\log_2 p$, $-\log_2 q$ formulation for information agrees with the intuitive concept.

The information carried by a group of 0 or 1 symbols should now be the sum of the bits of information carried by each appearance of 0 or of 1. If $p = 0.8$ and $q = 0.2$ and if p occurs 802 times in 1,000 possibilities, q occurring 198 times, the information content of the 1,000 appearances of a 0 or 1 is

$$H = -(802 \log_2 0.8 + 198 \log_2 0.2)$$
$$\doteq -1,000(0.8 \log_2 0.8 + 0.2 \log_2 0.2)$$
$$= -1,000(p \log_2 p + q \log_2 q)$$

(The dot over the equal sign indicates "approximately equal to.")

The information content of a longer message made up of many 0's and 1's thus depends on $p \log_2 p + q \log_2 q$, the information in bits per occurrence of a 0 or 1, times the relative frequency of occurrence of 0 or 1.

Generalizing for this case of two possible signals, we again consider a time interval T seconds long, subdivided into intervals τ seconds long. There are then $m = T/\tau$ possibilities for a 0 or 1 to occur. On the average $(m \gg 1/p$ and $1/q)$ the 0 will appear $mp = (T/\tau)p$ times, the 1, $mq = (T/\tau)q$ times in the T-second interval. (Remember again that q and p represent, respectively, the probability or relative frequency of occurrence of a 1 and a 0.)

The information content of a message T seconds long is thus, on the average,

$$H = m(-p \log_2 p - q \log_2 q) \qquad \text{bits in } T \text{ seconds} \qquad (1\text{-}10)$$

The average information per interval τ seconds long is

$$H_{av} = \frac{H}{m} = -p \log_2 p - q \log_2 q \qquad \text{bits/interval} \qquad (1\text{-}11)$$

A communication system capable of transmitting this information should thus have an average capacity

$$C_{av} \geq \frac{H_{av}}{\tau} = \frac{1}{\tau}(-p \log_2 p - q \log_2 q) \qquad \text{bits/s} \qquad (1\text{-}12)$$

As a check consider the two possibilities, 0 or 1, equally likely. Then $p = q = 0.5$, and

$$H_{av} = \log_2 2 = 1 \text{ bit/interval}$$

or

$$\frac{T}{\tau} \log_2 2 = \frac{T}{\tau} \qquad \text{bits in } T \text{ seconds}$$

(Note that $n = 2$ here.)

As a further check, $p = 1$, $q = 0$ or $q = 1$, $p = 0$ gives $H_{av} = 0$. This corresponds to the case of a completely determined message (all 0's or 1's), which should carry no information according to our previous intuitive ideas.

Since $q = 1 - p$ in this simple case, H_{av} may be plotted as a function of p to give the curve of Fig. 1-9.[5] H_{av} reaches its maximum value of 1 bit per interval when $p = q = \frac{1}{2}$, or the two possibilities are equally likely. This is of course also in accord with our intuitive ideas that the maximum information should be transmitted when events are completely random, or equally likely to occur.

---

[5] Shannon, "A Mathematical Theory of Communication," *op. cit.*

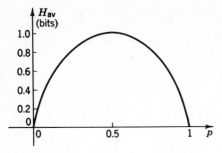

Figure 1-9 Average information per interval with two possible signals. (From C. E. Shannon, "A Mathematical Theory of Communication," *Bell System Tech. J.,* vol. 27, p. 394, July 1948, fig. 7. Copyright, 1948, The American Telephone and Telegraph Co., reprinted by permission.)

We can now generalize to the case of n possible signals or signal levels in any one signal interval τ seconds long. This could be the n levels of Sec. 1-2, the 26 possible letters in the English alphabet, any group of symbols or numbers of which one appears at a time, etc. We again seek an expression for the information content of a message T seconds long ($T > \tau$) and the average information in bits for the interval τ seconds long.

Let the relative frequency of occurrence, or probability, of each possible signal level or symbol be $P_1, P_2, \ldots, P_n$, respectively,

$$P_1 + P_2 + \cdots + P_n = 1$$

and, for the jth symbol, $0 \le P_j \le 1$. If we again assume that occurrences in adjacent intervals are independent (i.e., that any particular symbol occurring in any one interval does not affect the relative rate of occurrence of any of the symbols in any other interval), we can find the information carried by a particular selection in any one interval, the total information (on the average) in T seconds or $m = T/\tau$ intervals, and the average per τ-second interval, as before.

As before, if level (or symbol) j appears in any interval, it carries $-\log_2 P_j$ bits of information. In m intervals j will appear, on the average, mP_j times. The information in bits contributed, on the average, by each symbol appearing mP_j times in m intervals is then summed to give

$$H = -m \sum_{j=1}^{n} P_j \log P_j \qquad \text{bits in } m \text{ intervals}$$

$$= -\frac{T}{\tau} \sum_{j=1}^{n} P_j \log P_j \qquad \text{bits in } T \text{ seconds} \tag{1-13}$$

The *average* information per single symbol interval (τ seconds long) of a message with n possible symbols or levels, of probability P_1 to P_n, respectively, is then

$$H_{\text{av}} = -\sum_{j=1}^{n} P_j \log P_j \qquad \text{bits/interval} \tag{1-14}$$

With τ-second intervals the rate of transmission of information is $1/\tau$ symbols per second. The capacity required of a system to transmit this information would thus be

$$C_{\text{av}} \geq -\frac{1}{\tau}\sum_{j=1}^{n} P_j \log P_j \qquad \text{bits/s} \tag{1-15}$$

As a check, let $P_1 = P_2 = \cdot \ \cdot \ \cdot = P_n = 1/n$ (equally likely events). Then

$$H = m \log_2 n \qquad \text{bits in } m \text{ intervals} \tag{1-16}$$

as before.

As an example of the application of these results, we again consider the problem of determining the information content of a typical message of English speech. This is of course important knowledge required in determining the capacity of a communication system to be used for transmission of messages in English.

If we assume that the occurrence of any letter in the English alphabet is independent of preceding letters or words (a gross approximation), we may use a table of the relative frequency of occurrence of letters in the English alphabet to get an approximate idea of the information content of English speech or writing. This in turn can give us some estimate of the system capacity needed to transmit English at any specified rate (letters or words per second). Such a table appears in a book by Fletcher Pratt[6] and has been reprinted by S. Goldman[7] in his book *Information Theory*. C. E. Shannon[8] in the article already referred to reproduces some of the information available.

The relative frequency of occurrence of the letter e is found to be 0.131 (131 times in 1,000 letters), t occurs 0.105 of the time, a 0.086 of the time, etc., all the way down to z with a probability of 0.00077 (0.77 times in 1,000 letters). Equation (1-14) then gives, as a first approximation to the information content of English, 4.15 bits per letter. Had the letters been equally likely to occur ($P_j = \frac{1}{26}$ for all 26 letters), we would have had $H_{\text{av}} = \log_2 26 = 4.7$ bits per letter. The fact that some letters are more likely to occur than others has thus *reduced* the information content of English from 4.7 bits to 4.15 bits per letter. If six letters are to be transmitted every second, we require a system with a capacity of at least $6 \times 4.15 = 25$ bits/s. Doubling the number of letters to be transmitted doubles the required capacity as well.

The information content of English is actually much less than the 4.15 bits per letter figure because there is of course some dependence between successive letters, words, and even groups of words. Thus, if the letter q occurs, it is almost certain to be followed by a u. These two occurrences (q and then u) are thus not independent, and the uncertainty of a message with a q occurring is reduced, and with it the message information content. Similarly, t is frequently

[6] Fletcher Pratt, *Secret and Urgent*, Doubleday, Garden City Books, New York, 1942.
[7] S. Goldman, *Information Theory*, Prentice-Hall, Englewood Cliffs, N.J., 1953.
[8] Shannon, "A Mathematical Theory of Communication," *op. cit.*

followed by an h, r, or e and almost never by a q or an x. Certain patterns of letters in groups of two thus occur much more frequently than others. The same holds true for groups of three letters. (The group *ter* frequently occurs in that order, while *rtn*, as an example, rarely appears.) Patterns also exist for four- and even five-letter combinations (*mani-*, *semi-*, etc.). In addition, there is dependence between successive words. The word *the* is almost always followed by a noun or adjective, a noun is frequently followed by a verb, etc. Various other words commonly occur together.

All these constraints on different letter and word combinations in English tend to reduce its information content. This reduction in information content of a message from the maximum possible (equally likely and independent symbols) is called the *redundancy* of the message. The redundancy of English, as an example, has been estimated to be considerably more than 50 percent.[9]

To calculate the information content of English more accurately, one would have to consider the influence of the different letters and words on one another. This requires additional knowledge of the statistics of the language, for example, the probability that e occurring would be followed by a or b or any of the other letters, the probability that a *th* would be followed by a or b, etc. These probabilities can also be calculated from the relative frequencies of occurrence of the different combinations. They are called *conditional probabilities* because they relate the occurrence of an event to the previous occurrence of another event.

The existence of redundancy in message transmission can also be demonstrated very simply in the case of TV pictures. Here one signal element or interval corresponds to one spot on the screen, and the time taken for the electron beam to move one element corresponds to the time interval τ in our previous example. The number of levels, n, then corresponds to the number of intensity levels from white to gray to black that can be distinguished. In the case of TV one does not expect adjacent elements to change drastically very often from black to white as the beam sweeps across the screen (usually there is a gradual change from black through gray to white). In addition, a particular element will not be expected to change very much from one sweep interval to the next ($\frac{1}{30}$ s). More than likely a given picture will persist for a while, backgrounds may remain the same for long intervals, small areas of black will remain black for a while, etc. The signal message considered as a time sequence, with various voltage levels corresponding to the different brightness levels and τ to the time to move across one element, will thus not consist of equally likely voltage levels, with the possibility of completely independent changes from one τ interval to the next. These constraints reduce the different number of signal combinations possible and thus quite markedly the average information content of a TV picture. (The limiting case again corresponds to the one in which the picture remains unchanged indefinitely. This implies no information content: you might as well turn off your set and go to bed.) Televi-

---

[9] Goldman, *op. cit.*, p. 45.

sion scenes have a high percentage of redundancy and for this reason require (at least theoretically) much less system capacity than under the assumption of equally likely and independently varying signal levels. (Various estimates have indicated the redundancy of a typical TV pattern to be as high as 99.9 percent!)

PROBLEMS

1-1 In facsimile transmission, 2.25×10^6 square picture elements are needed to provide proper picture resolution. (This corresponds to 1,500 lines in each dimension.) Find the maximum information content if 12 brightness levels are required for good reproduction.

1-2 An automatic translator, with a capacity of 15×10^3 bits/s, converts information from one coding system to another. Its input is a train of uniformly spaced variable-amplitude pulses, 2.71×10^5 pulses occurring each minute. Its output is another uniformly spaced variable-amplitude pulse train, with one-fifth the number of possible amplitude levels as in the input. Find the repetition rate of the output pulses.

1-3 (*a*) Find the capacity in bits per second that would be required to transmit TV picture signals if 500,000 picture elements were required for good resolution and 10 different brightness levels were specified for proper contrast. Thirty pictures per second are to be transmitted. All picture elements are assumed to vary independently, with equal likelihood of occurrence.

(*b*) In addition to the above requirements for a monochrome system a particular color TV system must provide 30 different shades of color. Show that transmission in this color system requires almost $2\frac{1}{2}$ times as much capacity as the monochrome system.

1-4 Refer to Prob. 1-3*b*. If 10 of the 30 color shades require only 7 brightness levels instead of 10, what is the capacity of the system? How many times greater is this capacity than that required for the monochrome system described in Prob. 1-3*a*?

1-5 Express the following decimal numbers in the binary system of notation: 6, 16, 0, 33, 1, 63, 127, 255, 117.

1-6 A system can send out a group of four pulses, each of 1 msec width, and each equally likely to have a height of 0, 1, 2, or 3 V. The four pulses are always followed by a pulse of height -1, to separate the groups. A typical sequence of groups is shown in Fig. P1-6. What is the average rate of information in bits per second that is transmitted with this system?

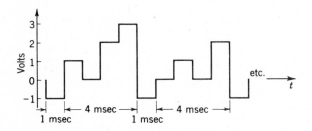

Figure P1-6

1-7 In the system of Prob. 1-6 the zero voltage level occurs one-half of the time on the average, the 1-V level occurs one-fourth the time on the average, and the remaining two levels occur one-eighth of the time each on the average. Find the average rate of transmission of information.

1-8 An alphabet consists of the letters *A, B, C, D.* For transmission each letter is coded into a sequence of two binary (on-off) pulses. The *A* is represented by 00, the *B* by 01, the *C* by 10, the *D* by 11. Each individual pulse interval is 5 ms.

(*a*) Calculate the average rate of transmission of information if the different letters are equally likely to occur.

(*b*) The probability of occurrence of each letter is, respectively, $P_A = 1/5$, $P_B = 1/4$, $P_C = 1/4$, $P_D = 3/10$. Find the average rate of transmission of information in bits/s.

1-9 Repeat Prob. 1-8 with the letters coded into single pulses of 0, 1, 2, or 3 V amplitude and of 10 ms duration.

1-10 A communication system is used to transmit one of 16 possible signals. Suppose that the transmission is accomplished by encoding the signals into binary digits.

(*a*) What will be the pulse sequence for the thirteenth symbol; for the seventh symbol?

(*b*) If each binary digit requires 1 μs for transmission, how much information in bits does the system transmit in 8 μs? Assume that the signals are equally likely to occur.

(*c*) If the symbols are sent directly without encoding, it is found that each symbol requires 3 μs for transmission. What is the information rate in bits/s in this case?

FREQUENCY RESPONSE OF LINEAR SYSTEMS

We indicated in Chap. 1 that the system capacity, or rate of information transmission through a communication system, is related to the rapidity with which signals may change with time. From studies of the transient behavior of networks we know that in all networks with energy-storage elements (L and C) currents, or voltages as the case may be, cannot change instantaneously with time. A specified length of time is required (depending on the network) to reach a desired amplitude level.

In all networks inherent capacitance and inductance limit the time response. In many networks additional limitations are purposely imposed by adding filtering circuits which include inductance and capacitance. From our previous studies we also know that the time, or transient, response is inherently related to the familiar frequency, or steady-state sine-wave, response.

Since frequency concepts are widely used in radio and communication practice, and since frequency analysis of networks frequently simplifies the study of a system, we shall review and extend, in detail, the relation between frequency and time response. The importance of these concepts can be seen from some typical examples:

1. Radio-broadcasting stations are required to operate at their assigned frequency with very tight tolerances. Channels are spaced every 10 kHz in the amplitude-modulation (AM) broadcast band. This 10-kHz spacing is specified in order to prevent overlapping of stations and to allow as many stations as possible to be "squeezed into" the available frequency spectrum. As we shall see later, these severe restrictions limit the maximum rate of information transmission.
2. Telephone cables are limited in their transient response (or, alternatively, in their frequency response). For a given cable to accommodate as many signal

channels as possible, a limitation must be placed upon the frequency extent, or bandwidth, of each channel. This again limits the rate of information transmission in a specified channel.

3. Television stations are limited to 6-MHz bandwidth, again to conserve available frequency space. This in turn imposes limitations on information-transmission capabilities.

Note that in all these simple examples it is the *frequency response* of the network that is specified. This has become common practice, and so it is important to study in detail the relation between frequency and time response, relating both in turn to the particular networks involved.

Consider the simplest set of examples of familiar time functions,

$$f(t) = a_1 \sin \omega_0 t, \ a_2 \sin 2\omega_0 t, \ a_3 \sin 3\omega_0 t, \ \ldots, \ a_n \sin n\omega_0 t \qquad (2\text{-}1)$$

The first two functions are shown in Fig. 2-1. As n increases, the rates of variation with time become more rapid. This may also be seen easily by comparing the different derivatives,

$$\frac{df(t)}{dt} = a_n n \omega_0 \cos n\omega_0 t \qquad (2\text{-}2)$$

As n increases, the maximum rate of change of $f(t)$ increases.

We can represent these functions in a different way by plotting the amplitude a_n versus angular frequency, as in Fig. 2-2. As the frequency increases, the time function varies more rapidly.

Now consider a simple example of an amplitude-modulated signal,

$$f(t) = A(1 + \cos \omega_m t) \cos \omega_0 t \qquad \omega_m \ll \omega_0 \qquad (2\text{-}3)$$

The amplitude varies slowly (as compared with $\cos \omega_0 t$ time variations) between 0 and $2A$. Its rate of variation is given by ω_m, the modulating frequency. ω_0 is the carrier frequency. Figure 2-3 is a plot of one cycle of this function. The amplitude variations form the envelope of the complete signal and represent any information being transmitted. (In a practical situation the envelope would actually be a more complex function of time.)

A simple trigonometric manipulation of Eq. (2-3) gives

$$A(1 + \cos \omega_m t) \cos \omega_0 t$$

$$= A \cos \omega_0 t + \frac{A}{2} [\cos (\omega_0 - \omega_m)t + \cos (\omega_0 + \omega_m)t] \qquad (2\text{-}4)$$

The complete function can thus be represented as the sum of three sinusoidal functions and be plotted on an amplitude-frequency graph (Fig. 2-4). The two smaller lines are called the sideband frequencies; the larger, central line, the carrier. As the amplitude-modulating signal varies more rapidly, ω_m increases and the sideband frequencies move farther away from the carrier. We therefore again have the notion that more rapid variations correspond to wider frequency swings.

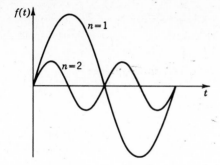

Figure 2-1 Sinusoidal time functions.

Figure 2-2 Amplitude-frequency plot.

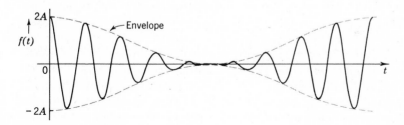

Figure 2-3 Amplitude-modulated wave.

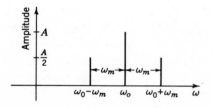

Figure 2-4 Amplitude-frequency plot, modulated carrier.

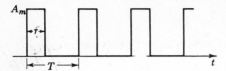

Figure 2-5 Periodic pulse sequence.

These simple examples have all been cases of sinusoidal variations. As such, they are presumed to exist for all time and so carry no information. (Information-carrying signals are continuously varying in an unpredictable manner.) They are valuable in studying networks and systems, however, since any physical time function existing over a finite time interval can be expanded into a Fourier series of sinusoidal functions. Such a series will then represent that function over the interval desired. (Because of the periodic nature of sinusoidal functions the Fourier series representation will then repeat itself regularly outside the specified time interval.)

It is this concept that makes the frequency approach so useful in the communication field. We are quite familiar with the steady-state sinusoidal response of networks. It is also a fact that sinusoidal analysis is frequently much simpler than a time, or transient, analysis. The ability to relate time to frequency response will thus simplify the solution of many problems.

The two examples previously discussed indicated that as the rate of time variation increased the frequency increased also. To generalize this concept somewhat, consider the series of periodic rectangular pulses shown in Fig. 2-5. Note that these pulses are related to the binary pulse sequence introduced in Sec. 1-1. We shall be dealing with such rectangular pulse sequences throughout the entire book because of their widespread occurrence in digital communications. Such a sequence of pulses can of course be expanded into a Fourier series. We ask the question: What is the amplitude-frequency plot of this sequence? This will further fix the time-frequency connection and will then enable us to go further into an examination of the effect of networks on time functions. Before treating this particular problem we review the Fourier-series concept.

2-1 REVIEW OF FOURIER SERIES

Let $f(t)$ be a periodic function of time with period T. Then $f(t)$ may be expanded into the following Fourier series[1] (we shall not deal here with the mathematical

---

[1] Many forms of the series may be written. For example,

$$f(t) = A_0 + \sum_{n=1}^{\infty} (A_n \cos \omega_n t + B_n \sin \omega_n t)$$

is commonly used. This is the same as Eq. (2-5) with

$$A_0 = \frac{a_0}{T} \qquad A_n = \frac{2a_n}{T} \qquad B_n = \frac{2b_n}{T}$$

conditions necessary):

$$f(t) = \frac{a_0}{T} + \frac{2}{T} \sum_{n=1}^{\infty} (a_n \cos \omega_n t + b_n \sin \omega_n t) \qquad \omega_n = \frac{2\pi n}{T} \qquad (2\text{-}5)$$

To find the constants a_n, we multiply through by $\cos \omega_n t$ and integrate over the period. All terms on the right-hand side vanish except the a_n term, since

$$\int_{-T/2}^{T/2} \cos \omega_j t \cos \omega_n t \, dt = 0 \qquad j \ne n$$

$$\int_{-T/2}^{T/2} \sin \omega_j t \cos \omega_n t \, dt = 0 \qquad \text{all } j$$

This gives us

$$\int_{-T/2}^{T/2} f(t) \cos \omega_n t \, dt = \frac{2a_n}{T} \int_{-T/2}^{T/2} \cos^2 \omega_n t \, dt = \frac{2a_n}{T} \frac{T}{2} = a_n \qquad (2\text{-}6)$$

[Remember that $\cos^2 \omega_n t = (1 + \cos 2\omega_n t)/2$.] Thus

$$a_n = \int_{-T/2}^{T/2} f(t) \cos \omega_n t \, dt \qquad n = 1, 0, 2, 3, \ldots \qquad (2\text{-}7)$$

Similarly, $\qquad b_n = \int_{-T/2}^{T/2} f(t) \sin \omega_n t \, dt \qquad n = 1, 2, \ldots \qquad (2\text{-}8)$

The amplitude-frequency plot is proportional to a plot of $\sqrt{a_n^2 + b_n^2}$ versus ω_n. [The phase-frequency characteristic is a plot of $\tan^{-1} (-b_n/a_n)$ versus ω_n.] This plot of $\sqrt{a_n^2 + b_n^2}$ versus frequency will be referred to as the amplitude *spectrum* of the function.[2]

We know from our circuit analysis that, if the voltage across a 1-Ω resistor is given by

$$v_n(t) = A_n \cos \omega_n t + B_n \sin \omega_n t$$

the average power dissipated in the resistor is $(A_n^2 + B_n^2)/2$ watts. Alternatively, if $A_n = 2a_n/T$ and $b_n = 2b_n/T$, the average power dissipated can be written as

$$\left(\frac{2}{T} \right)^2 \frac{a_n^2 + b_n^2}{2}$$

The square of the amplitude spectrum is thus a measure of the power dissipated in a 1-Ω resistor at the different frequencies ($n = 0, 1, 2, \ldots$). By adding the power dissipated at each frequency we get the total average power dissipated when a periodic voltage is impressed across a resistor.

In our work to follow we shall be more interested in the amplitude spectrum $\sqrt{a_n^2 + b_n^2}$ and the phase angle $\tan^{-1} (-b_n/a_n)$ than in the individual Fourier coefficients a_n and b_n. (Note that in general two quantities must be

---

[2] Note that this differs from the amplitude-frequency plot by the constant T.

specified at each frequency in order completely to specify the Fourier series.) This implies writing our Fourier series in the form

$$f(t) = \frac{a_0}{T} + \frac{2}{T} \sum_{n=1}^{\infty} \sqrt{a_b^2 + b_n^2} \cos(\omega_n t + \theta_n) \qquad \theta_n = \tan^{-1} \frac{-b_n}{a_n} \qquad (2\text{-}9)$$

Equation (2-9) can of course be obtained from Eq. (2-5) by a simple trigonometric manipulation.

Since we shall frequently be interested in the amplitude and phase characteristics $\sqrt{a_n^2 + b_n^2}$ and θ_n of a periodic function, it would be much simpler to obtain these directly from $f(t)$, rather than by first finding a_n and b_n. We shall show that this can be done very simply by using yet another form of the Fourier series, the complex exponential form. This alternative form of the series may be written as

$$f(t) = \frac{1}{T} \sum_{n=-\infty}^{\infty} c_n e^{j\omega_n t} \qquad (2\text{-}10)$$

The Fourier coefficient c_n is then a complex number defined as

$$c_n \equiv a_n - jb_n = \sqrt{a_n^2 + b_n^2}\, e^{j\theta_n} = \int_{-T/2}^{T/2} f(t) e^{-j\omega_n t}\, dt \qquad (2\text{-}11)$$

$|c_n| = \sqrt{a_n^2 + b_n^2}$ is thus the desired amplitude spectrum and

$$\theta_n = \tan^{-1} \frac{-b_n}{a_n}$$

represents the phase characteristic. The coefficient c_n gives the complete frequency spectrum.

Equations (2-10) and (2-11) are completely equivalent to Eqs. (2-5), (2-7), and (2-8). Not only does Eq. (2-11) give the amplitude and phase characteristics directly, but Eqs. (2-10) and (2-11) represent a much more compact form of the Fourier series,

$$f(t) = \frac{1}{T} \sum_{n=-\infty}^{\infty} c_n e^{j\omega_n t} \qquad (2\text{-}10a)$$

$$c_n = \int_{-T/2}^{T/2} f(t) e^{-j\omega_n t}\, dt \qquad (2\text{-}11a)$$

In rewriting Eqs. (2-5), (2-7), and (2-8) in this new form use has been made of the exponential form for the sine and cosine. Thus, in Eq. (2-5) let

$$\cos \omega_n t = \frac{e^{j\omega_n t} + e^{-j\omega_n t}}{2}$$

$$\sin \omega_n t = \frac{e^{j\omega_n t} - e^{-j\omega_n t}}{2j}$$

Regrouping terms in Eq. (2-5),

$$f(t) = \frac{a_0}{T} + \frac{1}{T} \sum_{n=1}^{\infty} [e^{j\omega_n t}(a_n - jb_n) + e^{-j\omega_n t}(a_n + jb_n)] \qquad (2\text{-}12)$$

If $c_n \equiv a_n - jb_n$, $c_n^* \equiv a_n + jb_n$, where c_n^* is the complex conjugate of c_n. But

$$a_n - jb_n = \int_{-T/2}^{T/2} f(t)(\cos \omega_n t - j \sin \omega_n t) \, dt = \int_{-T/2}^{T/2} f(t)e^{-j\omega_n t} \, dt \qquad (2\text{-}13)$$

from Eqs. (2-7) and (2-8). Since $\omega_n = 2\pi n/T$, $e^{-j\omega_n t} = e^{-j(2\pi nt/T)}$ and

$$e^{+j\omega_n t} = e^{-j(2\pi/T)(-n)t} = e^{-j\omega_{-n} t}$$

Therefore,

$$c_n^* = a_n + jb_n = \int_{-T/2}^{T/2} f(t)e^{j\omega_n t} \, dt = \int_{-T/2}^{T/2} f(t)e^{-j\omega_{-n} t} \, dt \qquad (2\text{-}14)$$

[Remember again that $\omega_n = 2\pi n/T$, $\omega_{-n} \equiv 2\pi(-n)/T$.] From Eq. (2-14), then, assuming $f(t)$ real, $c_n^* = c_{-n}$. (That is, replacing n by $-n$ in c_n gives c_n^*.) Equation (2-12) can now be rewritten as

$$f(t) = \frac{a_0}{T} + \frac{1}{T} \sum_{n=1}^{\infty} (e^{j\omega_n t} c_n + e^{-j\omega_n t} c_{-n}) \qquad (2\text{-}15)$$

But summing over $-n$ from 1 to ∞ is the same as summing over $+n$ from -1 to $-\infty$ ($c_{-3} \equiv c_n$, $n = -3$). Also,

$$c_0 = \int_{-T/2}^{T/2} f(t)e^{j0} \, dt = a_0$$

Equation (2-15) can thus be further simplified to

$$f(t) = \frac{1}{T} \sum_{n=-\infty}^{\infty} c_n e^{j\omega_n t} \qquad (2\text{-}10)$$

$$c_n = \int_{-T/2}^{T/2} f(t)e^{-j\omega_n t} \, dt \qquad (2\text{-}11)$$

Although "negative" frequencies seem to appear in Eq. (2-10), they are actually fictitious. For if Eq. (2-10) is rewritten in real form, Eq. (2-9) is obtained, in which the only frequencies appearing ($\omega_n = 2\pi n/T$) are positive. This is very simply shown by writing $c_n = |c_n|e^{j\theta_n}$. Then Eq. (2-10) becomes

$$f(t) = \frac{1}{T} \sum_{n=-\infty}^{\infty} |c_n|e^{j(\omega_n t + \theta_n)}$$

$$= \frac{1}{T} \sum_{n=1}^{\infty} [|c_n|(e^{j(\omega_n t + \theta_n)} + e^{-j(\omega_n t + \theta_n)})] + \frac{c_0}{T}$$

$$= \frac{a_0}{T} + \frac{2}{T} \sum_{n=1}^{\infty} |c_n| \cos (\omega_n t + \theta_n)$$

(Remember again that $-\omega_n = \omega_{-n}$, $-\theta_n = \theta_{-n}$.)

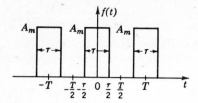

Figure 2-6 Fourier analysis of periodic pulses.

As an example of the utility of this complex form of the Fourier series, consider the series of pulses of Fig. 2-6. (The origin has been chosen to coincide with the center of one pulse.) Then

$$c_n = \int_{-\tau/2}^{\tau/2} A_m e^{-j\omega_n t}\, dt = -\left. \frac{A_m}{j\omega_n} e^{-j\omega_n t} \right]_{-\tau/2}^{\tau/2}$$

$$= A_m \frac{e^{j\omega_n \tau/2} - e^{-j\omega_n \tau/2}}{j\omega_n} = \frac{2A_m}{\omega_n} \sin \frac{\omega_n \tau}{2}$$

This may be written in the form

$$c_n = \tau A_m \frac{\sin (\omega_n \tau/2)}{\omega_n \tau/2} \tag{2-16}$$

If we define a normalized and dimensionless variable, $x = \omega_n \tau/2$,

$$c_n = \tau A_m \frac{\sin x}{x}$$

The $(\sin x)/x$ function will be occurring in many problems in the future and should be carefully studied. Note that it has its maximum value at $x = 0$, where $\sin x \to x$, $(\sin x)/x \to 1$. It approaches zero as $x \to \infty$, oscillating through positive and negative values. If x is a continuous variable, $(\sin x)/x$ has the form of Fig. 2-7.

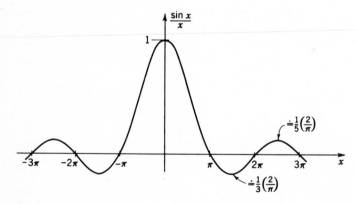

Figure 2-7 $[(\sin x)/x]$ versus x.

In our particular problem n has discrete values only, ω_n takes on discrete values (harmonics of $\omega_1 = 2\pi/T$), and the normalized parameter x is thus also defined only at discrete points. The *envelope* of the plot of c_n will be exactly the curve of Fig. 2-7. The plot of c_n itself is shown in Fig. 2-8 ($\tau \ll T$). Since c_n is for this example a real number (alternately positive and negative), there is no need to find $|c_n|$ and θ_n and plot each separately. (Note again that the "negative" frequencies shown are just a mathematical artifice, since $\omega_n = 2\pi n/T$ and $\omega_{-n} = -2\pi n/T$.) The spacing between the successive lines is

$$\Delta\omega_n = \frac{2\pi}{T}(n+1) - \frac{2\pi n}{T} = \frac{2\pi}{T}$$

or just the fundamental angular frequency. All lines of the frequency spectrum shown thus occur at multiples of this fundamental frequency. The "dc component" ($\omega_n = 0$) is of course just T times the average value.

Time-Frequency Correspondence

Although the periodic function of Fig. 2-6 contains frequency components at all integral multiples of the fundamental frequency, the envelope of the amplitude decreases at higher frequencies. Note, however, that as the fundamental period T *decreases* (more pulses per second), the frequency lines move out farther. Again a *more rapid variation* in the time function corresponds to *higher-frequency* components. Alternatively, as T increases, the lines crowd in and ultimately approach an almost smooth frequency spectrum. Since the lines concentrated in the lower-frequency range are of higher amplitude, we note that most of the energy associated with this periodic wave is confined to the lower frequencies. As the function varies more rapidly (T decreases), the relative amount of the energy contained in the higher-frequency range increases.

Figure 2-8 and Eq. (2-16) emphasize another interesting phenomenon that will be very useful to us in later work. As the pulse width τ *decreases*, the

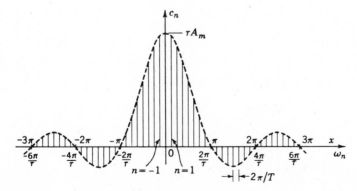

Figure 2-8 Frequency spectrum, rectangular pulses ($\tau \ll T$).

frequency content of the signal extends out over a larger frequency range. The first zero crossing, at $\omega_n = 2\pi/\tau$, moves out in frequency. There is thus an *inverse relationship between pulse width, or duration, and the frequency spread of the pulses*.

If $\tau \ll T$ (that is, very narrow pulses), most of the signal energy will lie in the range

$$0 < \omega_n < \frac{2\pi}{\tau}$$

The first zero crossing is frequently a measure of the frequency spread of a signal (assuming, of course, that the envelope of the amplitude spectrum decreases with frequency). In keeping with the notation for networks we can talk of the bandwidth of the signal as being a measure of its frequency spread.

As in the case of the frequency response of networks the bandwidth occupied by the signal cannot be uniquely specified unless the signal is "band-limited" (i.e., occupies a finite range of frequencies with no frequency components beyond the range specified). However, some arbitrary (and frequently useful) criterion for bandwidth may be chosen to specify the range of frequencies in which most of the signal energy is concentrated. As an example, if the bandwidth B is specified as the frequency extent of the signal from zero frequency to the first zero crossing,

$$B = \frac{1}{\tau} \tag{2-17}$$

where $\tau \ll T$. Any other criterion for bandwidth would still retain the inverse time-bandwidth relation, and, in general,

$$B = \frac{k}{\tau} \tag{2-18}$$

with k a constant depending on the choice of criterion. We shall return to this important concept later.

Power Considerations

We noted earlier that the average power dissipated in a 1-Ω resistor with a voltage

$$v_n(t) = A_n \cos \omega_n t + B_n \sin \omega_n t$$

impressed across it is just given by $(A_n^2 + B_n^2)/2$. If one deals with the complex exponential

$$v_n(t) = c_n e^{\omega_n t}$$

instead, it is apparent that the average power is given by

$$\bar{P} = \frac{1}{T}\int_0^T |v_n(t)|^2 \, dt = |c_n|^2$$

This may be simply generalized to the case of an arbitrary periodic function $f(t)$ by using the Fourier-series expansion. Thus we recall that with an arbitrary (and possibly complex) voltage $f(t)$ impressed across a 1-Ω resistor the instantaneous power is just $|f(t)|^2$ and the average power is given by

$$\bar{P} = \frac{1}{T}\int_0^T |f(t)|^2 \, dt \qquad (2\text{-}19)$$

[If $f(t)$ is real, as in most of the examples of this book, $|f(t)|$ is simply replaced by $f(t)$.] Substituting for $f(t)$ its Fourier-series expansion of Eq. (2-10), we get

$$\bar{P} = \frac{1}{T^3}\int_0^T \left[\sum_m \sum_n c_m^* c_n e^{j(\omega_n - \omega_m)t} \right] dt \qquad (2\text{-}19a)$$

Interchanging the order of summation and integration and noting that

$$\int_0^T e^{j(\omega_n - \omega_m)t} \, dt = T \qquad \omega_n = \omega_m$$
$$= 0 \qquad \text{elsewhere}$$

(this is simply checked by expanding in sines or cosines or by noting that $e^{j\omega t}$, although complex, is periodic, and averages to zero over a period), we get

$$\bar{P} = \frac{1}{T}\int_0^T |f(t)|^2 \, dt = \sum_{n=-\infty}^{\infty} \left| \frac{c_n}{T} \right|^2 \qquad (2\text{-}19b)$$

The average power is thus found by summing the power contribution at all frequencies in the Fourier-series expansion.

As an example, if we again consider the rectangular test pulses of Fig. 2-6, we have

$$\bar{P} = \frac{A_m^2 \tau}{T} = \sum_{n=-\infty}^{\infty} \left(\frac{\tau A_m}{T} \right)^2 \frac{\sin^2 (\omega_n \tau/2)}{(\omega_n \tau/2)^2} \qquad (2\text{-}20)$$

This expression enables us to determine the relative power contributions at the various frequencies. For example the dc power is just $(c_0/T)^2$ or $(\tau A_m/T)^2$ in the rectangular pulse case. (This is of course easily checked by noting that the dc power is $\left[1/T \int_0^T f(t) \, dt \right]^2$.) The power in the first harmonic, that at $\omega_1 = 2\pi/T$, is just

$$2 \left(\frac{\tau A_m}{T} \right)^2 \frac{\sin^2 (\omega_1 \tau/2)}{(\omega_1 \tau/2)^2} \qquad \text{etc.}$$

(The factor of 2 is due to the equal power contributions at ω_1 and $\omega_{-1} = -\omega_1$.)

Periodic Impulses

A special case of the rectangular pulse train, of great utility in both digital-communication analysis and modern digital computation, is obtained by letting

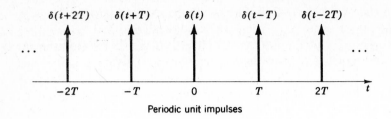

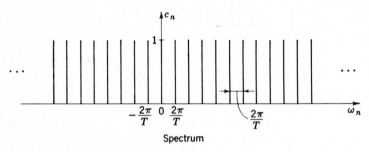

Figure 2-9 Spectrum of periodic impulses.

the pulse width τ in Fig. 2-6 go to zero and the amplitude $A_m \to \infty$, with $A_m\tau = 1$. This results in the extremely useful set of periodic impulses of infinite height, zero width, and unit area. These unit impulses are shown sketched in Fig. 2-9. The symbol usually adopted for an impulse of unit area and centered at $t = \tau$ is $\delta(t - \tau)$. It is represented by an arrow, as indicated in Fig. 2-9. If the area $(A_m\tau)$ is some number k, one writes $k\delta(t - \tau)$.

We shall have more to say about the impulse function later, but note now that all spectral lines have the same height $A_m\tau$. The "bandwidth" thus approaches infinity. This agrees of course with the inverse time-bandwidth relation. As the width τ of the pulse $\to 0$, its bandwidth $1/\tau \to \infty$.

2-2 FOURIER INTEGRAL AND ITS PROPERTIES

The discussion thus far has been restricted to the case of periodic time functions representable by Fourier series. Although such functions are commonly used for test purposes in many system studies, in practice they do not really represent the time functions occurring in communications practice. For as noted in Chap. 1, periodic functions carry no information. A closer approximation to the actual signals used in practice is provided by nonperiodic time functions. (A still better model will be the random signals discussed in later chapters.)

Here too we shall find the frequency response of signals playing a key role in the discussion. In particular the inverse time-frequency relationship developed in the preceding section for periodic functions will still be found valid for nonperiodic functions.

An extension of the time-frequency correspondence to nonperiodic time functions requires the introduction of the Fourier integral. This is simply done by recognizing that any time function (subject of course to certain broad mathematical definitions and restrictions) defined only over a specified time interval T seconds long may be expanded in a Fourier series of base period T. The time function is then artificially made to repeat itself outside the specified time interval. As the time interval of interest becomes greater, the Fourier period is correspondingly increased. Ultimately, as the region of interest is made to increase beyond bound, the resultant Fourier series becomes, in the limit, the Fourier integral.

Consider a periodic function $f(t)$,

$$f(t) = \frac{1}{T} \sum_{n=-\infty}^{\infty} c_n e^{j\omega_n t} \qquad \omega_n = \frac{2\pi n}{T} \tag{2-21}$$

$$c_n = \int_{-T/2}^{T/2} f(t) e^{-j\omega_n t} \, dt \tag{2-22}$$

A typical amplitude-spectrum plot would appear as in Fig. 2-10. The spacing between successive harmonics is just

$$\Delta\omega = \omega_{n+1} - \omega_n = \frac{2\pi}{T} \tag{2-23}$$

Equation (2-21) may be written as

$$f(t) = \frac{1}{2\pi} \sum_{n=-\infty}^{\infty} c_n e^{j\omega_n t} \, \Delta\omega \tag{2-24}$$

Now consider the limiting case as $T \to \infty$. Then $\Delta\omega \to 0$, the discrete lines in the spectrum of Fig. 2-10 merge, and we obtain a continuous-frequency spectrum. Mathematically, the infinite sum in Eq. (2-24) becomes the ordinary Riemann integral. c_n is now defined for *all* frequencies, not merely integral multiples of $2\pi/T$. In the limit, as $T \to \infty$, $\omega_n \to \omega$ and c_n becomes a continuous function $F(\omega)$.

$$F(\omega) = \lim_{T \to \infty} c_n \tag{2-25}$$

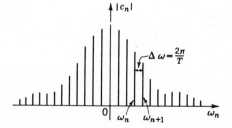

Figure 2-10 Amplitude spectrum, periodic function.

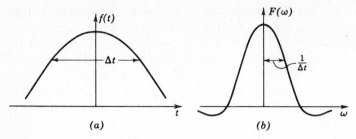

Figure 2-11 A typical time function and its spectrum. (a) $f(t)$. (b) $F(\omega)$.

In the place of the Fourier series of Eq. (2-21) we now obtain as the Fourier-integral representation of a nonperiodic function $f(t)$

$$f(t) = \frac{1}{2\pi} \int_{-\infty}^{\infty} F(\omega)e^{j\omega t}\, d\omega \tag{2-26}$$

with

$$F(\omega) = \int_{-\infty}^{\infty} f(t)e^{-j\omega t}\, dt \tag{2-27}$$

from Eq. (2-22). $F(\omega)$ is, in general, a complex function of ω and may be written

$$F(\omega) = |F(\omega)|e^{j\theta(\omega)} \tag{2-28}$$

A typical time function and its Fourier spectrum $F(\omega)$ are shown in Fig. 2-11. The periodic pulses previously considered serve as a good example of the transition from the Fourier series to the Fourier integral. The pulses are shown in Fig. 2-12a and the corresponding spectrum in Fig. 2-12b.

The frequency spectrum of the periodic pulses is of course a plot of the Fourier coefficient c_n (normally amplitude and phase plots).

$$c_n = V\tau \frac{\sin (\omega_n \tau/2)}{\omega_n \tau/2} \qquad \omega_n = \frac{2\pi n}{T} \tag{2-29}$$

As $T \to \infty$, all the pulses except for the one centered at $t = 0$ move out beyond bound and we are left, in the time plot, with a single pulse of amplitude V and width τ seconds.

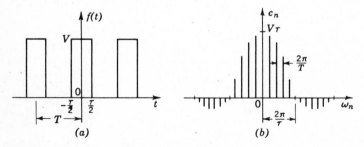

Figure 2-12 Periodic pulses and their spectrum.

In the frequency plot $\omega_n \to \omega$ as $T \to \infty$, the lines move together and merge, and the spectrum becomes a continuous one (Fig. 2-13). Thus

$$F(\omega) = \lim_{T \to \infty} c_n = V\tau \frac{\sin (\omega\tau/2)}{\omega\tau/2} \qquad (2\text{-}30)$$

The single pulse of Fig. 2-13a has the continuous-frequency spectrum of Fig. 2-13b, defined for all frequencies.

Equation (2-30) can of course be obtained directly from the defining relation for $F(\omega)$ [Eq. (2-27)].

$$F(\omega) = \int_{-\infty}^{\infty} f(t)e^{-j\omega t}\, dt$$

For a single pulse

$$f(t) = V \qquad |t| < \frac{\tau}{2}$$

$$f(t) = 0 \qquad |t| > \frac{\tau}{2} \qquad (2\text{-}31)$$

Then $F(\omega) = V \displaystyle\int_{-\tau/2}^{\tau/2} e^{-j\omega t}\, dt = \frac{V}{-j\omega}\,(e^{-j\omega\tau/2} - e^{j\omega\tau/2}) = V\tau\,\frac{\sin (\omega\tau/2)}{\omega\tau/2} \quad (2\text{-}32)$

Note the inverse time-frequency relationship between the rectangular pulse and its spectrum (Fig. 2-13), exactly as pointed out earlier: as the pulse width τ decreases, the spectral spread moves out as $1/\tau$. The bandwidth B in hertz, measured to the first zero crossing, is again $B = 1/\tau$.

The Fourier integral $F(\omega)$ is frequently called the *Fourier transform* of $f(t)$. The two defining relations, (2-27) for the Fourier transform of $f(t)$ and (2-26) for the inverse Fourier transform, represent a Fourier transform pair. Note the complete duality between the integrals, particularly if $d\omega/2\pi$ is written as df, a differential frequency in units of hertz. If the two functions $f(t)$ and $F(\omega)$ are interchanged, their respective transforms interchange as well. For example, a rectangular spectrum $F(\omega)$ should have as its inverse transform a $(\sin x)/x$ function in time. We shall explore this point later, in more detail.

The introduction and use of the Fourier transform $F(\omega)$ enables us to readily determine the time response of linear networks. We shall find ourselves

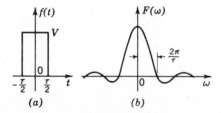

(a)   (b)

Figure 2-13 Rectangular pulse and its spectrum. (*a*) Time plane. (*b*) Frequency plane.

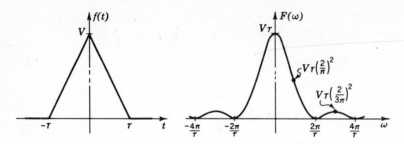

Figure 2-14 Triangular pulse and its spectrum.

continually referring to the "frequency or spectral content" of signals and the "frequency properties" of communication systems. Communication engineers commonly refer to the "bandwidth" of a system with which they may be working. They have found the concept of frequency response to be an extremely useful one in the design and analysis of communication systems. We shall in fact be continually referring to spectrum and bandwidth throughout this book. With such continuous usage the concept of the "frequency property" of a network will tend to take on a physical meaning. But we must keep in mind that these frequency and spectral considerations are in reality mathematical abstractions. The frequency content of a signal $f(t)$ is obtained from the Fourier transform $F(\omega)$, an integral representation of $f(t)$.[3]

Before proceeding to discuss the properties of Fourier integrals and the response of linear systems to aperiodic time functions using the frequency concept, it is worthwhile examining some additional examples of Fourier integral pairs. The five examples that follow all focus on pulse-type signals in time, because of their common occurrence in communication practice. The calculation of some of the transforms is left for the reader as an exercise. Others are developed, in a somewhat different manner, later, using the properties of Fourier transforms.

1. *Triangular pulse* (*Fig. 2-14*). A triangular pulse of height V and base width 2τ has as its spectrum

$$F(\omega) = (V\tau)\left|\left[\frac{\sin\,(\omega\tau/2)}{\omega\tau/2}\right]^2\right.$$
(2-33)

Note that this pulse has the same amplitude and area as the rectangular pulse of Fig. 2-13. Its width, as measured at the half-amplitude points, is just that of the rectangular pulse. This triangular pulse and its spectrum obey the same inverse time-frequency relationship discussed previously: as the pulse width τ decreases, the bandwidth B, in hertz, measured to the first zero crossing, or any other comparable measure, increases as $1/\tau$.

---

[3] The discrete Fourier transform is often used nowadays to calculate the spectrum of a signal, because of the advent of high-speed digital processing. See M. Schwartz and L. Shaw, *Signal Processing: Discrete Spectral Analysis, Detection, and Estimation*, McGraw-Hill, New York, 1975, chap. 2.

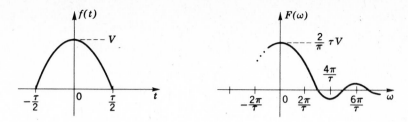

Figure 2-15 Cosine pulse and its Fourier transform.

Note, however, that the spectrum in this case is more tightly centered about the origin than in the previous rectangular pulse case. The spectrum here decreases as $1/f^2$ (or $1/\omega^2$ in radian measure), as contrasted with the $1/f$ decrease of the $(\sin x)/x$ spectrum of the rectangular pulse. The latter has a discontinuity in the function itself. The triangular pulse is a more smoothly varying function: it is continuous in the function itself and discontinuous in the derivative. We shall demonstrate further, by example, that the smoother the function, as exemplified by the continuity of higher and higher-order derivatives, the more rapidly the spectrum decreases with frequency, packing more of the frequency content into a specified bandwidth. Integration of the triangle of Fig. 2-14 would result in a spectrum decreasing as $1/f^3$ with frequency.

  2. *Cosine pulse* (*Fig. 2-15*). Here

$$f(t) = V \cos \frac{\pi t}{\tau} \qquad |t| \le \frac{\tau}{2} \;=\; T \tag{2-34}$$

$$= 0 \qquad\qquad \text{elsewhere}$$

One then finds that $\;=\; V \cos \frac{\pi t}{2T}$

$$F(\omega) = \frac{2\tau V}{\pi}\, \frac{\cos (\omega\tau/2)}{1 - (\omega\tau/\pi)^2} \tag{2-35}$$

Note that the spectrum here decreases as $1/f^2$ (or $1/\omega^2$) with frequency. The cosine pulse is also continuous in time, while its first derivative is discontinuous at $\pm\tau/2$. The pulse width is again τ, while the first zero crossing, at $\omega\tau = 3\pi$, if chosen as a measure of the pulse "bandwidth," corresponds to

$$B = \frac{3}{2\tau}$$

in hertz. The same inverse time-frequency relation again exists here.

  3. *Raised cosine pulse* (*Fig. 2-16*)

$$f(t) = \frac{V}{2}\left(1 + \cos \frac{\pi t}{\tau} \right) \qquad |t| \le \tau$$

$$= 0 \qquad\qquad \text{elsewhere} \tag{2-36}$$

$$F(\omega) = V\tau\, \frac{\sin \omega\tau}{\omega\tau[1 - (\omega\tau/\pi)^2]} \tag{2-37}$$

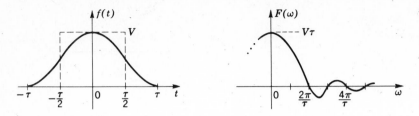

Figure 2-16 Raised cosine pulse and its Fourier transform.

The spectrum here decreases as $1/f^3$ for large frequency, since the first derivative is continuous at all values of t. There is a discontinuity in the second derivative at $t = \pm\tau$. Note here that although the full pulse width is 2τ, its width to the half-amplitude points (indicated in Fig. 2-16 by the dashed lines), is τ. The bandwidth to the first zero crossing is $B = 1/\tau$.

4. *Gaussian pulse* (*Fig. 2-17*)

$$f(t) = Ve^{-t^2/2\tau^2} \tag{2-38}$$

$$F(\omega) = \sqrt{2\pi}\,\tau Ve^{-\tau^2\omega^2/2} \tag{2-38a}$$

Note that here τ is one possible measure of the width of the pulse. The width of $F(\omega)$, defined in an identical manner, is then $1/\tau$. The bandwidth of the signal thus increases inversely with its width, as expected. The Gaussian pulse is the smoothest of all those considered. All its derivatives exist and are continuous at all values of time. The rate of drop-off of the transform $F(\omega)$ is thus the greatest, dropping exponentially with ω^2 at frequencies higher than the bandwidth. Note, however, that the pulse is neither time-limited nor band-limited. Both the width in time and the width in frequency have to be arbitrarily defined. [As an interesting sidelight notice that $f(t)$ and $F(\omega)$ in this case have exactly the same shapes: a Gaussian pulse in time gives rise to a Gaussian pulse in frequency.]

5. (*Sin x*)/*x pulse* (*Fig. 2–18*). Here

$$f(t) = V\frac{\sin(\pi t/T)}{\pi t/T} \tag{2-39}$$

This is just the dual of the rectangular pulse considered in Fig. 2-13. Since the Fourier transform pairs are reciprocal with respect to one another, it is apparent

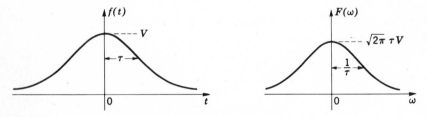

Figure 2-17 Gaussian pulse and its Fourier transform.

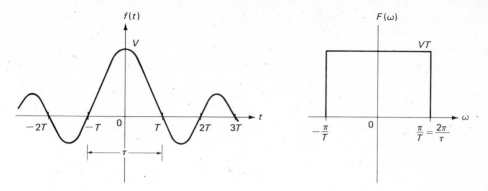

Figure 2-18 (Sin *x*)/*x* pulse and its transform.

as noted earlier that this pulse must give rise to the rectangular transform shown in Fig. 2-18. Details are left to the reader. Note that in this example the frequency spectrum is precisely limited in frequency, although the function itself is not time-limited. If one takes as a measure of the pulse width τ the width between the two first zero crossings, as shown, the radian bandwidth is $\pi/T = 2\pi/\tau$, and the bandwidth, in hertz, is $B = 1/\tau$. Equation (2-39) can then also be written, with specific emphasis on bandwidth, in the form

$$f(t) = V \frac{\sin 2\pi Bt}{2\pi Bt} \qquad (2\text{-}39a)$$

In all the examples considered thus far there exists an inverse time-bandwidth relation: if a pulse width τ is defined, the bandwidth B in hertz is $\propto (1/\tau)$. This is exactly the result noted earlier in our discussion of Fourier series and the spectra of periodic time functions [Eqs. (2-17) and (2-18)]. Thus, if the pulse width $\tau = 1\ \mu s$, $B = 1/\tau = 1$ MHz; if $\tau = 1$ ms, $B = 1/\tau = 1$ kHz; if $\tau = 10$ s, $B = 1/\tau = 0.1$ Hz. For simplicity's sake, and because this serves as a good rule of thumb, we shall often take $B = 1/\tau$ in this book. Although this generally serves as a good measure of the bandwidth, note that it is far from precise. The specific bandwidth in any particular case depends on the definition desired, or on the application itself. It may sometimes be defined as the frequency spacing between the half-power, or 3-dB points; it may be related to the zero crossings of the frequency spectrum as indicated in some of the examples considered here; it may be measured in terms of rms deviation about the center frequency (this is generally the case with the Gaussian pulse of Fig. 2-17). Other definitions are possible as well. Some of these will be considered elsewhere in this book. The concept of bandwidth will be particularly important in discussing the transmission of signals through linear systems. In Chap. 3, in discussing the transmission of digital symbols through telephone-type band-limited channels, bandwidth will play a particularly significant role in determining appropriate bit rates to be allowed through these channels. We shall then discuss signal shaping to pack as many bits per second through a given channel. But no matter what the definition of bandwidth, we shall always find that an inverse time-bandwidth relation exists. The particular definition used simply changes the proportionality constant k in Eq. (2-18): $B = k/\tau$.

The fact that there exists an inverse time-bandwidth relationship is due specifically to the Fourier integral definition of Eq. (2-27). Thus it is readily shown for *any* time function that if the time scale is reduced by τ, the frequency is increased correspondingly by $1/\tau$. To demonstrate this relationship in the general case, consider time to be normalized to a parameter τ, written as t/τ. The Fourier integral of a function $f(t/\tau)$ is then given by

$$F'(\omega) = \int_{-\infty}^{\infty} f(t/\tau)e^{-j\omega t}\, dt$$

$$= \tau \int_{-\infty}^{\infty} f(x)e^{-j(\omega\tau)x}\, dx$$

$$= \tau F(\tau\omega) \tag{2-40}$$

after introducing a dummy variable $x = t/\tau$, and recognizing that the resultant integral is again a Fourier transform, but with a radian frequency scale given by $\tau\omega$.

Equation (2-40) states, exactly as noted above, that if the time scale is *reduced* by a factor τ, the frequency scale is *increased* by $1/\tau$. If the pulse width in time is reduced from ms to μs, the bandwidth increases correspondingly, by a factor of 10^3, from kHz to MHz. The relationship (2-40) may be stated in a concise way by representing the Fourier transform–inverse Fourier transform relationship in terms of a doubleheaded arrow. Thus if we have

$$f(t) \leftrightarrow F(\omega) \tag{2-41}$$

then
$$f(t/\tau) \leftrightarrow \tau F(\tau\omega) \tag{2-42}$$

These expressions may be simply read as meaning that $F(\omega)$ is the Fourier transform of $f(t)$, and $f(t)$ the inverse transform of $F(\omega)$, or more simply, that the two form a Fourier transform pair.

Other properties of the Fourier integral may be deduced from the examples given above. Note for instance that in the case of both the rectangular pulse (Fig. 2-13) and the triangular pulse (Fig. 2-14), the dc content $F(0) = V\tau$, the *area* under each of the pulses. This is simply proven from Eq. (2-27), by setting $\omega = 0$:

$$F(0) = \int_{-\infty}^{\infty} f(t)\, dt \tag{2-43}$$

It is left to the reader to show that the other examples considered (see Figs. 2-14 to 2-16) do in fact obey this property.

The examples selected thus far have been *even* functions of time. $[f(t) = f(-t)]$. The resultant Fourier transforms are all real and *even* functions of frequency. That this is a general property of even time functions is readily proven from Eq. (2-27). Thus, if $f(t) = f(-t)$, we have

$$F(\omega) = \int_{-\infty}^{\infty} f(t)e^{-j\omega t}\, dt = \int_{-\infty}^{\infty} f(t)(\cos \omega t - j \sin \omega t)\, dt$$

$$= \int_{-\infty}^{\infty} f(t) \cos \omega t \qquad f(t) \text{ even} \tag{2-44}$$

since the product of the even $f(t)$ and odd $\sin \omega t$ integrate to zero. Hence

$$F(\omega) = F(-\omega) \qquad f(t) \text{ even} \tag{2-45}$$

It is apparent as well that if $f(t)$ is an *odd* function in time, with $f(t) = -f(-t)$, $F(\omega)$ is imaginary and *odd* in frequency. In this case,

$$F(\omega) = -j \int_{-\infty}^{\infty} f(t) \sin \omega t\, dt \qquad f(t) \text{ odd} \tag{2-46}$$

and $$F(\omega) = -F(-\omega) \qquad f(t) \text{ odd} \tag{2-47}$$

As an example, it is left for the reader to show that if $f(t)$ is given by

$$f(t) = e^{-at} \qquad t > 0$$
$$= -e^{at} \qquad t < 0 \tag{2-48}$$

as shown in Fig. 2-19, its Fourier transform is given by

$$F(\omega) = \frac{-2j\omega}{a^2 + \omega^2} \tag{2-49}$$

This is of course imaginary and an odd function of frequency, as also shown in Fig. 2-19.

Other useful Fourier transform properties may be readily deduced from the

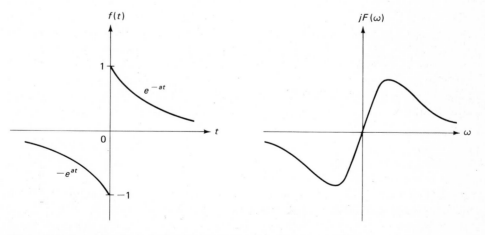

Figure 2-19 Example of an odd function in time and its Fourier transform.

defining Eqs. (2-26) and (2-27) of the Fourier transform pairs. We discuss just a few of these below. Others will be developed where needed in this book.

1. *Fourier transform of a derivative*. Let $f(t)$ be a differentiable function. The Fourier transform of $df(t)/dt$ is then $j\omega F(\omega)$. Thus, if

$$f(t) \leftrightarrow F(\omega)$$

then
$$\frac{df(t)}{dt} \leftrightarrow j\omega F(\omega) \qquad (2\text{-}50)$$

To demonstrate this we simply differentiate Eq. (2-26) with respect to time. Then

$$\frac{df(t)}{dt} = \frac{1}{2\pi} \int_{-\infty}^{\infty} j\omega F(\omega)e^{j\omega t} \, d\omega$$

and Eq. (2-50) follows.

Equation (2-50) implies several things. Multiplication by ω increases $F(\omega)$ at the higher frequencies. Hence one can say that differentiation increases the high-frequency content of a signal. This generalizes the statement made earlier that successive discontinuities in a function and its derivatives reduce the rate of drop-off of the frequency spectrum. A pulse is essentially the derivative of a triangle. The latter has a frequency spectrum dropping off as $1/f^2$, the former as $1/f$. This will be noted again below in discussing some examples of the application of Eq. (2-50). Before we consider these examples, note also that the derivative of an even function in time must be odd. Hence the transform of the derivative must be odd and imaginary. This is of course borne out by Eq. (2-50).

If higher-order derivatives of $f(t)$ exist, we can proceed in a similar manner to find the transforms of those higher-order derivatives. For example, it is apparent from Eq. (2-50) that we also have

$$\frac{d^2 f(t)}{dt^2} \leftrightarrow -\omega^2 F(\omega) \qquad (2\text{-}51)$$

So as a general rule, successive differentiation increases the higher-frequency content of a signal. This observation is particularly important in following a signal through a linear system through which it propagates.

Equation (2-50) enables us to rapidly derive some other Fourier transform pairs, using some of the examples introduced earlier. Three such examples of pulses, their derivatives, and the resultant Fourier transforms, appear in Fig. 2-20. Notice, as already pointed out, that the frequency spectrum of the derivative of the triangle drops off as $1/f$ with frequency. The corresponding time function has a discontinuity at the origin. The same holds true for the derivative of the cosine pulse in Fig. 2-20b. The raised cosine of Fig. 2-20c has a continuous first derivative. The transform of the raised cosine itself drops off as $1/f^3$ at high frequencies, while the transform of its derivative drops off as $1/f^2$.

It is instructive to differentiate the derivative of the raised cosine once more. The resultant cosinusoidal pulse appears in Fig. 2-21. It is shown there

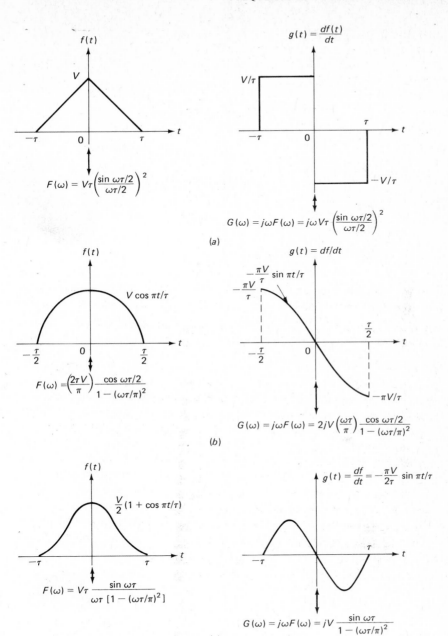

Figure 2-20 Fourier transforms of derivatives.

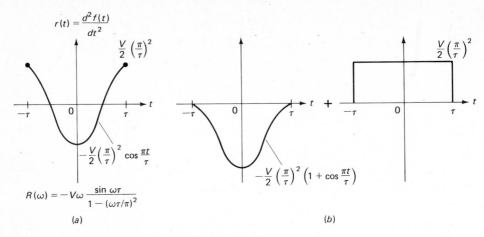

Figure 2-21 Decomposition of a cosinusoidal pulse.

that this may be considered as the superposition of a negative raised cosine pulse and a rectangular pulse of width 2τ. Multiplying the transform of the raised cosine pulse by $-\omega^2$, one gets the transform $R(\omega)$ shown in Fig. 2-21. As a check, we can simply add the transforms of the two pulses shown in Fig. 2-21b. It is left to the reader to show that these are given by

$$R(\omega) = -V\tau \left(\frac{\pi}{\tau}\right)^2 \frac{\sin \omega\tau}{\omega\tau[1 - (\omega\tau/\pi)^2]} + V\tau \left(\frac{\pi}{\tau}\right)^2 \frac{\sin \omega\tau}{\omega\tau}$$

$$= -V\omega \frac{\sin \omega\tau}{[1 - (\omega\tau/\pi)^2]}$$

2. *Shifting theorem.* Let the origin of time be shifted from $t = 0$ to $t = t_0$. This then corresponds to a phase shift of $-j\omega t_0$ in $F(\omega)$. More precisely,

$$f(t - t_0) \leftrightarrow e^{-j\omega t_0} F(\omega) \tag{2-52}$$

The proof of this relation is left to the reader. [As a hint, use $f(t - t_0)$ in Eq. (2-27). Introduce a dummy variable $x = t - t_0$; Eq. (2-52) then follows.]

As examples, consider the two rectangular pulses of Fig. 2-22a and b. The two Fourier transforms are then, respectively,

$$F_1(\omega) = Ve^{-j\omega(\tau/2)} \frac{\sin (\omega\tau/2)}{\omega\tau/2}$$

and

$$F_2(\omega) = Ve^{j\omega(\tau/2)} \frac{\sin (\omega\tau/2)}{\omega\tau/2}$$

As a check, the sum of the two pulses is given by the rectangular pulse of Fig. 2-22c. We thus have

$$F(\omega) = F_1(\omega) + F_2(\omega) = 2V\tau \frac{\sin \omega\tau}{\omega\tau}$$

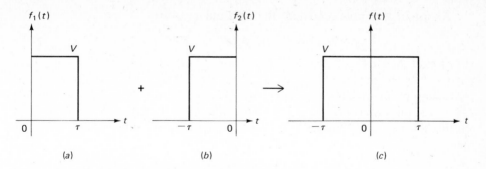

Figure 2-22 Examples of application of shifting theorem.

using the double-angle formula of trigonometry. Note that this result agrees, of course, with the Fourier transform of the pulse of Fig. 2-22c, written by inspection.

As another application of the shifting theorem, consider the function $f(t)$ of Fig. 2-23. This is obviously the superposition of the *difference* of the two pulses in Fig. 2-22a and b. Hence

$$f(t) = f_2(t) - f_1(t)$$

and

$$F(\omega) = F_1(\omega) - F_1(\omega)$$

$$= V\tau \frac{\sin (\omega\tau/2)}{\omega\tau/2} \underbrace{(e^{j\omega\tau/2} - e^{-j\omega\tau/2})}_{2j \sin (\omega\tau/2)}$$

$$= 2jV\tau \frac{\sin^2 (\omega\tau/2)}{\omega\tau/2}$$

Note that this agrees with the results obtained for the transform of the derivative of the triangle in Fig. 2-20b. Since $f(t)$ in this case is odd, $F(\omega)$ is imaginary, and odd as well.

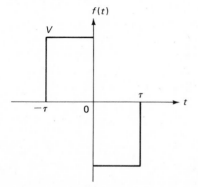

Figure 2-23. Another example of application of shifting theorem.

As a final example, consider the sinusoidal pulse

$$g(t) = -(\pi V/2\tau) \sin (\pi t/\tau), \quad -\tau \le t \le \tau$$

shown sketched in Fig. 2-20c. Recall that this is the derivative of the raised cosine pulse shown in the same figure. A little thought will indicate that this sinusoidal pulse may be considered the difference of two cosine pulses, one centered at $t = -\tau/2$, the other at $t = \tau/2$, the amplitude of both being $\pi V/2\tau$. Using the shifting theorem, the Fourier transform of the difference of these two pulses is then

$$F(\omega) = \frac{2\tau}{\pi} \left(\frac{\pi V}{2\tau}\right) \frac{\cos(\omega\tau/2)}{[1 - (\omega\tau/\pi)^2]} \underbrace{(e^{j\omega\tau/2} - e^{-j\omega\tau/2})}_{2j \sin \omega\tau/2}$$

$$= jV \frac{\sin \omega\tau}{[1 - (\omega\tau/\pi)^2]}$$

in agreement with the transform obtained by multiplying the raised cosine transform by $j\omega$ (Fig. 2-20c). Use has again been made of the trigonometric double-angle formula.

3. *Frequency shifting theorem.* This is the dual of the shifting theorem in time just described. Specifically, if $F(\omega)$ is shifted by ω_0 rad/s, $f(t)$ is multiplied by $e^{j\omega_0 t}$. More precisely, we have

$$f(t)e^{j\omega_0 t} \leftrightarrow F(\omega - \omega_0) \tag{2-53}$$

[compare with Eq. (2-52)]. This is easily proven from Eq. (2-27) by writing

$$F(\omega - \omega_0) = \int_{-\infty}^{\infty} f(t)e^{-j(\omega-\omega_0)t} dt = \int_{-\infty}^{\infty} [f(t)e^{j\omega_0 t}]e^{-j\omega t} dt$$

We shall see in our discussion of modulation in Chap. 4 that it is this simple relation that accounts for the frequency shift incurred when a signal $f(t)$ is modulated onto a carrier. Specifically, we say that amplitude modulation of a carrier signal $\cos \omega_0 t$ by an information-bearing signal $f(t)$ is given by the simple multiplicative expression

$$f(t) \cos \omega_0 t$$

$f(t)$ is then called the *baseband* or *modulating* signal, and $f_0 = \omega_0/2\pi$ the carrier frequency. Writing $\cos \omega_0 t$ as the sum of two complex exponentials, and using Eq. (2-53) on each resultant term separately, we get $\frac{1}{2}[F(\omega - \omega_0) + F(\omega + \omega_0)]$ as the Fourier transform of the modulated signal. The effect of the multiplication by $\cos \omega_0 t$, the carrier signal, is to shift the spectrum by f_0 hertz. If $f(t)$ has a spectrum centered about 0, the modulated signal has its spectrum centered about the carrier, f_0. An example appears in Fig. 2-24.

4. *Symmetry of amplitude and phase spectra.* In general $F(\omega)$ is a complex function of frequency. Just as in the case of the Fourier spectra discussed earlier, $F(\omega)$ may be decomposed into an amplitude spectrum and a phase spectrum. For real time functions $f(t)$ (except for complex exponentials and the discussion in Chap. 4 dealing with representations of single-sideband signals

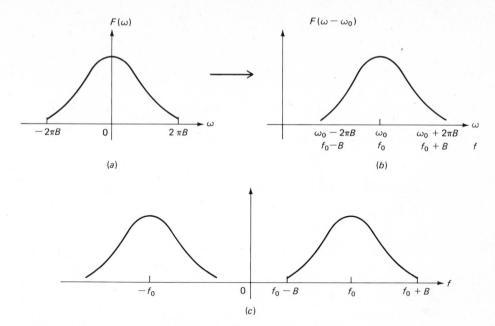

Figure 2-24 Illustration of frequency shift. (*a*) Original signal spectrum. (*b*) Shifted spectrum. (*c*) Spectrum of $f(t) \cos \omega_0 t$.

this is the only kind we shall be considering in this book), one finds that the *amplitude* spectrum is an *even* symmetrical function of frequency, and the phase spectrum is an *odd* symmetrical function of frequency.

To prove this statement we again invoke the defining Fourier integral relation (2-27). As noted previously in Eq. (2-28), we can write the generally complex $F(\omega)$ in the magnitude-phase complex form

$$F(\omega) = |F(\omega)|e^{j\theta(\omega)} \qquad (2\text{-}28)$$

The magnitude term $|F(\omega)|$, written as a function of frequency, is just the amplitude spectrum; the variation of the phase angle $\theta(\omega)$ with frequency represents the phase spectrum. From Eq. (2-27),

$$F(\omega) = \int_{-\infty}^{\infty} f(t) \cos \omega t \, dt - j \int_{-\infty}^{\infty} f(t) \sin \omega t \, dt$$

Then
$$|F(\omega)|^2 = \left[\int_{-\infty}^{\infty} f(t)\cos \omega t \, dt\right]^2 + \left[\int_{-\infty}^{\infty} f(t) \sin \omega t \, dt\right]^2 \qquad (2\text{-}54)$$

is an even function of frequency, and

$$\theta(\omega) = -\tan^{-1}\left[\int_{-\infty}^{\infty} f(t) \sin \omega t \, dt \bigg/ \int_{-\infty}^{\infty} f(t) \cos \omega t \, dt\right] \qquad (2\text{-}55)$$

is an odd function of frequency. (Why?)

As an example, consider the exponential time function of Fig. 2-25. It is readily shown that $F(\omega) = 1/(a + j\omega)$. Then $|F(\omega)| = 1/\sqrt{(a + \omega)^2}$, and $\theta(\omega) =$

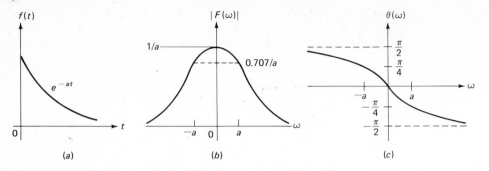

Figure 2-25 Amplitude and phase spectra of a signal. (*a*) Time function. (*b*) Amplitude spectrum. (*c*) Phase spectrum.

$-\tan^{-1}(\omega/a)$. Both $|F(\omega)|$ and $\theta(\omega)$ are sketched in Fig. 2-25. Notice the even symmetry of the amplitude spectrum and the odd symmetry of the phase spectrum. The parameter a represents the 3-dB *radian* bandwidth (the 3-dB bandwidth in hertz is then $a/2\pi$), since $|F(a)|/|F(0)| = 0.707$. The phase spectrum ranges between $\pm\pi/2$. The signal has zero phase angle at $\omega = 0$, has a $-\pi/4$-radian phase angle at $\omega = a$, and approaches $-\pi/2$ radians at large frequencies, $\omega \gg a$. The phase spectrum is generally nonlinear, but in the range $-a < \omega < a$ may be represented by the linear expression $\theta(\omega) \doteq -\omega/a$.

In discussing the concept of bandwidth earlier we focused exclusively on even symmetrical positive pulse-type time functions for which the Fourier transform is real and even, and exactly equal to the amplitude spectrum. (The phase spectrum is then zero for all ω. If the time function is negative, and even symmetrical, the phase spectrum is taken by definition to be π radians for negative frequency, and $-\pi$ radians for positive frequency. This keeps the phase spectrum odd symmetrical.) We then discussed various definitions of bandwidth in terms of the transform $F(\omega)$. More generally, the bandwidth is defined in terms of the amplitude spectrum, and everything noted earlier still holds true. The bandwidth may be taken as the 3-dB point of the amplitude spectrum as in the example of Fig. 2-25, it may be defined to be the frequency at which the amplitude spectrum first goes to zero, etc. In all cases, however, the same inverse time-frequency relation still holds. This is apparent from the exponential time function of Fig. 2-25: $1/a$ is a measure of the duration of the exponential, or its effective time constant. Its 3-dB bandwidth in hertz is then $a/2\pi$.

The Fourier transform relations discussed above may be summarized as follows:

Given
$$F(\omega) = \int_{-\infty}^{\infty} f(t)e^{-j\omega t}\,dt \tag{2-27}$$

$$f(t) = \frac{1}{2\pi}\int_{-\infty}^{\infty} F(\omega)e^{j\omega t}\,d\omega \tag{2-26}$$

or
$$f(t) \leftrightarrow F(\omega) \qquad (2\text{-}41)$$

Then

$$\frac{df(t)}{dt} \leftrightarrow j\omega F(\omega) \qquad (2\text{-}50)$$

$$f(t - t_0) \leftrightarrow e^{-j\omega t_0}F(\omega) \qquad (2\text{-}52)$$

$$f(t)e^{j\omega_0 t} \leftrightarrow F(\omega - \omega_0) \qquad (2\text{-}53)$$

$$F(\omega) = |F(\omega)| e^{j\theta(\omega)} \qquad (2\text{-}28)$$

$$|F(\omega)| \quad \text{even in } \omega$$

$$\theta(\omega) \quad \text{odd in } \omega$$

Other relations will be developed at appropriate places in the book.

2-3 SIGNALS THROUGH LINEAR SYSTEMS: FREQUENCY RESPONSE

The use of the Fourier transform and the bandwidth or spectral occupancy concept developed from it enables us to readily assess the effect of passing communication signals through systems. We commented on this point earlier. It is one of the reasons the frequency concept is so widely used in communication work. In this section we discuss the frequency response of linear systems in detail, focusing on the generally distorting effect of systems on signals passed through them. This distortion may be desired, as in the case of filters designed to produce certain controlled signal shapes at their output. The distortion may be undesirable, yet unavoidable, as in the transmission of various signal waveforms over their communication paths. (Both cases were first introduced in Sec. 1-1, Fig. 1-4, in discussing the transmission of a digital message over a communication path.)

In this section and throughout most of the book we take the system through which the signals are passed to be *linear*. This simplifies the analysis and enables us to readily assess the effect of the system on the signals in terms of frequency response. In later chapters, particularly Chaps. 4 and 5, we shall have occasion to deal with nonlinear responses as well. There too frequency analysis plays a useful role, but the analysis is much more complex and general results are hard to come by.

The simplest example of a linear system is the *RC* filter of Fig. 2-26. It is

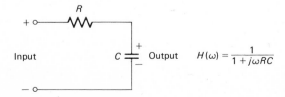

$$H(\omega) = \frac{1}{1 + j\omega RC}$$

Figure 2-26 *RC* filter as example of linear system.

well known from elementary circuit analysis that a linear system has a fre- quency transfer function $H(\omega)$. This is a measure of the (complex) response to the complex exponential time function $e^{j\omega t}$. Depending on the input applied (voltage or current in electric circuits), and the output measured (again voltage or current in electric circuits), this transfer function may be the complex impe- dance, complex admittance, or a transfer ratio. For example, in the RC filter of Fig. 2-26, $e^{j\omega t}$ applied at the input produces $H(\omega)e^{j\omega t}$ at the output. With the input-output ports indicated, $H(\omega)$, the voltage transfer ratio, is the ratio of the complex impedance $1/j\omega C$ at the output to the input impedance $R + 1/j\omega C$. This then gives

$$H(\omega) = \frac{1}{1 + j\omega RC} \tag{2-56}$$

as indicated in the figure. If the desired response were the voltage produced with a current $e^{j\omega t}$ applied at the input, $H(\omega)$ would be the input impedance $R + 1/j\omega C$. More generally, for more complex networks or systems, one would have to *measure* the transfer function. One technique is to apply a fixed ampli- tude and phase sinusoidal signal at the input, and measure both amplitude and phase at the same frequency, as the input frequency is varied over the desired frequency range.

The response of a linear system to any arbitrary input signal $f(t)$ can be readily related to the transfer function, and hence to the frequency response of the network, because of the concept of superposition. For recall that by the inverse Fourier transform of Eq. (2-26) any time function $f(t)$ may be written as the sum of an infinitely dense number of complex exponentials. Each one of these exponentials gives rise to an exponential of the same frequency at the output, multiplied by the appropriate transfer function. Superposing these out- puts, we find the desired output signal $g(t)$.

The concept of superposition is in fact the defining relation of a linear network, and it is because of this property that we confine ourselves to linear networks in this section. We also must invoke the assumption of time invari- ance or stationarity in order for the analysis to be valid. (Both nonlinear and time-varying networks generate new frequencies if a fixed-frequency sinusoidal signal is applied to them. This point is discussed further in Chaps. 4 and 5.)

Summarizing then, we deal in this section with linear stationary systems only. The RC network of Fig. 2-26 is probably the simplest example. Because of their importance to our work, we state the two basic properties in terms of which a linear stationary system is defined more precisely:

1. The response to a sum of excitations is equal to the sum of the responses to the excitations acting separately.
2. The relations between input and output are time-invariant or stationary.

The first condition is of course the statement of superposition noted above. The second condition implies that the system elements do not change with time.

Our familiar linear operations are those of multiplication by a constant, addition, subtraction, differentiation, and integration. The combination of these operations gives rise to a time-invariant linear system governed by constant-coefficient linear differential equations. The RC network of Fig. 2-26 is a particularly simple example of such a system.

The first condition cited above as defining a linear system may be summarized by saying that if an input $f_1(t)$ gives rise to an output $g_1(t)$, while an input $f_2(t)$ produces $g_2(t)$ at the output, the input $af_1(t) + bf_2(t)$ results in the output $ag_1(t) + bg_2(t)$. This is shown diagrammatically in Fig. 2-27. The time-invariant condition says that an input $f_1(t - \tau)$ gives rise to an output $g_1(t - \tau)$. It is these simple conditions that enable Fourier analysis to play so significant a role in linear system analysis. For, as already noted, the system response to a number of exponentials is simply the sum of the responses to each individual exponential. The sum of the responses to each exponential is then just the Fourier-integral representation of the output function $g(t)$, providing a powerful tool in linear system analysis.

To find the output response $g(t)$ of a stationary linear system with transfer function $H(\omega)$ to an input signal $f(t)$, we proceed as follows: From the inverse Fourier transform, Eq. (2-26), it is apparent that $f(t)$ is the sum of terms of the form $(1/2\pi)F(\omega)e^{j\omega t}$. Each such term produces an output $[H(\omega)/2\pi]F(\omega)e^{j\omega t}$. By superposition, then, we must have, as the response to the infinitely dense set of complex exponentials representing $f(t)$,

$$g(t) = \frac{1}{2\pi} \int_{-\infty}^{\infty} \underbrace{H(\omega)F(\omega)}_{G(\omega)} e^{j\omega t} \, d\omega \qquad (2\text{-}57)$$

Hence we have the well-known transfer function relation for linear systems,

$$G(\omega) = H(\omega)F(\omega) \qquad (2\text{-}58)$$

Here $G(\omega)$ is the Fourier transform of the desired output $g(t)$:

$$g(t) \leftrightarrow G(\omega) \qquad (2\text{-}59)$$

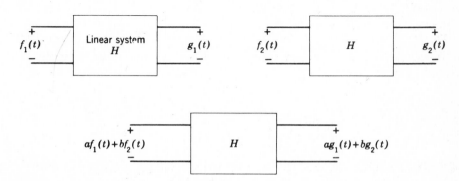

Figure 2-27 Linear system, defining relations.

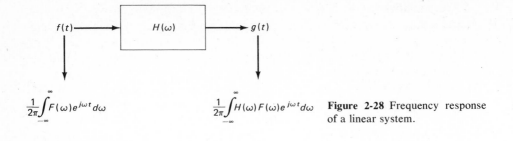

$$\frac{1}{2\pi}\int_{-\infty}^{\infty} F(\omega)e^{j\omega t}d\omega \qquad\qquad \frac{1}{2\pi}\int_{-\infty}^{\infty} H(\omega)F(\omega)e^{j\omega t}d\omega$$

Figure 2-28 Frequency response of a linear system.

This transfer relationship is shown schematically in Fig. 2-28. We have of course by no means proven Eqs. (2-57) and (2-58) rigorously. We shall return to such a proof later using the convolution integral and the network impulse response.

Equation (2-58) is an extremely useful relation, and accounts in part for the widespread use of frequency concepts in communication system analysis and design. For it enables us to determine exactly, if desired, the effect of a linear system on input signals applied to it, as well as to trace the changes in the signals (if any) as they move through the system. It also enables us to assess quite quickly the approximate effect of a linear network on signals passing through it.

It is this latter point that is quite important and particularly useful. We shall see that the concept of bandwidth plays a particularly significant role here. Using measured amplitude and phase spectra plots, one can quite readily determine the distorting effect of a network, or specify filters needed to obtain desired signal shapes. We shall in fact attempt to develop intuitive ideas along these lines throughout this book. To be more specific at this point, assume that $H(\omega)$ is centered about the origin. Say that $|H(\omega)|$, the amplitude spectrum of the system transfer function, has a characteristic width, measured with respect to the origin, that we call the system bandwidth B_s, in hertz, $2\pi B_s$ in rad/s. A simple example might be

$$H(\omega) = \frac{1}{1 + j\omega/\omega_0} \tag{2-60}$$

$$|H(\omega)| = \frac{1}{\sqrt{1 + (\omega/\omega_0)^2}} \tag{2-60a}$$

This is of course just the transfer function and amplitude spectrum of the RC network of Fig. 2-26, with $RC \equiv 1/\omega_0$. $|H(\omega)|$ for this example is shown sketched in Fig. 2-29. (Note that it is identical in form to the amplitude spectrum of the exponential signal of Fig. 2-25.) If the 3-dB point is again chosen as a bandwidth measure in this case, $2\pi B_s = \omega_0$ or $B_s = \omega_0/2\pi$ Hz.

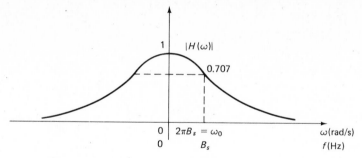

Figure 2-29 System bandwidth: an example.

Now let a pulse signal $f(t)$ with bandwidth B be applied at the input to $H(\omega)$. Consider three cases:

1. $B > B_s$
2. $B \sim B_s$
3. $B < B_s$

From the relation $G(\omega) = H(\omega)F(\omega)$ one would expect the output signal in case 3 to be a relatively undistorted replica of $f(t)$. In case 1 the output is highly distorted. In fact, in this case, with $B \gg B_s$, the characteristics of the input signal disappear, the output signal has a bandwidth = B_s, and the output signal shape is determined almost wholly by the characteristics of the linear system. Case 2 is of course intermediate between these. These three cases are shown schematically in Fig. 2-30, with the signal spectrum overlaid on the system spectrum in each case. Taking the product of the two spectra in each case, we see the reasoning behind the statements above.

It is thus apparent that to have relatively little distortion in a signal as it passes through a communication system, its bandwidth must be small compared to the system bandwidth. This idea is extremely intuitive and is of course well known to everyone: a hi-fi system, for example, should have an overall bandwidth of 20 kHz in order to reproduce signals with high fidelity. (This of course implies that all records or tapes have frequency components well below 20 kHz. We are neglecting here phase spectral effects on the signals. These may be quite significant in some cases. More will be said about this later.) Filters in telephone systems are purposely set to have cutoff bandwidths of 4 kHz. The human voice *or* music played over the telephone thus does not at all have a high-fidelity quality. AM receivers have notoriously poor frequency response. They sometimes have bandwidths as low as 2.5 kHz. It is this fact, not necessarily the transmission bandwidth of AM signals, that accounts for the relatively poor tone of AM reception.

Some special examples will make these points more quantitative. As the

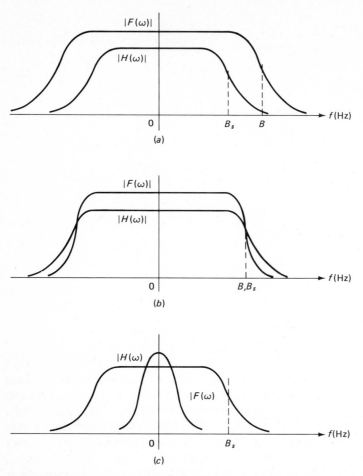

Figure 2-30 Effect of varying signal bandwidth. (a) $B > B_s$, high distortion. (b) $B \sim B_s$. (c) $B < B_s$, little distortion.

first example say that the measured characteristic of a linear system is of the form

$$H(\omega) = \frac{\sin (\omega/2B_s)}{\omega/2B_s}\, e^{-j\omega/2B_s} \tag{2-61}$$

As shown in Fig. 2-31, $|H(\omega)|$ has a bandwidth of B_s hertz, measured to the first zero crossing. Note that this is the same $\sin x/x$ function encountered earlier as the Fourier transform of a rectangular pulse (Fig. 2-13). Now consider a rectangular pulse signal $f(t)$ of width τ applied at the input to the same linear system. This is shown in Fig. 2-31. The bandwidth, in hertz, of the rectangular pulse, is $B = 1/\tau$, using the first zero crossing as a measure. $|F(\omega)|$ has the same $\sin x/x$ shape as $|H(\omega)|$, but with its bandwidth controlled by the pulse width τ. We now consider the three cases noted above:

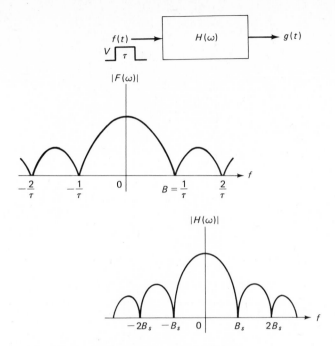

Figure 2-31 Response of linear system to a pulse.

1. $B \gg B_s$, or $\tau \ll 1/B_s$. The signal bandwidth is in this case much larger than the system bandwidth. The input pulse $f(t)$ is then very narrow compared to the characteristic time response of the system, which is the order of $1/B_s$. (As an example, say that $B_s = 100$ kHz and $\tau = 1$ μs. Then $B = 1$ MHz. The system time response is the order of 10 μs.) In this case the product $H(\omega)F(\omega)$ is dominated by the system transfer function [$|F(\omega)|$ is effectively constant and equal to $V\tau$ over most of the significant range of $|H(\omega)|$]. To a good approximation, then,

$$G(\omega) \doteq V\tau H(\omega) \qquad B \gg B_s \qquad (2\text{-}62)$$

and the output signal $g(t)$ is then the rectangular pulse shown in Fig. 2-32a. [Actually, $g(t)$ is not quite rectangular but has finite rise and fall times, as indicated in Fig. 2-32a, related to τ, the width of the input pulse. This will be discussed further in later sections.] The *shape* of $g(t)$ is to a good approximation determined by $H(\omega)$ only, although its amplitude $V\tau B_s$ depends on the input pulse $f(t)$ as well. As far as the system is concerned, $f(t)$ appears very much as an impulse and $g(t)$ is then the *impulse response*. We shall have more to say about the impulse response of linear systems later, but note that a simple condition for a real excitation to approximate an impulse is exactly the relation assumed above:

$$B \gg B_s \qquad \text{or} \qquad 1/\tau \gg B_s$$

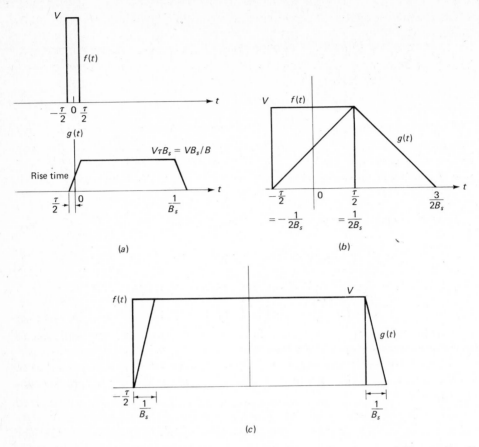

Figure 2-32 Effect of varying pulse width and bandwidth. (a) $B \gg B_s$ ($\tau \ll 1/B_s$). (b) $B = B_s$ ($\tau = 1/B_s$). (c) $B \ll B_s$ ($\tau \gg 1/B_s$).

The excitation [$f(t)$ in this example] must have a frequency response that is wide compared to the system response.

2. $B = B_s$ or $\tau = 1/B_s$. In this case the signal and system bandwidths are the same. One would expect the output to be a distorted version of the input. In fact, in this case, because of the particular form of the transfer function as-- sumed, we have

$$G(\omega) = H(\omega)F(\omega) = V\tau \left[\frac{\sin (\omega/2B_s)}{\omega/2B} \right]^2 e^{-j\omega/2B_s} \tag{2-63}$$

From our previous discussion [Eqs. (2-33) and (2-52)] this represents the transform of a triangle centered at $\tau = 1/2B_s$. Both the input $f(t)$ and the output $g(t)$ for this case are sketched in Fig. 2-32b. Note that the distorting effect of the linear system has been to broaden the rectangular pulse out to the triangular shape shown. Alternatively, the rise time $\tau = 1/B_s$ is just half the width of the pulse.

3. $B \ll B_s$ or $\tau \gg 1/B_s$. In this case $|F(\omega)|$ is much more narrow than $|H(\omega)|$, all frequencies in the range $f < B_s$ are passed essentially unchanged, and $g(t)$ looks very much like the input $f(t)$:

$$G(\omega) \doteq F(\omega) \qquad B \ll B_s \qquad (2\text{-}64)$$

The resultant input and output pulses are shown sketched in Fig. 2-32c. They again differ only in the rise and fall portions. The output pulse has a rise time shown of about $1/B_s$ seconds. More-quantitative considerations relating to pulses passing through linear systems appear in the next section.

2-4 RESPONSE OF IDEALIZED NETWORKS

The example used in the previous section for determining the effect of linear filtering on a pulse-type signal was useful in showing the results of three extreme cases: the signal bandwidth much greater than the filter bandwidth, equal to the filter bandwidth, and much smaller than the filter bandwidth. It is less useful in intermediate cases because of the cumbersome mathematics involved in determining the output pulse response $g(t)$. More generally, one would like to *quantitatively* evaluate the response of a linear system or filter to a pulse-type signal. For this purpose numerical evaluation of the defining inverse transform relation (2-57) must be resorted to.[4] We repeat that relationship here for convenience:

$$g(t) = \frac{1}{2\pi} \int_{-\infty}^{\infty} \underbrace{H(\omega)F(\omega)}_{G(\omega)} e^{j\omega t}\, d\omega \qquad (2\text{-}57)$$

This is the equation used to determine the effect of measured network amplitude and phase characteristics on signals.

Many examples of measured characteristics could be used for this purpose. It is more useful for us, since we are primarily interested in developing insight into the effect of filtering on signals, to consider instead an idealized network characteristic. This serves as a useful model of the filtering characteristics of many real networks, it enables us to demonstrate quantitatively the effect of filtering on a signal, and it focuses on the variation of only one parameter, the system bandwidth B_s. By varying B_s, we can thus see the effect of the network on the signal passing through it.

The example we shall study in detail is the ideal low-pass filter of Fig. 2-33. For simplicity of notation we define the amplitude characteristic of a linear system to be $|H(\omega)| = A(\omega)$. Hence we have, in general,

---

[4] In practice one often deals with discrete Fourier transforms instead, as already noted. See Schwartz and Shaw, *op. cit.*

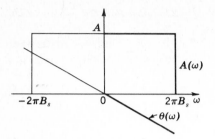

Figure 2-33 Ideal low-pass filter.

$$H(\omega) = A(\omega)e^{j\theta(\omega)} \qquad (2\text{-}65)$$

For the ideal low-pass filter of Fig. 2-33, the amplitude characteristic is taken as a constant value A for all radian frequencies below the cutoff frequency $2\pi B_s$. Its bandwidth is thus precisely $2\pi B_s$ in radian measure, or B_s hertz. We thus have

$$\begin{aligned} A(\omega) &= A & |\omega| &\leq 2\pi B_s \\ A(\omega) &= 0 & |\omega| &> 2\pi B_s \end{aligned} \qquad (2\text{-}66)$$

The phase shift $\theta(\omega)$ is assumed linearly proportional to frequency.

$$\theta(\omega) = -t_0\omega \qquad (2\text{-}67)$$

(t_0 is a constant of this ideal network. Negative frequencies must again be introduced in order to use the Fourier integrals, defined also for negative ω.) This amplitude response is physically unattainable. The linear-phase-shift characteristic assumed is also physically impossible for finite lumped-constant networks. (Smooth transmission lines can have linear phase shift.) In addition, the amplitude and phase characteristics of a given network are connected together by the pole-zero plot of the network. They are thus normally not chosen independently, as was done here. (There do exist *all-pass* networks, synthesized from lattice structures, which provide phase variation with constant-amplitude response. These networks have their poles and zeros symmetrically arranged on either side of the $j\omega$ axis. Such networks can be included in an overall network to provide independent choice of amplitude and phase.)

The use of these idealizations to investigate the response of physical networks could thus lead to absurdities unless we are careful in interpreting our results.

We now investigate in detail the response of the ideal low-pass filter to a pulse. We shall vary the filter bandwidth B_s and determine the output signal $g(t)$ for several values of B_s. The results will of course be in agreement with those of the last section (Fig. 2-32), but more quantitatively specified.

If the input signal is a single rectangular pulse,

$$F(\omega) = V\tau \frac{\sin(\omega\tau/2)}{\omega\tau/2} \qquad (2\text{-}68)$$

(The time origin is chosen at the center of the pulse.) The transform of the output signal is

$$G(\omega) = V\tau \frac{\sin (\omega\tau/2)}{\omega\tau/2} Ae^{-jt_0\omega} \qquad -2\pi B_s < \omega < 2\pi B_s$$

$$G(\omega) = 0 \qquad\qquad\qquad\qquad \text{elsewhere} \tag{2-69}$$

Then $g(t)$ will be

$$g(t) = \frac{AV\tau}{2\pi} \int_{-2\pi B_s}^{2\pi B_s} \frac{\sin (\omega\tau/2)}{\omega\tau/2} e^{j\omega(t-t_0)} d\omega \tag{2-70}$$

To evaluate this integral, we recall that

$$e^{j\theta} = \cos \theta + j \sin \theta$$

Then

$$\int_{-2\pi B_s}^{2\pi B_s} \frac{\sin (\omega\tau/2)}{\omega\tau/2} e^{j\omega(t-t_0)} d\omega$$

$$= \int_{-2\pi B_s}^{2\pi B_s} \frac{\sin (\omega\tau/2)}{\omega\tau/2} [\cos \omega(t - t_0) + j \sin \omega(t - t_0)] d\omega$$

$$= \int_{-2\pi B_s}^{2\pi B_s} \frac{\sin (\omega\tau/2)}{\omega\tau/2} \cos \omega(t - t_0) d\omega$$

$$+ j \int_{-2\pi B_s}^{2\pi B_s} \frac{\sin (\omega\tau/2)}{\omega\tau/2} \sin \omega(t - t_0) d\omega \tag{2-71}$$

The integrand of the first integral is an even function of ω. The integral is then just twice the integral from 0 to $2\pi B_s$. The integrand of the second integral is an odd function, and the integral, between equal negative and positive limits, vanishes. Equation (2-71) can thus be written

$$2 \int_0^{2\pi B_s} \frac{\sin (\omega\tau/2)}{\omega\tau/2} \cos \omega(t - t_0) d\omega$$

$$= \int_0^{2\pi B_s} \left[\frac{\sin \omega(t - t_0 + \tau/2)}{\omega\tau/2} - \frac{\sin \omega(t - t_0 - \tau/2)}{\omega\tau/2} \right] d\omega \tag{2-72}$$

using the trigonometric relation for sum and difference angles.

Breaking the integral up into two integrals and changing variables [$x = \omega(t - t_0 + \tau/2)$ in the first integral, $x = \omega(t - t_0 - \tau/2)$ in the second], we get finally for $g(t)$

$$g(t) = \frac{AV}{\pi} \int_0^{2\pi B_s(t-t_0+\tau/2)} \frac{\sin x}{x} dx - \frac{AV}{\pi} \int_0^{2\pi B_s(t-t_0-\tau/2)} \frac{\sin x}{x} dx \tag{2-73}$$

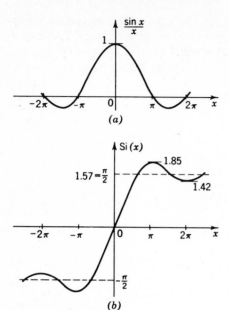

Figure 2-34 (a) $(\sin x)/x$ versus x. (b) Si x versus x.

Unfortunately, $\int_0^a [(\sin x)/x]\, dx$ cannot be evaluated in closed form but must be evaluated by expanding $(\sin x)/x$ in a power series in x and integrating term by term. Tables are available for the integral,[5] however, and it is called the sine integral of x,

$$\text{Si } x \equiv \int_0^x \frac{\sin x}{x}\, dx \qquad (2\text{-}74)$$

Equation (2-73) can thus be written

$$g(t) = \frac{AV}{\pi} \left\{ \text{Si}\left[2\pi B_s \left(t - t_0 + \frac{\tau}{2} \right) \right] - \text{Si}\left[2\pi B_s \left(t - t_0 - \frac{\tau}{2} \right) \right] \right\} \qquad (2\text{-}75)$$

The sine integral appears very frequently in the literature pertaining to pulse transmission through idealized networks. It represents the area under the $(\sin x)/x$ curve plotted previously and reproduced in Fig. 2-34a. It thus has its maxima and minima at multiples of π (the points at which $\sin x$ changes sign). The first maximum is 1.85 at $x = \pi$. The curve is odd-symmetrical about $x = 0$ and approaches $\pi/2 = 1.57$ for large values of x.

Since $(\sin x)/x$ is an even function and has zero slope at $x = 0$, the initial slope of Si x is linear and Si $x \doteq x$, $x \ll 1$. The sine integral is plotted in Fig. 2-34b.

[5] E. Jahnke and F. Emde, *Tables of Functions,* Dover, New York, 1945.

The response of an idealized low-pass filter to a rectangular pulse of width τ sec is given by Eq. (2-75) in terms of the sine integral. Figure 2-35 shows Eq. (2-75) plotted for different filter bandwidths B_s.

1. $$B_s = \frac{1}{5\tau} \qquad \left(B_s \ll \frac{1}{\tau} \right)$$

2. $$B_s = \frac{1}{\tau}$$

3. $$B_s = \frac{5}{\tau} \qquad \left(B_s \gg \frac{1}{\tau} \right)$$

The three cases are shown superimposed and compared with the rectangular-pulse input.

What conclusions can we draw from the curves of Fig. 2-35?

1. All three output curves are displaced t_0 seconds from the input pulse and are symmetrical about $t = t_0$. The negative-linear-phase characteristic assumed for the filter has thus resulted in a *time delay* equal to the slope of the filter phase characteristic. This of course agrees with the shifting theorem of Eq. (2-52).
2. The curves bear out the filter-bandwidth–pulse-width relations developed previously.
   (a) $B_s \ll 1/\tau$. With the filter bandwidth much less than the reciprocal of the pulse width the output is much broader than the input and peaks only slightly, i.e., is a grossly distorted version of the input. This approximates the impulse response of the filter.
   (b) $B_s = 1/\tau$. Here the output is a recognizable pulse, roughly τ sec in width, but far from rectangular. It is close to the triangle of Fig. 2-32b. The rise time is approximately half the pulse width.

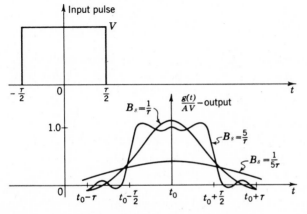

Figure 2-35 Response of low-pass filter.

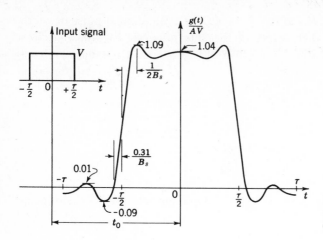

Figure 2-36 Signal transmission through ideal low-pass filter $(B_s \gg 1/\tau)$.

(c) $B_s \gg 1/\tau$. The output resembles the input closely and has approximately the same pulse width. There are several marked differences between output and input, however. To point these out, the curve for this case is replotted on an expanded time scale in Fig. 2-36. Note that the output pulse, although a delayed replica of the input, has a nonzero rise time. This was first noted in the previous section. This rise time is inversely proportional to the filter bandwidth, as was to be expected. In particular, if the rise time here is defined as the time for a pulse to rise from zero to its maximum value of $1.09\,AV$,

$$\text{Rise time} = 0.8\,\frac{1}{B_s} \qquad (2\text{-}76)$$

Alternatively, if the rising curve is approximated by its tangent at the point $g(t) = 0.5AV$, the rise time of the resultant straight line (0 to AV) is

$$\text{Rise time} = 0.5\,\frac{1}{B_s} \qquad (2\text{-}77)$$

(Note that these results are valid only if $B_s \gg 1/\tau$, as is the case here.)

These results are in agreement with those obtained in the previous section for the $(\sin x)/x$ filter. If the object is merely to produce an output pulse which has about the same width as the input pulse, with fidelity unimportant, then the filter bandwidth required is approximately the inverse of the pulse width. This is the situation, for example, in a search radar system, where a recognizable signal pulse is required but its shape is of secondary interest.[6] If fidelity is

---

[6] Noise considerations lead to the same bandwidth requirement, as will be seen later on. Search radar specifications thus indicate that the overall receiver bandwidth should be at least the reciprocal of the pulse width. (Actually the i-f bandwidth is normally specified and is $2B_s$.)

required, then the bandwidth specified must be at least several times the reciprocal of the pulse width and is actually determined by rise-time considerations. For example, in tracking radars the time of arrival of individual pulses must be accurately known. The output pulse of the radar receiver must rise quite sharply so that the leading edge of the pulse may be accurately determined. The same considerations hold for loran navigational systems, where the time of arrival of each pulse, or its leading edge, must be accurately known. In some pulse-modulation systems the bandwidth is also determined by the specified pulse rise time. (Note that ordinarily i-f bandwidths, rather than a low-pass equivalent, are involved. As will be shown later, the i-f bandwidth is twice the low-pass equivalent, or $2B_s$. From Eqs. (2-76) and (2-77), then, rise time is $1/B_s$ or $1.6/B_s$, depending on the rise-time definition in this case. Defining rise time as the time required for a pulse to rise from 10 to 90 percent of its peak value leads essentially to the same result.)

These results, of course, account for the rule of thumb used in practice in most pulse system work that the system bandwidth should be the reciprocal of the pulse rise time.

Some other interesting conclusions may be drawn from a study of Fig. 2-36. Note that the output pulse *overshoots,* or oscillates with damped oscillations, about the flat-top section of the pulse. This phenomenon is characteristic of filters with sharply cutoff amplitude response.

Figure 2-36 indicates that the output pulse actually has nonzero value for $t < -\tau/2$, *before* the input pulse has appeared. In fact, Eq. (2-75) gives nonzero values for negative time as well as positive time (although centered about $t = t_0$). This appearance of an output before the input producing it has appeared is obviously physically impossible and is due specifically to the nonphysically realizable filter characteristics assumed. Thus the rectangular amplitude characteristic of the idealized low-pass filter can never be realized with physical circuits. (It can be approached closely, but the number of elements required increases as the approximation becomes better. An exact fit requires an infinite number of elements theoretically. The filter phase-shift constant t_0 then becomes infinite, and the filter produces the amplitude of the pulse after infinite delay.) As pointed out previously, however, network idealizations are valuable since they frequently provide insight into system performance and enable general conclusions as to network response to be drawn.

In this section we have focused on the effect of amplitude spectral limiting on pulse transmission through a network. We chose an idealized low-pass filter with bandwidth B_s hertz and constant amplitude at frequencies $|f| < B_s$. We assumed the phase spectrum was linear with frequency.

It is left for the reader to show that a flat amplitude characteristic over all frequencies with linear phase shift results in a delayed replica of the input, although of possibly different magnitude. The delay is, of course, just given by the slope of the linear phase characteristic from the shifting theorem of Eq. (2-52). Thus if $\theta(\omega) = -\omega t_0$, the output pulse is delayed t_0 seconds with respect to the input one. A system in which the output is a delayed version of the input represents distortionless transmission, and is often very desirable in practice. It is again left for the reader to show that for distortionless transmission to occur, the amplitude-phase characteristic must be flat and linear with frequency, respectively, for all frequencies. In this section we investigated the effect of a non-flat-amplitude characteristic by assuming that the amplitude cuts off sharply at a prescribed bandwidth. In the next section we discuss briefly the effect of nonlinear phase variation on the signal transmission.

2-5 EFFECT OF PHASE VARIATION ON SIGNAL TRANSMISSION

We noted in the previous section that a distortionless-transmission system would require flat (uniform- or constant-) amplitude and linear-phase-shift characteristics over the range of frequencies covered by the signal.

In practice all linear systems introduce a certain amount of signal distortion because of bandwidth limitations and nonlinear phase characteristics. As pointed out previously, the amplitude and phase response of a given network are interrelated, and to treat the two characteristics separately in studying signal transmission could become an academic approach. We can gain some further insight into the effect of network characteristics on signal response, however, by treating amplitude and phase separately and attempting to superimpose the results.

We considered previously the effect of amplitude limiting on pulse transmission and arrived at the basic relation between system bandwidth and pulse rise time. The phase characteristic assumed was a linear one, and it is of interest to explore the question of network phase response further.

Note first that, although the effect of bandwidth limiting on an input pulse was to distort the pulse, the *distortion* was found to be *symmetrical:* the output waves were symmetrical about a delayed time $t = t_0$ (Fig. 2-35). This symmetry in the output pulses is a direct consequence of the linear-phase-shift characteristic assumed for the network. It is a general property of networks that where amplitude distortion occurs (as in band limiting) the output transient is symmetrically related to the input if the network phase shift is linear. Phase linearity in networks thus gives rise to the symmetrical pulse response and symmetrical step response and is particularly desirable in TV and radar systems.[7] Amplitude and phase distortion of a pulse input are illustrated in Fig. 2-37.[8]

The symmetrical-output property due to linear phase shift can be easily demonstrated. Assume an idealized linear network with frequency transfer function

$$H(\omega) = A(\omega)e^{-jt_0\omega} \tag{2-78}$$

[7] T. Murakami and M. S. Corrington, "Applications of the Fourier Integral in the Analysis of Color TV Systems," *IRE Trans. Circuit Theory,* vol. CT-2, no. 3, September 1955.

[8] C. Cherry, *Pulses and Transients in Communication Circuits,* Chapman & Hall, London, 1949; Dover, New York, 1950.

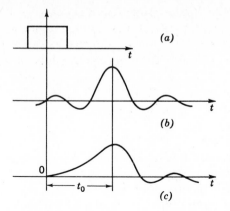

Figure 2-37 Network amplitude and phase distortion. (*a*) Input pulse. (*b*) Output response, amplitude distortion, linear phase shift (symmetrical). (*c*) Phase distortion (nonsymmetrical). (From C. Cherry, *Pulses and Transients in Communication Circuits*, Chapman & Hall, London, 1949, fig. 60, p. 147; Dover Publications, New York, 1950.)

The phase-shift characteristic is thus linear, with the amplitude characteristic $A(\omega)$ arbitrary. The input signal is $f(t)$, and the output response is $g(t)$. Then

$$G(\omega) = F(\omega)H(\omega) \qquad (2\text{-}79)$$

with $F(\omega)$ the Fourier integral of $f(t)$ and $G(\omega)$ the Fourier integral of $g(t)$. We assume that $f(t)$ is a symmetrical function in time, either odd or even. We have previously shown that if $f(t)$ is an even function, $F(\omega)$ is real and even in ω, while if $f(t)$ is an odd-symmetric function, $F(\omega)$ is imaginary and odd-symmetric in ω.

We should now like to prove that either type of symmetrical function produces a symmetrical function of the same type when passed through the idealized network with linear phase shift.

Case 1: $f(t)$ even. Passing $f(t)$ through the idealized network, we obtain

$$g(t) = \frac{1}{2\pi} \int_{-\infty}^{\infty} A(\omega)e^{-jt_0\omega} F(\omega)e^{jt\omega} \, d\omega$$

$$= \frac{1}{2\pi} \int_{-\infty}^{\infty} A(\omega)F(\omega)[\cos \omega(t - t_0) + j \sin \omega(t - t_0)] \, d\omega \qquad (2\text{-}80)$$

But $A(\omega)$ must be an even function of ω since it represents the magnitude of $H(\omega)$. $F(\omega)$ is even for $f(t)$ even. Therefore,

$$A(\omega)F(\omega) \cos \omega(t - t_0)$$

is an even function, and

$$A(\omega)F(\omega) \sin \omega(t - t_0)$$

is an odd function. Or, from Eq. (2-80),

$$g(t) = \frac{1}{\pi} \int_{0}^{\infty} A(\omega)F(\omega) \cos \omega(t - t_0) \, d\omega \qquad (2\text{-}81)$$

an even function in *time* and symmetrical about $t = t_0$.

Case 2: f(t) odd.   From Eq. (2-80), with $F(-\omega) = -F(\omega)$ and imaginary,

$$g(t) = \frac{1}{2\pi} \int_{-\infty}^{\infty} A(\omega)F(\omega)[\cos \omega(t - t_0) + j \sin \omega(t - t_0)] \, d\omega$$

$$= \frac{j}{\pi} \int_{0}^{\infty} A(\omega)F(\omega) \sin \omega(t - t_0) \, d\omega \tag{2-82}$$

$g(t)$ is thus symmetrical about $t = t_0$. [$\sin \omega(t - t_0) = -\sin \omega(t_0 - t)$.]

Any signal input $f(t)$ beginning at time $t = 0$ can be shown to be de-composable into a sum of an even and an odd function. The foregoing results may then be superimposed. The odd component produces an odd component of the output, the even component an even component of the output. The over-all output function $g(t)$ is thus also symmetrical although symmetrical about its average value. For example, if $f(t)$ is a step, it may be decomposed into the odd and even functions shown in Fig. 2-38b. The response of the filter with linear phase shift to the unit step is symmetrical about $g(t) = 0.5$ as shown in Fig. 2-38c.

Although the linear-phase network is an idealization, the phase shift of a given network may be made very nearly linear by the addition of phase-correction networks. (The all-pass network mentioned previously may be used for this purpose.)

As an example of the effect of a nonlinear phase characteristic on signals consider the step response of the simple RC low-pass filter of Fig. 2-39. The step response is of course

$$v_0(t) = 1 - e^{-t/RC} \qquad v_i(t) = u(t) \tag{2-83}$$

This is obviously *not* symmetrical about the $v_0(t) = 0.5$ point. The non-linearity of the filter phase characteristic is simply shown by writing the fre-quency transfer function:

$$H(\omega) = \frac{1}{1 + jRC\omega} = \frac{1}{\sqrt{1 + (\omega RC)^2}} e^{-j \tan^{-1} \omega RC} \tag{2-84}$$

The phase characteristic is thus, as noted previously,

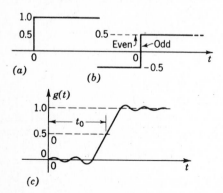

(a)

(b)

(c)

Figure 2-38 Response of linear-phase network to a unit step. (*a*) Unit step. (*b*) Even and odd com-ponents. (*c*) Output response [symmetrical about $g(t) = 0.5$].

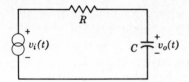

Figure 2-39 *RC* network.

$$\theta(\omega) = -\tan^{-1} \omega RC \tag{2-85}$$

A linear-phase network with the same amplitude characteristics would have as its transfer function

$$H(\omega)_{\text{linear phase}} = \frac{1}{\sqrt{1 + (\omega RC)^2}} e^{-j\omega t_0} \tag{2-86}$$

The step response of a network with this phase characteristic has been calculated by Murakami and Corrington, in the paper previously referred to, and is shown plotted in Fig. 2-40, together with the step response of the ordinary *RC* network. The slope of the phase-shift curve, t_0, is chosen so that both curves coincide at the 50 percent point. Note the symmetrical response about the 50 percent point of the linear-phase network and the precursor or anticipatory transient ($t < 0$), the result of "noncompatible" amplitude and phase characteristics. The 10 to 90 percent rise time is 5 percent better for the linear-phase case.

2-6 IMPULSES AND IMPULSE RESPONSE OF A NETWORK

In Sec. 2-3 we commented briefly on a pulse approximation to an impulse. We now pursue the concept of impulses and impulse response in more detail.

Specifically, let the input signal $f(t)$ to a network be the unit impulse function $\delta(t)$. Its Fourier transform is then

$$F(\omega) = \int_{-\infty}^{\infty} \delta(t)e^{-j\omega t} \, dt = 1 \tag{2-87}$$

since

$$\int_{-\infty}^{\infty} \delta(t)r(t) \, dt = r(0) \tag{2-88}$$

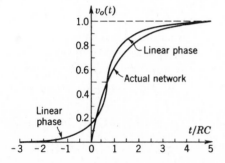

Figure 2-40 Step response of *RC* network: actual and linear phase. (From T. Murakami and M. S. Corrington, "Applications of the Fourier Integral in the Analysis of Color TV Systems," *IRE Trans. Circuit Theory*, vol. CT-2, no. 3, September 1955, by permission.)

is the defining relation of the unit impulse function. An impulse function thus has a *flat* frequency characteristic, providing equal amplitude at all frequencies. If this impulse is applied to a linear system, it is equivalent to exciting the system with *all* frequencies simultaneously. The system selects the frequencies to be outputted according to its own characteristic transfer function $H(\omega)$ and outputs their weighted sum as the impulse response $h(t)$:

With $G(\omega) = H(\omega)F(\omega) = H(\omega)$,

$$g(t) = h(t) = \frac{1}{2\pi} \int_{-\infty}^{\infty} H(\omega)e^{j\omega t} \, d\omega \tag{2-89}$$

The impulse response and the system transfer function represent a Fourier transform pair. This is an extremely useful relation in measuring the transfer function of any linear system. Instead of applying an oscillator and varying its frequency continuously, one can equally well excite the system with an impulse, measure the response $g(t) = h(t)$, and then take its Fourier transform $H(\omega)$:

$$H(\omega) = \int_{-\infty}^{\infty} h(t)e^{-j\omega t} \, dt \tag{2-90}$$

Any pulse-type signal whose bandwidth $B \gg B_s$ can serve as a good approximation to an impulse. Alternatively, all we need is a pulse of width $\tau \ll 1/B_s$. For example,

if $B_s = 1$ MHz $\qquad \tau \ll 1\ \mu s$
if $B_s = 10$ kHz $\qquad \tau \ll 100\ \mu s$
if $B_s = 10$ Hz $\qquad \tau \ll 0.1$ s

Two examples of the impulse response of a network have already been discussed. If $H(\omega)$ is of the $(\sin x)/x$ form, as shown in Fig. 2-31, $h(t)$ must be a rectangular pulse of width $1/B_s$ s (see Fig. 2-32a). If $H(\omega)$ is the ideal low-pass filter of Fig. 2-33, its impulse response must be of the $(\sin x)/x$ type:

$$h(t) = 2B_sA \, \frac{\sin 2\pi B_s(t - t_0)}{2\pi B_s(t - t_0)} \tag{2-91}$$

(see Fig. 2-18). Details are left to the reader.

Convolution

Using the impulse response $h(t)$ and its Fourier transform $H(\omega)$, one can actually prove the transfer function product form developed intuitively previously. Thus assume that we have a time-invariant linear system with impulse response $h(t)$. We can apply the convolution integral to obtain the response $g(t)$ to any input $f(t)$:

$$g(t) = \int_{-\infty}^{\infty} h(\tau)f(t - \tau) \, d\tau \tag{2-92}$$

Take the Fourier transform of both sides of this equation. One then gets

$$G(\omega) = \int_{-\infty}^{\infty} \left[\int_{-\infty}^{\infty} h(\tau)f(t - \tau)\, d\tau \right] e^{-j\omega t}\, dt$$

Interchanging the order of integration on the right-hand side and using Fourier transform relations and the shifting theorem, one finds

$$G(\omega) = H(\omega)F(\omega) \tag{2-93}$$

where
$$H(\omega) \leftrightarrow h(t) \tag{2-94}$$

and
$$F(\omega) \leftrightarrow f(t) \tag{2-95}$$

As a check on the usefulness of the convolution integral and its direct connection to our frequency concepts, take the case where both $f(t)$ and $h(t)$ are rectangular pulses starting at $t = 0$, and lasting τ seconds and $1/B_s$ seconds, respectively. This then corresponds to the example discussed earlier at the end of Sec. 2-3 (see Fig. 2-31). The convolution for the same three cases is carried out graphically in Fig. 2-41. Note the agreement with the previous results. Note also how the rise times arise naturally in this case.

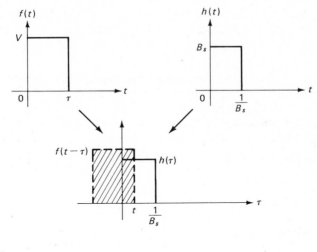

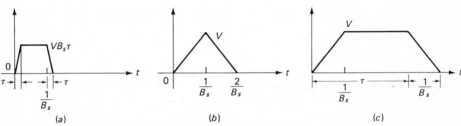

Figure 2-41 Convolution applied to example of Figs. 2-31 and 2-32. (a) $\tau < 1/B_s$. (b) $\tau = 1/B_s$. (c) $\tau > 1/B_s$.

Relations (2-92) and (2-93) may be written in more compact form by using an asterisk to represent the process of convolution. We then have the following equivalent Fourier transform pair representing the two equations:

$$g(t) = h(t) * f(t) \leftrightarrow G(\omega) = H(\omega)F(\omega) \tag{2-96}$$

A little thought will indicate that because of the symmetry in the Fourier transform–inverse Fourier transform relations there must exist an equivalent convolution in frequency when two time functions are multiplied together. The specific transform pair may be written concisely as follows:

$$g(t) = f_1(t)f_2(t) \leftrightarrow G(\omega) = \frac{1}{2\pi} F_1(\omega) * F_2(\omega) \tag{2-97}$$

The shorthand asterisk form of the right-hand side of (2-97) implies, as above, the following convolution integral:

$$G(\omega) = \frac{1}{2\pi} \int_{-\infty}^{\infty} F_1(x)F_2(\omega - x) \, dx \tag{2-98}$$

The proof of (2-97) is very similar to that of (2-96). One takes the inverse transform of both sides of (2-98), interchanges the order of integration, applies the shifting theorem in frequency, and recognizes that the product of the inverse transforms of $F_1(\omega)$ and $F_2(\omega)$ results. Details are left to the reader.

This convolution in frequency is a very useful concept. We shall find applications in Chap. 3 in discussing sampling of time functions, and in later chapters as well in discussing the frequency response of nonlinear and time-varying linear circuits. As an example, say that we apply a communications signal $f(t)$ to a square-law device. Such a device simply outputs the square of $f(t)$. We would like to know the spectrum at the output of the device. We shall encounter such devices in Chap. 4 in discussing the generation and detection of modulated signals. Specifically, then, let

$$g(t) = f^2(t) \tag{2-99}$$

From (2-97) we have

$$G(\omega) = \frac{1}{2\pi} \int_{-\infty}^{\infty} F(x)F(\omega - x) \, dx \tag{2-100}$$

As an example, let $f(t)$ be a $(\sin x)/x$ pulse in time. Its bandwidth is of course confined to B hertz, as shown at the input to the device in Fig. 2-42. The convolution of a rectangle with itself must then produce the triangular spectrum shown at the output in Fig. 2-42. Notice the characteristic spectral *broadening* due to the nonlinear device. Recall that linear time-invariant devices alter the amplitude and phase of the spectrum by multiplying by the transfer function $H(\omega)$. No new frequency terms are produced, however. This is the direct consequence of the superposition property. Nonlinear devices, and time-varying devices as well (as will be seen later in Chaps. 3 and 4), do produce new frequency terms. It is precisely this property that requires their use as mod-

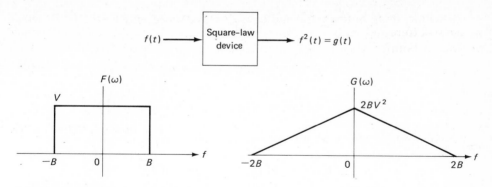

Figure 2-42 Spectrum at output of nonlinear device.

ulators and demodulators (or detectors) in communication systems. [As a simple check on the result of Fig. 2-42, let $f(t) = \cos \omega_0 t$. Then $f^2(t) = \frac{1}{2} + \frac{1}{2} \cos 2\omega_0 t$. The single frequency term has been shifted down to dc and up to twice the frequency by the process of squaring. Figure 2-42 could in fact be obtained by approximating $f(t)$ as the dense sum of cosinusoidal terms, squaring the sum, and then using the trigonometric sum and difference formulas, term by term. This is just an approximate form of evaluating Fourier transforms.]

Periodic Signals

Impulse functions can be used to represent periodic signals, and from this representation, Fourier-series spectra can be represented naturally in terms of discrete spectrum impulse functions. With this approach both periodic and aperiodic signals can be incorporated in a common Fourier-transform framework. To demonstrate this idea, recall that

$$\delta(t) \leftrightarrow 1 \tag{2-101}$$

and, from the shifting theorem,

$$\delta(t - t_0) \leftrightarrow e^{-j\omega t_0} \tag{2-102}$$

But recall as well the symmetry of the transforms in frequency and in time. Thus, let $F(\omega) = \delta(\omega - \omega_0)$. Then

$$f(t) = \frac{1}{2\pi} \int_{-\infty}^{\infty} F(\omega)e^{j\omega t} \, d\omega = \frac{1}{2\pi} e^{j\omega_0 t}$$

using the defining relation (2-88) for impulse functions. We thus have the following added transform pairs:

$$e^{j\omega_0 t} \leftrightarrow 2\pi\delta(\omega - \omega_0) \tag{2-103}$$

and, in a similar manner,

$$e^{-j\omega_0 t} \leftrightarrow 2\pi\delta(\omega + \omega_0) \tag{2-104}$$

In particular, then, both $\sin \omega_0 t$ and $\cos \omega_0 t$ may be presented by their Fourier transforms as well:

$$\sin \omega_0 t = \frac{e^{j\omega_0 t} - e^{-j\omega_0 t}}{2j} \leftrightarrow -j\pi[\delta(\omega - \omega_0) - \delta(\omega + \omega_0)] \qquad (2\text{-}105)$$

$$\cos \omega_0 t = \frac{e^{j\omega_0 t} + e^{-j\omega_0 t}}{2} \leftrightarrow \pi[\delta(\omega - \omega_0) + \delta(\omega + \omega_0)] \qquad (2\text{-}106)$$

The $\sin \omega_0 t$ amplitude spectrum thus consists of two impulse functions, each of weight π, at $\omega = -\omega_0$ and $\omega = +\omega_0$. Its phase spectrum is $\pi/2$ radians at $\omega = -\omega_0$ and $-\pi/2$ at $\omega = \omega_0$. These are shown in Fig. 2-43.

In general, we know that a periodic signal is representable by the complex Fourier series of Eq. (2-10). Taking Fourier transforms term by term and using (2-103) and (2-104), we get

$$f(t) = \frac{1}{T} \sum_{n=-\infty}^{\infty} c_n e^{j\omega_n t} \leftrightarrow F(\omega) = \frac{2\pi}{T} \sum_{n=-\infty}^{\infty} c_n \delta(\omega - \omega_n) \qquad (2\text{-}107)$$

Any periodic signal may thus be represented by a discrete spectrum Fourier transform consisting of impulses at the harmonics of the periodic signal. This is of course the same as the Fourier-series spectrum, except that weighted impulses are used here in place of the lines used previously. This is extremely useful in describing the Fourier transform of combined periodic and aperiodic signals. One simply invokes the impulses at the harmonics of the periodic signals to represent the periodic contribution.

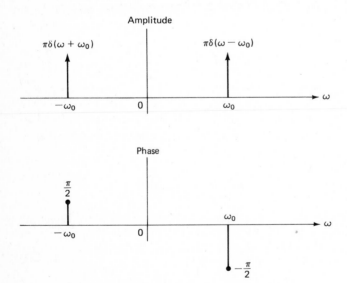

Figure 2-43 Fourier transform spectrum of $\sin \omega_0 t$.

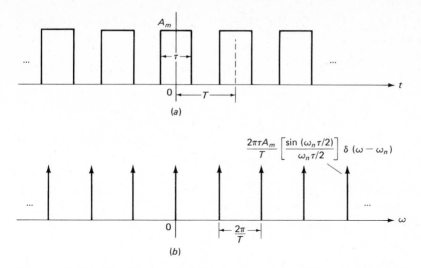

Figure 2-44 Fourier transform of rectangular pulse train. (*a*) Pulses. (*b*) Spectrum.

As a simple example consider the periodic rectangular pulse train of Fig. 2-5. Taking these pulses to be centered at the time origin, we have

$$F(\omega) = \frac{2\pi\tau A_m}{T} \sum_{n=-\infty}^{\infty} \left[\frac{\sin(\omega_n\tau/2)}{\omega_n\tau/2} \right] \delta(\omega - \omega_n) \qquad \omega_n = \frac{2\pi n}{T} \qquad (2\text{-}108)$$

This is shown sketched in Fig. 2-44. As a special case, let $\tau \to 0$, $\tau A_m = 1$. The periodic pulses become a train of periodic impulses. For this case we get complete symmetry: periodic impulses in time are representable by a set of periodic impulses in frequency. Thus we have

$$f(t) = \sum_{n=-\infty}^{\infty} \delta(t - nT) \leftrightarrow F(\omega) = \frac{2\pi}{T} \sum_{n=-\infty}^{\infty} \delta(\omega - \omega_n) \qquad (2\text{-}109)$$

This Fourier-transform pair is shown sketched in Fig. 2-45. The reader is asked to compare this result with the equivalent result of Fig. 2-9. Note that the spectrum there is given by the Fourier-series coefficients. The spectrum of Fig. 2-45, on the other hand, is that of the Fourier integral or transform. It is left to the reader to show that the two results are identical.

2-7 TELEVISION BANDWIDTH REQUIREMENTS

Before concluding this chapter on signal transmission through linear networks, we give one further example—the calculation of the bandwidth needed to transmit a typical TV test pattern.[9]

[9] S. Deutsch, *Theory and Design of Television Receivers*, McGraw-Hill, New York, 1951.

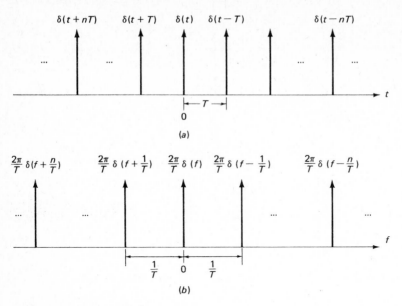

Figure 2-45 Fourier transform of periodic impulses. (a) f(t): periodic impulses in time. (b) F(ω): periodic impulses in frequency.

The pattern we choose consists of a series of alternating black and white spots. The pattern is 6 in high by 8 in wide, and the spots are 0.0121 in high and 0.0188 in wide. (These dimensions have been chosen as a compromise between the ability of the eye to resolve detail and the frequency bandwidth required.) There are then 6/0.0121 = 495 horizontal lines that must be covered by a scanning beam of electrons. The equivalent of 30 additional lines is allowed to enable the beam to retrace from the final line back to the first again. (Actually, interlaced scanning is used in which even-numbered lines are first covered, then odd-numbered lines. This increases the time allotted for scanning the picture. The persistence of the eye is then relied on to give the effect of a more rapid scanning rate.) The total number of lines scanned is thus 525 (the time for the 30 additional lines is blanked out). The standard scanning rate is 30 frames per second, so that each line is scanned in

$$\frac{1}{30 \times 525} = 63.5 \ \mu s$$

This represents the time in which the beam must sweep from left to right through the 8 in of the pattern and return to the left side again. Allowing 10 μs for the return, or horizontal, trace interval leaves 53.5 μs for the actual sweep. In this time 8/0.0188 = 425 alternating black and white spots will be swept through. The electrical output will thus consist of a series of square waves of 53.5/425 = 0.125 μs width.

Our problem is to determine how much bandwidth is needed to resolve these square-wave pulses of 0.125 μs duration, separated by the same time

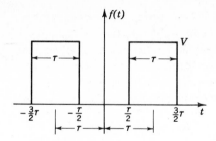

Figure 2-46 Two test pulses.

interval. We can relate this to the problem of resolving two pulses of width τ, separated by τ seconds (τ is 0.125 μs). These pulses are shown in Fig. 2-46.

The response of the idealized low-pass filter to a single pulse was given by Eq. (2-75). Since the system is assumed linear, superposition may be applied. The two pulses applied to the low-pass filter with bandwidth B_s hertz then produce a response at the output given by

$$g(t) = \frac{AV}{\pi} \left\{ \text{Si} \left[2\pi B_s(t - t_0) + \frac{3\tau}{2} \right] - \text{Si} \left[2\pi B_s(t - t_0) + \frac{\tau}{2} \right] \right.$$
$$\left. + \text{Si} \left[2\pi B_s(t - t_0) - \frac{\tau}{2} \right] - \text{Si} \left[2\pi B_s(t - t_0) - \frac{3\tau}{2} \right] \right\} \quad (2\text{-}110)$$

This equation has been plotted by S. Goldman in his book,[10] and some of these curves are reproduced in Fig. 2-47. (t_0 is assumed zero so that the input and output curves are superimposed for comparison.) These results indicate that for resolution of (the ability to separate) the two pulses

$$B_s \geq \frac{1}{2\tau} \quad (2\text{-}111)$$

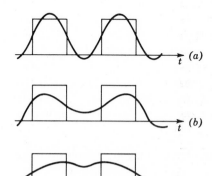

Figure 2-47 Low-pass filter response to TV test pattern (two pulses). (a) $B_s = 1/\tau$. (b) $B_s = 1/2\tau$. (c) $B_s = 1/3\tau$. (From S. Goldman, *Frequency Analysis, Modulation, and Noise*, McGraw-Hill, New York, 1948).

[10] S. Goldman, *Frequency Analysis, Modulation, and Noise*, McGraw-Hill, New York, 1948.

Note that this is a different situation from the one previously stressed in this chapter, in which it was desired to reproduce the pulse faithfully ($B_s \gg 1/\tau$). Here the problem is to distinguish between black and white spots or between an on and off signal. The signal detail as such is not important in this application.

If $B_s < 1/2\tau$, the two pulses merge into one and the resolution disappears. If $B_s > 1/2\tau$, the sides of the pulses are sharpened, the rise times decreasing.

For the TV case outlined $\tau = 0.125 \ \mu s$, and

$$B_s \geq 4 \ \text{MHz}$$

It will be shown later that in order to transmit these signals at high frequencies bandwidths of $2B_s = 8$ MHz will be required. In practice some bandwidth compression is utilized (vestigial sideband transmission), and 6-MHz bandwidths are prescribed for home TV receivers.

2-8 SUMMARY

Major emphasis has been placed in this chapter on developing, through different examples, the inverse frequency-time relationship of signal transmission.

We demonstrated, first through the use of the Fourier-series representation for periodic functions and then through the Fourier-integral representation of nonperiodic functions, that more rapid time variations in a signal give rise to the higher-frequency components in the signal spectrum.

A general conclusion drawn from an analysis of signal transmission through linear networks was that the system bandwidth had to be approximately the reciprocal of the signal duration in order to produce at the system output a signal of the same general form as the input. For a symmetrical output signal the system phase characteristics had to be linear.

High-fidelity signal reproduction, or reproduction of detail, required bandwidths in excess of the reciprocal of the signal duration. The bandwidths needed were found to be about the reciprocal of the rise time, or the time taken for the signal to change from one level to another.

These conclusions are of importance in the design of practical system circuitry and are also of considerable theoretical importance in the study of information transmission through communication systems.

We recall that in Chap. 1 we discussed in a qualitative way the two system limitations on the amount of information per unit time (system capacity) a system could transmit:

1. Inability of the system to respond instantaneously to signal changes (due to the presence of energy-storage devices)
2. Inability of the system to distinguish infinitesimally small changes in signal level (due to inherent voltage fluctuations or noise)

These two limitations were tied together in a simple expression developed for system capacity,

$$C = \frac{1}{\tau} \log_2 n \qquad \text{bits/s} \qquad (2\text{-}112)$$

where τ was the minimum time required for the system to respond to signal changes and n the number of distinguishable signal levels.

In this chapter we have found that this minimum response time is proportional to the reciprocal of the system bandwidth (two alternative ways of referring to the same phenomenon). System capacity thus could be written

$$C = B \log_2 n \qquad \text{bits/s} \qquad (2\text{-}113)$$

where B is the system bandwidth in hertz.

In the next chapters we shall discuss some common communication systems and their bandwidth requirements. We shall then return to a discussion of the second limitation on the system capacity—inherent noise.

PROBLEMS

2-1 (*a*) Find the Fourier-series representations of each of the pulse trains in Fig. P2-1. Choose the time origin so that a cosine series is obtained in each case.

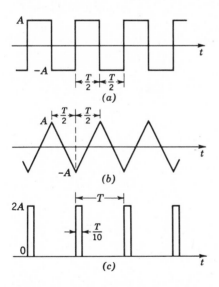

Figure P2-1

(*b*) Plot the first 10 Fourier coefficients vs. frequency for each pulse train. Compare the plot of Fig. P2-1*a* with each of the other two, paying particular attention to the rate of decrease of the higher-frequency components. Note that Fig. P2-1*b* represents in form the integral of Fig. P2-1*a* and has no discontinuities in the function. The pulses of Fig. P2-1*c* are much narrower than those of Fig. P2-1*a*.

2-2 (*a*) Find the cosine Fourier-series representation of the half-wave rectified sine wave of Fig. P2-2.

(*b*) Compare the successive Fourier coefficients and their rate of decrease with those of Fig. P2-1*a*.

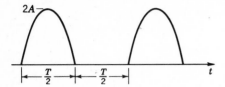

Figure P2-2

2-3 Find the complex Fourier series for the two pulse trains of Fig. P2-3. Plot and compare the two amplitude spectra. What is the significance of the term $e^{-j\omega_n t_0}(\omega_n = 2\pi n/T)$ in the expression for the complex Fourier coefficient for the rectangular pulses?

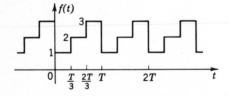

Figure P2-3

2-4 Find the complex Fourier series for the periodic function of Fig. P2-4. *Hint:* Use superposition and the result of Prob. 2-3.

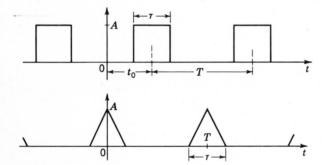

Figure P2-4

2-5 Consider the three cases of a rectangular, a triangular, and a half cosine pulse train, all with $\tau/T = \frac{1}{10}$ (Fig. P2-5). Plot the amplitude spectra to the first zero crossing in each. What is the percent of the total power in the first 10 frequency components?

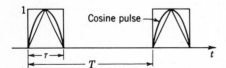

Figure P2-5

2-6 Find the complex Fourier series for the periodic function of Fig. P2-6. Find the percent power in the first six components.

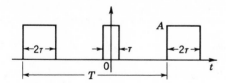

Figure P2-6

2-7 Plot the frequency spectra for the pulses of Fig. P2-7.

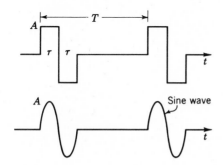

Figure P2-7

2-8 Find the complex Fourier series for the two periodic functions of Fig. P2-8.

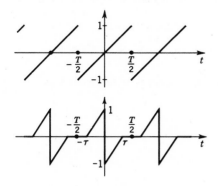

Figure P2-8

2-9 $f(t) = \sin \omega t$ for $0 \le t \le \pi/\omega$. It is undefined for values of t outside this interval.

(*a*) Express $f(t)$ inside the given interval as a sum of cosine terms with fundamental radian frequency ω.

(*b*) Express $f(t)$ in the same interval as a sum of sine terms with fundamental radian frequency ω.

2-10 T is the period of expansion in the Fourier-series representation for the function $f(t)$.

(*a*) Show that if $f(t) = f(t + T/2)$ the Fourier series will contain no odd harmonics.

(*b*) Show that if $f(t) = -f(t + T/2)$ the Fourier series will contain no even harmonics.

2-11 Each of the pulse trains shown in Fig. P2-11 represents a voltage $v(t)$ appearing across a load resistor of 1 Ω.

(*a*) Find the total average power dissipated in the resistor by each voltage.

(*b*) Find the percentage of the total average power contributed by the first-harmonic (fundamental) frequency of each voltage.

(*c*) Find the percentage of the total average power due to the first 10 harmonics of each voltage.

(*d*) Find the percentage of the total average power due to those harmonics of the voltage within the first zero-crossing interval of the amplitude spectrum.

(*e*) Add a fixed-bias (dc) level E volts to each of the pulse trains. Sketch the spectrum of each as E is increased from zero to 10 V.

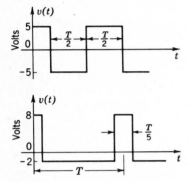

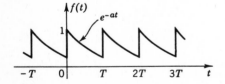

Figure P2-11

2-12 The duty cycle of a train of rectangular pulses is defined to be the ratio of time on to time off, or τ/T in the notation of this book. Sketch the pulse trains and their corresponding frequency spectra for the following cases:

(*a*) 0.1 duty cycle; (1) pulse width τ of 1 μs; (2) pulse width of 10 μs. Compare the frequency spectra.

(*b*) Repetition period T is 1 ms; (1) τ is 10 μs; (2) τ is 1 μs. What is the effect of varying τ, with T fixed, on the frequency scale? Compare the time and frequency plots.

(*c*) 10-μs pulses; (1) 0.1 duty cycle; (2) 0.001 duty cycle. What is the effect of varying T, with τ fixed, on the frequency spectrum?

2-13 Find the complex Fourier coefficient c_n, and write the Fourier series for the function shown in Fig. P2-13. Leave the coefficients in complex form.

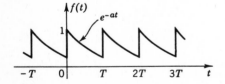

Figure P2-13

2-14 (*a*) Show that the complex Fourier coefficient c_n of the pulse train of Fig. P2-14 is given by $c_n = 2A\tau[\sin{(\omega_n\tau/2)}/(\omega_n\tau/2)]\cos{\omega_n\tau}$. *Hint:* Use the result of Prob. 2-3 and the principle of superposition.

(*b*) Sketch the spectrum envelope and indicate the location of the spectral lines for $\tau/T = 0.1$.

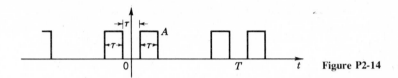

Figure P2-14

2-15 The teletype printer uses a code consisting of seven units or time intervals per character (letter). This includes start and stop pulses during the first and last time interval corresponding to each letter. During the remaining five intervals the signal may be either on (a 1) or off (a 0). The average character has the form shown in Fig. P2-15.

Assuming that the average word in the English language contains five letters and that one character is needed to transmit a space between words, there are on the average six characters per word. The rate of transmission is 25 words per minute.

(*a*) Find the length of one time interval.

(*b*) Find the bandwidth needed to transmit the first five harmonics of the pulse train of Fig. P2-15. What bandwidth would be needed if the rate of transmission were increased to 100 words per minute? Why are these bandwidths the maximum required to transmit a telegraph signal?

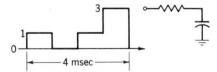

Figure P2-15

2-16 A pulse-code-modulation (PCM) system produces symmetrical trapezoidal-shaped pulses, 5 μs wide at the top and 5.5 μs wide at the base. The average spacing between pulses is 125 μs.

(*a*) What is the approximate system bandwidth required?

(*b*) If the average pulse spacing were halved, what bandwidth would be required?

(*c*) The pulses are passed through a low-pass filter cutting off at 200 kHz. Sketch the waveshape at the output of the filter.

2-17 (*a*) At what frequency does the amplitude spectrum of the four-pulse group of Fig. P2-17 first go to zero?

(*b*) The pulse group in (*a*) is put through a single low-pass RC section having as its half-power frequency the frequency found in (*a*). Sketch the output waveshape.

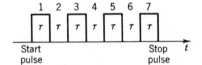

Figure P2-17

2-18 (*a*)What is the bandwidth in which 90 percent of the power in the trapezoidal pulses of Fig. P2-18 is contained?

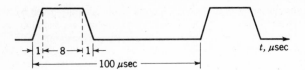

Figure P2-18

(*b*) What is the effect of passing these pulses through a single *RC* filter with a 3-dB (half-power) frequency of 318 kHz?

2-19 Find the Fourier transform of each of the pulses shown in Fig. P2-19. Compare with the complex Fourier coefficients of the corresponding pulse trains obtained in Prob. 2-5.

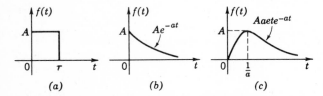

Rectangular pulse   Half-cosine pulse   Triangular pulse   **Figure P2-19**

2-20 The Fourier transform of $f(t)$ is $F(\omega)$, and the Fourier transform of $g(t)$ is $G(\omega)$. For each of the following cases, find $G(\omega)$ in terms of $F(\omega)$:

(*a*) $g(t) = f(3t)$.

(*b*) $g(t) = f(t + a)$.

2-21 (*a*) Show that for each of the pulses of Prob. 2-19 any "bandwidth" definition (e.g., the first zero crossing of the spectrum) would give $B = K/\tau$, K a constant.

(*b*) Superimpose sketches of $|F(\omega)|$ for each of these pulses, and compare the spectrum curves. Focus attention particularly on the l-f and h-f ends of the spectra.

2-22 Find the Fourier transform of each of the pulses of Fig. P2-22.

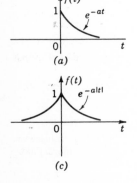

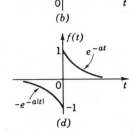

(*a*)   (*b*)   (*c*)   **Figure P2-22**

2-23 Find the Fourier transform of each of the functions of Fig. P2-23. The last two time functions are even and odd, respectively. What, therefore, is to be expected of their Fourier transforms?

(*a*)   (*b*)

(*c*)   (*d*)   **Figure P2-23**

2-24 Find the Fourier transforms of the pulses of Figs. 2-14 to 2-16 and show they are given by (2-33), (2-35), and (2-37), respectively. *Hint:* Represent the cosine as the sum of two exponentials. Superimpose sketches of the spectra. Check the comments made in the text on the locations of the first zero crossings and variations of the spectra with frequency. Check that the dc content $F(0)$ is in each case given by the area under the curve [Eq. (2-43)].

2-25 Check that the inverse Fourier transform of the rectangular spectrum of Fig. 2-18 is given by the $(\sin x)/x$ time function of Eq. (2-39).

2-26 Use the Fourier transform relation of Eq. (2-50), connecting the derivative of a function to the Fourier transform of that function, to verify the Fourier transforms of Figs. 2-20 and 2-21.

2-27 Prove the shifting theorem of Eq. (2-52).

2-28 Consider the delayed exponential $f(t) = 2e^{-5(t-10)}U(t - 10)$ shown in Fig. P2-28 [$U(t)$ is the unit-step function]. Find its Fourier transform $F(\omega)$. Sketch both $|F(\omega)|$ and $\angle F(\omega)$.

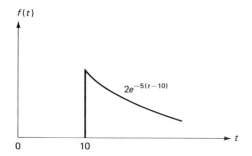

Figure P2-28

2-29 Find and sketch the spectrum of a cosinusoidal pulse, $A \cos \omega_0 t$, $-\tau/2 \le t \le \tau/2$; $= 0$, elsewhere. Compare with the spectrum of a rectangular pulse of width τ, centered about $t = 0$. *Hint:* Show the cosinusoidal pulse is represented by a rectangular pulse $\times \cos \omega_0 t$.

2-30 Find the Fourier transform of $f(t) = e^{-a|t|} \cos \omega_0 t$.

2-31 Find the frequency spectrum of the output voltage of Fig. P2-31. Leave the answer in complex form.

 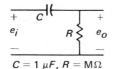 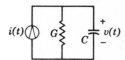

Figure P2-31   **Figure P2-32**

2-32 For the circuit and the input current shown in Fig. P2-32, find the capacitor voltage as a function of time. Leave the result in integral form.

2-33 "Integration smoothes out time variations; differentiation accentuates time variations." Verify this statement for the two circuits shown in Fig. P2-33, by comparing the input and output spectra with an arbitrary input $v_i(t)$.

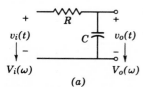 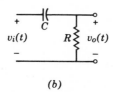

(a)   **(b)**

Figure P2-33 (*a*) Integrator $(R \gg 1/j\omega C)$. (*b*) Differentiator $(R \ll j\omega C)$.

2-34 A delay line and integrating circuit, combined as shown in Fig. P2-34, are one example of a "holding circuit" commonly used in radar work, sampled-data servo systems, and pulse-modulation systems.

(a) Tracing through the circuit step by step, show that the frequency transfer function is $H(\omega)$ $= e^{-j\omega\tau/2} \sin(\omega\tau/2)/(\omega\tau/2)$.

(b) If $v_i(t)$ is a rectangular pulse of width τ seconds, show that the output is a triangular pulse of width 2τ seconds, by (1) actually performing the successive operations indicated in the figure; (2) using the spectrum approach, the results of (a), and Eq. (2-33).

2-35 A pulse of the form $e^{-at}u(t)$ with $u(t)$ the unit-step function is applied to an idealized network with transfer characteristic $H(\omega) = Ae^{-jt_0\omega}$, $|\omega| \le \omega_c$; $H(\omega) = 0$, $|\omega| > \omega_c$.

(a) Show that the output time response is given by

$$g(t) = \frac{A}{\pi} \int_0^{\omega_c} \left[\frac{a \cos \omega(t - t_0)}{a^2 + \omega^2} + \frac{\omega \sin \omega(t - t_0)}{a^2 + \omega^2} \right] d\omega$$

(b) Show that, for $\omega_c \gg a$,

$$g(t) = \frac{A}{\pi} \text{Si } \omega_c(t - t_0) + \frac{A}{2} e^{-a|t-t_0|}$$

$$\text{Si } x \equiv \int_0^x \frac{\sin x}{x} dx$$

Note that $\int_0^\infty [(\cos mx)/(1 + x^2)] dx = (\pi/2)e^{-a|m|}$.

Differencer

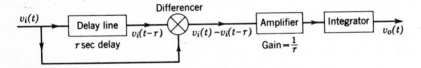

Figure P2-34 Holding circuit, $(\sin x)/x$ filter.

2-36 Using the results of Prob. 2-35, show that the *unit-step response* of an idealized network with amplitude A, cutoff frequency ω_c, and phase constant t_0 is given by

$$A(t) = A\left[\frac{1}{2} + \frac{1}{\pi} \text{Si } \omega_c(t - t_0) \right]$$

Sketch $A(t)/A$.

2-37 The unit-step response of the idealized network of Prob. 2-36 can also be obtained by first finding the response to a pulse of unit amplitude and width τ. As τ is allowed to get very large, the pulse approaches a unit step. Show that the network response approaches the unit-step response of Prob. 2-36.

2-38 A filter has the following amplitude and phase characteristics:

$$A(\omega) = 1 - \alpha + \alpha \cos 2\pi n \frac{\omega}{\omega_c} \qquad |\omega| \le \omega_c$$

$$A(\omega) = 0 \qquad |\omega| > \omega_c$$

$$\theta(\omega) = -t_0\omega$$

(This represents a filter with ripples in the passband.)

(a) Show that the step response has the form

$$A(t) = \frac{1}{2} + \frac{1}{\pi}\left\{(1-\alpha)\text{ Si }\omega_c(t-t_0) + \frac{\alpha}{2}\text{ Si }[\omega_c(t-t_0) + 2\pi n] + \frac{\alpha}{2}\text{ Si }[\omega_c(t-t_0) - 2\pi n]\right\}$$

(Use the approach of either Prob. 2-36 or Prob. 2-37.)

(b) Sketch $A(t)$ if $\alpha = \frac{1}{4}$, $n = 2$. Use the straight-line approximation for Si x indicated in Fig. P 2-38.

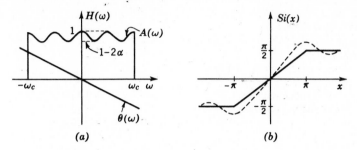

(a)   (b)

Figure P2-38 (a) Low-pass filter with ripples. (b) Linear approximation to Si x.

2-39 In the filter characteristic of Prob. 2-38 let $2n = 1$ and $\alpha = \frac{1}{2}$. Sketch the amplitude characteristic for this case. Find the step response of this filter, and compare with the step response of the idealized filter of Prob. 2-36. This shows the effect of rounding off the sharp corners of the ideal low-pass filter.

2-40 Two students were asked the effect of passing the square wave of Fig. P2-40 through the high-pass RC network shown. John reasoned that since the fundamental frequency of the square wave was four octaves above the 3-dB falloff frequency of the filter its attenuation and that of all the harmonics should be negligible. The output should thus look like the input. Conrad disagreed and to prove his point set up an experiment and obtained the oscilloscope trace for the output shown in the lower part of Fig. P2-40.

Explain the fallacy (or fallacies) in John's argument. Describe the characteristics of a network to be connected in tandem with the RC network so that no overall distortion will result.

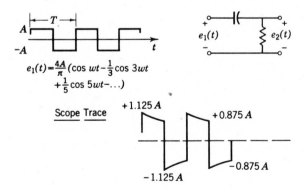

$$e_1(t) = \frac{4A}{\pi}\left(\cos wt - \frac{1}{3}\cos 3wt + \frac{1}{5}\cos 5wt - \ldots\right)$$

Figure P2-40

2-41

$$H(\omega) = \frac{1}{a + j\omega} = |H(\omega)|e^{j\theta(\omega)}$$

(a) Sketch $|H(\omega)|$ and $\theta(\omega)$.
(b) Find and sketch the impulse response of a network with this transfer function.

(*c*) Show a simple *RC* circuit that has a transfer function of this type. Compare the impulse response of the *RC* circuit with the result of (*b*).

2-42 Show that the impulse response of the ideal low-pass filter of Fig. 2-33, with linear phase characteristic defined as in Eq. (2-67), is given by (2-91). Sketch this response and relate the time difference between successive zero crossings to the bandwidth B_s.

2-43 Prove Eq. (2-97), relating convolution in frequency to the multiplication of two time functions. Use this transform pair, combined with (2-106), the Fourier transform of cos $\omega_0 t$, to prove the frequency shift property of $f(t)$ cos $\omega_0 t$.

2-44 Refer to Fig. P2-44, $RC = 1$ ms. Sketch the response $v_0(t)$ of the *RC* circuit to the excitation $f(t)$ over a 3-ms interval for four cases:

    (*a*) $f(t)$ is a unit-amplitude rectangular pulse 1 ms wide.
    (*b*) $f(t)$ is a unit-amplitude pulse 1 μs wide.
    (*c*) $f(t)$ consists of two unit-amplitude pulses, each 1 μs apart wide, spaced 1 μs apart.
    (*d*) $f(t)$ consists of two unit-amplitude pulses, each 1 μs wide, spaced 1 ms apart.
    In which of these cases may $f(t)$ be approximated by an impulse function? Explain.

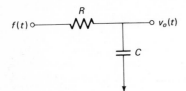

Figure P2-44

2-45 Repeat Prob. 2-44 if instead of the *RC* circuit an ideal low-pass filter of bandwidth $B_s = 1$ kHz is used.

2-46 Find the Fourier transforms of the two pulse trains of Prob. 2-3.

2-47 A time function $f(t)$ is multiplied by a periodic set of unit impulses, as shown in Fig. P2-47. The Fourier transform $F(\omega)$ is band-limited to (has no frequency components above) B hertz as shown. Use frequency convolution to find the spectrum at the multiplier output. Sketch for the three cases $1/T = 4B$, $1/T = 2B$, and $1/T = B$.

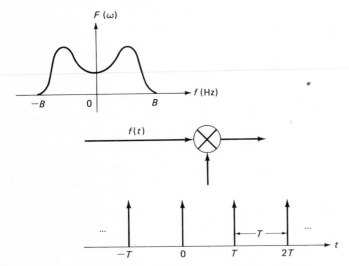

Figure P2-47

THREE

DIGITAL COMMUNICATION SYSTEMS

3-1 INTRODUCTION

In this book we are basically interested in investigating the information-handling capabilities of different communication systems. We showed in Chap. 2 that one important parameter determining the system information capability is the system bandwidth. In Chap. 5 we shall discuss the limitations on system performance due to noise.

We pause at this point in our discussion to describe some typical digital communication systems, to compare them on the basis of bandwidth requirements, and to develop further the inverse frequency-time relationship of Chap. 2.

The stress here will be on digital systems because of their paramount importance in modern-day technology and because the concepts important to information transmission are so easily developed through the study of digital systems. This was first pointed out in Chap. 1, where the transmission of digital signals step by step through a typical system was discussed.[1]

---

[1] Good overall references on digital systems, covering much of the material of this chapter in more detail, are the books *Data Transmission,* W. R. Bennett and J. R. Davey, McGraw-Hill, New York, 1965 and *Principles of Data Communication,* R. W. Lucky, J. Salz, and E. J. Weldon, McGraw-Hill, New York, 1968. The emphasis in these books is on digital systems developed primarily for telephone usage. Various aspects of digital space systems are considered in S. W. Golomb (ed.), *Digital Communications with Space Applications,* Prentice-Hall, Englewood Cliffs, N.J., 1964; A. V. Balakrishnan (ed.), *Space Communications,* McGraw-Hill, New York, 1965; E. L. Gruenberg (ed.), *Handbook of Telemetry and Remote Control,* McGraw-Hill, New York, 1967; and J. J. Spilker, Jr., *Digital Communications by Satellite,* Prentice-Hall, Englewood Cliffs, N.J., 1977. A more general book is the one by W. C. Lindsey and M. K. Simon, *Telecommunications System Engineering,* Prentice-Hall, Englewood Cliffs, N.J., 1973. Pulse code modulation (PCM) is covered in detail in the book by K. W. Cattermole, *Principles of Pulse Code Modulations,* Illiffe Books Ltd., London, 1969; American Elsevier, New York, 1975.

The current widespread use of digital signaling is the result of many factors.

1. The relative simplicity of digital circuit design and the ease with which one can apply integrated circuit techniques to digital circuitry
2. The ever-increasing use and availability of digital processing techniques
3. The widespread use of computers in handling all kinds of data
4. The ability of digital signals to be coded to minimize the effects of noise and interference

Although some communications signals are inherently digital in nature—e.g., teletype data, computer outputs, pulsed radar and sonar signals, etc., many signals are analog, or smooth, functions of time. If these signals are to be transmitted digitally, they first have to be sampled at a periodic rate and then further converted to discrete amplitude samples by quantization. The sampling procedure in this analog-to-digital (A/D) conversion process will be the first topic treated in this chapter. Speech, TV, facsimile, and telemetered data signals are all examples of analog signals that are often transmitted digitally.

We shall discuss time multiplexing of digital signals, in which signals from different sources are sequentially transmitted through the system. (Sine-wave carrier modulation by these signals for remote transmission will be discussed in Chap. 4.) Some of the problems arising in the use of digital systems will also be discussed. These include time synchronization, intersymbol interference, quantization noise, transmission bandwidth requirements, etc. Examples will be given of various types of digital systems in use.

3-2 NYQUIST SAMPLING

Consider a continuous varying signal $f(t)$ that is to be converted to digital form. We do this simply by first sampling $f(t)$ periodically at a rate of f_c samples per second. Although in practice this sampling process would presumably be carried out electronically by gating the signal on and off at the desired rate, we show the sampling process conceptually in Fig. 3-1 using a rotating mechanical switch.

Assume that the switch remains on the $f(t)$ line τ seconds, while rotating at the desired rate of $f_c = 1/T$ times per second ($\tau \ll T$). The switch output $f_s(t)$ is then a sampled version of $f(t)$. A typical input function $f(t)$, the sampling times,

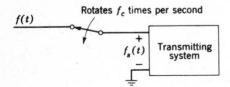

Figure 3-1 Sampling of analog signal.

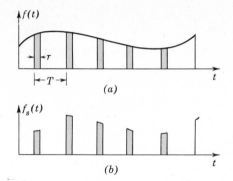

Figure 3-2 The sampling process. (τ = sampling time; $T = 1/f_c$: sampling interval.) (*a*) Input $f(t)$. (*b*) Sampled output $f_s(t)$.

and the sampled output $f_s(t)$, are shown in Fig. 3-2. f_c is called the sampling rate, and T is the sampling interval.

The question that immediately arises is: What should the sampling rate be? Are there any limits to the rate at which we sample? One might intuitively feel that the process of sampling has irretrievably distorted the original signal $f(t)$. The process of sampling has been introduced to convert the signal $f(t)$ to a digital form for further processing and transmission. Yet eventually, and at a remote location, in most cases, one would like to retrieve $f(t)$ again. Have we lost valuable information in the sampling process?

The answer to this last question, and from this, to the other questions asked, is that under a rather simple assumption (closely approximated in practice), the sampled signal $f_s(t)$ contains within it *all* information about $f(t)$. Further, $f(t)$ can be uniquely extracted from $f_s(t)$! This rather amazing and not at all obvious result can be demonstrated through the use of the Fourier analysis of Chap. 2.

We first assume the signal $f(t)$ is *band-limited* to B hertz. This means there are absolutely no frequency components in its spectrum beyond $f = B$. The Fourier transform $F(\omega)$ of such a signal is shown in Fig. 3-3. Physically occurring signals generally do not have the sharp frequency cutoff implied by the concept of bandlimitedness. Except for a few singular cases, examples of which were given in Chap. 2, real signals contain frequency components out to all frequencies. Yet we know from our discussion in Chap. 2 that the frequency content drops rapidly beyond some defined bandwidth. This approximation of real signals by band-limited ones introduces no significant error in the analysis and will therefore henceforth be assumed. (In practice, sharp-cutoff low-pass

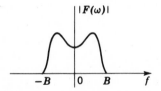

Figure 3-3 Band-limited signal.

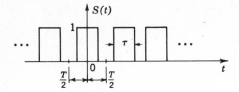

Figure 3-4 Periodic switching function.

filters are frequently introduced before the sampling process to ensure that the band-limited condition is obeyed to the approximation desired.)

With the signal $f(t)$ band-limited to B hertz, it is then readily shown that sampling the signal does *not* destroy any information content, provided that the sampling rate $f_c \geq 2B$. The minimum sampling rate of $2B$ times per second is called the *Nyquist*[2] sampling rate and $1/2B$ the Nyquist sampling interval.

To demonstrate this result through Fourier analysis we use a simple strategem. It is apparent that the sampled signal $f_s(t)$ may be represented in terms of $f(t)$ by the simple relation

$$f_s(t) = f(t)S(t) \tag{3-1}$$

with $S(t)$ a periodic series of pulses of unit amplitude, width τ, and period $T = 1/f_c$. This *switching* or *gating function* is sketched in Fig. 3-4. (Note the recurrence of our familiar periodic pulses of Chap. 2.) We use the relationship of (3-1) to derive the spectrum $F_s(\omega)$ of the sampled signal $f_s(t)$. We demonstrate two ways of doing this, both of which rely on Fourier-analysis relations developed in Chap. 2. First note that the periodic switching function $S(t)$ may be expanded in its Fourier series. It is apparent that the sampled signal $f_s(t)$ can be written in the form

$$f_s(t) = df(t) \left(1 + 2 \sum_{n=1}^{\infty} \frac{\sin n\pi d}{n\pi d} \cos 2\pi nf_c t \right) \tag{3-2}$$

Here $d = \tau/T$, the *duty cycle*. Note the typical term $f(t) \cos 2\pi nf_c t$ appearing. By the shifting theorem of Chap. 2, the Fourier transform $F_c(\omega)$ of this term represents $F(\omega)$, the transform of $f(t)$, shifted positively and negatively by $n\omega_c$, $\omega_c = 2\pi f_c$. Thus

$$F_c(\omega) = \tfrac{1}{2}F(\omega - n\omega_c) + \tfrac{1}{2}F(\omega + n\omega_c) \tag{3-3}$$

If $F(\omega)$ was originally centered about 0, $F_c(\omega)$ is centered at $\pm n\omega_c$. As pointed out in Chap. 2, the amplitude modulation of $\cos n\omega_c t$, the *carrier frequency*, by $f(t)$, the modulating signal, results in a new signal centered about the carrier.

Repeating this process for each value of n, weighting by the appropriate factors of (3-2), and then surperposing all the terms, we have for the Fourier transform of $f_s(t)$,

---

[2] After H. Nyquist of the Bell Telephone Laboratories.

$$F_s(\omega) = dF(\omega) + d \sum_{\substack{n=-\infty \\ n\neq 0}}^{\infty} \frac{\sin n\pi d}{n\pi d} F(\omega - n\omega_c) \tag{3-4}$$

This sum of individual Fourier transforms, each centered at a multiple of the sampling frequency, is shown sketched in Fig. 3-5. Note that the amplitude of each successive component decreases as $(\sin n\pi d)/n\pi d$. The effect of sampling $f(t)$ has thus been to shift its spectrum up to all harmonics of the sampling frequency. Alternatively, we may say that the effect of multiplying the periodic sampling function $S(t)$ by the nonperiodic $f(t)$ has been to broaden its discrete-line spectrum into a continuous spectrum symmetrically situated about the original frequency lines.

Equation (3-4) for the spectrum of the sampled signal $f_s(t)$ may be derived in an alternative instructive fashion from the product relation of (3-1) by use of the frequency convolution theorem of Chap. 2. Recall from (2-97) that the product of two time functions results in a Fourier transform given by the convolution of the respective Fourier transforms. Specifically, we have, from (3-1),

$$F_s(\omega) = \frac{1}{2\pi} F(\omega) * S(\omega)$$

$$= \frac{1}{2\pi} \int_{-\infty}^{\infty} F(x)S(\omega - x) \, dx \tag{3-5}$$

But the Fourier transform of the periodic sampling function $S(t)$ is, from (2-108), given by an infinite set of equally spaced impulse functions in frequency:

$$S(\omega) = 2\pi d \sum_{n=-\infty}^{\infty} \left(\frac{\sin n\pi d}{n\pi d} \right) \delta(\omega - n\omega_c) \qquad d = \frac{\tau}{T} \qquad \omega_c = \frac{2\pi}{T} \tag{3-6}$$

Inserting (3-6) in (3-5), and noting that

$$\int_{-\infty}^{\infty} F(x)\delta(\omega' - x) \, dx = F(\omega')$$

we get (3-4). Alternatively, the reader is asked to carry out a graphical convolution by plotting some arbitrary band-limited $F(x)$ and then sliding the impulse train of (3-6) past it. The spectrum of Fig. 3-5 results.

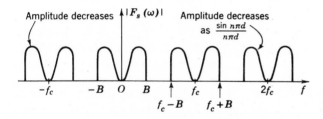

Figure 3-5 Amplitude spectrum, sampled input.

Consider Fig. 3-5 now. As long as the different spectra in Fig. 3-5 are separated, it is apparent that $f(t)$ can be filtered out from $f_s(t)$. This requires a low-pass filter, passing $F(\omega)$ but cutting off sharply before reaching the frequency spectrum component centered at f_c. This answers the question previously raised about the possibility of retrieving $f(t)$, *undistorted,* from its sampled version $f_s(t)$. Systems transmitting these sampled values of the signal $f(t)$ are commonly called *sampled-data* or *pulse-modulation* systems.

We now visualize the switch rotation rate slowing down. The frequency f_c and all its harmonics start closing in on one another. It is apparent that eventually the spectral components in Fig. 3-5 will overlap and merge. The component $F(\omega - \omega_c)$ centered about f_c will in particular merge with the unshifted $F(\omega)$ term, centered about the origin. It is then impossible to separate out $F(\omega)$, and hence $f(t)$, from $f_s(t)$. This phenomenon of overlapping spectra due to samples that are too widely spaced, and the distortion that results, is termed *aliasing*. The limiting frequency at which $F(\omega)$ and $F(\omega - \omega_c)$ merge is, from Fig. 3-5, given by $f_c - B = B$, or

$$f_c = 2B \tag{3-7}$$

which is just the Nyquist sampling rate introduced earlier.

Generally, one samples at a somewhat higher rate to ensure separation of the frequency spectra and to simplify the problem of low-pass filtering to retrieve $f(t)$. As an example, speech transmitted via telephone is generally filtered to $B = 3.3$ kHz. The Nyquist rate is thus 6.6 kHz. For digital transmission the speech is normally sampled at an 8-kHz rate, however. As another example, if the signal $f(t)$ to be sampled contains frequency components as high as 1 MHz, at least 2 million samples per second are needed.

Sampling Theorem—Further Discussion

The minimum Nyquist sampling rate, $2B$ samples per second, for signals band-limited to B hertz, is so important a concept in digital communications that it warrants further discussion and extension. In particular, we shall demonstrate rather simply that any $2B$ independent samples per second (not necessarily periodically obtained) suffice to represent the signal uniquely. One may thus sample aperiodically, if so desired; one may also sample a signal $f(t)$ and its successive derivatives, etc.

The lower, Nyquist, limit on the sampling rate in the case of periodic sampling is highly significant. Once we have satisfied ourselves that the sampled values of a signal $f(t)$ do in fact contain complete information about $f(t)$ we might logically feel that there must be a *minimum* value of f_c in order not to lose information or to be able eventually to reconstruct the continuous input signal. For we note that if we sample at too low a rate, the signal may change radically between sampling times. We thus lose information and eventually produce a distorted output. A sampling rate too low for the signal involved is shown in Fig. 3-6. In order not to lose the signal dips and rises, additional sampling pulses must be added as shown.

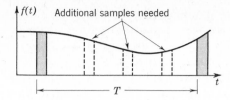

Figure 3-6 Sampling frequency too low.

There is obviously a relation between the rate at which a signal varies and the number of pulses needed to reproduce it exactly. The rate at which a signal varies is of course related to its maximum frequency component, or bandwidth, B. Equation (3-7) tells us that at least $2B$ uniformly spaced samples are needed every second in order eventually to reproduce the signal without distortion.

This statement, arising quite naturally out of a consideration of the frequency spectrum of a periodically sampled signal, is the famous *sampling theorem*[3] of modern communication theory.

The theorem is of course particularly important in sampled-data and pulse-modulation systems, where sampling is inherent in the operation of the system. But it has deep significance in the modern concepts of information theory. For any measure of the information content of a specified signal must be related to the number of *independent* quantities needed to describe that signal completely (see Chap. 1). If the number is written in the form of binary units, the information content is measured in bits.

Although our statement of the sampling theorem, arising from Eq. (3-7), relates to the periodic sampling of a band-limited signal, the theorem can be generalized to *any* group of independent samples. Thus the more general theorem states that *any $2B$ independent samples per second will completely characterize a band-limited signal. Alternatively, any $2BT'$ unique (independent) pieces of information are needed to completely specify a signal over an interval T' seconds long.*

These statements should come as no surprise to us, for we use similar results constantly in applying the Taylor's-series expansion of calculus: a function, obeying certain conditions, may be expressed and completely specified at any point in terms of the value of the function and its successive derivatives at another point. All polynomials, for example $y = t, t^2, t^3$, etc., possess a finite number of nonzero derivatives, and so only a finite number of pieces of information are needed to describe these functions. Here the number of independent pieces of information are measured in the vicinity of one point and are not periodically spaced samples.

---

[3] H. S. Black, *Modulation Theory,* chap. 4, Van Nostrand, Princeton, N.J., 1953; A. Papoulis, *The Fourier Integral and Its Applications,* McGraw-Hill, New York, 1962, and *Systems and Transforms with Applications in Optics,* McGraw-Hill, New York, 1968. A tutorial paper stressing the history of the sampling theorem concept appears in A. J. Jerri, ''The Shannon Sampling Theorem—Its Various Extensions and Applications: A Tutorial Review,'' *Proc. IEEE,* vol. 65, no. 11, pp. 1565–1596, November 1977.

The proof of the more general form of sampling theorem, that *any* $2BT'$ pieces of information are needed to characterize a signal over a T'-second interval, follows readily from an application of the Fourier series. Thus assume that we are interested in a band-limited function $f(t)$ over an interval T' seconds long. We may then expand $f(t)$ in a Fourier series with T' as the base period. But with $f(t)$ band-limited to B hertz, we get only a finite number of terms in the Fourier series,

$$f(t) = \frac{c_0}{T'} + \frac{2}{T'} \sum_{n=1}^{BT'} |c_n| \cos(\omega_n t + \theta_n) \qquad \omega_n = \frac{2\pi n}{T'} \tag{3-8}$$

or
$$f(t) = \frac{c_0}{T'} + \frac{2}{T'} \sum_{n=1}^{BT'} (a_n \cos \omega_n t + b_n \sin \omega_n t) \tag{3-9}$$

with $|c_n| = \sqrt{a_n^2 + b_n^2}$, $\theta_n = -\tan^{-1}(b_n/a_n)$. [Since B is the maximum-frequency component of $f(t)$, ω_n has a maximum value

$$2\pi B = \frac{2\pi n}{T'}$$

The maximum value of n is thus BT'.]

The c_0 term is the dc term. It merely serves to shift the level of $f(t)$ and does not provide any new information. (Information implies signals *changing* with time, as pointed out in Chap. 1.) There are thus $2BT'$ independent Fourier coefficients (the sum of the c_n's and θ_n's or a_n's and b_n's), and any $2BT'$ independent samples of $f(t)$ are needed to specify $f(t)$ over the T'-second interval.

It is of interest to note at this point the connection between the sampling theorem just discussed and the equation for channel capacity developed in Chap. 1 and rephrased at the end of Chap. 2,

$$C = B \log_2 n \qquad \text{bits/s} \tag{3-10}$$

where B is the channel bandwidth in hertz. In T' seconds the channel will allow the transmission of

$$BT' \log_2 n \qquad \text{bits}$$

This expression for system capacity includes possible effects of noise ($\log_2 n$), but, aside from noise considerations, it says simply that the information that can be transmitted over a band-limited system is proportional to the product of bandwidth times the time for transmission. The concept embodied in this last sentence was first developed by R. V. L. Hartley of Bell Laboratories in 1928 and is called *Hartley's law*.[4] Modern communication theory has extended Hartley's law to include the effects of noise, giving rise to Eq. (3-10), the expression for system capacity.

Hartley's law and the sampling theorem (as developed by Nyquist in 1928) are in essence the same. For a band-limited signal may be viewed as having

---

[4] Black, *op. cit.*

been "processed," or emitted, by a band-limited system. The information carried by this signal must thus be proportional to BT' by Hartley's law. This is the same result as that obtained from the sampling theorem. (The factor of 2 that appears to distinguish the expression for channel capacity and the results of the sampling theorem can be accounted for by using $\tau = 1/2B$ instead of $1/B$ in the development of the channel-capacity expression of Chap. 2.)

Demodulation of Sampled Signals

If $2BT'$ samples completely specify a signal, it should be possible to recover the signal from the samples. This is the demodulation process required for sampled-data or pulse-modulation systems. How do we accomplish it?

Note that the sampled output was expressed by Eq. (3-2) in the form

$$f_s(t) = df(t) \left(1 + 2 \sum_{n=1}^{\infty} \frac{\sin n\pi d}{n\pi d} \cos \frac{2\pi nt}{T} \right) \qquad (3\text{-}11)$$

As noted earlier, the simplest way to demodulate this output signal would be to pass the sampled signal through a low-pass filter of bandwidth B hertz. This is shown in Fig. 3-7.

If we sample at exactly the Nyquist rate ($f_c = 2B$), the filter required must have infinite-cutoff characteristics, as shown in Fig. 3-7a. This requires an ideal filter, an impossibility in practice. A practical low-pass filter with sharp-cutoff characteristics could of course be used, with resulting complexity in filtering and some residual distortion (part of the lower sideband about f_c would be transmitted). This situation can of course be relieved somewhat by sampling at a higher rate, as shown in Fig. 3-7b. A guard band is thus made available, and filter requirements are less severe. The filter must cut off between B and $f_c - B$, its attenuation at $f_c - B$ being some prescribed quantity measured with respect to the passband.

In the example given previously, that of transmitting speech signals digitally, it was noted that the voice transmission is commonly band-limited to 3.3

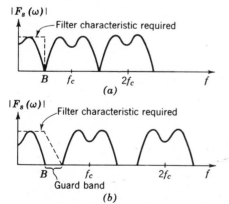

Figure 3-7 Sample-data demodulation using low-pass filter. (a) $f_c = 2B$. (b) $f_c > 2B$.

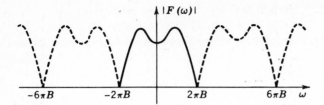

Figure 3-8 $F(\omega)$ represented as a periodic function.

kHz. The Nyquist sampling rate is then 6.6 kHz. A sampling rate of 8 kHz is most frequently used, however, so that the filter guard band is 1.4 kHz (from 3.3 to $8.0 - 3.3 = 4.7$ kHz).

It is instructive to consider a simple proof for the low-pass filter demodulation of periodic samples.[5] This will fill out in a more quantitative way our rather qualitative discussion based on Eq. (3-11) and Fig. 3-7. It also helps to clarify further the actual mechanism of filtering.

Thus assume that a signal $f(t)$ band-limited to B hertz has been sampled at intervals of $1/2B$ seconds. We shall first show that $f(t)$ may be reconstructed from these samples (a necessary result from the sampling theorem) and shall then demonstrate that an ideal low-pass filter is called for in the reconstruction or demodulation process.

To show that the given samples suffice to reproduce $f(t)$, take the Fourier transform $F(\omega)$ of $f(t)$,

$$F(\omega) = \int_{-\infty}^{\infty} f(t)e^{-j\omega t}\, dt \tag{3-12}$$

Then

$$F(\omega) = 0 \qquad |\omega| > 2\pi B \tag{3-13}$$

by virtue of the band-limited assumption on $f(t)$.

$F(\omega)$ can be arbitrarily made periodic with a period of $4\pi B$, as shown in Fig. 3-8. It can then be expanded in a Fourier series of period $4\pi B$ to be used within the interval $|\omega| \le 2\pi B$. Thus

$$F(\omega) = \frac{1}{4\pi B} \sum_{-\infty}^{\infty} c_n e^{j(2\pi n/4\pi B)\omega}$$

$$= \frac{1}{4\pi B} \sum_{n=-\infty}^{\infty} c_n e^{jn\omega/2B} \qquad |\omega| < 2\pi B \tag{3-14}$$

$$F(\omega) = 0 \qquad\qquad\qquad |\omega| > 2\pi B$$

with c_n defined by

$$c_n = \int_{-2\pi B}^{2\pi B} F(\omega)e^{-jn\omega/2B}\, d\omega \tag{3-15}$$

[5] B. M. Oliver, J. R. Pierce, and C. E. Shannon, "Philosophy of PCM," *Proc. IRE,* vol. 36, no. 11, p. 1324, November 1948.

But, since $F(\omega)$ is the Fourier transform of $f(t)$, $f(t)$ can be written

$$f(t) = \frac{1}{2\pi} \int_{-\infty}^{\infty} F(\omega)e^{j\omega t}\, d\omega = \frac{1}{2\pi} \int_{-2\pi B}^{2\pi B} F(\omega)e^{j\omega t}\, d\omega \qquad (3\text{-}16)$$

In particular, at time $t = -n/2B$

$$f\left(-\frac{n}{2B}\right) = \frac{1}{2\pi} \int_{-2\pi B}^{2\pi B} F(\omega)e^{-jn\omega/2B}\, d\omega = \frac{c_n}{2\pi} \qquad (3\text{-}17)$$

from Eq. (3-15).

This means that if we are given $f(t)$ at the various sampling intervals (for example, $t = -3/2B,\ -2/2B,\ -1/2B,\ 0,\ 1/2B,$ etc.), we can find the corresponding Fourier coefficient c_n. But knowing c_n, we can in turn find $F(\omega)$ from the Fourier series of Eq. (3-14). Knowing $F(\omega)$, we find $f(t)$ for *all* possible times by Eq. (3-16). A knowledge of $f(t)$ at sampling intervals $1/2B$ seconds apart thus suffices to determine $f(t)$ at all times. This completes the first part of our proof. We have demonstrated that $f(t)$ may be reproduced completely, solely from a knowledge of $f(t)$ at the periodic sampling intervals. How do we now actually reconstruct $f(t)$ from these samples?

If we substitute the Fourier-series expansion for $F(\omega)$ [Eq. (3-14)] into Eq. (3-16), the Fourier-integral representation of $f(t)$, we get

$$f(t) = \frac{1}{2\pi} \int_{-2\pi B}^{2\pi B} F(\omega)e^{j\omega t}\, d\omega$$

$$= \frac{1}{2\pi} \int_{-2\pi B}^{2\pi B} \frac{1}{4\pi B} \left(\sum_n c_n e^{jn\omega/2B} \right) e^{j\omega t}\, d\omega \qquad (3\text{-}18)$$

If the order of integration and summation are now interchanged (the integral is finite and no difficulties can arise because of improper integrals; the c_n coefficients also approach zero for n large enough), the resulting integral may be readily evaluated. We get

$$f(t) = \sum_n \frac{c_n}{2\pi} \frac{1}{4\pi B} \int_{-2\pi B}^{2\pi B} e^{j\omega(t+n/2B)}\, d\omega$$

$$= \sum_n \frac{c_n}{2\pi} \frac{\sin 2\pi B(t + n/2B)}{2\pi B(t + n/2B)} \qquad (3\text{-}19)$$

But $c_n/2\pi = f(-n/2B)$, from Eq. (3-17). Therefore,

$$f(t) = \sum_{n=-\infty}^{\infty} f\left(-\frac{n}{2B}\right) \frac{\sin 2\pi B(t + n/2B)}{2\pi B(t + n/2B)}$$

$$= \sum_{n=-\infty}^{\infty} f\left(\frac{n}{2B}\right) \frac{\sin 2\pi B(t - n/2B)}{2\pi B(t - n/2B)} \qquad (3\text{-}20)$$

since all positive and negative values of n are included in the summation. Mathematically, then, Eq. (3-20) indicates that we are to take each sample, multiply it by a $(\sin x)/x$ weighting factor centered at the sample's time of occurrence, and sum the resultant terms. This is exactly what is done, however, when we pass the samples through an ideal low-pass filter cutting off at B hertz.

We can demonstrate this very simply in the following manner: Assume $f(t)$ again limited to B hertz. We sample $f(t)$ for a length of time τ (the sampling time) periodically at intervals of $1/2B$ seconds. If $\tau \ll 1/2B$, $f(t)$ may be assumed very nearly constant over the sampling time (Fig. 3-9).

The individual sample $f(n/2B)$ has a Fourier transform (or frequency spectrum) given by

$$F_n(\omega) = \int_{-\infty}^{\infty} f(t)e^{-j\omega t}\, dt \doteq \tau f\left(\frac{n}{2B}\right) e^{-jn\omega/2B} \qquad (3\text{-}21)$$

since $f(t)$ is assumed constant over the τ-second interval and zero elsewhere.

But note that this is just the Fourier transform of an impulse (delta function) of amplitude $\tau f(n/2B)$ and located at $t = n/2B$ in time. [See Eq. (2-87) plus the shifting theorem.]

The amplitude factor $\tau f(n/2B)$ represents the area under the curve of the individual sample. By assuming the sample duration τ to be very small, we have effectively approximated the sample by an impulse of the same area.

Assume that this impulse is now passed through an ideal low-pass filter of bandwidth B hertz with zero phase shift and unity amplitude assumed for simplicity. (The idealized linear phase shift just serves to delay the occurrence of the output signal, as pointed out in Chap. 2.) The response of this idealized filter to an impulse $K\delta(t - t_0)$ is, from (2-91), found to be given by

$$2KB\, \frac{\sin 2\pi B(t - t_0)}{2\pi B(t - t_0)}$$

The output response $g_n(t)$ of the ideal low-pass filter to the individual sample $f(n/2B)$ must thus be given by

$$g_n(t) = 2\tau f\left(\frac{n}{2B}\right) B\, \frac{\sin 2\pi B(t - n/2B)}{2\pi B(t - n/2B)} \qquad (3\text{-}22)$$

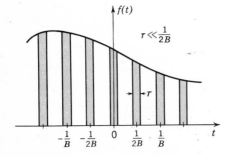

Figure 3-9 The sampling process.

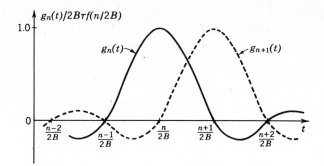

Figure 3-10 Filter response to sampled inputs.

This result can of course also be obtained directly from Eq. (3-21) and the assumed filter characteristic. Thus, for an ideal low-pass filter with zero phase shift and unit amplitude, the output response to the sample $f(n/2B)$ applied at the input is

$$g_n(t) = \frac{1}{2\pi} \int_{-2\pi B}^{2\pi B} F_n(\omega) e^{j\omega t} \, d\omega$$

$$= f\left(\frac{n}{2B}\right) \frac{\tau}{2\pi} \int_{-2\pi B}^{2\pi B} e^{j\omega(t - n/2B)} \, d\omega \tag{3-23}$$

Integrating, as indicated, we get for $g_n(t)$

$$g_n(t) = 2B\tau f\left(\frac{n}{2B}\right) \frac{\sin 2\pi B(t - n/2B)}{2\pi B(t - n/2B)}$$

as in Eq. (3-22). $g_n(t)$ is plotted in Fig. 3-10. Note that $g_n(t)$ has its maximum value at $t = n/2B$ (the filter was assumed to have zero phase shift), and precursors again appear because of the idealized filter characteristics assumed. The output is a maximum at the given sampling point, $t = n/2B$, and zero at all the other sampling points. At the sampling point $g_n(n/2B) = 2B\tau f(n/2B)$, or g_n is just the input $f(n/2B)$ to within a constant. The next sample, occurring $1/2B$ seconds later, or at $t = (n + 1)/2B$, likewise produces an output given by Eq. (3-22), but delayed $1/2B$ seconds, and proportional to $f[(n + 1)/2B]$.

The peak of each $(\sin x)/x$ term occurs at a sampling point where all other outputs are zero (Fig. 3-10), and the output at each of the sampling points is exactly proportional to the magnitude of the input sample at that point.

Since we have been assuming a linear ideal filter, the complete output is just the superposition of the individual sample outputs, or

$$g(t) = \sum_{n=-\infty}^{\infty} g_n(t)$$

$$= 2B\tau \sum_{n=-\infty}^{\infty} f\left(\frac{n}{2B}\right) \frac{\sin 2\pi B(t - n/2B)}{2\pi B(t - n/2B)}$$

$$= 2B\tau f(t) \tag{3-24}$$

by comparison with Eq. (3-20). The output of the low-pass filter, $g(t)$, is thus identically proportional to the original signal $f(t)$ at *all* instants of time, not only at the sampling points. [Had we included a linear-phase-shift characteristic, we would of course have obtained a constant time delay for the output function. The output $g(t)$ would thus be a delayed replica of $f(t)$.]

The original input $f(t)$ may thus be reproduced from the samples by passing them through an ideal low-pass filter of bandwidth B hertz. The $2B\tau$ proportionality factor represents the ratio of the filter and sampling pulse bandwidths. Since we assumed $\tau \ll 1/2B$, and since the effective pulse "bandwidth" is proportional to $1/\tau$ (Chap. 2), the pulse spectral width is much greater than that of the filter. [We in fact assumed the pulse spectrum to be constant over the filter bandwidth, as shown by Eq. (3-21).] Most of the energy of the input sample thus lies outside the filter spectrum, as shown in Fig. 3-7. If the pulse width is so narrow that its spectrum is almost constant (again the justification for representing it by an impulse of the same area), the output amplitude is reduced by the ratio of bandwidths, or just $2B\tau$.

These considerations of filter operation from a frequency point of view of course agree with the time approach. For, as shown by Eq. (3-11), an ideal low-pass filter of unit amplitude and bandwidth B would produce $g(t) = df(t)$ at the output. But the pulse duty cycle d is simply τ/T, or $2B\tau$, with $T = 1/2B$, agreeing with the result of Eq. (3-24).

The ideal filter is of course a mathematical artifice and can only be approximated in practice. Actual filter outputs will thus only approximate the actual input $f(t)$.

3-3 TIME MULTIPLEXING OF SIGNALS: AN INTRODUCTION

We have indicated in the previous section that one may convert band-limited analog data to a sampled form by sampling at least at the Nyquist rate. These sampled values of the signal carry the original intelligence, and demodulation may be carried out by low-pass filtering. A system transmitting these sampled values of the signal is commonly called a *pulse-amplitude-modulation* (PAM) system. For the sequence of samples may alternatively be visualized as a periodic sequence of pulses (the *carrier*) whose amplitude is modulated (or varied) in accordance with the intelligence to be transmitted. This is in fact apparent from the form of the sampled data expression $f_s(t) = f(t)S(t)$. The switching function $S(t)$ represents the unmodulated pulse carrier and $f(t)$ the intelligence modulating the carrier. This concept of amplitude modulation will be pursued further in Chap. 4.

Most pulse communication systems in use transmit many signals simultaneously, rather than just one. It is apparent from the sampling process, with a very narrow sample τ seconds wide taken every T seconds, that much of the time no information is being transmitted through the system (see Fig. 3-2). It is thus possible to transmit other information from other sources in the vacant intervals. The transmission of samples of information from several signal chan-

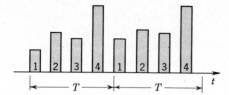

Figure 3-11 Time multiplexing.

nels simultaneously through one communication system with different channel samples staggered in time is called *time-division multiplexing* or *time multiplexing* for short.

Most time-multiplexed systems in use now are digital. This implies that analog signals are converted to digital format before transmission in multiplexed form. Digital signals (from data terminals, computers, printers, or any other digital source) are of course already in the form needed for multiplexing. In this introductory treatment we choose to separate the two ideas of time-multiplexing and analog-to-digital conversion. In this section we discuss the basic idea of multiplexing of signals, without indicating whether they are digital or analog in form. In the next section we discuss the analog-to-digital (A/D) operation and its application to PCM systems. In Sec. 3-8 we return to time multiplexing, focusing there on *digital* systems only. We then describe the multiplexing process in more detail, drawing on examples of multiplexing techniques used in practice. In the digital multiplexer case it is immaterial whether the signals to be multiplexed were digital to begin with, were analog and then converted using A/D techniques to digital format, or are combinations of both. All digital signals are handled in the same way once they are in that form.

In a typical time-multiplexing scheme, the various signals to be transmitted are sequentially sampled and combined for transmission over the one channel. Figure 3-11 shows a time diagram of a time-multiplexed signal system having four information carriers. It is apparent that all signals to be multiplexed must either be of the same bandwidth, or, if such a scheme is to be used, sampling must take place at a rate determined by the maximum bandwidth signal. (Alternatively, relatively low-bandwidth signals may be first combined before sampling, as will be noted later.) A sampler time-multiplexing a multichannel input sequentially into the transmission channel is shown in Fig. 3-12. A mechanical switch is again shown for simplicity, but in practice electronic switching would normally be used.

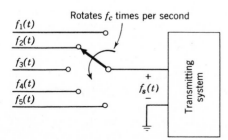

Figure 3-12 Sampler for time multiplexing.

Time multiplexing has been widely used in the radio, teletype, and telephone fields, and for telemetering purposes. Experimental data from space probes are commonly sampled and transmitted sequentially, for example. A space vehicle transmitting back vital measured information such as temperature, electron density, magnetic intensity, and a multitude of other sensor outputs need only utilize one transmission channel, providing the sampling is carried out rapidly enough to accommodate the most rapidly varying signal to be transmitted.

If the various signals to be multiplexed have widely differing bandwidths, or, as is commonly the case with data sources, have widely different data rates, two approaches may be used. Proportionately more samples of the wider-bandwidth or higher-data-rate signals may be taken and combined with samples of the more slowly varying signals, or the more slowly varying signals may first be combined into a single wider-bandwidth analog signal by *frequency-multiplexing* techniques. This latter approach is discussed in Chap. 4. The former case, the direct time multiplexing of data sources with different data rates, is discussed in detail in the context of some examples in Secs. 3-6 and 3-8. In this section we assume, for simplicity, that all signals are sampled and combined at the same rate, as dictated by the most rapidly varying signal.

As is true with all engineering developments, one pays a price for introducing time multiplexing into a system. First it is apparent that the necessary transmission bandwidth increases with the number of signals multiplexed, for the transmission bandwidth is proportional to the reciprocal of the width of the pulses transmitted (Chap. 2). Thus in the case of the single-signal pulses of Fig. 3-2, the bandwidth required to transmit the pulses shown would be approximately $1/\tau$ hertz. However, it is possible to widen the pulses substantially, just to the point where they begin to overlap say, and so require approximately $1/T$-hertz bandwidth. The pulses of Fig. 3-11 can only be widened to the point where they begin to interfere with the next adjacent pulses, however. In this case the minimum bandwidth is four times that of the single-signal case of Fig. 3-2. In general, then, if N signals are time-multiplexed, the required transmission bandwidth is N times that for single-signal samples.

As an example, assume that 10 voice channels, each band-limited to 3.2 kHz, are sequentially sampled at an 8-kHz rate and time multiplexed on one channel. The successive pulses are then spaced 12.5 μs apart, as shown in Fig. 3-13. The bandwidth required to transmit these pulses is roughly 80 kHz. (A more accurate determination of the filtering and bandwidth necessary to prevent pulse overlap will be considered later in Sec. 3-9 in the discussion of intersymbol interference.) The filter shown in Fig. 3-13 is used to widen the pulses as required. Filtering could be incorporated in the sampling operation as well, or it could be carried out further along the transmission path.

The other problem introduced by time multiplexing involves proper synchronization and registration of the successive pulses at the receiver. For it is apparent that the successive pulses must on reception be delivered to the appropriate destination. This implies that a switch is available at the receiver, is

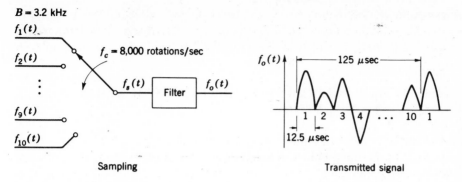

Sampling   Transmitted signal

Figure 3-13 Time-multiplexed voice channels.

synchronized to the original transmitter switch, and deposits each sample in its appropriate signal channel. This is no mean task with high-speed data systems, and with receiver and transmitter located thousands or millions of miles apart. Figure 3-14 shows transmitter and receiver switching in a time-multiplexed system. The registration problem, with sample 1 placed in the 1 line at the destination and not in line 2 or 3, is particularly acute. (Although you might not mind eavesdropping on someone else's conversation, you in turn would not like to have your conversation picked up by a stranger!)

Various techniques have been utilized in practice to perform synchronization and registration of signals. The techniques have included: (1) the use of special marker pulses, tagged to be easily differentiable from regular signal pulses, and sent periodically at prescribed intervals; (2) continuous sine waves of known phase and frequency which can be filtered out at the receiver to provide the necessary timing information; and (3) schemes which derive timing information from the transmitted signal pulses themselves by averaging over long periods of time, etc. Reference is made to the literature for details of

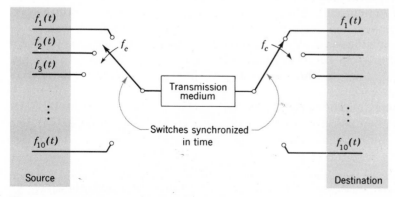

Figure 3-14 Switching in a time-multiplexed system.

various synchronizing techniques.[6] Some specific methods adopted for particular systems are included in Sec. 3-6 in the description of some actual operating systems. Other methods are described in detail in Sec. 3-8.

3-4 ANALOG-TO-DIGITAL CONVERSION: APPLICATION TO PULSE CODE MODULATION

Very commonly in modern communication technology the sampled analog signals are further digitized before transmission. The digital signals may then be encoded into any equivalent form desired. Systems embodying the transmission of digitized and coded signals are commonly called *pulse-code-modulation* (PCM) systems. Binary digital systems constitute the most frequently encountered form of PCM systems.

There are many advantages to using PCM systems.

1. The signals may be regularly reshaped or regenerated during transmission since information is no longer carried by continuously varying pulse amplitudes but by discrete symbols.
2. All-digital circuitry may be used throughout the system.
3. Signals may be digitally processed as desired.
4. Noise and interference may be minimized by appropriate coding of the signals, etc.

Some of these factors will be described in detail later in the book.

The process of digitizing the original analog signals is called the *quantization* process. It consists of breaking the amplitudes of the signals up into a prescribed number of discrete amplitude levels. The resultant signals are said to be *quantized*. Unlike the sampling process this results in an irretrievable loss of information since it is impossible to reconstitute the original analog signal from its *quantized* version. However, as first noted in the discussion of the concept of information in Chap. 1, there is actually no need to transmit all possible signal amplitudes. Because of noise introduced during transmission and at the receiver, the demodulator or detector circuit will not be able to distinguish fine variations in signal amplitude. In addition, the ultimate recipient of the information—our ears in the case of sound or music, our eyes in the case of a picture—is limited with regard to the fine gradation of signal it can distinguish.

This ultimate limitation in distinguishing among all possible amplitudes thus makes quantization possible. In a specific system the sampled pulses may be quantized, or both quantization and sampling may be performed simulta-

---

[6] See Bennett and Davey, *op. cit.*, chap. 14; J. J. Stiffer, *Theory of Synchronous Communications*, Prentice-Hall, Englewood Cliffs, N.J., 1971; as well as the other references cited at the beginning of this chapter.

neously. This latter process is portrayed in Fig. 3-15. The total amplitude swing of $A_0 = 7$ V is divided into equally spaced amplitude levels $a = 1$ V apart. There are thus $M = A_0/a + 1$ possible amplitude levels, including zero. In Fig. 3-15 samples are shown taken every second and the nearest discrete amplitude level is selected as the one to be transmitted. The resultant quantized and sampled version of the signal of the smooth signal of Fig. 3-15a thus appears in Fig. 3-15b. (The signal of 0.3 V at 0 s is transmitted as 0 V, etc.)

Although the level separation is shown here as uniform, the separation is often tapered in practice to improve the system noise performance. In particular, the spacing of levels is decreased at low amplitude levels. This is done by a technique called *compression*. We shall discuss one common technique and its effect on performance in the next section. We shall assume equal spacing for simplicity, however, in this section.

Obviously, the quantizing process introduces some error in the eventual reproduction of the signal. The demodulated signal will differ somewhat from the derived signal, as noted earlier. The overall effect is as if additional noise had been introduced into the system. (In the case of sound transmission this

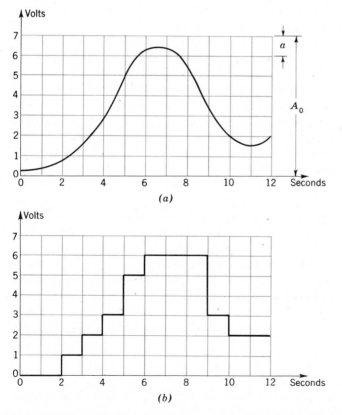

Figure 3-15 Quantization and sampling. (*a*) Given signal. (*b*) Quantized and sampled version.

manifests itself as a background cackle. In the case of picture transmission the continuous gradation of grays from black to white is replaced by a discrete number of grays, and the picture also looks somewhat noisy.) This *quantization noise* may of course be reduced by decreasing the level separation *a* or by increasing the number of levels *M* used. Experiment has shown[7] that 8 to 16 levels are sufficient for good intelligibility of speech. (Two-level PCM is even understandable, although quite noisy![8])

If the quantized signal samples were transmitted directly as pulses of varying (although quantized) heights, the resultant system would simply be quantized PAM. (This is also often referred to as *M*-ary PAM.) But with discrete or numbered voltage levels each level can be coded in some arbitrary form before transmission. It is this possibility of coding discrete sample levels that gives quantized signals much greater flexibility in transmission than the continuous-varying pulses of PAM.

Most commonly the quantized and sampled signal pulse is encoded into an equivalent group or packet of binary pulses of fixed amplitude. This finally provides the binary signals first encountered in Sec. 1-1, typical examples of which are shown in Fig. 1-2. The encoding of amplitude levels into binary form can be done in various ways. One procedure is to follow the usual decimal-to-binary conversion, tabulated for the numbers 0 to 7 in Sec. 1-3, and for the numbers 0 to 15 in Table 3-1.

One difficulty with the normal decimal-to-binary conversion (labeled binary code in Table 3-1) is that in moving from one adjacent decimal digit to another the binary code changes by a variable number of binary digits. This is particularly apparent in changing from 3 to 4 and 7 to 8 in Table 3-1. In the first case three binary digits change, in the second case four digits change. In 7-bit PCM, handling levels 0 to 127, as many as seven binary digits may change. This makes the binary code highly susceptible to error in the A/D conversion. One would prefer a code in which only one binary digit at a time changed as the corresponding decimal digit changed by one level. The Gray code shown in Table 3-1 is an example of such a code.

The Gray code is easily obtained from the binary code by the following conversion equations. Say in general that an n-digit binary code is used. (This corresponds to 2^n possible decimal digits, including 0.) Call the successive bits in order, from the most significant to the least significant, $b_1 b_2 b_3 \cdots b_n$. (The digits in Table 3-1 are then $b_1 b_2 b_3 b_4$, as shown.) Let the corresponding Gray code digits be $g_1 g_2 g_3 \cdots g_n$. The reader can then check, using Table 3-1 as an example, that the following equations convert the binary code to the Gray code:

$$g_1 = b_1$$
$$g_k = b_k \oplus b_{k-1} \qquad k \geq 2 \qquad\qquad (3\text{-}25)$$

---

[7] H. F. Mayer, "Pulse Code Modulation," summary chapter in L. Martin (ed.), *Advances in Electronics,* vol. III, pp. 221–260, Academic Press, New York, 1951.

[8] J. S. Mayo, "Pulse Code Modulation," *Electro-Technol.,* pp. 87–98, November 1962.

Table 3-1 Decimal-to-binary conversion

Digit	Binary code				Gray code			
	b_1	b_2	b_3	b_4	g_1	g_2	g_3	g_4
0	0	0	0	0	0	0	0	0
1	0	0	0	1	0	0	0	1
2	0	0	1	0	0	0	1	1
3	0	0	1	1	0	0	1	0
4	0	1	0	0	0	1	1	0
5	0	1	0	1	0	1	1	1
6	0	1	1	0	0	1	0	1
7	0	1	1	1	0	1	0	0
8	1	0	0	0	1	1	0	0
9	1	0	0	1	1	1	0	1
10	1	0	1	0	1	1	1	1
11	1	0	1	1	1	1	1	0
12	1	1	0	0	1	0	1	0
13	1	1	0	1	1	0	1	1
14	1	1	1	0	1	0	0	1
15	1	1	1	1	1	0	0	0

The symbol $\oplus$ represents modulo-2, or exclusive-or addition of the binary numbers. ($0 \oplus 0 = 0$; $1 \oplus 0 = 1$; $0 \oplus 1 = 1$; $1 \oplus 1 = 0$.) Note that the procedure adds the current digit, b_k, to the next-most-significant one, b_{k-1}. Conversion from the Gray code to the binary code is easily accomplished by reversing (3-25): (In mod-2 addition, subtracting two numbers is the same as adding them.)

$$b_1 = g_1$$
$$b_k = g_k \oplus b_{k-1} \qquad k \geq 2 \qquad (3\text{-}26)$$

For simplicity's sake we shall generally assume the usual decimal-to-binary conversion (the binary-code column of Table 3-1, for example) in discussing binary coding in this chapter. The appropriate conversion to the Gray code can then always be made using (3-25). In Chap. 6 we consider further encoding of these binary digits, with additional *parity bits* added for error detection and/or correction purposes.

It is apparent that in either case, the usual binary code or the Gray code, the number of binary digits used depends on the number of quantization levels used. Thus 8-level PCM requires three binary digits for transmission, 16-level PCM requires four binary digits, and 128 levels require 7 binary digits or bits for the representation of each level. Since these binary digits must be transmitted in the sampling interval originally allotted to one quantized sample, the binary pulse widths are correspondingly narrower and the transmission bandwidth goes up proportionately to the number of binary pulses needed. An example of the binary encoding process is shown in Fig. 3-16. Here three binary pulses are transmitted in the original sampling interval, so that the bandwidth is increased by a factor of 3. (Note that if this represented a time-multiplexed system, the

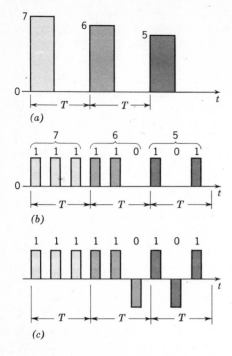

Figure 3-16 Binary coding of samples. (*a*) Given signal (already sampled and quantized). (*b*) Coded samples. (*c*) Another form of binary code: polar signals.

three samples of amplitudes 7, 6, and 5, respectively, would represent three separate signals.)

Two examples of binary signals are shown in Fig. 3-16: (1) an on-off signal, in which the 1 bit is represented by the presence of a pulse, the 0 bit by its absence; and (2) a signal with the 1 represented by a positive pulse, the 0 by a negative pulse.[9]

A binary code is just one special case of the coding theoretically possible in a PCM system. In general, any one quantized signal sample may be coded into a group of m pulses, each with n possible amplitude levels. These m pulses must be transmitted in the original sampling interval allotted to the quantized sample. Since the information carried by these m pulses is equivalent to the information carried by the original M amplitude levels, the number of possible amplitude combinations of the m pulses must equal M. Thus

$$M = n^m \tag{3-27}$$

(Recall from Chap. 1 that m pulses, each with n possible heights, may be combined in n^m different ways. Each combination must correspond to one of the original M levels.) If there are two possible levels, $n = 2$, and we have the binary code just mentioned. With $M = 8$, three binary pulses are necessary. If n

---

[9] Different binary PCM waveforms are used in practice. Two examples, the nonreturn-to-zero (NRZ) and the bipolar signal sequences, are discussed later in this chapter.

= 3, a ternary code results. Obviously, if $m = 1$, $n = M$ and we are back to our original uncoded but quantized samples. As the number n of levels chosen for the coded pulses increases, m decreases, as does the bandwidth required for transmission.

This ability to code back and forth is one of the reasons for the increased usage of PCM systems. Although most PCM systems in use are binary in form, so that the bandwidth required for transmission is wider than that required before binary encoding, bandwidth *reduction* schemes have been suggested in which successive M-level samples would be further collapsed into much higher-level samples. This is a reversal of the binary encoding process. For example, two successive binary pulses provide a total of four possible combinations, and so need four numbers to represent them. Similarly, 2 eight-level samples require 64 numbers to represent them. If combined into one *wider* pulse, covering the two adjacent time slots, the required bandwidth could be reduced by a factor of 2, but the one pulse used for transmission would take on 64 possible levels. This process of combining or collapsing successive pulses into one much wider pulse with many more amplitude levels can of course be continued indefinitely. [The m in Eq. (3-27) then represents the number of adjacent pulses, each with n amplitude levels, that are collapsed into one pulse with M levels. The bandwidth is then reduced by a factor of m.] There is one major difficulty, however; if the spacing between levels remains fixed, the required peak power goes up *exponentially* with the number of pulses combined. Alternatively, if the peak power or amplitude swing is to remain fixed, the levels must be spaced closer and closer together. This then makes it easier for noise in the system to obscure adjacent levels. Such bandwidth reduction techniques are thus only possible in a relatively low-noise environment.

The binary form of transmission provides, on the other hand, the most noise immunity, and this is therefore the digital communication scheme with which we shall concern ourselves most in the chapters to follow. For, as indicated by the examples of Fig. 3-16, the information carried by the binary signals is represented either by the absence or presence of a pulse, or by its polarity. All the receiver has to do is to then correspondingly recognize the absence or presence of a pulse, or the polarity (plus or minus) of a pulse, and then decode into the original quantized form to reconstruct the signal. The pulse shape or exact amplitude is not significant as in the case of the original analog signal, or a PAM signal. By transmitting binary pulses of high-enough amplitude, we can ensure correct detection of the pulse in the presence of noise with as low an error rate (or possibility of mistakes) as required. We shall have more to say about this noise-improvement capability of PCM systems in later chapters of the book.[10]

---

[10] In Chap. 6 we consider encoding blocks of m successive binary digits into $M = 2^m$ possible signal symbols or code words for the purpose of reducing the probability of error. The resultant bandwidths then generally *increase* with m, or at best remain fixed. These encoding schemes for error reduction should not be confused with the binary to M-level amplitude transmission discussed here for bandwidth reduction.

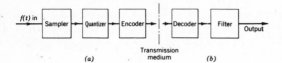

Figure 3-17 One-channel PCM system. (a) Transmitter. (b) Receiver.

As noted earlier, binary PCM lends itself readily as well to signal reshaping at periodic intervals during transmissions. This is a common practice in digital communication over telephone circuits. The reshaping of signals at these intermediate *repeaters* enhances the signal decisions when finally received at the receivers.[11]

It is now of interest to tie together the processes of sampling, quantization, and binary encoding, as well as their reverse processes at the receiver. Figure 3-17 shows these various operations in block diagram form for a complete one-channel PCM system. (Synchronizing circuitry, as well as modulation and demodulation circuits to be discussed later, are not shown.) In Fig. 3-18 we take a 10-channel PCM system and show the timing and pulse widths required as we progress through the system. Again, for simplicity's sake, necessary synchronizing pulses are not shown. (The insertion of additional synchronizing bits would of course require narrower pulses and hence wider bandwidths.) Thus if the original analog channels each have 3.2-kHz bandwidth, as shown, and a sampling rate of 8,000 samples per second is used, the time-multiplexed pulses at the sampler output [point (1), part (a)] appear at 12.5-μs intervals. Assuming an eight-level quantizer, with integer steps for simplicity, the quantized output pulses also appear at 12.5-μs intervals, but with integer amplitudes only as shown in part (c), point (2), of Fig. 3-18. Each quantized pulse is then encoded into three binary pulses, each occupying a 4.2-μs time slot, as shown.

If the pulses were allowed to occupy the full time slots, as shown in the figure, the bandwidths required at the three points shown would be approximately

1. 80 kHz (1/12.5 μs)
2. 80 kHz
3. 240 kHz

(1) is the PAM bandwidth; (3) is the binary PCM bandwidth. (As noted earlier, we shall discuss requisite bandwidths more quantitatively later after a discussion of intersymbol interference.)

3-5 QUANTIZATION NOISE AND COMPANDING IN PCM[12]

As noted in the previous section, a continuous or analog signal to be coded into digital form must first be quantized into discrete steps of amplitude. Once

[11] Bennett and Davey, *op. cit.*, pp. 128–131. Since most telephone systems provide relatively low-noise transmission, multilevel PCM rather than binary is often used to provide a higher rate of information transmission over the channel. Section 5-3 gives a brief analysis of repeaters.

[12] A detailed treatment appears in Cattermole, *op. cit.*, chap. 3.

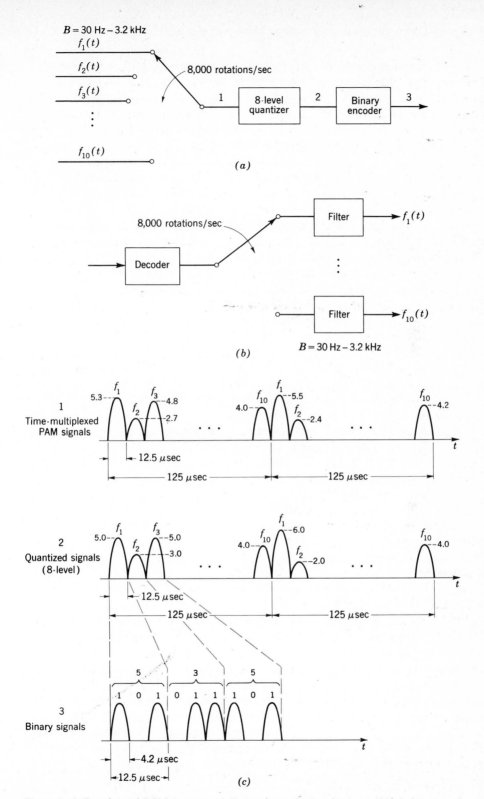

Figure 3-18 Ten-channel PCM system. (*a*) Transmitter. (*b*) Receiver. (*c*) Signal shapes.

quantized, the instantaneous values of the continuous signal can never be restored exactly. This, as we have pointed out previously, gives rise to random error variations called *quantization noise*. This noise can be reduced to any desired degree by choosing the quantum steps or level separations fine enough. There is a trade-off here, however. The larger the number of quantum steps used, the greater the number of binary digits or bits required to represent the signal and hence the wider the bandwidth needed for transmission. One then normally chooses as few quantization levels as possible consistent with the objective of the transmission. It has already been noted that for speech transmission via PCM, subjective tests indicate that 128 levels, or 7-bit PCM, suffice to ensure high-quality transmission. For other applications the number may differ. For the transmission of data via telemetry, the accuracy possible or required in the transmission determines the number of levels used.

For example, assume that the temperature of a refining process at a remote plant is to be transmitted back to a centralized computer site as part of the control process. If temperature readings to within 1°C only are required, and the expected range is 20 to 80°C, only 60 levels are required, so one would presumably choose 64 levels, or 6-bit PCM. If the temperature sensors used can only read to 2°C accuracy, 30 levels are available, and 5-bit PCM could be used. If readings to within 0.5°C accuracy are required and sensors capable of this accuracy are available, 128 levels and 7-bit PCM would be used.

These rather qualitative remarks are useful to provide a first, intuitive understanding of the trade-offs among accuracy, quantization noise, and the number of quantization levels used. A more sophisticated treatment suitable for design purposes requires a quantitative approach, however. We have tacitly assumed to this point that the analog signal to be quantized has definite minimum and maximum values between which it ranges, and that it is equally likely to occupy any value in this range. We have thus taken the quantization levels to be equally spaced. For speech transmission via PCM telephony, this is far from the true situation, however. There one has talkers of different power and timbre. Some people shout, others speak very quietly on the telephone. If the 128 quantization levels assigned were to be equally spaced and cover the range from a soft whisper to a shout, it is apparent that good reproduction for all types of speakers would not be possible. One thus has to take into account the *statistics* of the signal and design the quantization scheme on this basis. The same problem arises in a telemetry situation: lower temperature readings may occur more frequently than high readings and may be required to be known with greater accuracy. One thus needs a *tapered* quantization level structure that is finer at the lower readings, and more coarse at the high readings.

To quantify these comments in this section we shall introduce the idea of a signal-to-quantization noise ratio. This can be defined in various ways, as we shall see, but basically it establishes the idea, to be repeated many times throughout this book in discussing other kinds of noise, that the effect of noise on the performance of a system depends on the amount of noise relative to the signal. In a speech system quantization noise cackle, as heard by a listener, is only objectionable if it is noticeable compared to the signal intensity of the speaker. A given noise level is more objectionable when a quiet speaker is speaking than when a loud one is communicating. We shall be calculating a mean-squared quantization noise, or its square root, the rms noise level, but shall relate it to some measure of the signal level: mean-squared value or peak value squared, to determine the quantization noise performance of a PCM system.

We shall also introduce a statistical model of the signal and see how tapering the quantization-level structure improves the performance compared to the equal-level case. This is the first example in this book of the use of statistical concepts to develop design relations. Similar approaches will be used quite extensively in Chaps. 5 and 6. We shall focus primarily on the speech telephony application because of the availability of design procedures in this area, and because of its technological importance: the use of PCM in telephone transmission has increased dramatically since its introduction by the Bell System in the United States in the early 1960s. The ideas and concepts introduced are of course applicable to other application areas as well.

We first assume equally spaced quantization levels and calculate the signal-to-quantization noise ratio for this case. In practice, rather than use tapered quantization levels to match the signal characteristics the signals are first nonlinearly *compressed* in amplitude to force all signals to lie within a specified range. This compression characteristic is typically of a logarithmic form. Uniform quantization levels are then applied to this compressed signal. The effect is to provide proportionately more quantization levels at the smaller signal levels, as if the quantization-level spacing had been reduced at the lower signal levels. At the receiver the signal is *expanded* to its original amplitude through an inverse logarithmic form. The combination of compression and expansion is referred to as *companding* for short. Using some typical signal statistics for the speech example we shall show that the effect of companding (or nonuniform tapering of quantization levels) is to provide a uniform signal-to-quantization noise ratio over a much wider dynamic range of signals than in the noncompressed case.

Quantization Noise, Equal-Level Spacing

To calculate the rms quantization noise in this case, prior to setting up an expression for the signal-to-quantization noise ratio, let the signal at the transmitter be quantized into a total of M levels, with a the spacing in volts between adjacent levels. With a maximum plus-minus signal excursion of P volts, or a maximum excursion positively or negatively of V volts (see Fig. 3-19),

$$a = \frac{P}{M} = \frac{2V}{M} \qquad (3\text{-}28)$$

(The continuous signal is assumed to have 0 average value, or no dc component.) The quantized amplitudes will be at $\pm a/2, \pm 3a/2, \ldots, \pm(M-1)(a/2)$, and the quantized samples will cover a range

$$A = (M-1)a \qquad \text{volts}$$

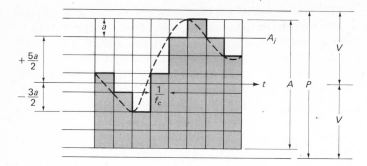

Figure 3-19 Quantized approximation to a signal: eight levels.

As noted earlier, the quantization process introduces an irreducible error, since a sample appearing at the receiver output at quantized voltage A_j volts could have been due to any signal voltage in the range $A_j - a/2$ to $A_j + a/2$ volts. This region of uncertainty is shown in Fig. 3-20. As far as the ultimate recipient of the message is concerned, this region of uncertainty could just as well have been due to additive noise masking the actual signal level. The one difference is that additive noise, as we shall see in Chap. 5, can theoretically take on all possible voltage values. The quantization noise, on the other hand, is limited to $\pm a/2$ volts. The distinction between these two will become clearer in Chap. 5.

We can calculate a mean-squared-error voltage due to quantization. To do this, assume that over a long period of time all voltage values in the region of uncertainty eventually appear the same number of times. The instantaneous voltage of the signal will be $A_j + \epsilon$, with $-a/2 \leq \epsilon \leq a/2$. ϵ represents the error voltage between the instantaneous (actual) signal and its quantized equivalent. Under our assumption all values of ϵ are equally likely. The mean-squared value of ϵ will then be

$$E(\epsilon^2) = \frac{1}{a} \int_{-a/2}^{a/2} \epsilon^2 \, d\epsilon = \frac{a^2}{12} \tag{3-29}$$

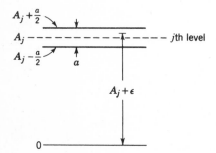

Figure 3-20 Region of uncertainty at system output.

with the symbol $E(\cdot)$ representing statistical expectation. The average value of the error is zero with the assumption made. The rms error is then $a/\sqrt{12} = a/(2\sqrt{3})$ volts, and this represents the rms "noise" at the system output.

We now introduce the idea of the signal-to-quantization noise ratio (SNR). We do this two ways, by defining two different SNR ratios, one in terms of the peak signal, V volts, the other in terms of the mean signal power, $S_o = (M^2 - 1)a^2/12$.

Peak signal. Since $V = aM/2$ is the peak signal excursion, the ratio of peak signal voltage to rms noise will be

$$\frac{S_{ov}}{N_{ov}} = \frac{V}{a/(2\sqrt{3})} = \sqrt{3}\, M \tag{3-30}$$

The corresponding power ratio is

$$\frac{S_o}{N_o} = 3M^2 \tag{3-31}$$

or, in decibels,

$$\left(\frac{S_0}{N_0}\right)_{dB} = 4.8 + 20\log_{10} M \tag{3-32}$$

The power ratio thus goes up as the square of the number of levels. The SNR decibel improvement with M is given in Table 3-2. Also indicated is the relative bandwidth as obtained from the discussion following.

Since M, the number of levels used, determines the number of pulses into which the quantized signal is encoded before transmission, increasing M increases the number of code pulses and hence the bandwidth. We can thus relate SNR to bandwidth. This is easily done by noting that $M = n^m$, with m the number of pulses in the code group and n the number of code levels. With this relation, Eqs. (3-31) and (3-32) become, respectively,

$$\frac{S_o}{N_o} = 3n^{2m} \tag{3-33}$$

Table 3-2 Quantization SNR improvement with number of levels

S_0/N_0, dB	M	Relative bandwidth
11	2	1
17	4	2
23	8	3
29	16	4
35	32	5
41	64	6
47	128	7

and
$$\left(\frac{S_o}{N_o}\right)_{\text{dB}} = 4.8 + 20m \log_{10} n \qquad (3\text{-}34)$$

In particular, for a binary code ($n = 2$),

$$\left(\frac{S_o}{N_o}\right)_{\text{dB}} = 4.8 + 6m \qquad (3\text{-}35)$$

Since the bandwidth is proportional to m, the number of pulses in the code group, the output SNR increases exponentially with bandwidth. The decibel SNR increases linearly with bandwidth [Eq. (3-35)].

For a 128-level system $S_o/N_o = 47$ dB, and seven-pulse binary-code groups are transmitted, requiring a sevenfold bandwidth increase.

Mean signal power. Essentially similar results are obtained upon defining a mean power SNR. With a quantized level spacing of a volts and signal swings of $\pm V$ volts the mean signal power is readily found to be

$$S_o = \tfrac{1}{12}(M^2 - 1)a^2$$

assuming all signal levels equally likely. Details are left to the reader as an exercise.

Since $N_o = a^2/12$, the mean power output SNR is

$$\frac{S_o}{N_o} = M^2 - 1 \qquad (3\text{-}36)$$

For $M \gg 1$ this differs only by a constant from the peak S_o/N_o relation given by Eq. (3-31). For a system with 128 levels the quantization SNR is 42 dB. A binary-code group requires seven pulses ($2^7 = 128$), or seven times the bandwidth of the original quantized signal.

Signal-to-Noise Ratios with Companding

In the previous paragraph we tacitly assumed that a known peak signal amplitude V (or peak-to-peak excursion $P = 2V$) existed and chose the M quantization levels equally spaced to cover the total signal excursion. For many classes of signals there is no specified peak value and the signal level may in fact change in a random manner. The most common example is that of speech transmission, with different speakers using the same transmission facilities. The range of speech intensity may vary as much as 40 dB in going from the whisper of a quiet speaker to the bellowing tones of a powerful speaker. It is apparent that to cover this dynamic range effectively nonuniform quantization-level spacing, or its equivalent, signal compression, must be used. If this is not done and equally spaced levels are chosen to cover the widest signal variation expected, the soft speakers will be penalized. The same problem obviously arises in the PCM transmission of any analog signals expected to cover a wide dynamic range.

We first demonstrate the effect of dynamic variation of the signal power on the SNR with uniform spacing of quantization levels by repeating the analysis of the previous paragraph somewhat differently. Say that the quantizer is again designed to accept a peak-to-peak signal excursion of $P = 2V$ volts, but that this corresponds to the maximum level of the highest-intensity signal expected. Let the actual signal appearing at the quantizer input have a mean power (mean-squared value) of σ^2. This should obviously be significantly less than V^2 to be accommodated by the quantizer. (As an example, if the signal is statistical in nature and follows the familiar gaussian probability distribution σ^2 is the variance of the distribution. This is discussed in detail in Chaps. 5 and 6. Theoretically a gaussian random variable, in this case the signal amplitude, can take on any value whatsoever. There is no theoretical maximum. There is a 99.99 percent probability that the variable will lie within the range $\pm 4\sigma$, however. One can thus safely pick $V = 4\sigma$ in this case. In this example, then, $\sigma^2 = V^2/16$ is the maximum intensity signal that can be accommodated by the quantizer. If the signal covers the range $\pm V$ uniformly, however, $\sigma^2 = V^2/3$ is the maximum signal power that can be accommodated.)

As previously, we have for the mean-squared quantization noise

$$E(\epsilon^2) = \frac{a^2}{12}$$

$$= \frac{V^2}{3M^2} \tag{3-37}$$

using the quantizer characteristic $a = 2V/M$. With an input signal power of σ^2 we have as the signal-to-noise ratio

$$\text{SNR} = \frac{\sigma^2}{E(\epsilon^2)} = 3M^2 \left(\frac{\sigma^2}{V^2} \right) \tag{3-38}[13]$$

Since the quantization noise is fixed, independent of σ^2, in the case of uniformly spaced levels, the SNR is proportional to σ^2. As a speaker reduces his intensity the SNR reduces correspondingly. The quantization noise becomes that much more noticeable. This is the problem alluded to above. To mitigate this and obtain a relatively fixed SNR over a wide dynamic range of signals it is necessary to introduce quantization-level tapering. Alternatively, as pointed out previously, it is simpler in practice to nonlinearly compress the signal and then apply uniform-level spacing to the compressed output signal. At the receiver the signal is then expanded following an inverse nonlinear characteristic. It is

---

[13] This definition of signal-to-noise ratios differs somewhat from those of the previous paragraph because it is the ratio of input signal power to mean-squared quantization noise. Previously, we used the output, quantized, power for the calculation. It is more convenient for our purpose now to refer to the input power, since it is this quantity that we will be varying. If the number of quantization levels $M \gg 1$, the difference between the input and output signal power, with equally spaced levels, is slight. Notice, for example, that $M^2 - 1$ appears in (3-36) rather than the M^2 factor in (3-38).

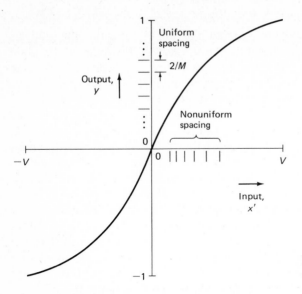

Figure 3-21 Nonlinear compressor characteristic.

apparent that this is exactly equivalent to nonuniform spacing of the levels. An example appears in Fig. 3-21. x' represents the input signal and y the output. For the characteristic chosen, the equivalent input levels move farther and farther apart as the input amplitude approaches $\pm V$. This is due to the compression of the higher input values into a correspondingly smaller range of output values.

A typical compression characteristic has a logarithmic form. A particularly common form implemented in practice for speech telephony is the *μ-law companding*. This has the specific form

$$y(x') = \frac{\ln (1 + \mu x'/V)}{\ln (1 + \mu)} \qquad 0 \le x' \le V \tag{3-39}$$

It is odd symmetric with this characteristic about the $x' = 0$ point. $y(x')$ thus ranges between ± 1. The parameter μ appearing can be varied to obtain a variety of characteristics. Note that for $x' \ll V/\mu$ the characteristic is almost linear:

$$y(x') \doteq \frac{\mu x'}{V \ln (1 + \mu)} \qquad x' \ll \frac{V}{\mu} \tag{3-40}$$

As x' increases beyond the point V/μ, the logarithmic characteristic takes over. For $\mu \ll 1$, $y(x') \doteq x'/V$, nonlinear compression disappears, and the uniform spacing of the output y corresponds to uniform spacing of the input x'. The Bell System in the United States has adopted a $\mu = 255$ companding law for its digital carrier systems.[14] This particular characteristic is sketched, for $x' \ge 0$,

[14] Bell Telephone Laboratories, *Transmission Systems for Communications*, 4th ed., 1970, pp. 571–583.

in Fig. 3-22. We shall carry out a quantization noise and SNR analysis for the μ-law compression characteristic. For this purpose it is useful to define a normalized input signal $x = x'/V$. In terms of this normalized signal, the μ-law compressor characteristic is given by

$$y(x) = \frac{\ln (1 + \mu x)}{\ln (1 + \mu)} \qquad 0 \le x \le 1 \qquad (3\text{-}41)$$

The normalized input form is the one actually sketched in Fig. 3-22. Our analysis will depend on the normalized signal power (or variance) $\sigma_x^2 = \sigma^2/V^2$. In terms of this parameter the SNR for uniform quantization levels (no compression) is given by

$$\text{SNR} = 3M^2\sigma_x^2 \qquad (3\text{-}42)$$

We shall compare the SNR with compression with this expression.

To analyze the effect of compression or nonlinear tapering on the quantization noise and hence the SNR, we note that the output signal y has values equally spaced $2/M$ units apart, from -1 to $+1$, as shown in Fig. 3-21. We focus on the analysis for positive x only. This uniform spacing projects into a nonuniform spacing Δ_j $(j = 1, 2, \ldots, M/2)$ at the input which depends on the compressor characteristics. Consider a particular spacing Δ_j centered at x_j as shown in Fig. 3-23. It is apparent from the figure that for Δ_j small enough $(2/M \ll 1)$, $2/M \doteq (dy/dx)|_{x_j} \Delta_j$, or

$$\Delta_j \doteq \frac{2}{M} \bigg/ \frac{dy}{dx}\bigg|_j \qquad (3\text{-}43)$$

All values of x in the range Δ_j centered at x_j will, after quantization of the compressed signal, correspond to one output value. Hence the mean-squared quantization error due to these values can be found by appropriately averaging over Δ_j. Assume that now the input signal is random with a known probability

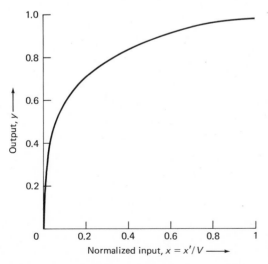

Figure 3-22 Compressor characteristic, μ-law companding, $\mu = 225$ (positive quadrant only).

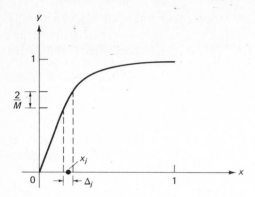

Figure 3-23 Compressor analysis.

density function $f(x)$. The mean-squared variation about x_j is then, by definition, given by

$$E(\epsilon_j^2) = \int_{x_j-\Delta_j/2}^{x_j+\Delta_j/2} (x - x_j)^2 f(x)\, dx \tag{3-44}$$

Since we have been assuming the number of quantization levels $M \gg 1$, it is a
reasonable approximation that all values of x in the range Δ_j have the same
probability of occurring. This corresponds to taking $f(x)$ constant over that
range. We thus have

$$E(\epsilon_j^2) \doteq f(x_j) \int_{x_j-\Delta_j/2}^{x_j+\Delta_j/2} (x - x_j)^2\, dx$$

$$= f(x_j)\, \frac{\Delta_j^3}{12} \tag{3-44a}$$

Note that this is similar to the analysis carried out in the previous paragraph,
except that there the level spacing was a fixed value $\Delta_j = a$ [see Eq. (3-29)].
From (3-43) we can now write

$$E(\epsilon_j^2) \doteq \frac{1}{3M^2}\, \frac{f(x_j)}{[y'(x_j)]^2}\, \Delta_j \tag{3-44b}$$

where we have used the shorthand notation $y'(x)$ to represent dy/dx. The total
mean-squared quantization noise $E(\epsilon^2)$, including negative x_j, is twice the sum
of all $M/2$ contributions, $E(\epsilon_j^2), j = 1, 2, \ldots, M/2$:

$$E(\epsilon^2) = 2 \sum_{j=1}^{M/2} E(\epsilon_j^2) \doteq \frac{2}{3M^2} \sum_{j=1}^{M/2} \frac{f(x)}{[y'(x)]^2}\bigg|_j \Delta_j \tag{3-45}$$

For M large enough we can approximate the sum of (3-45) by the equivalent
integral, and get

$$E(\epsilon^2) \doteq \frac{2}{3M^2} \int_0^1 \frac{f(x)}{[y'(x)]^2}\, dx \tag{3-45a}$$

It is this expression that we use to evaluate the effect of the compression law. Note, as a check, that if there is no compression $y = x$, $y'(x) = 1$, and $E(\epsilon^2) = 1/3M^2$, as found previously. [See (3-37). Recall again that we are dealing here with the normalized input $x = x'/V$.] The effect of the compressor is thus given by the square of the derivative appearing in the denominator of the integral in (3-45a).

For the μ-law compander specifically, we have, from (3-41),

$$y'(x) = \frac{\mu}{\ln (1 + \mu)} \frac{1}{1 + \mu x} \tag{3-46}$$

Inserting this in the denominator of (3-45a), we get

$$E(\epsilon^2) = \frac{2}{3M^2} \left(\frac{\ln (1 + \mu)}{\mu} \right)^2 \int_0^1 f(x)(1 + \mu x)^2 \, dx$$

$$= \left[\frac{\ln (1 + \mu)}{\mu} \right]^2 \frac{1}{3M^2} (1 + \mu^2 \sigma_x^2 + 2\mu E[|x|]) \tag{3-47}$$

where

$$\sigma_x^2 \doteq \int_{-1}^{1} x^2 f(x) \, dx \tag{3-48}$$

is the variance of the signal distribution to be quantized [$f(x)$ is assumed symmetrical about zero and concentrated in the range -1, 1], and

$$E[|x|] \doteq 2 \int_0^1 x f(x) \, dx \tag{3-49}$$

Equation (3-47) enables us to calculate the mean-squared quantization noise for the μ-law compressor characteristic for any set of signal statistics. The signal statistics required are the two parameters σ_x^2 and $E[|x|]$. The actual form of the signal probability density function $f(x)$ is not too critical. The ratio $E[|x|]/\sigma_x$ does not vary very much from one distribution to another and hence $E(\epsilon^2)$ will be almost the same for various density functions. Three examples are tabulated below. The corresponding density functions are shown sketched in Fig. 3-24.

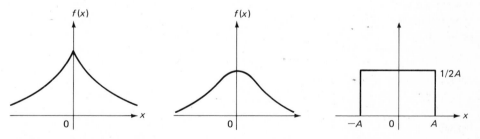

Figure 3-24 Examples of signal probability density functions. (*a*) Laplacian. (*b*) Gaussian. (*c*) Uniform.

1. *Laplacian signal:*

$$f(x) = \frac{1}{\sqrt{2}\,\sigma_x}\,e^{-\sqrt{2}|x|/\sigma_x} \tag{3-50}$$

$$\frac{2E[|x|]}{\sigma_x} = \sqrt{2} = 1.414 \tag{3-51}$$

2. *Gaussian signal:*

$$f(x) = \frac{e^{-x^2/2\sigma_x^2}}{\sqrt{2\pi\sigma_x^2}} \tag{3-52}$$

$$\frac{2E[|x|]}{\sigma_x} = 2\sqrt{\frac{2}{\pi}} = 1.6 \tag{3-53}$$

3. *Uniformly distributed signal:*

$$f(x) = \frac{1}{2A} \qquad -A \leq x \leq A \tag{3-54}$$

$$\sigma_x^2 = \frac{A^2}{3}$$

$$\frac{2E[|x|]}{\sigma_x} = \sqrt{3} = 1.732 \tag{3-55}$$

The particular values of $E[|x|]$ given in each case are obtained by integrating (3-49) using the appropriate form of $f(x)$. It turns out that the laplacian density function of (3-50) is a relatively good model for speech,[15] so we will focus on it in making calculations of the SNR. The other density functions give almost identical results, however. The gaussian density function is the one most commonly appearing in the theory of probability and we shall encounter it regularly in Chaps. 5 and 6 when modeling additive noise. The quadratic form in the exponential causes $f(x)$ to drop very rapidly for $|x| > \sigma_x$. It thus models signals that have a relatively low probability of exceeding σ_x. The laplacian density function serves as a good model of signals that have a relatively higher probability of attaining higher values. It is left to the reader to show, by integrating (3-50) from $4\sigma_x$ to ∞ and $-4\sigma_x$ to $-\infty$, that the probability that $|x|$ exceeds $4\sigma_x$, in that case, is 0.0035, contrasted with the equivalent probability for the gaussian case of 10^{-4}.

The signal-to-quantization noise ratio is now obtained, as previously, by defining it to be the mean-squared signal power σ_x^2 divided by $E(\epsilon^2)$. We thus get, for the μ-law compander,

$$\text{SNR} = \frac{\sigma_x^2}{E(\epsilon^2)} = \frac{3M^2}{[\ln(1+\mu)]^2}\,\frac{1}{1 + 2\,E[|x|]/\mu\sigma_x^2 + 1/\mu^2\sigma_x^2} \tag{3-56}$$

[15] *Ibid.*

Table 3-3 Signal-to-quantization noise, 7-bit code, μ-law companding, $\mu = 255$
SNR vs. input signal power

σ_x^2, dB, relative to maximum input signal	SNR, dB		
	Uniform spacing	Companded case	
		Laplacian	Gaussian
−60	−13.1	18.6	18.5
−56	−9.1	21.8	21.6
−50	−3.1	25.7	25.5
−46	0.9	27.6	27.3
−40	6.9	29.7	29.5
−26	20.9	31.5	31.5
−20	26.9	31.8	31.8
−14	32.9	32	32
−8	38.9	32	32
0	46.9	32	32

Using this equation and various models for the signal statistics one can obtain the SNR for various compander laws. For $\mu \to 0$ it is easy to see, as a check, that SNR $= 3M^2\sigma_x^2$, just the result for uniform-level spacing obtained earlier [Eq. (3-42)]. In Table 3-3 we tabulate SNR in decibels ($10 \log_{10}$ SNR) as a function of σ_x^2, in decibels also, for the $\mu = 255$ compander law and a 7-bit ($M = 128$) system. This is shown for both the laplacian and gaussian cases. Note that the results for the two are comparable, as expected from the discussion above. Also shown for comparison is the uniform-level spacing case, obtained from (3-42). Since $\sigma_x^2 < 1$ because of the normalization, its value is indicated as dB, relative to the maximum input signal. These results are sketched as well in Fig. 3-25. The important point to note is that over a broad range of input signals, the SNR is almost constant for the compander case. It is only when σ_x^2 reaches -40 dB, that the SNR begins to drop appreciably. The uniform level SNR shows wide variations, however. For $\sigma_x^2 > -15$ dB, i.e., for higher-amplitude signals, it begins to exceed the compander result. But for all the other values shown (lower-level signals), it drops lower and lower with respect to the compander result. Uniform spacing thus favors the higher-amplitude signals at the expense of the lower-amplitude signals. The companded SNR is of course almost constant over almost a 40-dB range of input amplitudes.

Other logarithmic companding laws have been proposed in addition to the μ-law characteristic discussed here. One example is the A law,[16] with the following characteristic (the input signal x is again normalized, relative to the

[16] *Ibid.*

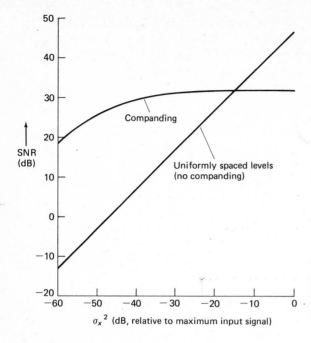

Figure 3-25 Signal-to-quantization noise, with companding, vs. input signal power, M = 128 levels, μ = 225.

maximum input signal):

$$y = \frac{1 + \ln Ax}{1 + \ln A} \qquad \frac{1}{A} \le x \le 1$$

$$= \frac{Ax}{1 + \ln A} \qquad 0 \le x \le \frac{1}{A} \qquad (3\text{-}57)$$

(The full characteristic is odd symmetric about $x = 0$.) A value of $A = 100$ is typical of practical companders using this characteristic. The SNR characteristic for this value of A is found to provide a somewhat wider dynamic range than the μ-law compander with $\mu = 255$, but its output SNR is somewhat smaller.[17]

Note that the A-law characteristic is defined to be linear for small x and logarithmic for large x. Both the μ-law and A-law characteristics are implemented in practice by piecewise linear approximations.[18]

3-6 EXAMPLES OF PAM AND PCM SYSTEMS

In this section we present some examples of specific PAM and PCM systems. The examples have been chosen to further focus attention on the elements of time multiplexing, quantization, and A/D conversion discussed earlier, as well

[17] *Ibid.*, p. 582, fig. 25-11.
[18] *Ibid.*, pp. 580, 581.

as to present some typical engineering solutions to problems raised earlier. The emphasis here, as in previous sections, is on analog signals to be transmitted using digital techniques. Other examples of digital communication systems in use are described briefly in some of the sections following.

In Sec. 3-8 we stress the time multiplexing of signals already in digital format. These could be data terminal or computer signals, or they could be analog signals converted to digital form.

The applications described in this section include an older PAM system typical of those used for many years in radio telemetry, a PCM system developed for a meteorological satellite, a deep-space telemetry system, and the Bell System T1 PCM system in common use for short-haul telephone communications. The techniques used in generating the time-multiplexed PAM or PCM signals in these examples are representative of those used in a broad variety of application areas requiring the telemetering of data. These include industrial telemetry for process control purposes, automatic monitoring of power-generation plants, biomedical telemetry, telemetry for geoseismic and geophysical exploration, etc. In the older systems the various sampling, A/D, and multiplexing operations were carried out using hard-wired logic. More recently, with the advent of programmable logic and microprocessor-controlled systems, the various operations have been brought under software control. This results in a flexibility of operation not possible with hard-wired logic. It often results in cheaper, more compact, and less power consuming systems as well. Since the principles of multiplexing, commutation, and synchronization are independent of specific implementation, both older and newer systems can be described side by side to obtain insight into the operation of PAM and PCM systems. This is particularly true of the simple radio telemetry PAM system to be described first. Although PCM systems have become much more common in telemetry, and have in many cases displaced older PAM systems, the techniques used to carry out the sampling and multiplexing operations are common to both PAM and PCM systems. This is apparent from our discussion in Sec. 3-3.

Examples of systems other than those described here may be found in the current literature.[19]

A Radio Telemetry PAM System[20]

The PAM system to be discussed is typical of those used for radio telemetry purposes. It is of interest because it provides for the multiplexing of many data channels of greatly varying data rates or bandwidths. This is done by performing the multiplexing operations in two steps: low-data-rate channels are first

---

[19] Journals and conference proceedings commonly publishing papers on digital communication systems include the *IEEE Transactions on Communications,* the *IEEE Transactions on Aerospace and Electronic Systems,* and the annual proceedings of both the IEEE National Telecommunications Conference and the IEEE International Communications Conference.

[20] E. L. Gruenberg (ed.), *Handbook of Telemetry and Remote Control,* McGraw-Hill, New York, 1967, pp. 9–11 to 9–18.

Table 3-4 Multiplexing

Group	Number of data channels	Bandwidth per channel	Sampling rate, samples/ second	Required accuracy, %	Main multiplexer position
1	3	2 kHz	5,000	10	2 and 10, 4 and 12, 5 and 13
2	2	1 kHz	2,500	5	3, 6
3	5	100 Hz	312.5	2	7
4	28	25 Hz	78	2	8
5	55	5 Hz	39	1	9
6	115	5 Hz	19.5	2	14
7	110	1 Hz	19.5	2	15

time-multiplexed to form composite data channels of much wider bandwidths; the composite channels are then in turn time-multiplexed with wider-bandwidth channels to form the main multiplexed signal. Other techniques are discussed in Sec. 3-8.

Specifically, in a typical application 318 different data channels are to be sampled, time-multiplexed, and transmitted via radio. The data rates involved range from a bandwidth of 1 Hz to one of 2 kHz. The main multiplexer is designed to handle 16 channels, with 2,500 samples per second for each. The low-data-rate signals must thus be combined in groups to provide the 2,500 samples per second required for any one channel. The various data channels, and their bandwidths, sampling rates, and group assignments, where necessary, are indicated in Table 3-4.[21]

The two 1-kHz channels, sampled at a 2,500 samples per second rate, obviously can be tied directly to the main multiplexer. The channels in groups 3 to 7 must be combined, however. For ease in timing this combining is done by sampling at binary submultiples of the final 2,500 samples per second rate. The required sampling rates can then be obtained by successively dividing down by factors of 2 from a master clock. As an example, the 312.5 samples per second rate is obtained by dividing down by 8 from 2,500, and the 19.5 samples per second rate by dividing down by 128. Figure 3-26 shows the preliminary multiplexing at the 312.5 sample rate. Note that five data channels only are combined, leaving three time slots available for synchronization, calibration, or inclusion of additional data channels later, if so desired.

It is apparent that such spare capacity exists in each of the low data rate groupings. Thus a sampling rate of 78 samples per second makes 32 time slots available. In this application there are only 28 channels (group 4) requiring this sampling rate. The 55 data channels of 5-Hz bandwidth (group 5) are grouped together and shown sampled at almost four times the Nyquist rate to meet the

[21] *Ibid.*, p. 9–11.

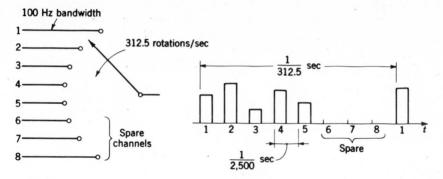

Figure 3-26 Preliminary multiplexing, group 3, PAM system.

higher accuracy requirements in that case. Here nine spare time slots are available. The 110 channels of 1-Hz bandwidth are obviously highly oversampled, but system simplicity results.

The three high-bandwidth channels indicated as group 1 require 5,000 samples per second each. Two time slots, spaced $\frac{1}{5000}$ s apart, must be thus set aside, of the 16 available in the main multiplexer, for each of the three channels. These then use six time slots. Adding the two used by the two 1-kHz channels, and the five time slots required by the premultiplexed groups 3 to 7, we have 13 used in all. Two additional time slots in the final multiplexed signal are used for synchronization purposes, and one is left available as a spare. The main multiplexer assignments are indicated in the table.

The final multiplexed signal, following these assigned time slots, is shown in Fig. 3-27.[22] The slots occupied by the three wide-bandwidth channels in group 1 are indicated by the labels 1A, 1B, and 1C, respectively, while the two 1-kHz channels are indicated by 2A and 2B. The entire sequence of 16 time slots shown is called a *frame*. In the format of Fig. 3-27 the composite data samples occupy 50 percent of each time slot. To provide synchronization all data samples are placed on a pedestal of about 20 percent of the full-scale data level. The absence of pulses in slots 1 and 11 then provides the necessary synchronization. Zero data corresponds to the pedestal height, as shown by the spare data slot 16. (In other systems synchronization may be provided by a specified digital code word inserted in the appropriate time slot.[23])

The block diagram of the entire PAM system (transmitter only) is shown in Fig. 3-28.[24] The programmer shown provides the clock pulses for all multiplexers, starting with a clock rate of 40 kHz, and counting down by 2's, as indicated earlier.

The receiver for this PAM radio telemetry system performs the necessary

[22] *Ibid.*, p. 9–13, fig. 12*a*.
[23] *Ibid.*, p. 9-13, fig. 12; pp. 5-34 and 5-35, fig. 5*a, b, c, d*.
[24] *Ibid.*, p. 9-12, fig. 11.

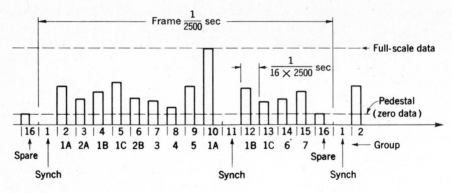

Figure 3-27 Composite signal format.

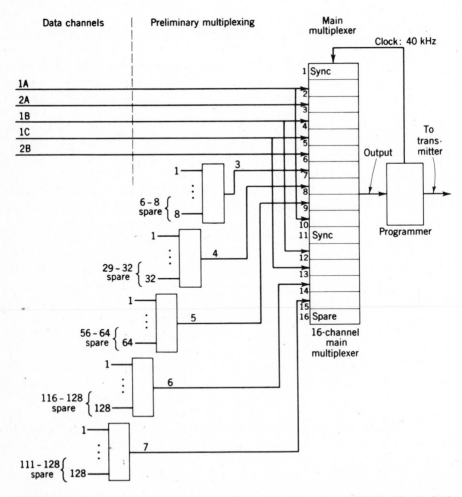

Figure 3-28 PAM system for radio telemetry. [From E. L. Gruenberg (ed.), *Handbook of Telemetry and Remote Control*, McGraw-Hill, New York, 1967, chap. 9, fig. 11, simplified.]

functions of demultiplexing the data samples, low-pass filtering them, and sending them to the appropriate receiving channel. We shall not discuss the demultiplexer here, but refer the reader to the reference for a detailed discussion.[25]

Commutation

The sampling and time-multiplexing operations represented conceptually here and in previous sections by a rotating mechanical switch represent the heart of PAM and PCM systems. Many electronic schemes have been devised to perform these operations. The process of sequentially sampling a group of input data channels is frequently referred to as a *commutation* process, and various electronic commutators have been devised to perform this function.[26] The commutator is commonly divided into two parts: channel gates that sequentially open to allow one channel at a time through, performing the sequential sampling operation, and a sequential gate control generator, operated by a clock, that actually operates the gates. One common example is shown in Fig. 3-29a.[27] The sequential gate control generator here consists of a chain of flip-flops. The master pulse turns on the first element of the chain. The next channel pulse turns off the first flip-flop, and this in turn turns on the second element. This stepping continues to the end of the chain, when the master pulse again turns on the first element.

The counter-matrix arrangement of Fig. 3-29b provides another common form of commutation. The outputs shown open the corresponding gates sequentially. To perform the necessary sequencing, counter waveforms are combined appropriately. As an example consider the binary waveforms of Fig. 3-30. These correspond respectively to the appropriate binary counters of Fig. 3-29b. Thus the basic binary sequence 1 and its phase-shifted version 1′ correspond to 1 and 1′, respectively, of Fig. 3-29b. Dividing down successively by 2, as indicated in Fig. 3-29b, we get the waveforms shown in Fig. 3-30. The three pairs of waveforms shown provide the $2^3 = 8$ possible connections needed for the eight channels indicated. Specifically, note from Fig. 3-30 that outputs 1, 2, and 3 are all simultaneously positive during the first $\frac{1}{8}$ of the switching cycle of Fig. 3-30. Connecting these outputs to channel 1 as shown in Fig. 3-29b, channel 1 is gated on during the first $\frac{1}{8}$ of the cycle only, being turned off during the remainder of the cycle. It is apparent from Fig. 3-30 that connecting the other seven channels appropriately, as shown in Fig. 3-29b, three simultaneously positive pulses are obtained in time sequence, providing the desired sequential gating action.

As noted above, sampling, multiplexing, and commutating functions just described can be implemented on LSI chips, or, with the addition of RAM and ROM memories, carried out using programmable logic. Microprocessor-controlled versions can be developed as well. In a deep-space telemetry system

[25] *Ibid.*, pp. 9-13 to 9-18.
[26] *Ibid.*, pp. 4-53 to 4-92.
[27] *Ibid.*, fig. 78a, p. 4-79.

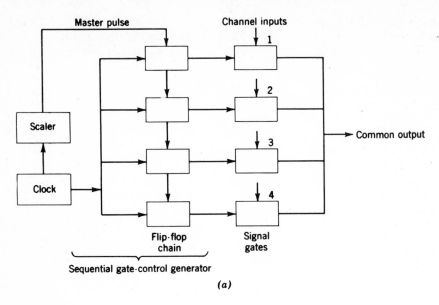

(a)

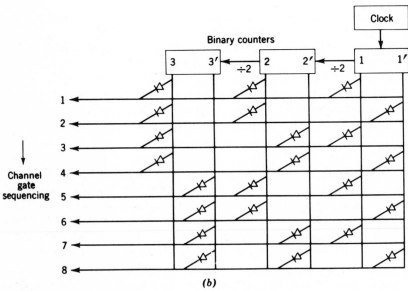

(b)

Figure 3-29 Typical electronic commutators. (*a*) Broken-ring commutator: four data channels. (*b*) Counter-matrix gate-control generator: eight channels. [From E. L. Gruenberg (ed.), *Handbook of Telemetry and Control,* McGraw-Hill, New York, 1967, chap. 4, figs. 78(a) and 79, simplified.]

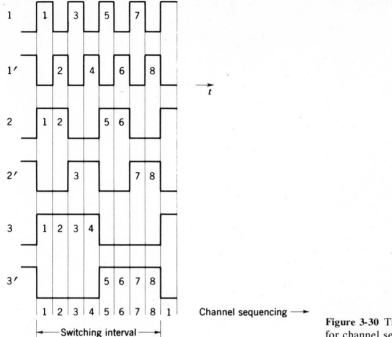

Channel sequencing ⟶

Figure 3-30 Timing waveforms for channel sequencing.

to be described later, programmed control is in fact used to carry out the multiplexing operations.

PCM Telemetry for the Nimbus Satellite

We turn to an example of a PCM system developed for space applications. The particular example we have chosen to describe very briefly is the telemetering system used aboard the NASA second-generation meteorological satellite, the Nimbus spacecraft. This spacecraft is designed to provide appropriate meteorological data (cloud measurements, distribution of rainfall, the earth's outgoing thermal radiation, heat balance, etc.) of the entire earth at least once a day.[28]

Two independent PCM telemetering systems are used. One, the so-called A system, records continuously on an endless loop tape recorder; the other, the B system, may be commanded at any time. In the A system 544 inputs are multiplexed for recording and ultimate transmission on command. Thirty-two of these input channels, in two groups of 16 channels each, are sampled once per second; the remaining 512 channels, in two groups of 256 channels each, are

---

[28] R. Stampfl, in A. V. Balakrishnan (ed.), *Space Communications,* McGraw-Hill, New York, 1963, chap. 18. The telemetry systems are described on pp. 370–376 and 404–406.

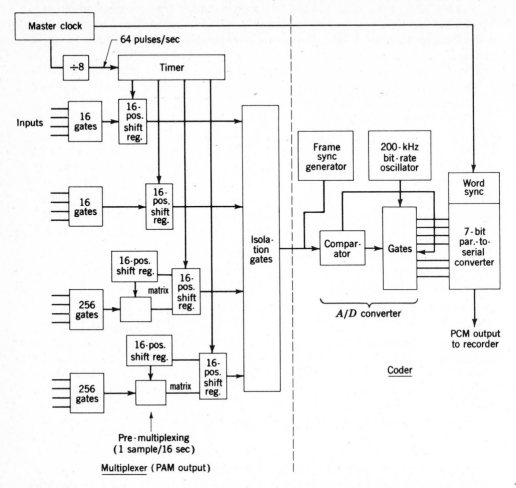

Figure 3-31 "A" telemeter, Nimbus PCM system, simplified. [Adapted from A. V. Balakrishnan (ed.), *Space Communications,* McGraw-Hill, New York, 1963, fig. 18-4, pp. 372–373.]

sampled once every 16 s. Premultiplexing, just as in the PAM systems described in the previous paragraph, must thus be carried out on the two 256 channel groups, combining them first before combining with the 16 channel groups. A simplified block diagram of the system is shown in Fig. 3-31.[29]

Sampling and multiplexing are first carried out by means of the timer and 16-position shift registers shown. A pulse of the 64 pulses per second generator shown triggers the first position of the first (uppermost) shift register shown, gating on the first of the 16 upper channels. The second pulse, $\frac{1}{64}$ s later, triggers the first position of the second shift register, gating on the first of the 16 chan-

---

[29] *Ibid.,* fig. 18-4, pp. 372, 373.

nels in the next group. One of the 256 multiplexed groups is then gated on, etc. The fifth pulse returns to the first shift register, opening its second position. The cycle thus continues for the full 64 pulses, returning to the first of the 16 upper channels 1 s later, as desired.

Premultiplexing of the 256 channel groups is done by using an additional 16-position shift register for each group to form a matrix with the shift registers mentioned above.

Each of the 64 outputs of the multiplexer constitutes a word $\frac{1}{64}$ s long. The 64 consecutive words in turn make up a frame 1 s long. Each word is quantized into $2^7 = 128$ levels and converted into a 7-bit sequence. A word sync bit (a zero) is added and the coded PCM stream fed out to the recorder. The complete frame 1 s long thus has $64 \times 8 = 512$ bits.

The A/D conversion is carried out by comparing the incoming (PAM) signal to a sequence of binary weighted reference voltages. The binary-coded signal is stored in a buffer where the word sync bit is added. Frame sync is also provided at the A/D converter input.

The Mariner Jupiter/Saturn 1977 Telemetry System[30]

This telemetry system was developed to handle the telemetry requirements of NASA's spacecraft missions to Jupiter and Saturn (1.5×10^9 km from the earth). The telemetry system is particularly interesting since its requirements vary with the location of the spacecraft: during the launch period, during the interplanetary cruise period, during the planetary encounter period. Playback of data recorded on tape is also used when earth communications are interrupted.

Both engineering data and science data from a variety of instruments (including imaging data of the planets), at varying data rates, are being, or will be telemetered back, as the vehicle moves through space. As an example, 1,200-bits/s high-rate engineering data were to be transmitted during the launch period, during periods of memory readout, and for anomaly diagnosis. During the encounter phase the engineering data rate is set at a normal rate of 40 bits/s. Science and engineering instrument data rates range from 80 to 2,560 bits/s during the cruise phase. Real-time engineering and science transmission during the encounter phase is transmitted at 7.2 kbits/s, while image information is transmitted at a maximum rate of 108 kbits/s. When this high rate of transmission is not possible, one of four selectable lower bit rates can be used, or the imaging data can be stored for retransmission later.

To accommodate this variety of data rates, computer control is used. Data from the science instruments and from engineering sensors are sampled and formatted at the proper rate and in the proper sequence by a special-purpose computer, called the Flight Data Subsystem. Data within a particular format

[30] G. E. Wood and T. Risa, "Design of the Mariner Jupiter/Saturn 1977 Telemetry System," *Proceedings, International Telemetry Conference,* Los Angeles, vol. 10, pp. 606–615, 1974.

are grouped into 8-bit bytes, or multiples thereof. More than 35 different programs are required to handle the different telemetry modes, depending upon the vehicle location and communication path conditions. Each of two 4,096 sixteen-bit memories in the computer can store at least two separate programs. As the mission progresses, new programs required are transmitted to the spacecraft. The sequencing between the various instruments and sensors and data-format generation are controlled by these programs.

Note how the use of computer control makes adaptive telemetry relatively easy to handle. A 32-bit synchronization word and a nonambiguous time identification word are used to provide the appropriate timing and synchronization.

The T1 System: The Bell Telephone PCM System for Short-Haul Telephone Communication

The Bell System in the United States pioneered in the early 1960s by its introduction of a PCM system for digital voice communication over short-haul distances of 10 to 50 mi. The T1 system, as it is called, has found widespread adoption throughout the United States, Canada, and Japan. It is the basis for a complete hierarchy of higher-order multiplexed systems, used for either longer-distance transmission or for transmission in heavily populated urban areas, to be described in Sec. 3-8. It forms the basis for many purely digital transmission systems as well. The 8-kHz sampling rate used, as well as the 8 bits per sample quantization currently in use, form the basis for most digital PCM voice systems in use throughout the world, both for terrestrial and satellite communications. In the T1 system 24 telephone channels are time-multiplexed, sampled, and coded into 1.544 Mbits/s PCM for carrier transmission, or for further multiplexing for longer-distance transmission.

Since the development of the T1 system, worldwide standards calling for 30-channel time-division-multiplexed PCM systems with 2.048 Mbits/s transmission, have been developed and implemented in many parts of the world.[31] Because of the earlier development of the 24-channel T1 system and its subsequent widespread adoption in certain parts of the world, two *de facto* standards now exist. Appropriate interfaces must be employed for PCM telephone signals moving from a 24-channel to a 30-channel system. The 30-channel PCM system is described briefly in Sec. 3-8 in a discussion of time-division-multiplexing hierarchies using this system as a base.[32]

The T1 system, as originally conceived, used $2^7 = 128$ levels of quantization for each of the 24 voice channels multiplexed. More recently, 256 levels have been adopted, leading to a quieter system with less distortion. The signaling format for the system groups twenty-four 8-bit PCM words, each corre-

---

[31] The 30-channel system is the recommended CCITT standard. The CCITT is the international standards agency for telephone communications.

[32] Details of the 30-channel system appear in D. W. Davies and D. L. A. Barber, *Communication Networks for Computers*, Wiley, London, 1973, chaps. 2 and 7.

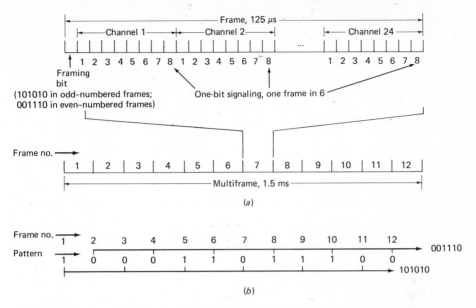

Figure 3-32 Signaling format, T1 system. (*a*) Frame structure, T1 system. (*b*) Framing bit pattern.

sponding to a coded PAM sample, sampled at a rate of 8,000 per second, plus a 1-bit framing bit, into a frame 125 μs long. The resultant transmission rate, corresponding to 193 bits transmitted in 125 μs, is thus 1.544 Mbits/s. Every sixth frame the eighth, least significant bit in each channel is deleted or "robbed" and a signaling bit put into its place. The sequence of signaling bits, one every six frames, appearing thus at a rate of about 1,330 bits/s, is used to transmit dial pulses, as well as telephone off-hook/on-hook signals. The frame format thus has the form shown in Fig. 3-32*a*.

The set of framing bits, one at the beginning of each frame, constitutes a frame bit pattern that is used to maintain frame synchronization. A 1010 . . . sequence, alternating every other frame, is used for this purpose. A 6-bit subframe bit pattern, 000111, at the beginning of the other frames, is used to identify the occurrence of the one frame in six containing signaling bits. The resultant framing pattern appears in Fig. 3-32*b*. The 8-bit PCM word in each channel is coded as a *bipolar signal*.[33] In this type of signal, one of many that could be used, binary 0's are represented by an off position (no pulse), while binary 1's alternate in polarity. This results in a zero dc component, a suitable condition for transmission through the telephone plant. A typical 8-bit bipolar word is shown in Fig. 3-33.

The T1 system consists of two basic components: the PCM terminal, called the D3 channel bank, and a T1 line for transmitting the composite 1.544-Mbits/s data. The channel bank is the device that carries out the sampling,

---

[33] W. R. Bennett and J. R. Davey, *Data Transmission*, McGraw-Hill, New York, 1965, p. 27.

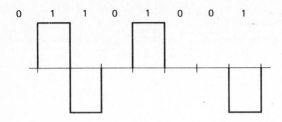

Figure 3-33 Bipolar signal (alternating polarity 1's).

multiplexing, and nonlinear quantization operations required to form the transmitted T1 data stream, as well as the receiver operations, in the reverse direction, that reconstitute an analog voice signal on each of the 24 voice channels.

Some of the PCM operations in the D3 channel bank are carried out on a per channel basis, for each of the 24 channels; the remainder are carried out in common.[34] Specifically, filtering for band limiting, and sampling through the use of sampling gates to generate PAM signals, are done on a per channel basis. A sampling rate of 8,000 samples per second is used. Each channel is sampled in sequence, and the PAM samples are passed on to the transmit portion of the common equipment. Here each sample is held momentarily while a PCM coder approximates it with an 8-bit digital code word. This thus represents the A/D process. Nonlinear quantization, using the μ-law compression characteristic described in the last section, is utilized. The common equipment generates the bipolar signal format, adds framing bits at the end of a frame, and inserts signaling bits in place of framing bits every sixth frame.

On reception a common receive unit processes the incoming bit pattern to extract clock information and develop frame synchronization, decodes the PCM words into PAM speech samples using an inverse nonlinear expansion characteristic, and then directs the PAM samples to the appropriate channel. Low-pass reconstruction filters are used in each channel to smooth the sequence of incoming PAM samples and to output the desired analog signal.

A simplified block diagram of the D3 channel bank units appears in Fig. 3-34.

3-7 DELTA MODULATION

It has been pointed out in the foregoing two sections that PCM techniques have been widely applied in the digital transmission of speech (telephony) as well as to the transmission of various types of telemetry signals. They have been used for the transmission of images, both fixed frame and TV, as well as for other

[34] W. B. Gaunt and J. B. Evans, Jr., "The D3 Channel Bank," *Bell Lab. Rec.,* vol. 50, pp. 329–333, August 1972; J. B. Evans, Jr., and W. B. Gaunt, "The D3 Channel Bank," *Conference Record, International Conference on Communications,* Minneapolis, Minn., June 1974, pp. 7D-1 to 7D-5.

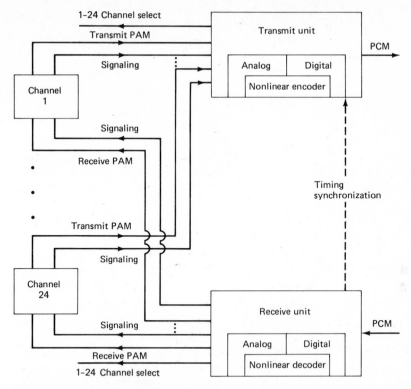

Figure 3-34 Simplified block diagram, D3 channel bank. (From J. B. Evans, Jr., and W. B. Gaunt, "The D3 Channel Bank," *Conference Record, International Conference on Communications,* Minneapolis, Minn., June 1974, fig. 2, p. 7D-4; reprinted by permission.)

forms of analog signals converted to digital format. Note that with widespread digital transmission and processing of signals, using computers, it becomes very natural to carry out the A/D operations and convert the signals to the equivalent binary format. Digital transmission of signals generally results in bandwidth expansion, however: as the number of quantization levels or, equivalently, the number of bits per sample increases, the transmission bandwidth required goes up accordingly.

Alternative methods of converting analog signals to digital format have been proposed for some applications, with the hope of reducing the bandwidth required, improving the performance, or reducing the cost. Delta modulation is one such scheme that has been adopted for some applications involving the transmission of speech or images. In those applications there is a great deal of redundancy in the information to be transmitted. Past information should thus allow current information to be predicted fairly well, so that new signals need only be transmitted if a significant change in the signal occurs.

Consider the block diagram of a delta-modulation system shown in Fig. 3-35. Just as in the usual A/D conversion discussed earlier the analog signal $x(t)$

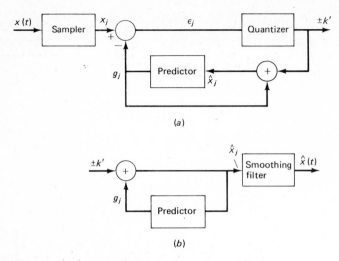

Figure 3-35 Delta modulation. (a) Transmitter. (b) Receiver.

must first be sampled periodically. These sampled values, designated x_j, $j =$ . . . , $-2, -1, 0, 1, 2, 3,$ . . . are compared with a predicted sample value, designated g_j, and the *difference* ϵ_j is then passed to a quantizer. Obviously, if ϵ_j is small most of the time, so that the prediction is good, few bits will be needed to represent this difference signal. A delta modulator uses a two-level quantizer, so that one bit only is used to represent the signal. The two quantizer levels in Fig. 3-35 are designated $\pm k'$. At the receiver the quantized difference signal is added to the predictor output to obtain a discrete estimate $\hat{x}_j$ of the desired sampled signal x_j. These discrete estimates are then passed through a low-pass filter, for smoothing, to generate the desired estimate $\hat{x}(t)$.

It is apparent that improved performance could be obtained by using more than two levels for the quantizer. This more general scheme is called *differential pulse code modulation* (DPCM). Delta modulation is thus a special case of DPCM, with two quantizer levels used.

The predictor used is generally a weighted sum of a number of past sample estimates. Specifically, with x_{j-1} the previous sample and $\hat{x}_{j-1}$ its estimate, $\hat{x}_{j-2}$ the estimate two time samples back, etc., the general form of a linear predictor may be written

$$g_j = \sum_{l=1}^{K} h_l \hat{x}_{j-l} \tag{3-58}$$

The h_l coefficients are the weighting factors. The simplest predictor is one that uses the previous sample estimate only as an estimate of the current sample x_j. For this case we have

$$g_j = h_1 \hat{x}_{j-1} \tag{3-59}$$

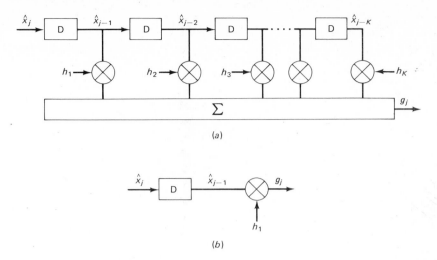

(a)

(b)

Figure 3-36 Predictor for delta modulation. (a) General case (nonrecursive digital filter). (b) Previous example estimator.

The coefficients $h_l, l = 1, 2, \ldots, K$, may be chosen to reduce some measure of the estimate errors to as small a value as possible. For example, one common measure of performance is the mean-squared error, averaged over many samples, or averaged statistically over the known statistics of the incoming samples x_j.

The linear predictor of (3-58) is an example of a transversal or nonrecursive digital filter.[35] It is readily implemented using shift-register elements. A conceptual block diagram appears in Fig. 3-36a. The D boxes shown represent delay elements that delay the incoming samples the required sampling interval. A block diagram for the special case of the previous sample estimator of (3-59) appears in Fig. 3-36b.

How well does the delta modulator perform? This is rather difficult to determine in general since the performance depends on the type of input signal, the sampling rate, the quantizer levels, and the form of predictor used. Although the device is conceptually simple, it is difficult to analyze because of the feedback implicit in its operation. Many studies of delta modulator performance have been carried out in recent years. We shall focus here on some gross characteristics only, referring the interested reader to the literature for detailed studies of performance.

As is the case with PCM systems, noise is introduced by the use of delta modulation. Because the difference signal is quantized to two levels, granular or *quantization noise* similar to PCM quantization noise appears at the receiver

[35] M. Schwartz and L. Shaw, *Signal Processing*, McGraw-Hill, New York, 1975.

output. Unlike PCM systems, with this noise reduced by using more quantization levels, the noise here can only be reduced by sampling more often. One
thus finds delta-modulation systems using sampling rates much higher than the
Nyquist rate used with PCM. This means that the ultimate bit rate is higher than
originally expected through the use of two-level quantization. In some cases the
bit rate required may even be higher than that of PCM. For this reason adaptive
delta-modulation schemes, to be described briefly later, have come into use.

In addition to quantization noise, another type of noise is encountered. This
is called *overload noise* and occurs if the quantization levels $\pm k'$ are too small to
track a rapidly changing signal. To demonstrate the occurrence of both types of
noise, consider a typical input signal and its predicted signal samples as shown
in Fig. 3-37. For simplicity's sake previous sample estimation is used. At sampling time $j-1$ the predicted value g_{j-1} is found to be smaller than the actual
signal sample x_{j-1}. Since $\epsilon_{j-1} = x_{j-1} - g_{j-1}$ is positive, the quantizer outputs
$+k'$. This is then added to g_{j-1} to provide the estimate $\hat{x}_{j-1}$. The predicted signal
at sampling time j is then $g_j = h_1\hat{x}_{j-1}$. Assuming that h_1 is somewhat less
than 1, one gets g_j as shown in the figure. Since $g_j > x_j$, the actual signal sample
shown in the figure, $\epsilon_j < 0$, and $-k'$ is outputted. The resultant signal estimate
is $\hat{x}_j = g_j - k'$, as shown. As one proceeds sample by sample, the sample
estimates $\hat{x}_j$ tend to follow the curve of $x(t)$, sometimes lying above, sometimes
lying below. The result is a received signal that tracks the original signal, but
introduces quantization noise.

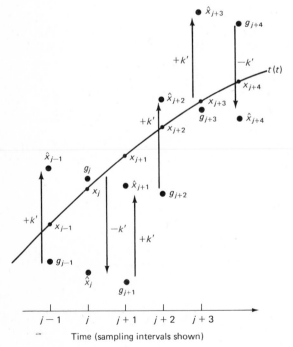

Time (sampling intervals shown)

Figure 3-37 Operation of delta modulator.

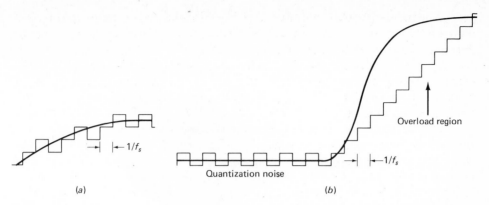

Overload region

Quantization noise

(a) (b)

Figure 3-38 Sources of noise in delta modulation. (*a*) Quantization noise. (*b*) Overload noise.

If a simple hold circuit is used at the receiver to provide the filtering, the resultant output signal appears as in Fig. 3-38*a*. It is apparent that the quantization noise introduced is proportional to the step size k'. By reducing k' the noise is reduced. Too small a value of k' results in the overload noise mentioned above, however. This is due to the inability of the modulator to track large changes of the input signal $x(t)$ in a small interval of time. An example appears in Fig. 3-38*b*. It is thus apparent that an optimum step size k' must exist: this depends on the characteristics of the input signal $x(t)$, the sampling rate, and the total noise that can be tolerated.

The overload region of Fig. 3-38*b* occurs because the step size k' sets a maximum limit on the input signal slope the modulator can follow. Specifically, let the sampling rate be f_s samples per second. The time between samples is thus $1/f_s$ seconds, as shown in Fig. 3-38. The maximum slope the system can follow corresponds to k' units of amplitude in $1/f_s$ seconds or $k'f_s$. This is apparent from the overload region of Fig. 3-38*b*. But increasing k' to reduce the overload noise increases the quantization noise as well. This thus gives rise to the optimum choice of k' noted above.

This optimum choice of step size k' is shown graphically in Fig. 3-39, depicting a typical performance curve for delta modulation. The performance

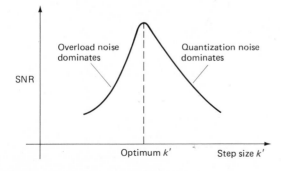

Overload noise dominates

Quantization noise dominates

SNR

Optimum k' Step size k'

Figure 3-39 Typical performance curve for delta modulation.

here is measured by the reciprocal of the noise, or the signal-to-noise ratio which is proportional to this quantity. Obviously, as the noise is reduced, the performance improves.

These concepts may be made more quantitative by considering, as an example, a sine-wave test signal. Say that the signal $x(t)$ is given simply by

$$x(t) = A \sin \omega_m t \tag{3-60}$$

Its maximum slope, at $t = 0$ and times spaced multiples of half a period ($1/2f_m$ seconds) from this point, is just $A\omega_m = 2\pi A f_m$. This is shown in Fig. 3-40. Now let this signal be sampled *many* times in a period at a rate of f_s samples per second. Thus

$$f_s \gg f_m \tag{3-61}$$

As an example, we might use 10 to 20 samples in a period. This is of course a much higher sampling rate than that required by the Nyquist sampling theorem. With delta modulation we now compare the sampled values of $x(t)$ with the predicted values and transmit either of two levels, $+k'$ or $-k'$, as noted earlier. The delta modulator will track changes in $x(t)$ if they differ by no more than k' units in $1/f_s$ seconds. Hence for this sine wave, overload can be avoided if $k'f_s$ is greater than or equal to the maximum slope:

$$k'f_s \geq 2\pi A f_m \tag{3-62}$$

This is often shown in normalized form by defining a relative step size $k \equiv k'/A$, i.e., the ratio of the step size k' to the maximum signal level. For more realistic signals, often represented by randomly varying signals, as noted in Sec. 3-5, a definite maximum may not exist. As in that section, then, in discussing nonuniform quantization, one often arbitrarily selects as the maximum signal level a value four standard deviations, 4σ, from the average. The relative step size would then be defined as $k \equiv k'/4\sigma$.

In terms of the relative step-size parameter k, then, one can avoid overload noise for the sine-wave test signal by selecting

$$kf_s \geq 2\pi f_m \tag{3-63}$$

But, as noted previously, large values of k' (or k) result in significant quantization noise. For example, if one chooses for the sine-wave test signal $kf_s = 2\pi f_m$, the minimum step size to eliminate overload noise, and then picks $f_s = 10f_m$, as noted earlier, one has $k \doteq 0.6$. The two quantization levels transmitted are each more than half the peak amplitude. This could lead to significant quantization

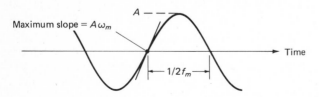

Figure 3-40 Sinusoidal test signal.

noise. The noise can be reduced, however, by sampling at a higher rate. Or, since it is the *total* noise that counts, the quantization noise can be reduced, at the expense of introducing overload noise, by reducing k below the critical value, for a sine wave, of $2\pi f_m/f_s$. This is the trade-off noted earlier, as exemplified by Fig. 3-39.

Another procedure, adopted in practice, is to introduce *adaptive delta modulation*. In this scheme the quantization step sizes are kept small, to reduce quantization noise to a tolerable level, until slope overload begins to manifest itself, as exemplified by a string of output values of the same polarity. The step size is then increased. If the output values alternate in polarity for a specified number of sampling intervals, denoting a signal not changing very much in time, the step size is reduced. A series of step-size changes may be used to increase the adaptivity of the technique.

We can gain somewhat more insight into the slope-overload effect by actually calculating the mean-squared overload noise for the sine-wave test signal under some restricted conditions. We shall then quote results from the literature for overload noise calculations for random-type signals more characteristic of signals encountered in practice. To do this we first define two normalizing parameters that allow a simplification of notation. The ratio of sampling rate f_s to sine-wave frequency f_m plays a key role in the determination of overload noise. For more general signals it is the ratio of f_s to the bandwidth B that plays the same role. We accordingly define a normalized sampling parameter

$$F_s \equiv \frac{f_s}{B}$$

For the sine-wave case B is just f_m. It is apparent from (3-63) that the quantity $kf_s/2\pi f_m = kF_s/2\pi$ plays a key role in the analysis of overload noise. This is a normalized slope parameter that we shall label α:

$$\alpha \equiv \frac{kF_s}{2\pi} = \frac{kf_s}{2\pi B} = \frac{k'f_s}{2\pi AB}$$

Although developed for a sine-wave test signal, this parameter is found to govern the delta-modulation overload-noise performance with other test signals as well. A little thought will indicate why this is so. As noted throughout Chap. 2, and as is exemplified by the simple sine-wave test signal, the bandwidth B is a measure of the rate of change of a signal. The larger B is (f_m for the sine wave), the more rapidly the signal may change. Specifically, the minimum time required for a signal to change a "significant" amount is the order of $1/B$. This is shown in Fig. 3-41. (We cannot be more precise than this, as noted in Chap. 2, because of the variety of measures of bandwidth. But recall that the pulse rise time discussed in Chap. 2 was proportional to $1/B$.) If we let this "significant" amount be the amplitude A or the 4σ deviation noted earlier, the maximum slope of *any* signal must be proportional to $4\sigma B = AB$. The maximum slope the delta modulator can follow is $k'f_s$. The parameter α is proportional to the ratio

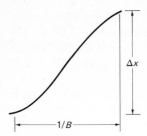

Figure 3-41 Rate of change of typical signal.

of these quantities. As α increases, the slope overload noise decreases; as α decreases this noise increases.

Now consider the sine-wave test signal again. Say that the parameter $\alpha < 1$. From Eq. (3-63) overload noise will result. A typical case appears in Fig. 3-42. The sine wave is sketched there as a function of normalized time $\theta \equiv \omega_m t$. Assume that the quantization levels are chosen small enough so that quantization noise is negligible. A delta-modulator output thus tracks, or follows, the sine wave until the slope of the sine wave at some point $-\theta_1$ equals the value $k'f_s$. Beyond this point the modulator output can only rise at the maximum rate of $k'f_s$. This is shown by the straight line of the figure. The sine-wave slope exceeds this value for $\theta > -\theta_1$, and the overload-noise region is entered. The region ends at the point θ_2 when the modulator output intersects the signal. As $k'f_s$ increases θ_1 and θ_2 both decrease, and the overload-noise region is reduced correspondingly. From the figure θ_1 is defined by the expression $\alpha = \cos \theta_1$. In the limit when $\alpha \equiv k'f_s/\omega_m A = 1$, there is no overload noise, as indicated earlier. For simplicity of analysis we shall restrict θ_1 to the range $\leq \pi/4$. This corre-

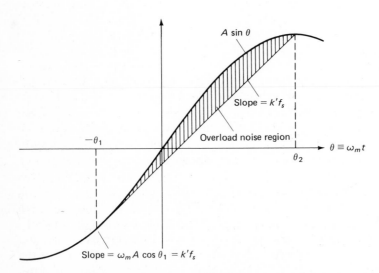

Figure 3-42 Calculation of overload noise, sine-wave signal.

sponds to $\alpha \geq 0.707$. For this range it is left to the reader to show that $\theta_2 \doteq 2\theta_1$, the approximation improving as α increases and θ_1 decreases.

The slope overload noise N_o is now defined as the average of the square of the difference between the true signal and the modulator output. Since the overload region of Fig. 3-42 repeats itself twice each cycle of the sine wave we need only average over the one region shown to determine the overload noise. It is left to the reader to show that this is given by

$$N_o = \frac{1}{\pi} \int_{-\theta_1}^{\theta_2} \left\{ A \sin\theta - \left[\frac{k'f_s}{\omega_m} (\theta + \theta_1) - A \sin\theta_1 \right] \right\}^2 d\theta \qquad (3\text{-}64)$$

or $\qquad \dfrac{\pi N_o}{A^2} = \displaystyle\int_{-\theta_1}^{\theta_2} [(\sin\theta - \alpha\theta) + \sin\theta_1 - \alpha\theta_1]^2\, d\theta \qquad (3\text{-}65)$

This may be integrated and evaluated to find N_o as a function of α. Doing this, and using the approximation $\theta_2 = 2\theta_1$, one finds, with $\theta_1 \leq \pi/4$, as assumed,

$$\frac{2N_o}{A^2} \doteq 0.32\theta_1^5 \doteq 1.8(1 - \alpha)^{5/2} \qquad (3\text{-}66)$$

(For this range of θ_1, $\alpha = \cos\theta_1 \doteq 1 - \theta_1^2/2$.) Note that the mean-squared signal itself is $A^2/2$. The expression in (3-66) is then just the reciprocal of the signal power to overload noise power, or the signal-to-noise ratio with overload noise only considered. Calling this $\text{SNR}|_0$, we get, for the sine-wave signal, with $0.7 \leq \alpha \leq 1$,

$$\text{SNR}|_0 \doteq 0.56(1 - \alpha)^{-5/2} \qquad (3\text{-}67)$$

As an example, for $\alpha = 0.707$, $\text{SNR}|_0$ is 10.8 dB. For $\alpha = 0.95$, $\text{SNR}|_0 = 30$ dB. In the first case overload noise dominates. In the second case the signal-to-noise ratio is quite high for overload noise. One would thus expect quantization noise to be the dominant noise.

To calculate the quantization noise for the sine-wave text signal we assume the crudest of models: we first take $\alpha \geq 1$ to avoid overload noise. The delta modulator thus tracks the sine-wave signal. We then assume that on the average the modulator output signal $\hat{x}(t)$ will overshoot the signal with equal likelihood in either direction. Although the successive *signal* samples are closely correlated (i.e., predictable or highly redundant) with the samples closely spaced ($F_s \gg 1$), the signal at the output of the modulator is assumed to fluctuate randomly between the maximum values $\pm k'$ about the true signal being tracked. This is the picture shown in Fig. 3-38a and b. (We are neglecting the smoothing by the filter at the modulator output, as shown in the block diagram of Fig. 3-35.) But with this crude model of quantization the mean-squared quantization noise is exactly that calculated for PCM in Sec. 3-5. This is just the mean-squared variation about the desired level given by (3-29). The level spacing a there is replaced by $2k'$ here. Calling N_q the quantization noise we thus have

$$N_q = \frac{(2k')^2}{12} = \frac{k'^2}{3} \qquad (3\text{-}68)$$

It is apparent, as noted previously in Sec. 3-5 and implicitly assumed in the discussion of overload noise here, that it is only the noise relative to the signal that has significance. For it is the perturbation in the signal that we call noise. We thus normalize to the mean signal power $A^2/2$ to obtain the relative quantization noise. This is exactly what we did in (3-66) with the overload noise. We thus have

$$\frac{2N_q}{A^2} = \frac{2}{3} k^2 \tag{3-69}$$

using the relative step size parameter k introduced previously.

It is often assumed that with the quantization noise and overload noise each small enough, one can decouple their effects. They are assumed independent of one another. The total noise in the system is then taken as the sum of the two, or

$$N = N_o + N_q \tag{3-70}$$

The signal-to-noise ratio for the delta modulator, with a sine-wave test signal at the input and invoking the various assumptions made up to this point, may then be written as

$$\text{SNR} = \frac{A^2}{2N} = \frac{A^2}{2(N_o + N_q)} \tag{3-71}$$

Using (3-66) and (3-69) for N_o and N_q, respectively, one may evaluate the SNR as a function of k or kF_s for various values of the normalized sampling frequency F_s. Figure 3-43 shows the results of such a calculation. Note that this is of the general form indicated previously in Fig. 3-39. There is a narrow range of values of kF_s (or, equivalently, α) for which the signal-to-noise ratio is maximum. For smaller values the quantization noise is reduced but overload noise increases rapidly. For larger values the overload noise disappears but quantization noise begins to increase. The optimum value of kF_s is about 6 for the sine-wave test signal, under the assumptions made, for a value of α slightly less than 1. The curves for $F_s = 16$, 32, and 64, respectively, appear separate in the quantization-noise region because the SNR is plotted versus $kF_s = 2\pi\alpha$. With the assumptions made previously, quantization noise is independent of the sampling rate. It depends only on the quantization level k. (More accurate analyses do indicate a dependence on F_s, as will be noted later.) There is an indirect improvement with sampling rate F_s, however, since, as noted earlier, increased sampling reduces the quantization level k required to maintain a specified overload-noise level. This in turn reduces the quantization noise.

More accurate analyses have been carried out for various classes of random signals. As an example, assume that the input signal $x(t)$ fed into the delta modulator is a gaussian random signal (this is then the model used in Sec. 3-5 in discussing nonuniform quantization in PCM) with a flat frequency spectrum out

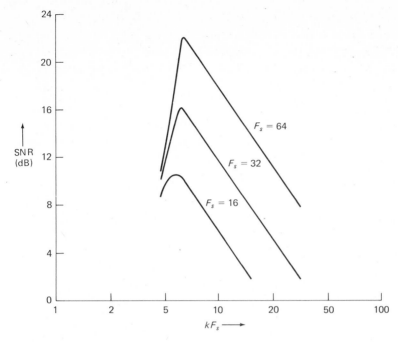

Figure 3-43 Delta modulator SNR, sine-wave signal.

to a bandwidth of B hertz.[36] The overload-noise power is then found to be given by[37]

$$N_o = 0.83\alpha^{-5}e^{-1.5\alpha^2} \tag{3-72}$$

with $\alpha \equiv kF_s/2\pi$ and $F_s = f_s/B$, as previously. This noise is actually normalized to the input variance, or signal power, σ_x^2. Alternatively, one may look on this as the overload-noise power, with $\sigma_x^2 = 1$. The step-size parameter k is the actual step size k' normalized to $4\sigma_x$, as noted previously. The overload noise thus drops extremely rapidly with increasing α.

[36] The spectral properties of random signals, as contrasted to the *deterministic* signals discussed in Chap. 2, will be treated in detail in Chap. 5.

[37] There are a large number of papers in the field of delta modulation and DPCM in which both overload and quantization noise are analyzed. Examples include J. E. Abate, "Linear and Adaptive Delta Modulation," *Proc. IEEE,* vol. 55, no. 3, pp. 290–308, March 1967; J. B. O'Neal, "Delta Modulation Quantizing Noise, Analytical and Computer Simulation Results for Gaussian and Television Input Signals," *Bell System Tech. J.,* vol. 45, pp. 117–141, January 1966; J. B. O'Neal, "Predictive Quantizing Systems (Differential PCM) for the Transmission of Television Signals," *Bell System Tech. J.,* vol. 45, pp. 1023–1036, May–June 1966; N. S. Jayant, "Digital Coding of Speech Waveforms: PCM, DPCM, and DM Quantizers," *Proc. IEEE,* vol. 62, no. 5, pp. 611–632, May 1974. The book by James J. Spilker, Jr., *Digital Communications by Satellite,* Prentice-Hall, Englewood Cliffs, N.J., 1977, reproduces these derivations as well.

The normalized quantization noise N_q is found to be given approximately by the expression

$$N_q \doteq \frac{2}{3} \frac{k^2}{F_s} \tag{3-73}$$

Note that this is of the form of our previous crude expression for N_q, (3-69), with the addition of the inverse dependence on the normalized sampling rate F_s. This reflects the correlation, or dependence, of successive input samples on one another, ignored in our previous assumption of completely random fluctuation of successive samples at the modulator output. The quantization noise here is also normalized to the input signal power, or variance, $\sigma_x{}^2$.

If we now assume, as we did earlier, that the overload noise occurs in short bursts, while the quantization noise is small enough in the vicinity of overload noise to be negligible, we can decouple the two noises and assume that they occur independently. The total noise power is then $N = N_o + N_q$. The signal-to-noise ratio is then $1/N = 1/(N_o + N_q)$, since $\sigma_x{}^2 = 1$ in writing (3-72) and (3-73). The resultant SNR, in dB, is shown plotted as a function of the normalized step size $kF_s = 2\pi\alpha$ in Fig. 3-44.[38] Note how similar the curve is to that for the sine-wave test signal discussed earlier (Fig. 3-43). The relative signal-to-noise ratios are different (both quantization and overload noise are smaller in the gaussian signal case), but the main point made earlier still holds: there exists a narrow region of kF_s or α, for which the SNR is maximum. For larger values of kF_s or α quantization noise dominates. For smaller values of kF_s, the slope overload noise dominates. In the quantization noise region an increase in sampling rate will reduce the quantization noise. This is apparent from the form of N_q in (3-73). The optimum value of kF_s varies from 10 for $F_s = 8$ to 15 for $F_s = 64$.

It is of interest, of course, to compare the performance of delta modulation with PCM. To do this assume that single-channel PCM is used, with uniform quantization (no companding). For a signal bandwidth of B hertz, the minimum sampling rate is of course $2B$ samples per second. With $M = 2^n$ quantization levels used, the *bit* rate is $2nB$ bits/s transmitted. This corresponds to the sampling rate f_s used with delta modulation. For comparison with delta modulation the *effective* normalized sampling rate is then $F_s = 2n$. From (3-38) the PCM signal-to-noise ratio with no companding is

$$\text{SNR} = 3M^2 \frac{\sigma_x{}^2}{V^2} \tag{3-38}$$

V is the maximum input signal intensity and $\sigma_x{}^2$ its average power, or variance. As noted several times previously, there is no theoretical maximum signal for gaussian random signals, the signal model assumed here. It is customary, as in the delta modulation case just noted, to choose $V = 4\sigma_x$ quite arbitrarily. For gaussian statistics this is the signal intensity exceeded only 0.01 percent of the time, on the average. For this choice of V,

---

[38] O'Neal, "Delta Modulation Quantizing Noise," *op. cit.*, p. 123, Fig. 4.

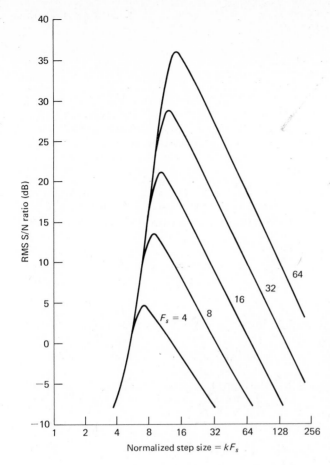

Figure 3-44 SNR for band-limited flat gaussian signals. (From J. B. O'Neal, "Delta Modulation Quantizing Noise, Analytical and Computer Simulation Results for Gaussian and Television Input Signals," *Bell System Tech. J.*, vol. 45, p. 123, January 1966, fig. 4. Copyright, 1966, The American Telephone and Telegraph Co., reprinted by permission.)

$$ \text{SNR} = \frac{3M^2}{16} \tag{3-74} $$

Converting to decibels, letting $M = 2^n$, and then letting $F_s = 2n$, we have

$$ \text{SNR}|_{\text{dB}} = 3F_s - 7.3 \tag{3-75} $$

This has been plotted in Fig. 3-45, together with two curves for delta modulation.[39] The lower curve, labeled flat signal, corresponds to the optimum SNR points in Fig. 3-44. The upper curve, a more accurate model for television signals, assumes that the input signal has the *RC* spectral characteristic $1/(1 + f/\beta)^2$, up to a maximum frequency of $B = 8\pi\beta$. β is thus the 3-dB bandwidth of this signal. Note the improvement in delta modulation SNR of this type of smoothed input signal over the one with a flat spectral characteristic.

[39] *Ibid.*, p. 127, fig. 7.

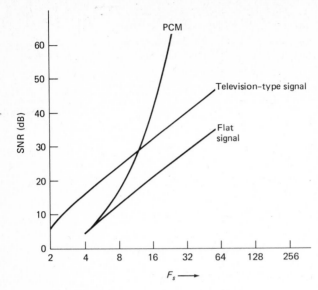

Figure 3-45 Comparison of delta modulation and standard PCM for band-limited gaussian signals. (From J. B. O'Neal, "Delta Modulation Quantizing Noise, Analytical and Computer Simulation Results for Gaussian and Television Input Signals," *Bell System Tech. J.,* vol. 45, p. 127, January 1966, fig. 7. Copyright, 1966, The American Telephone and Telegraph Co., reprinted by permission.)

For this class of signals delta modulation appears to provide an improvement in SNR at the lower sampling rates.[40] This corresponds to situations in which simplicity and low cost dominate over performance. Where high-quality speech transmission or very accurate telemetering of data is not a prime consideration, nonadaptive delta modulation with relatively low sampling rates offers advantages. For higher-quality transmission, adaptive delta modulation is used to improve the performance.

Various data-compression schemes have been used to reduce the data-rate requirements in the transmission of both speech and pictures. These schemes are often coupled with delta modulation or DPCM operating on the compressed data to provide still further reductions in data-transmission-rate requirements.[41]

An adaptive delta-modulation system for digitized voice transmission has been developed and put into operation by the Bell System in the United States for use in rural areas.[42] This Subscriber Loop Multiplex (SLM) system multiplexes 24 voice channels, each sampled at a rate of 57.2 k-samples/s. A typical system serves up to 80 telephone lines sharing the 24 channels. Assuming a telephone bandwidth of $B = 4$ kHz, $F_s = 57.2/4 = 14.3$ in this case. The adaptive step-size feature allows the step sizes to be changed, by factors of 2, over a large range from 1 to 128. In its simplest form the adaptive algorithm

---

[40] Additional substantial improvement in the delta-modulation SNR is possible through output smoothing. See Spilker, *op. cit.,* sec. 4-3.

[41] See, for example, the Special Issue on Image Bandwidth Compression, *IEEE Trans. Commun.,* vol. COM-25, no. 11, November 1977.

[42] I. M. McNair, Jr., "Digital Multiplexer for Expanded Rural Service," *Bell Lab. Rec.,* vol. 50, pp. 80–86, March 1972.

senses the polarity of strings of successive output samples. If 7 output bits in a row, for example, are of the same polarity, the step size is increased. In this case the original step size is presumed to have been too small, resulting in slope overload. If strings of alternate polarity appear, the step size is reduced. This is an indication that the signal is not changing very much, slope overload noise is not a problem, and the quantization noise can be safely reduced.

3-8 TIME-DIVISION MULTIPLEXING OF DIGITAL SIGNALS

Introduction

The concept of time multiplexing of signals was introduced earlier, in Secs. 3-3 and 3-4. There, for simplicity's sake, we indicated how a group of signals, analog and digital, might be time-multiplexed for transmission over a common line by sampling them sequentially at a common sampling rate. We now focus more deeply on the subject of multiplexing, specializing to the case of *digital* signals being multiplexed. This is not at all restrictive, however, since the vast majority of time-multiplexing applications today involve operations on digital signals, and the trend is moving to conversion of analog signals to digital form as early in the system as possible.

As noted in our prior discussion in Sec. 3-3, it does not matter what the source was of a digital signal to be multiplexed. The signal could be a data set output, computer output, a digitized voice signal (or group of voice signals), digital facsimile or TV information, telemetry information to be transmitted to a remote point, etc. The multiplexing operation treats them all alike. The object is to combine digital signals of possibly varying bit rates and feed the combined signal stream out sequentially, at a correspondingly higher bit rate, over one higher bandwidth/higher bit rate line. As an example, one might want to combine two 1,200 bits/s sources with three 2,400-bits/s sources and feed all five signals out sequentially at the combined nominal rate of 9,600 bits/s. Alternatively, one might want to combine several T1 signal streams into one correspondingly higher bandwidth/higher bit rate stream. A conceptual diagram of the multiplexing-demultiplexing operation appears in Fig. 3-46.

The one distinction made that does indirectly distinguish between data (alphanumeric symbols) and other digital signals that might be transmitted is

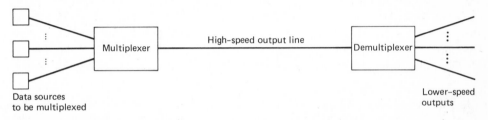

Data sources
to be multiplexed

Lower-speed
outputs

Figure 3-46 Multiplexing operation.

the interleaving of digital signals on either a character or a bit basis. Low-speed data terminals (up to 1,200 bits/s normally) generally transmit data asynchronously in character form, a character ranging from 5 to 10 bits in length, depending on the type of terminal and code used. Multiplexing of these terminals is generally carried out by interleaving characters. PCM signals, on the other hand, have no natural character forms and are multiplexed by interleaving on a bit-by-bit basis. Higher-speed data terminals (1,200 bits/s and up) are sometimes also multiplexed by interleaving bits. We shall have more to say about this distinction later, in discussing examples of multiplexers in current use.

No matter how the multiplexing of digital signals may be carried out, several basic points can be made:

1. Some form of framing structure must be incorporated, a frame representing the smallest unit of time in which all signals to be multiplexed are serviced at least once.
2. The frame is slotted into time allocations, uniquely assigned to each data source connected. A timing procedure, comparable to the simple periodic sampling of individual channels in Sec. 3-3, must thus be developed to sample each data source at the appropriate time in the frame. (In Sec. 3-3 we focused on the sequential sampling of a group of channels to be combined. It is the procedure used in combining channels of the same bandwidth or bit rate. With widely differing bit rates or bandwidths, channels are sampled at correspondingly different rates.)
3. Framing and synchronization bits must be appended to enable the receiving system to uniquely synchronize in time with the beginning of each frame, with each slot in the frame, and with each bit within a slot. These bits may be collectively called control bits.
4. Provision must be made for handling small variations in the bit rates of the incoming digital signals to be multiplexed.

Although some variable-frame multiplexing schemes are coming on the market,[43] we focus here on fixed-frame-size schemes only. A typical frame then might look as shown in Fig. 3-47. C_1 and C_2 represent sequences of control bits (in this example they are placed at the beginning and end of the frame, respectively) and four data sources are shown multiplexed. One of these transmits at twice the rate of the other three, so that it is allocated two slots per frame, as shown. As an example, say the basic slot contains 10 bits of data. Say control sequence C_1 is 3 bits, and C_2 is 2 bits long. The total frame length, in this example, is 55 bits long. Say units 2, 3, and 4 each transmit at 1,200 bits/s, while unit 1 transmits at 2,400 bits/s. The total data bit rate is then 6,000 bits/s.

[43] J. B. Van der Mey, "The Architecture of a Transparent Intelligent Network," *Conference Record, IEEE National Telecommunications Conference,* Dallas, Texas, December 1976, pp. 7.2.1–7.2.5.

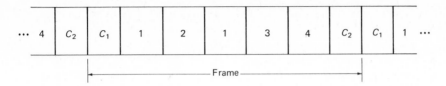

Figure 3-47 Typical framing structure.

Because of the control bits used in this example, however, the multiplexer must actually be transmitting at a rate of 6,600 bits/s. This is of course a fictitious example, but it illustrates the point.

Two major classes of multiplexing appear in practice. The first group comprises multiplexers designed to combine lower-speed data signals, up to 4,800 bits/s maximum bit rate, into one higher-speed multiplexed signal of up to 9,600 bits/s data rate. These techniques are used primarily to transmit data over *voice-grade channels* of a telephone network. ("Voice grade" simply refers to the fact that these channels are nominally of the bandwidth required to transmit a single analog voice signal. These channels are exactly those of 3.3-kHz nominal bandwidth referred to previously in Sec. 3-3 in first discussing Nyquist sampling, and then again in Sec. 3-6 in discussing the T1 system. Modems, devices required to convert the digital format to the analog format required to transmit signals over the telephone channels, will be discussed in Chapter 4.) A customer with several data terminals to be connected to a geographically distant computer, or to equivalent terminals remotely located, can thus share the cost of a single voice channel among his terminals by multiplexing the signals into a single voice channel. The use of a multiplexer obviates the need for individual telephone connections for each terminal.

Multiplexers designed for this purpose normally transmit at output rates of 1,200, 2,400, 3,600, 4,800, 7,200, or 9,600 bits/s, depending on the application, and on whether the voice channel used is privately leased and specially conditioned (in which case higher bit rates can be used) or dialed, as in any telephone call. The input rates handled generally vary from 75 bits/s to as high as 7,200 bits/s. We shall discuss the specific numbers in more detail later in this section in describing a typical commercial multiplexer designed for this purpose.

The second broad class of multiplexing occurs at much higher bit rates, and is part of the data transmission service generally provided by communication carriers. As an example, the Bell System in the United States has developed a digital hierarchy using the T1 carrier described in Sec. 3-6 as a base.[44] The hierarchy appears in schematic form in Fig. 3-48. Note that at any level a data signal at the input rate shown would be multiplexed with the other input signals

[44] R. T. James and P. E. Muench, "A.T.&T. Facilities and Services," *Proc. IEEE,* vol. 60, no. 11, pp. 1342–1349, November 1972.

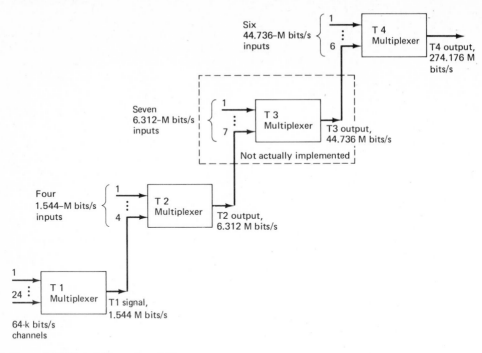

Figure 3-48 Digital hierarchy, ATT system.

at the same rate for more economical transmission as one combined higher-speed signal. The T1 multiplexer, although designed originally to handle 24 digitized voice circuits, is not restricted to multiplexing voice signals. Any 64 kbits/s signal, of the appropriate format, could be transmitted as one of the 24 input channels shown. Similarly, at a higher level in the hierarchy, not all inputs need to have been derived from a lower-level multiplexer. At the T3 level, for example, some of the 6.312-Mbits/s inputs could represent digitized TV inputs appearing directly at this bit rate; others could be multiplexed T1 signals in groups of four (T2 signals) transmitting voice information; others could be derived by multiplexing upward and combining appropriately some lower-speed data traffic. (The T3 level shown in Fig. 3-48 has not actually been implemented in AT&T systems. Transmission systems include T1, T2, and T4 only.)

The multiplexing of signals allows a given transmission channel to be shared by a number of users, reducing the cost. A similar hierarchy, but using different bit-rate levels, has been recommended as an international standard by the CCITT, the international committee for standards relating to telephony and telegraphy. This will presumably be adopted by many nations in the world outside North America and Japan. It is based on the lowest-level PCM international standard of 2.048 Mbits/s, which applies worldwide outside North Amer-

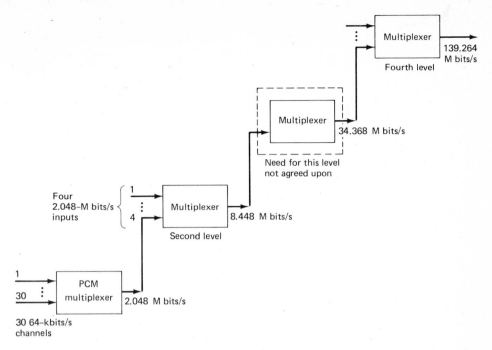

Figure 3-49 Digital hierarchy, CCITT recommendation.

ica and Japan. This PCM standard multiplexes 30 channels at 64 kbits/s each, with two additional channels used for signaling and other purposes. It thus differs from the North American T1 standard. The CCITT-recommended digital hierarchy appears in Fig. 3-49.[45] This hierarchy consists of four levels of multiplexing as shown. The third level shown in the figure is left optional, however.

These very high speed multiplexing hierarchies have been developed for use by the various national communications carriers. They are based historically on the PCM standards used to multiplex digitized voice channels, although other types of signals can be multiplexed as well.

Other multiplexing schemes derived specifically for data transmission come up with other numbers, of course. As an example, the DATRAN system, which was to have been a specialized communications carrier in the United States devoted exclusively to the transmission of data, but ceased operation in late 1976 because of financial difficulties, had developed its own digital hierarchy based on the common lower-speed data rates.[46] In this system the first-level

[45] T. Irmer, "An Overview of Digital Hierarchies in the World Today," *Conference Record, IEEE International Conference on Communications,* San Francisco, June 1975, pp. 16-1 to 16-4.

[46] F. T. Chen *et al.,* "Digital Multiplexing Hierarchy for an Integrated Data Transmission and Switching System," *Conference Record, IEEE National Telecommunications Conference,* New Orleans, December 1975.

multiplexer accepted data at a rate of 4.8, 9.6, or 19.2 kbits/s, and combined the various subscriber channels to form a high-speed output at either 56 or 168 kbits/s. Groups of these signals were in turn fed into a second-level multiplexer to form outputs at either 1.344 or 2.688 Mbits/s. These were in turn further multiplexed to the third level of 21.504 Mbits/s, capable of accommodating 3,927 4.8-kbits/s data channels, or smaller numbers of higher-speed data signals.

In the next three subsections we discuss the time-multiplexing operation in more detail. We first focus on bit-interleaved multiplexers, and discuss design questions involved in combining independent bit streams of possibly differing rates, with statistical fluctuations in their rates. We then describe, by example, the multiplexing process using character interleaving. Questions relating to frame loss and acquisition, applicable to all types of multiplexers, are introduced briefly in the final subsection.

Bit-interleaved Multiplexers and Bit Stuffing

It was noted earlier that among the basic problems arising in the multiplexing of independent digital streams are those of framing, synchronization, and rate adjustment to accommodate small variations in the input data rates. We discuss the question of rate adjustment, by bit stuffing, in this subsection. This technique is used in bit-interleaved multiplexers. In the next subsection we mention, by example, another scheme used with character-interleaved multiplexers.

In a bit-interleaved multiplexer the slots devoted to individual input channels (Fig. 3-47) are one or more bits in length. The multiplexer should presumably have the appropriate number of bits ready and available for transmission when an input data source's slot time arrives. Independent data sources will be expected to experience variations in their data rate, however. A nominal 2,400-bits/s input rate may occasionally drop to 2,390 bits/s, for example, or increase to 2,410 bits/s. A 9,600-bits/s multiplexer combining four such inputs must accommodate for such variations. Otherwise, severe misalignment and synchronization problems can arise. The problem is handled in the bit-interleaved case by, first, running the multiplexed output at a speed slightly higher than the sum of the maximum expected rates of the input channels. Second, to accommodate small reductions in the input rates, as well as to handle the nominal rates which have been designed to be somewhat below the multiplexer rate, *bit stuffing* on a per channel basis is often used.[47] The very fact that output pulses appear multiplexed at a higher rate than the expected input rate is equivalent to occasionally "stuffing" the output with additional non-information-carrying pulses.

---

[47] V. I. Johannes and R. H. McCullough, "Multiplexing of Asynchronous Digital Signals Using Pulse Stuffing with Added-Bit Signaling," *IEEE Trans. Commun. Tech.*, vol. COM-14, no. 5, pp. 562–568, October 1966.

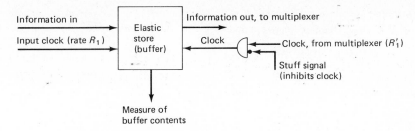

Figure 3-50 Elastic store for bit stuffing.

To accomplish this, each input data stream feeds a buffer or *elastic store* at the multiplexer. The contents of this buffer are fed out, at the higher rate, onto the outgoing line when the appropriate slot interval appears. A simple schematic of the elastic store operation appears in Fig. 3-50. The input is designated as R_1. The output rate is $R_1' > R_1$. If the input rate begins to drop, relative to the multiplexer output rate, the store contents decrease. The multiplexer monitors the store contents, as indicated in Fig. 3-50, and when the number of bits stored drops below a specified threshold the multiplexer disables readout of this store after a fixed time delay. The clock-inhibiting stuff signal shown in Fig. 3-50 is used for this purpose. This effectively inserts a "blank" into the corresponding slot position in the frame. This blank can be coded as a 1, if desired. When the store contents rise up above the threshold, sampling of the store contents at the appropriate slot time is again resumed.

An example of the bit-stuffing process, using two multiplexed signals, appears in Fig. 3-51. Both input channels are assumed to operate at the same

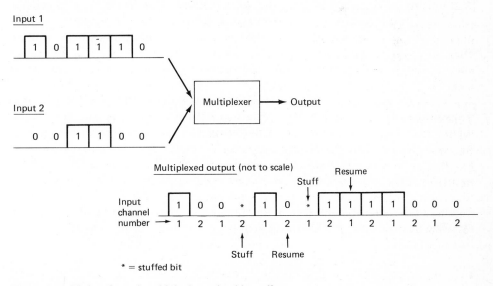

Figure 3-51 Bit-interleaved multiplexing using bit stuffing.

→ Order of transmission

M_0 (48) C_I (48) F_0 (48) C_I (48) C_I (48) F_1 (48)

M_1 (48) C_{II} (48) F_0 (48) C_{II} (48) C_{II} (48) F_1 (48)

M_1 (48) C_{III} (48) F_0 (48) C_{III} (48) C_{III} (48) F_1 (48)

M_1 (48) C_{IV} (48) F_0 (48) C_{IV} (48) C_{IV} (48) F_1 (48)

Figure 3-52 M12 multiplexer, frame format, Bell System.

nominal bit rate. They are thus interleaved, bit by bit, in order, as shown. In this example first channel 2 and then channel 1 slow down somewhat, relative to the output rate, and stuffed bits are inserted. These could equally well be 0's or 1's, following an agreed-upon convention.

At the demultiplexer at the other end of the line, the stuffed bits must obviously be removed from the data stream. This requires a method of identifying the stuffed bits. Various techniques are possible. As an example, we present the approach used with the Bell System M12 multiplexer, combining four T1 digital streams into one T2 stream.[48] This corresponds to the second level of the digital hierarchy of Fig. 3-48. This example also enables us to show one method of providing frame synchronization for a bit-interleaved system. The M12 frame contains 1,152 data bits and 24 control bits, for a total of 1,176 bits. Its efficiency (data bits/total bits) is thus 98 percent. The frame format appears in Fig. 3-52. The multiplexer uses bit-by-bit interleaving of the four inputs, until a total of 48 bits, 12 from each input, is accumulated. A control bit is then inserted. The 24 control bits thus appear spread out throughout the frame, enclosing sequences of 48 data bits. Three types of control bits are used, as indicated in Fig. 3-52. They are there labeled the M series, the C series, and the F series. The first line of the figure is transmitted, then the second, third, and fourth, in that order.

These three control signals are used to provide frame indication and synchronization, and to identify which of the four input signals has been stuffed. Only one stuffed bit/input channel is allowed per frame. As will be shown shortly, this is sufficient to accommodate expected variations in the input signal rate.

The subscripts on the M- and F-series control denote the actual bit (0 or 1) transmitted. The total F pattern appearing, 01010101, provides the main framing pattern. The multiplexer uses this to synchronize on the frame. After locking onto this pattern, the demultiplexer searches for the 0111 pattern given by the M series, which further breaks the frame down into four subframes, each corresponding to a line in Fig. 3-52. Bit stuffing for each input signal is restricted to the particular subframe (or line) corresponding to the subscripts of the C bits in that line. C_I thus stands for input channel I, C_{II} for input channel II, etc. The insertion of a stuffed pulse in any one subframe is denoted by setting all three C's in that line to 1. Three 0's in a line denote no stuffing. If a bit has been

---

[48] Bell Telephone Laboratories, *Transmission Systems for Communications,* 4th ed., 1970, p. 612.

stuffed it appears as the first information bit of the 12 for that input signal, following F_1 in the same subframe. An example of a typical bit sequence appears in Fig. 3-53. $C_1 = 1$ denotes one stuffed bit, as shown.

The 3-bit C sequence used to denote the presence or absence of a stuffed bit is necessary to reduce the chance of missing a stuffed bit to a tolerably low value. If a stuffed bit is mistakenly called an information bit and not deleted from the bit stream at the demultiplexer, all the bits for the entire frame for the T1 system in question will be in error. With three bits used to denote the stuffing, a single error in any one of the three bits will be recognized. Majority logic decoding is used. This simply means that the majority of the C bits, in this case two out of three, determine whether the all-zero or all-one sequence (denoting a stuffed bit present in the information sequence following) was transmitted. Any combination of two 1's and a 0, or the presence of three 1's, thus means that a stuffed bit is present. Similarly, any combination of two 0's and a 1, or three 0's present, means that no stuffing is used. An error in any one bit of the three is thus disregarded. (This is a very simple example of a single error-correcting code. More general codes of this type are discussed in Chap. 6.)

An error will be made if two or all three of the C bits in any subframe are misinterpreted or detected incorrectly. If bit errors occur independently from bit to bit, the probability of a stuffed bit error is the sum of the probabilities of the possible ways in which such an error can occur. If the probability p that an individual bit is detected incorrectly is very small, however, the chance that all three C bits will be detected incorrectly is extremely small. The chance of a stuffed bit error occurring is then dominated by the chance that two bits will be transposed. The probability of a stuffing error is then given, very nearly, by

$$P_s \doteq 3p^2 \tag{3-76}$$

As an example, if $p = 10^{-5}$ (the chance of a bit error is then one in 100,000 bits transmitted, on the average, a typical number for data transmission over

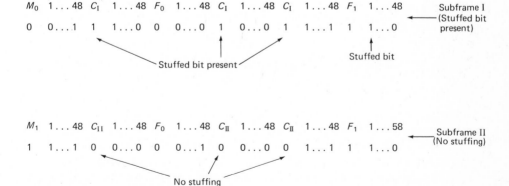

Figure 3-53 Example of the bit-stuffing process, M12 frame.

telephone lines),

$$P_s \doteq 3 \times 10^{-10}$$

Three frames in 10^{10}, on the average, of each of the T1 channels will thus be lost because of this problem. (Note that only the T1 frame of the 4 in an M12 frame that is directly affected will be lost.) Since the T2 frames are 1,176 bits long and data are transmitted at a 6.3-Mbits/s rate, it is apparent that this converts to an average time between such occurrences of 0.63×10^6 s, 10^4 min, or 175 h. If a *single* bit were used instead to denote bit stuffing in one of the four input signal streams in any one frame, the chance of an error event would be just $p = 10^{-5}$. With 5,300 frames/s transmitted, this would mean an error on the average of once every 19 s! The need for the longer bit sequence denoting a bit stuffed is apparent.

In general, for C bits used to denote bit stuffing in a bit-interleaved multiplexer, and majority decision decoding used to determine the presence of a stuffed bit, it is apparent that an error will be made if $(C + 1)/2$ or more bits are transposed. (C is assumed odd.) If bits may be independently transposed (converted from a 0 to a 1, or vice versa), it is left for the reader to show that the probability of a stuffing error is just the sum of binomial probabilities, or

$$P_s = \sum_{i = \frac{C+1}{2}}^{C} \binom{C}{i} p^i (1 - p)^{C-i} \tag{3-77}$$

For $p \ll 1$, the leading term in this sum is dominant [i.e., $(C + 1)/2$ errors have a much higher probability of occurring than $(C + 1)/2 + 1$ errors], and

$$P_s \doteq \binom{C}{\frac{C+1}{2}} p^{(C+1)/2} = \frac{C!}{\left(\frac{C-1}{2}\right)! \left(\frac{C+1}{2}\right)!} p^{(C+1)/2} \tag{3-78}$$

For $C = 3$, this gives just the expression $3p^2$ written previously. For $C = 5$, $P_s \doteq 10p^3$, showing the improvement made possible by just adding two bits per C series sequence. (Note, however, that all of this discussion assumes *independent* bit errors. Errors sometimes occur in bursts and this assumption is no longer valid in that case.)

It has been noted above that the Bell System M12 multiplexer we have been citing as an example of a bit-interleaved multiplexer allows only one bit per frame to be stuffed per input signal channel. Is this sufficient to accommodate expected variations in the input bit rates? Since a frame contains 288 bits per input channel (12 bits $\times$ 24 information sequence positions per frame), the maximum fractional change in input bit rate that can be accommodated is 1/288, or 0.36 percent. For a 1.544-Mbits/s T1 channel this represents a 5.4-kbits/s variation, far more than the change in input bit rates actually expected.

This discussion of bit stuffing can be made somewhat more precise and generalized to include the extent of expected variations in the input clock (data) rate by the following argument. Say that m input signal streams (we sometimes use the word channels synonymously with this), each of nominal bit rate R_1 bits/s, are multiplexed together. The frame consists of I information bits and X control bits. If the maximum expected fractional increase in R_1 is δ, the multiplexer output rate R_O required to accommodate the possible input clock rate increase is given by

$$R_O = mR_1(1 + \delta)\left(\frac{I + X}{I}\right) \tag{3-79}$$

It is apparent that even if all m input channels are operating at the nominal data rates, stuffing will occasionally occur to offset the higher output rate R_O. If at most one stuffed bit per frame can be used per input channel, the average bit-stuffing rate must be less than 1 bit in the I/m information bits transmitted per channel in any one frame. Call the average number of bits stuffed per channel in any frame S. Then $S < 1$ is desired. As a matter of fact, it is desirable to have $S \leq \frac{1}{2}$ to accommodate input clock rates reduced by a fraction δ *below* the nominal rate. At $S = \frac{1}{2}$, bit stuffing will occur, on the average, once every two frames with the input clocks operating at their nominal values. In particular, we would like to have

$$S = \delta\left(\frac{I}{m}\right) \leq \frac{1}{2} \tag{3-80}$$

Consider the M12 multiplexer example again. Here $m = 4$, $(I + X)/I = 49/48$ (one control bit appears for every 48 information bits), and $I/m = 288$ bits/frame/input signal stream. Then

$$R_O = 4R_1 \times \frac{49}{48} \times \left(1 + \frac{S}{288}\right) \tag{3-81}$$

With $R_O = 6.312$ Mbits/s (a design decision made to have the equivalent clock frequency be a multiple of 8 kHz[49]), and $R_1 = 1.544$ Mbits/s, $S = \frac{1}{3}$. This choice of output bit rate thus satisfies the condition $S \leq \frac{1}{2}$ with some spare stuffing capacity.

In a later subsection we shall return to the question of the appropriate choice of control bits for establishing and maintaining frame synchronization. Before doing that, however, we take up in the next paragraph the subject of character-interleaved multiplexers, used, as already noted, in multiplexing low-speed data sources for transmission over a higher-speed circuit. This circuit could be a telephone voice channel capable of accommodating up to 9,600 bits/s, or a higher-speed channel as well.

Character-interleaved Multiplexers

Many modern data terminals transmit at relatively low speeds such as 75, 110, 150, 300, and 1,200 bits/s. Most commonly they transmit alphanumeric sym-

[49] *Ibid.*, p. 613.

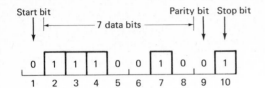

Figure 3-54 Typical asynchronous terminal character using ASCII code.

bols (letters of the alphabet, decimal numbers, and various control symbols) in character form. Teletypewriters and other keyboard terminals are typical examples.[50] Characters are transmitted asynchronously, one at a time. For this reason these devices are called asynchronous machines, and their transmission, as sequences of characters, is called *asynchronous transmission*. To be received properly each character must carry its own start and stop, or synchronization, information.

A host of such devices exist, with a variety of character sizes ranging from 5 to 8 bits of data, plus 2 to 3 bits for the start-stop function.[51] The majority of these devices in the United States have now standardized on a 10-bit character, consisting of a 0 start bit, 8 data bits (most commonly 7 data bits, plus a parity check bit of the type discussed in Chap. 6), and a 1 stop bit at the end. The data bits are based on the U.S. standard 7-bit ASCII code (American Standard Code for Information Interchange), which is in turn compatible with a corresponding international standard.[52] An example of a typical ASCII character appears in Fig. 3-54.

It is obviously simpler to multiplex these devices character by character, rather than bit by bit. This is particularly so when the terminals to be multiplexed may have different character sizes. For this reason character-interleaved time-division multiplexers have been widely adopted for these lower-speed applications. Since individual characters are sent out at a much higher rate than received from the terminal, the multiplexer must be able to buffer at least one character from each terminal connected to it. Start and stop bits are generally dropped from the character before buffering. These must then be reinserted by the demultiplexer at the receiving end, if the destination is another terminal. Synchronously transmitted data streams, generally transmitted at higher speeds and without the start-stop bits, are often multiplexed together with the asynchronous data. Examples of the procedure will be provided later.

The number of terminals (or data sources) that may be combined by a multiplexer depend on the output bit rate, the character size adopted for the multiplexer on the output line, the types of terminals to be accommodated, and the way in which synchronization is carried out. Most commercial multiplexers

---

[50] James Martin, *Telecommunications and the Computer,* 2nd ed., Prentice-Hall, Englewood Cliffs, N.J., 1976, chap. 3.

[51] *Ibid.*

[52] *Ibid.* Also, M. Schwartz, *Computer Communication Network Design and Analysis,* Prentice-Hall, Englewood Cliffs, N.J., 1977, chap. 14.

allow a wide range of terminal types and different combinations of terminals to access them. Some multiplexers use program control to change the multiplexing procedure if the number of types of terminals change. The multiplexers at both ends of the high-speed link must of course conform to the same multiplexing format.

Some examples will clarify the procedure. Say that a 2,400-bits/s line is available for the high-speed transmission. Eight-bit characters are sent out over this line. The line capacity, or the output speed of the multiplexer, is then 300 characters per second. Say that a synchronizing character is inserted as every tenth character. The data transmission rate nominally available is then 270 characters per second. Let 3 percent of this number, or 8 characters per second, be set aside to accommodate the maximum expected increase in input terminal rate. This allows 262 characters per second to be allocated to the terminals to be multiplexed. Say that one 300-bits/s and four 150-bits/s terminals, plus an undetermined number of 110-bits/s terminals, are to be multiplexed. How many such terminals can be handled? The 300- and 150-bits/s terminals output 10-bit characters, including the stop and start bits. They thus transmit at 30- and 15-characters per second rates, respectively. Their total requirements thus come to 90 characters per second. The 110-bits/s terminals output 11-bit characters, so that their character output rate is 10 characters per second. It is apparent that this system will allow 17 such terminals to be multiplexed together, in addition to the 300-bits/s and the four 150-bits/s terminals. A possible timing diagram appears in Fig. 3-55. Each slot represents one 8-bit output character. It is apparent that the 150-bits/s output appears only once for every two 300-bits/s characters output, while the 110-bits/s output appears only once for every three 300-bits/s characters. A blank must be inserted at the end of every third subframe of 10 characters, as shown, to smooth out the timing characteristic and make it as uniform as possible. Every six such subframes, the entire pattern repeats itself.

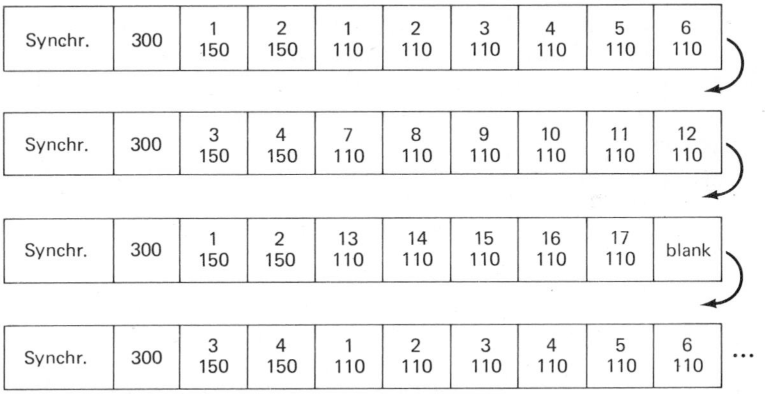

Synchr.	300	1 150	2 150	1 110	2 110	3 110	4 110	5 110	6 110

Synchr.	300	3 150	4 150	7 110	8 110	9 110	10 110	11 110	12 110

Synchr.	300	1 150	2 150	13 110	14 110	15 110	16 110	17 110	blank

Synchr.	300	3 150	4 150	1 110	2 110	3 110	4 110	5 110	6 110

Figure 3-55 Typical multiplexer format, 2,400-bits/s output.

A commercial time-division multiplexer, the Timeplexer,[53] requires a synchronization character once every 120 time slots only. This makes for a more efficient system than the example discussed above. The maximum character rate accommodated by the Timeplexer is the output line rate (in characters per second) less 3 percent to account for both synchronization, and possible increases in input terminal rates. Take, as an example, the same case considered above. Say the output line speed is again 2,400 bits/s, or 300 characters per second if 8-bit characters are used. 291 characters per second are thus available for terminal traffic. With 90 characters per second allocated to the 300-bits/s and four 150-bits/s terminals, it is apparent that up to twenty 110-bits/s terminals (10 characters per second each) can be accommodated. The allocation of slots to terminals is similar to that shown in Fig. 3-55, except that a synchronization character appears only once every 120 slots.

Various programming assignments of these 120 slots can be set up, depending on the terminal types and number of terminals to be multiplexed. Synchronous data channels may be multiplexed together with the asynchronous inputs. The synchronous input speeds must, however, be an integer fraction of the high-speed output rate. The asynchronous data rate available is then the high-speed rate less the sum of the synchronous input rates. For example, one 1,200-bits/s synchronous channel plus 1,200 bits/s of asynchronous data could be multiplexed onto a 2,400-bits/s line. (The 1,200 bits/s for asynchronous data must be converted to the characters per second form to find the asynchronous terminal capacity available.) Four 1,200-bits/s synchronous channels plus 2,400 bits/s of asynchronous data could be multiplexed onto a 7,200-bits/s line. Various other combinations are possible.

The way in which the input line scanning and subsequent multiplexing is actually carried out depends on the specific implementation. A brief description of one commercial character-interleaved time multiplexer, the Codex 920 TDM,[54] provides one example. This multiplexer will multiplex (and demultiplex, of course), up to 64 low-speed data channels onto a high-speed output line or trunk. Up to 6 of the 64 inputs may be synchronous input channels. The inputs are connected directly to multiplexer ports. The various asynchronous low-speed data rates that may be handled include 75, 110, 134.5, 150, 300, 600, and 1,200 bits/s. (The terminal types can use any character code of up to 8 bits length.) Synchronous data rates that can be handled include 1,200, 2,000, 2,400, 3,600, 4,800, and 7,200 bits/s. The high-speed trunk rates that can be used include 1,200, 2,400, 3,600, 4,800, 7,200, and 9,600 bits/s.

To accommodate this wide variety of possible inputs and outputs in one machine, the port scanning and multiplexing operations are controlled by instructions stored in two memories, labeled A and B memories. The A memory constitutes a table of addresses, scanned sequentially to determine the port, at a particular time in a framing cycle, whose character buffer contents are to be

[53] See various application notes, Timeplex, 100 Commerce Way, Hackensack, N.J. 07601.
[54] Operation manual, Codex 920 TDM, Codex Corp., Mansfield, Mass., May 1976.

outputted onto the high-speed trunk. The memory is divided into eight sections or subframes, each one of which has twenty-four 8-bit "slots." A conceptual diagram appears in Fig. 3-56. Depending on the speed of the output trunk, one or more of the subframes is scanned periodically. For a 1,200-bits/s trunk speed one subframe only is used and is scanned completely in 180 ms. If a 2,400-bits/s output line is used, two subframes are enabled, the two of them still being scanned in the same 180 ms. If a 9,600-bits/s trunk is used, all eight frames are scanned, again in the same interval of 180 ms.

Each of the 24 slots in a subframe holds a 6-bit address, as shown in Fig. 3-56, corresponding to the port whose waiting character is to be outputted onto the trunk, or read out on the low-speed port line, depending on the direction of transmission. (Two additional bits indicate whether the data are asynchronous or synchronous.) By controlling the addresses in the table, ports may be scanned the appropriate number of times per frame: higher-speed ports are addressed more often than lower-speed ports.

As an example, a maximum of twelve 110-bits/s terminals may be multiplexed over a 1,200-bits/s trunk. In this case each terminal's address would appear twice in the subframe. If four 300-bits/s terminals are multiplexed instead, each would be addressed six times in the subframe. The addresses in both examples would be spaced uniformly, to provide a desirable uniform sampling. If combinations of terminals are multiplexed, the number of slots per terminal vary accordingly: four 150-bits/s and two 300-bits/s terminals can be multiplexed, for example, over a 1,200-bits/s trunk. The 300-bits/s ports would

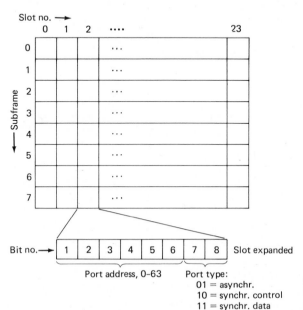

Figure 3-56 A-memory format. Codex 920 multiplexer. (From *Operational Manual,* May 1976, Codex Corporation, Newton, Mass., fig. 5-1; reprinted by permission.)

each be addressed six times in the subframe, while the 150-bits/s ports would be addressed three times each.

The characteristics of the terminal connected to the port whose address is being scanned at any time are stored in the second memory, the B memory. These characteristics include the terminal bit rate (speed), the character (word) length if an asynchronous terminal, the type of parity check used, and the length of the stop element at the end of the asynchronous character. (Terminals use lengths of 1, 1½, or 2 bits as the stop indicator.) The port characteristic table stored in the B memory has sixty-four 8-bit locations, one for each port address. The 8-bit word for any one address appears in Fig. 3-57. Note that 3 of the 8 bits are used to indicate the bit rate of the terminal connected to that port. The distinction between asynchronous and synchronous inputs is made by the last 2 bits ("port type") of the appropriate entry in the A memory (Fig. 3-56). The B-memory entries need therefore not duplicate this information. Two bits are used to designate the four different character (word) lengths (excluding start and stop bits) of the various asynchronous terminals that could be connected. Two bits are used to designate the type of parity checking used with the terminal code. (As will be discussed in Chap. 6, parity checking is the simplest form of bit error detection that can be implemented. The data bits are added modulo-2. An additional parity bit is then added to make the total sum, modulo-2, odd or 1, in the case of odd parity; or even or 0, if even parity is

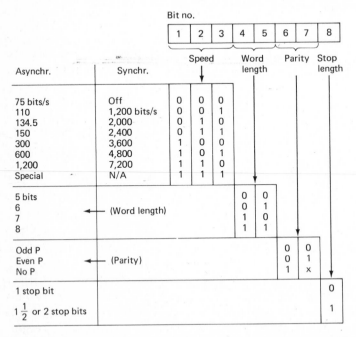

Figure 3-57 Port characteristic table, B-memory, Codex 920 multiplexer. (From *Operational Manual*, May 1976, Codex Corporation, Newton, Mass., fig. 5-8; reprinted by permission.)

used.) Finally, the stop length single-bit indicator is used to distinguish terminals with stop elements lasting 1 bit from those with stop elements longer than 1 bit. All of these characteristics are needed to cue the multiplexer into how to handle the incoming character at a port, or to properly convert high-speed into the appropriate low-speed characters fed out onto a low-speed line.

This multiplexer outputs 9-bit characters onto the high-speed line. It adds an even-parity bit to 8-bit asynchronous data bits, after stripping off the start and stop bits. Where the asynchronous data words are less than 8 bits in length, 0's are added to build them up to the desired 8 bits. Synchronous data bits are simply grouped into 9-bit characters, and transmitted directly, with no parity checking. On reception, from the high-speed line, the receiving multiplexer (or demultiplexer) can carry out both internal (bit 8) error checking, as well as bit-9 error checking on asynchronous data. All these tasks can be performed using the port characteristics stored in the B memory.

It is apparent that both the local (initiating) multiplexer and the remote device must have the same entries stored in the respective tables. They must also maintain synchronism with respect to one another. If configurations change frequently, RAM memories or alterable ROM memories can be used. On a command from the local site, the contents of both the A and B memories can be transmitted to the remote site. Synchronization signals are sent as well.

A specific ''handshaking'' procedure is used to reestablish synchronization whenever required. This includes the time of initial power-up, whenever a loss of frame is detected, when no framing is received from the remote site after a specified timeout, etc. The procedure consists of first sending a 255-bit pseudorandom binary sequence (known to both the local and the remote multiplexers) to be used to lock the remote multiplexer to the local one. This is followed by an 8-bit status word which indicates whether or not the A- and B-memory contents are to be transferred, and whether or not the A- and B-memory contents of the remote multiplexer are requested to be transferred to the local site. The memory transfer (2,048 bits) then follows, if so specified. This phase of the handshaking procedure is concluded with the transmission of a 25-bit check sequence to detect errors made during transmission. (This is again an example of error detection and correction procedures to be discussed in Chap. 6.) The remote multiplexer concludes the handshaking by transmitting the same sequence, less the memory transfer (unless specifically requested), back to the local site.

Once transmission is initiated, synchronism is maintained by a continual check of the asynchronous data characters, using the characteristics of the A and B memories, as already noted. Out-of-synchronization conditions can thus be recognized, initiating a reframing procedure to correct the condition. Since this is not possible with synchronous channels, at least one subframe must always be devoted to asynchronous use. A 1200-bits/s output trunk can thus not support synchronous transmission. A minimum of 2,400-bits/s output line capacity is needed if at least one of the channels to be multiplexed is synchronous. For the same reason, no more than 7,200 bits/s from synchronous channels can be transmitted over a 9,600-bits/s trunk—at least one subframe, accommodating 1,200-bits/s data transmission, must be left for asynchronous transmission.

Variations in the asynchronous terminal speeds are accommodated by the use of control characters and the transmission of 9-bit characters. The former corresponds to pulse stuffing in the bit-interleaved multiplexer, the latter to transmitting at higher rates to accommodate expected increases in the input terminal data rate. To see this, recall that ASCII terminals output 10-bit characters, of which 8 bits are data (Fig. 3-54). The 8 data bits only are transmitted by this multiplexer, with an even-parity check bit added. Nine bits are thus transmitted for every 10 arriving, allowing small increases in input data rate to be readily accommodated.

As an example, say that a 300-bits/s terminal is connected to one of the input ports. This uses up one-fourth of the capacity of a 1,200-bits/s subframe, for, as noted earlier, the terminal would be scanned 6 times of the 24 slots of the subframe. Since a complete scan is carried out in 180 ms, the terminal is scanned every 30 ms. In this time 9 bits are output, so that the output rate is the nominal 300-bits/s rate. However, 8 data bits only are output, so that the output *data* rate is actually 8 bits/30 ms = 267 bits/s. The terminal itself is inputting data at a 240-bits nominal rate, however (8 bits for every 10 bits in), so that 27 bits/s, or a 10 percent margin, is left available to accommodate increases in the terminal data rate.

Should the asynchronous port have no data to transmit, control characters are transmitted instead. This maintains the requisite synchronization. Control characters are denoted by transmitting as the first 3 bits of the character the sequence 010, and using an odd-parity bit as the ninth bit. (This latter feature is necessary to avoid confusing a data character beginning 010 with a control character. Recall that data characters used even parity.) Other nondata characters transmitted include transmit and receive garble characters, used to denote a parity error detected in a port character, an overload of data characters from a port, or an unrecognizable asynchronous character received from the remote multiplexer, and break characters, used to replace an asynchronous data character with no stop bits at its end (hence a framing error must have occurred at the asynchronous ports).

Frame Loss and Acquisition

As is apparent from the discussion in the previous subsections and from our earlier discussion of time multiplexing in this chapter, framing and synchronization must be maintained for the corresponding transmitting and receiving data sources to stay in step with one another. Framing ensures that slots in the received time sequence will be correctly associated with the appropriate receiver terminals. Synchronization implies that the transmitter and receiver clocks are locked to one another, so that bit integrity is maintained. The subjects of framing and synchronization are quite broad and detailed in their own right. We have earlier mentioned typical ways in which synchronization and framing are maintained without discussing the related design questions in detail. These include such questions as the length of the frame (in characters or bits) in fixed-frame systems, the length and type of framing pattern to be used, the design of the synchronization pattern, etc.

Rather than discuss these questions directly here, in this introductory text, we shall come at them indirectly, motivating them through a discussion of the way in which one maintains frame integrity. Specifically, we discuss in this concluding subsection of time-division multiplexing the questions of loss of frame and time required to acquire frame again, once a loss of frame has occurred. These will serve to introduce some of the quantitative aspects of synchronization and framing. The interested reader is referred to the literature for further study of the subject.[55]

Specifically, there is a probability that a frame will be lost, owing to errors in the framing pattern. The longer the framing pattern generally, the higher the probability of a loss. Yet a longer frame pattern is easier for synchronization. So there is a trade-off in pattern length. One can add error-correction capability to the pattern, hence allowing a longer length to be used, but this reduces the efficiency of the frame (more overhead is introduced, at the expense of data), and results in increased complexity as well. Alternatively, one can require a frame error or violation detected to be repeated several times in a row before a firm decision is made that the frame is lost. This reduces the chance that a bit received in error due to noise will throw the framing procedure off. Thus the more repetitions required, the more accurate the final decision will be, but it will take the system correspondingly longer to reach a decision. If the system *is* out of frame, proportionately more data will be lost.

To quantify these statements say that a frame is L bits long, of which n bits represent the framing pattern. Let the probability of incorrectly detecting a bit again be $p \ll 1$. Assume that these bit errors are due to noise encountered during transmission, and that errors occur randomly, from bit to bit. For the simplest framing scheme an error in any of the n framing bits will be assumed to produce a framing error. For the n-bit pattern the probability P_f of a framing error is then

$$P_f = 1 - (1 - p)^n \doteq np \qquad p \ll 1 \qquad (3\text{-}82)$$

Now require k frames in succession to be declared in error before a frame loss is declared present. The probability P_l of a loss is then

$$P_l = (P_f)^k \doteq (np)^k \qquad (3\text{-}83)$$

If F frames per second are transmitted, the average time, in seconds, between frame losses is approximately

$$T_l = \frac{1}{P_l F} \qquad (3\text{-}84)$$

---

[55] See, e.g., J. J. Stiffler, *Theory of Synchronous Communication,* Prentice-Hall, Englewood Cliffs, N.J., 1971.

As an example, say that $p = 10^{-5}$, $n = 4$, and $k = 3$. The framing pattern is thus 4 bits long, and three repeats of a detected error are required before framing information is declared lost. Then $P_f \doteq 4 \times 10^{-5}$, and

$$P_l \doteq 0.64 \times 10^{-13}$$

On the average, then, 1 frame in 1.6×10^{13} frames transmitted is lost. If the bit rate is 1.6×10^6 bits/s and frames are $L = 10^3$ bits long, the frame rate is $F = 1{,}600$ frames per second. A frame will thus be lost, on the average, once every 10^{10} s, clearly a tolerable number! Now let p increase to 10^{-3}. Then $P_f \doteq 4 \times 10^{-3}$, and $P_l = 0.64 \times 10^{-7}$. A frame will now be lost, on the average, once every 10^4 s, or 3 h. This may or may not be tolerable.

Once a loss of frame has occurred, it requires some time to reacquire frame. One would like to reduce this time to as small a value as possible. Again trade-offs exist. There are two components to this time. One is the time required to detect the out-of-frame condition. This is the order of k frames. Hence now we want k small, yet previously we wanted k large to prevent the loss from occurring too often. The other component is the time required to acquire frame again, once it has been declared lost.

To quantify these comments, focus first on the time required to detect a lost frame. If the frame is lost, the receiving system may still not notice this, since the data are random, and there is a probability 2^{-n} that an n-bit data pattern will look like the expected frame pattern. Hence the probability the loss *is* detected in any one frame is $(1 - 2^{-n})$. The probability the frame loss is detected is then $(1 - 2^{-n})^k = 1 - k \cdot 2^{-n} + \cdots$. We want this to be close to 1, and hence we want

$$k \cdot 2^{-n} \ll 1$$

If this condition is satisfied, the time to detect a lost frame is very nearly k frames. In the example used earlier, with $k = 3$ and $n = 4$, $k \cdot 2^{-n} = \frac{3}{16}$.

Summarizing the conclusions thus far:

1. We want $np \ll 1$. This suggests picking n small, as noted earlier.
2. We want k *large* to keep random errors from establishing a frame loss.
3. We want k *small* to reduce the time to detect a lost frame.
4. We want $k \cdot 2^{-n} \ll 1$ to prevent a random data pattern from being mistaken for a framing pattern. Hence we want n *large*.

Note the trade-offs in n and k required here, as pointed out earlier.

Assuming it takes k frames to detect a frame loss, how long does it take to reframe? This is the time required to search through the data bits, n at a time, until a framing pattern is definitively detected as such. One simple procedure is to require a framing pattern, once detected, to appear at the same position in a frame m times in a row. m is normally chosen to be greater than 1 to prevent an n-bit data pattern from being (mistakenly) labeled as a frame pattern. Once such a mistake is made, a minimum of k frames is required to detect it, and to break out of the erroneous pattern. Using $m > 1$ reduces the chance of this occurrence. On the other hand, $m < k$ is desired to reduce the reframe time.

To calculate the average reframe or acquire time, once a frame loss is detected, consider the worst possible case—the one in which one starts searching at the opposite end of the frame from the location of the true frame pattern. It thus takes a minimum of $(m + 1)$ frames to acquire frame. A typical picture appears in Fig. 3-58. The 4-bit sequence 0101 is the frame pattern to be detected in this example. In the worst case the search begins at the second bit (the 1 in this sequence). The correct sequence is thus L bits away.

As the search proceeds, there is always a probability 2^{-n} that a frame pattern will be detected erroneously. If this occurs, search is suspended until the next frame, to check to see if the same pattern again appears at the same location in the frame. The chance of the frame pattern appearing erroneously at the same location during the next frame is again 2^{-n}. For $n > 2$ or 3 it is apparent that, on the average, only one frame interval will be required to detect the false pattern, with search then being resumed. Say that on the average h such suspensions of search or *holds* are encountered during the L-bit search. The worst-case acquire time is then

$$T_a = m + 1 + h \text{ frames} \tag{3-85}$$

To calculate h we note that in this worst case, $(L + h)$ patterns must be examined. Of these, h are detected as frame patterns. The ratio $h/(L + h)$ must thus be the same as the probability of detecting a frame pattern, so that

$$\frac{h}{L + h} = 2^{-n}$$

and

$$h = \frac{L}{2^n - 1} \tag{3-86}$$

As an example, let $L = 1,000$ bits and $n = 4$ bits. Then $h = 1,000/15 \sim 66$ frames! This is obviously much too big. The implication is that with such a large frame there are many opportunities, on the average, to (erroneously) detect frame patterns among the data bits. If n is increased to 8 bits, $h = 1,000/255 \sim 4$ frames, a much more tolerable value.

The total average (worst case) time to detect a frame loss and acquire frame again can now be calculated as

$$T = k + m + 1 + \frac{L}{2^n - 1} \quad \text{frames} \tag{3-87}$$

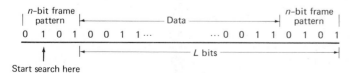

Figure 3-58 Worst-case search for n-bit frame pattern.

As an example, take $L = 1,000$ bits, $k = 3$, $m = 2$, and $n = 8$ bits, as in the example above. Then $T = 10$ frames to detect a frame loss and reacquire frame. This corresponds to 10,000 bits. At a 1.5-Mbits/s rate, this is 6 ms. At a 50-kbits/s rate, this is 0.2 s. At a 10-kbits/s rate, this is 1 s.

If in this example we must take $n = 8$ to reduce the acquire time, what happens to the probability of losing a frame due to noise? Using $p = 10^{-5}$ again as the probability of erroneous bit detection, $P_f \doteq 8 \times 10^{-5}$, and $P_l \doteq 5 \times 10^{-13}$. For a trunk bit rate of 1.5 Mbits/s, and a corresponding frame rate of 1,500 frames per second, a frame will be lost, on the average, approximately once every 10^9 s, clearly a tolerable number again. If $p = 10^{-3}$ now, a frame will be lost, on the average, once every 1,000 s, or once every 16 min. This may no longer be tolerable. Hence one would require the use of low-noise lines in such a case.

3-9 WAVESHAPING AND BANDWIDTH CONSIDERATIONS

In discussing digital communication systems thus far, we have ignored the shapes of the pulses used to transmit the information. In many examples considered we have simply shown the digital pulses to be rectangular in shape. In actual practice, where these pulses will modulate a carrier for transmission over relatively long distances, pulse shaping must be carried out. This is particularly true where constraints are placed on the channel bandwidth.

We have occasionally commented that the bandwidth is roughly given by the reciprocal of the time slot or interval in which a pulse is constrained to lie. Thus if 10 signals are sampled and time-multiplexed every 125 μs, each signal sample is constrained to lie within its 12.5-μs time slot. The bandwidth required to transmit the time-multiplexed signal train is then roughly $1/12.5$ μs, or 80 kHz. If eight-level quantization is now used, and binary signals sent, the bandwidth increases to 240 kHz. The T1 PCM system discussed in Sec. 3-6 transmits at the rate of 1.544 Mbits/s. If binary symbols are used, one would then argue that the bandwidth required for baseband transmission is roughly 1.5 MHz. As noted in Chap. 2, this reciprocal time-bandwidth relation represents a very useful rule of thumb. One has to be more precise, however, in dealing with pulse shaping to attain a specified bandwidth. We shall in fact find that up to $2B$ pulses per second can be transmitted over a channel with bandwidth B hertz. This is called the *Nyquist rate*, and is obviously related to the Nyquist sampling rate.

The technology has been moving to the transmission of higher and higher bit rates over given channels, using multilevel signaling. (We shall describe some of these techniques briefly in this section and the one following, and discuss them in more detail, in terms of modulation techniques, in Chap. 4.) This can result in considerable spillover of pulse energy into adjacent time slots, resulting in *intersymbol interference* if care is not taken about the shapes of the pulses transmitted. It is thus important to focus more deeply on the dual concepts of pulse shaping and the bandwidths necessary to transmit the shaped pulses.

To make these comments on pulse shaping, bandwidth, and Nyquist rate more precise, it is useful to discuss the concept of intersymbol interference in more detail. By choosing signal waveshapes to minimize or eliminate this phenomenon, we shall find the bandwidth necessary for their transmission.

Consider the sequence of pulses shown in Fig. 3-59. Although these are shown as binary pulses, they could equally well be pulses of identical shape, but of arbitrary height (PAM or quantized PAM). They are shown recurring at T-second intervals. T is the sampling interval in the PAM or quantized PAM case, the binary interval in the case of binary-encoded symbols. System filtering causes these pulses to spread out as they traverse the system, and they overlap into adjacent time slots as shown. At the receiver the original pulse message may be derived by sampling at the center of each time slot as shown, and then basing a decision on the amplitude of the signal measured at that point.

The signal overlap into adjacent time slots may, if too strong, result in an erroneous decision. Thus, as an example, in the case of Fig. 3-59 the 0 transmitted may appear as a 1 if the tails of the adjacent pulses add up to too high a value. (In practice there may be contributions due to the tails of several adjacent pulses rather than the one pair shown in Fig. 3-59.) This phenomenon of pulse overlap and the resultant difficulty with receiver decisions is termed *intersymbol interference*.

Note that this interference may be minimized by purposely widening the transmission bandwidth as much as desired. This is unnecessarily wasteful of bandwidth, however, and if carried too far may introduce too much noise into the system (see Chap. 1). Instead we seek a way of *purposely* designing the signal waveshapes and hence transmission filters used to minimize or eliminate this interference with as small a transmission bandwidth as possible. One obvious signal waveshape to use is one that is maximum at the desired sampling point, yet goes through zero at all adjacent sampling points, multiples of T seconds away. This ideally provides zero intersymbol interference. With such a waveshape, chosen at the receiver, it should then be possible to design the overall system, back to the original sampling point at the transmitter, to provide

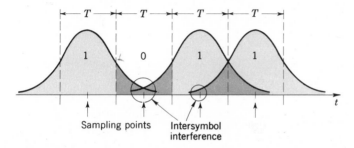

Figure 3-59 Intersymbol interference in digital transmission.

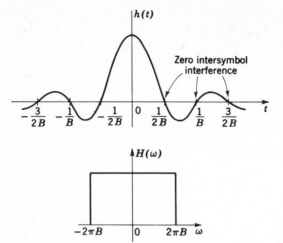

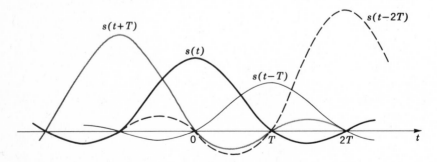

Figure 3-60 Pulse providing zero intersymbol interference.

this desired waveshape. (Recall that the original sampled pulses are essentially impulses if the time τ, during which the analog signal is sampled, is small compared to the sampling interval. The pulse broadening and shaping is then due to innate system filtering as well as filters purposely put in to achieve the final desired shape.)

One signal waveshape producing zero intersymbol interference is just the $(\sin x)/x$ pulse introduced in Chap. 2 as the impulse response of an ideal low-pass filter. Specifically, if the filter has a flat amplitude spectrum to B hertz, and is zero elsewhere, the impulse response is just $(\sin 2\pi Bt)/2\pi Bt$. This pulse is shown sketched in Fig. 3-60. Note that the pulse goes through zero at equally spaced intervals, multiples of $1/2B$ seconds away from the peak at the origin. If $1/2B$ is chosen as the sample interval T, it is apparent that pulses of the same shape and *arbitrary amplitude* that are spaced $T = 1/2B$ seconds apart will not interfere. This is shown in Fig. 3-61. $2B$ pulses per second may thus be transmitted over a bandwidth of B hertz if this waveshape is used. This is just the Nyquist rate noted earlier.

Figure 3-61 Sequence of pulses: zero intersymbol interference.

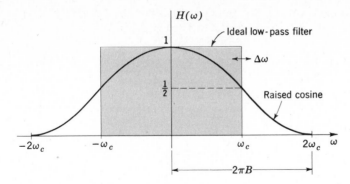

Figure 3-62 Raised cosine spectrum.

There are practical difficulties with this particular waveshape, however:

1. It implies that the overall characteristic between transmitter sampling and receiver decision point is that of an ideal low-pass filter. As noted in Chap. 2, this is physically unrealizable, and very difficult to approximate in practice because of the sharp cutoff in its amplitude spectrum at B hertz.
2. This particular pulse, if attainable, would require extremely precise synchronization. If the timing at the receiver varies somewhat from exact synchronization, the zero intersymbol interference condition disappears. In fact, under certain signal sequences, the tails of all adjacent pulses may add up as a divergent series, causing possible errors! Since some timing jitter will inevitably be present even with the most sophisticated synchronization systems, this pulse shape is obviously not the one to use.

It is, however, possible to derive from this waveshape related waveshapes with zero intersymbol interference that do overcome the two difficulties mentioned. They are much simpler to attain in practice, and the effects of timing jitter may be minimized. The particular class of such waveshapes we shall discuss is one of several first described by Nyquist.[56]

To attain this particular class we start with the ideal low-pass filter of Fig. 3-60, but modify its characteristics at the cutoff frequency to attain a more gradual frequency cutoff, which is hence more readily realized. In particular, if the new frequency characteristic is designed to have odd symmetry about the low-pass cutoff point, it is readily shown, following Nyquist, that the resultant impulse response retains the derived property of having zeroes at uniformly spaced time intervals. An example of such a spectrum often approximated in practice is the _raised cosine amplitude spectrum_. This particular spectrum and the original ideal low-pass spectrum are shown in Fig. 3-62. To avoid confusion

[56] H. Nyquist, "Certain Topics in Telegraph Transmission Theory," *Trans. AIEE,* vol. 47, pp. 617–644, April 1928. See W. R. Bennett and J. R. Davey, *Data Transmission,* McGraw-Hill, New York, 1965, chap. 5, for a detailed discussion.

with the transmission bandwidth B, we henceforth label the ideal low-pass cutoff frequency ω_c. The raised cosine spectrum is then given by

$$H(\omega) = \frac{1}{2}\left(1 + \cos\frac{\pi\omega}{2\omega_c}\right) \qquad |\omega| \leq 2\omega_c \qquad (3\text{-}88)$$
$$= 0 \qquad\qquad\qquad \text{elsewhere}$$

The bandwidth in this case is just $2\omega_c = 2\pi B$. If we now measure frequency with respect to ω_c by letting $\omega = \omega_c + \Delta\omega$ (see Fig. 3-62), we have

$$H(\omega) = \frac{1}{2}\left[1 + \cos\frac{\pi}{2}\left(1 + \frac{\Delta\omega}{\omega_c}\right)\right] = \frac{1}{2}\left(1 - \sin\frac{\pi}{2}\frac{\Delta\omega}{\omega_c}\right) \qquad (3\text{-}89)$$

Since the sine term has odd symmetry [$\sin(-x) = -\sin x$], it is apparent that the raised cosine spectrum displays the odd symmetry about ω_c noted above. This is also apparent from Fig. 3-62.

The impulse response of a filter with this frequency characteristic is readily shown to be given by

$$h(t) = \frac{\omega_c}{\pi}\frac{\sin\omega_c t}{\omega_c t}\frac{\cos\omega_c t}{1 - (2\omega_c t/\pi)^2} \qquad (3\text{-}90)$$

It has the $(\sin x)/x$ term of the ideal filter multiplied by an additional factor that decreases with increasing time. The $(\sin x)/x$ term ensures zero crossings of $h(t)$ at precisely the same equally spaced time intervals as the ideal low-pass filter. The additional factor multiplying the $(\sin x)/x$ term reduces the tails of the pulse considerably below that of the $(\sin x)/x$ term, however, so that such pulses when used in digital transmission are relatively insensitive to timing jitter.

Letting the sampling interval $T = 1/2f_c = \pi/\omega_c$, so that the zeros of the pulse occur at T-second intervals, as in the previous ideal low-pass case, we have the transmission bandwidth given by $B = 2f_c = 1/T$. This is just the bandwidth criterion adopted rather arbitrarily in previous sections. As an example, consider again an analog signal of 3.2 kHz sampled at an 8-kHz rate. If this signal only were transmitted by PAM, and the $(\sin x)/x$ ideal waveshape were used, the transmission bandwidth required would be $B = 1/2T = 4\text{kHz}$. Using a raised cosine spectrum, however, the bandwidth required would be $B = 1/T = 8$ kHz. If 10 signals were time-multiplexed, the bandwidth would increase by a corresponding factor of 10. Thus although the $(\sin x)/x$ signal shape theoretically allows transmission at very nearly the original analog signal bandwidth (it would be the original bandwidth if the Nyquist sampling rate were used), the use of more realistic waveshapes results in an increase of the required bandwidth. The raised cosine spectrum doubles the bandwidth required. Other shapes to be discussed reduce this requirement.

The fact that the raised cosine spectrum is just one example of a class of spectra with odd symmetry about ω_c providing zero crossings at equally spaced sampling intervals is demonstrated as follows.

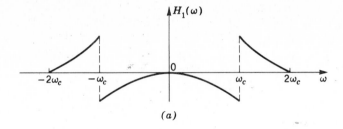

(a)

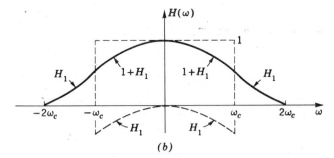

(b)

Figure 3-63 Nyquist filter. (a) Odd symmetry about ω_c. (b) Overall filter.

Assume a low-pass filter to have the characteristic

$$
\begin{aligned}
H(\omega) &= 1 + H_1(\omega) & |\omega| &< \omega_c \\
&= H_1(\omega) & \omega_c &< |\omega| < 2\omega_c \\
&= 0 & &\text{elsewhere}
\end{aligned}
\tag{3-91}
$$

(If H_1 is zero, we have just the ideal low-pass filter.) As an example, let

$$
\begin{aligned}
H_1(\omega) &= \frac{1}{2}\left(\cos \frac{\pi}{2} \frac{\omega}{\omega_c} - 1 \right) & |\omega| &< \omega_c \\
&= \frac{1}{2}\left(\cos \frac{\pi}{2} \frac{\omega}{\omega_c} + 1 \right) & \omega_c &< |\omega| < 2\omega_c
\end{aligned}
\tag{3-92}
$$

This is of course just the raised cosine case.

Assume now for simplicity's sake that the overall filter has zero phase shift. (A linear phase term of course results in a corresponding pulse time delay.) Assume further that $H_1(\omega)$ has *odd symmetry* about ω_c. Then

$$
H_1(\omega_c + \Delta\omega) = -H_1(\omega_c - \Delta\omega)
\tag{3-93}
$$

The raised cosine has this property. Another arbitrarily chosen example is shown in Fig. 3-63a, with the overall characteristic sketched in Fig. 3-63b.

Taking the Fourier transform of Eq. (3-91) to find the impulse response $h(t)$, we have, by superposition,

$$h(t) = \frac{\omega_c}{\pi} \frac{\sin \omega_c t}{\omega_c t} + h_1(t) \tag{3-94}$$

where
$$h_1(t) = \frac{1}{2\pi} \int_{-\infty}^{\infty} H_1(\omega) e^{j\omega t} \, d\omega \tag{3-95}$$

But $H_1(\omega)$, having zero (or linear) phase shift, must be even in ω (see Fig. 3-63a). (Recall that all *amplitude* spectra are symmetrical about $\omega = 0$.) Then

$$h_1(t) = \frac{1}{\pi} \int_0^{\infty} H_1(\omega) \cos \omega t \, d\omega$$

$$= \frac{1}{\pi} \int_0^{\omega_c} H_1(\omega) \cos \omega t \, d\omega + \frac{1}{\pi} \int_{\omega_c}^{2\omega_c} H_1(\omega) \cos \omega t \, d\omega \tag{3-96}$$

using the fact that $H_1 = 0$, $\omega > 2\omega_c$.

We now make use of the property of odd symmetry about ω_c by letting $\omega = \omega_c - x$ in the first integral of Eq. (3-96), and $\omega = \omega_c + x$ in the second integral. The new dummy variable x ranges between 0 and ω_c in both integrals, and the two may be combined into the following one integral, after using the odd-symmetry property:

$$h_1(t) = \frac{1}{\pi} \int_0^{\omega_c} H_1(\omega_c - x) \left[\cos (\omega_c - x) t - \cos (\omega_c + x) t \right] dx \tag{3-96a}$$

Now using the trigonometric identity

$$\cos (a - b) - \cos (a + b) = 2 \sin a \sin b$$

we finally obtain the following interesting result:

$$h_1(t) = \frac{2}{\pi} \sin \omega_c t \int_0^{\omega_c} H_1(\omega_c - x) \sin xt \, dx \tag{3-96b}$$

Note that independently of the precise value of the integral (this will depend on the particular characteristic chosen for H_1), the $\sin \omega_c t$ preceding *guarantees* that $h_1(t)$ will be *zero* at intervals spaced $T = \pi/\omega_c$ seconds apart. But this is just the original sampling interval T, so that $h(t)$ from Eq. (3-94) does go through zero at all intervals that are multiples of T away from the desired sampling

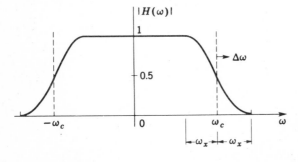

Figure 3-64 Sinusoidal roll-off spectrum.

point $t = 0$. This property of $h(t)$ is of course due to the odd symmetric choice $H_1(\omega)$.

An interesting practical example of a spectrum satisfying the odd-symmetric property of (3-93) that leads theoretically to pulses with zero intersymbol interference is the *sinusoidal roll-off* spectrum of Fig. 3-64.[57] Measuring frequency from ω_c, as shown in the figure, the amplitude spectrum for this example is specified by

$$|H(\Delta\omega)| = \frac{1}{2}\left(1 - \sin\frac{\pi}{2}\frac{\Delta\omega}{\omega_r}\right) \qquad |\Delta\omega| < \omega_r$$

$$= 0 \qquad\qquad\qquad |\Delta\omega| > \omega_r \qquad\qquad (3\text{-}97)$$

$$= 1 \qquad\qquad\qquad -\omega_c < \Delta\omega < -\omega_r$$

The design parameter ω_r represents the radian frequency by which the radian bandwidth exceeds ω_c. The ratio $r = \omega_r/\omega_c$ is called the *roll-off factor*. Smaller values of ω_r/ω_c lead to smaller bandwidth requirements, but require correspondingly tighter control on the design and more complex filter designs. The case of $\omega_r = 0$, or zero roll-off, is just the ideal low-pass filter case, and gives rise to the $(\sin x)/x$ pulse shape already discussed. Similarly, unity roll-off, with $\omega_r/\omega_c = 1$, results in the raised cosine spectrum discussed. By varying the roll-off factor a whole class of pulse shapes is generated. We shall have occasion to refer to examples of this spectral class later in the book in discussing spectral shaping for data transmission in satellite communications, as well as over telephone lines.

Note that in proving that the Nyquist odd-symmetry condition of (3-93) gives rise to a class of pulses capable (theoretically) of attaining zero intersymbol interference we assumed zero phase shift. More generally, linear phase shift, $\theta(\omega) = -\omega t_0$, over the range of frequencies of interest ($|\omega| < \omega_c + \omega_r$) must be assumed. This of course makes the filter design more difficult. It is left to the reader to show that the impulse response for the sinusoidal roll-off spectrum with linear phase shift is given by

$$h(t) = \frac{\omega_c}{\pi}\frac{\sin\omega_c(t - t_0)}{\omega_c(t - t_0)}\frac{\cos\omega_r(t - t_0)}{1 - [2\,\omega_r(t - t_0)/\pi]^2} \qquad (3\text{-}98)$$

This is just a special case of (3-94) and (3-96b), and can of course be obtained from those equations by appropriate integration.

The bandwidth/pulse rate trade-off using this sinusoidal roll-off spectrum (and other related spectra) depends of course on the roll-off factor chosen. What it does indicate is that if the desired rate of pulse transmission is $1/T$ pulses per second, the bandwidth B, in hertz, required is

$$B = \frac{1}{2T}\left(1 + \frac{f_r}{f_c}\right) = \frac{1}{2T}(1 + r) \qquad (3\text{-}99)$$

[57] Bennett and Davey, *op. cit.*, pp. 55, 56.

Alternatively, with B specified, the number of pulses per second that may be transmitted is given by

$$\frac{1}{T} = \frac{2B}{1 + f_x/f_c} = \frac{2B}{1 + r} \tag{3-100}$$

The number thus ranges from an unattainable maximum of $2B$ pulses per second to B pulses per second ($f_x/f_c = 1$), just the rule of thumb used up to now.

As an example, say that the allowable bandwidth is 2.4 kHz. (This is a number often used as a measure of the *available* bandwidth over telephone channels.) The maximum pulse rate over this channel is then $2B = 4800$ pulses per second. If a 25 percent roll-off factor is used, 3840 pulses per second may be transmitted. If the raised cosine spectrum is used, this reduces to 2400 pulses per second. In the chapters to come we shall often refer to the range B to $2B$ pulses per second, as the pulse rate that may be transmitted over a bandwidth of B hertz. Specific numbers in this range depend, as already noted, on design requirements, allowable costs, and various practical constraints introduced in any real application. As an example, timing jitter may play a prominent role in a particular application. If so, a roll-off factor close to zero may not be acceptable and a design close to the raised cosine spectrum may be necessary.

In addition, the discussion thus far has focused on eliminating intersymbol interference. The problem of minimizing intersymbol interference with digital transmission is a major one in the telephone plant where the effects of additive noise are generally minimal. In digital systems designed for space communications, however, noise considerations play a highly significant role. Receiver design and signal shaping must take the effects of noise into account. We shall see, in considering this problem later, in Chap. 5, that the minimization of errors due to additive noise leads to *matched-filter* receivers. In an environment involving the minimization of both intersymbol interference and errors due to noise, some compromise between the two effects is generally necessary. Yet our discussion later will show that the receiver and signal shaping problem is not too critical; in the matched-filter case, for example, we shall find that transmission bandwidths equal to the reciprocal of the interval T are close to optimum, so that one may simultaneously combat both intersymbol interference and noise without too much difficulty.[58]

For theoretical considerations it is often useful to use the maximum pulse rate, the Nyquist rate of $2B$ pulses per second, as the pulse rate to be expected

[58] The minimization of errors due to both intersymbol interference and noise is considered in R. W. Lucky, J. Salz, and E. J. Weldon, *Principles of Data Communication*, McGraw-Hill, New York, 1968, chap. 5. They show that if the channel may be assumed to introduce no distortion other than noise, the effect of both intersymbol interference and noise may be simultaneously minimized by splitting the Nyquist filter $H(\omega)$ equally into two parts, $H^{1/2}(\omega)$ at the transmitter, $H^{1/2}(\omega)$ at the receiver. As will be seen in Chap. 5, the receiver filter is then exactly the *matched filter* that minimizes errors due to noise.

with a bandwidth of B hertz. B is then referred to the Nyquist bandwidth. More commonly, rather than speak of *pulses per second,* as we have been doing to this point, one refers to the number of *symbols per second* that one may transmit over a channel of B hertz. The use of the term "symbol" rather than "pulse" will become clear in Chap. 4 when we discuss modulated signals. Alternatively, one refers as well to the number of *symbols per hertz* that may be accommodated. Thus the Nyquist rate, with ideal low-pass shaping, provides 2 symbols/Hz; the raised cosine spectrum provides 1 symbol/Hz. The concept of symbols/Hz is particularly useful since it enables us to compare various transmission schemes. We shall see an example of this in the next section, in discussing duobinary transmission, a technique used to specifically increase the symbol transmission rate over a specified channel.

As noted earlier, the ideal transmission rate of $2B$ symbols per second agrees with the Nyquist sampling rate of Sec. 3-2. There is a close connection between the two concepts, one involving the number of samples required to uniquely represent a band-limited signal, the other the maximum rate of transmission of symbols over a band-limited channel. It is no accident that $(\sin x)/x$ filtering for signal reconstruction in the first case, and $(\sin x)/x$ pulses in time, in the second, both appear. The maximum transmission rate of $2B$ symbols per second agrees as well with Hartley's law, mentioned in Sec. 3-2.

It is important at this point to distinguish between symbols/s and bits/s. We have already noted, in our discussion of the A/D or quantization process in Sec. 3-4, that a pulse may have a discrete number of amplitude levels. With the number of levels (generally a power of 2) specified, a unique conversion between a single multilevel pulse and the equivalent set of two-level, or binary, pulses exists. Specifically, a pulse of $M = 2^n$ levels is representable by n binary pulses. Generalizing to symbols, a multivalued symbol having $M = 2^n$ possible values is representable by n binary symbols. As an example, we shall see in Chap. 4, in discussing sine-wave signals, that amplitude, phase, and frequency can be varied independently. The discussion thus does not have to be restricted to *amplitude*-level variations. The symbol in that case is a sine wave with a particular amplitude, phase, and frequency.

More generally, different waveshapes can be used to transmit digital information. We use the word *symbol* to represent a specified waveshape which may or may not be of the simple pulse type assumed up to now, and refer generically to a class of M such specified waveshapes as the M symbols under consideration. One of these M symbols is assumed transmitted in an interval T seconds long, so that the transmission rate is 1 symbol/T seconds. If a bandwidth B hertz is available, the Nyquist rate is $2B$ symbols/s. Since each symbol may be uniquely coded into $n = \log_2 M$ bits, the equivalent *bit rate* is $2Bn$ bits/s, or $2n$ bits/Hz. Transmission of the M-valued symbols, one at a time, is referred to as *M-ary transmission*. We shall have occasion to refer to specific schemes of this type in Chaps. 4 and 6. The simplest example is the one already referred to—the transmission of shaped pulses of varying amplitude. We have referred previously, in Sec. 3-4, to the bandwidth trade-off possible by coding between binary signals and M-level signals in PCM. Thus a quantized PAM signal of M levels may be coded into binary PCM with $n = \log_2 M$ bits used to represent one of the M-level signals. The corresponding bandwidth is increased by a factor of n, since the binary digits each occupy $1/n$ of the original time. Conversely, n successive bits may be coded or collapsed into one symbol of n times the duration. The bandwidth is thus reduced by a factor of n. It is the latter concept that we have been exploiting in this current discussion. The bandwidth and *symbol* duration are inversely connected. If a symbol is one of $M = 2^n$ that could be transmitted in a specified interval, it represents $n = \log_2 M$ bits of information. By going to multivalued symbols one can increase the bit rate over a fixed bandwidth channel. An example of binary-to-M-ary conversion with $n = 3$ and $M = 8$ appears in Fig. 3-65.

Since it is the *symbol rate* that is determined by the bandwidth, one can theoretically increase the bit rate to be transmitted over a band-limited channel by going to higher- and higher-level M-ary schemes. In practice this is not always possible, however. As an example, take the case of M-level pulses. Figure 3-61 shows pulses of four different heights being transmitted in consecutive intervals T seconds long. In any real system some intersymbol interference must be present due to inaccuracies and tolerances in filter design, as well as to timing jitter. This is due to the tails of other pulses (those coming before and those coming after) having nonzero values at the sampling instants. It is apparent that the higher-level pulses in Fig. 3-61 will create more of a problem than will the lower-level ones. As the number of levels increases and the corresponding heights increase, the problem is exacerbated. One cannot indefinitely increase the pulse heights anyway because of power limitations. Trying to reduce the amplitude differences of the multilevel pulses is not helpful beyond a point, since ever-present noise means there should be some minimum amplitude separation or level spacing between the pulses. (The identical problem arises in attempting to distinguish between sinewaves of varying phase or frequency.) So limits do exist on the bit rate possible over a channel of given bandwidth. As an

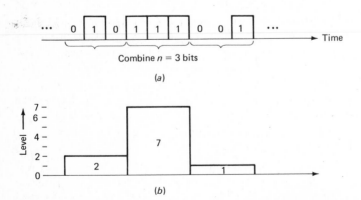

Figure 3-65 Binary M-ary conversion. (a) Binary input sequence. (b) Output M-ary sequence.

example, the current state of the art for digital transmission over a telephone channel of 2,400 Hz bandwidth is 9,600 bits/s. Sixteen different sine waves, of differing amplitude and phase, are used for this purpose. Details of the scheme used will be presented in Chap. 4. The combination of bandwidth, signal power, and noise in limiting the rate of transmission of signals over a given channel has already been noted briefly in Chaps. 1 and 2. In this section we have attempted to quantify the bandwidth limitation. In later chapters, after discussing the effects of noise on digital transmission, we shall come up with specific quantitative expressions that show how both bandwidth and noise impact on the rate of data transmission. In Chap. 5 we shall develop a bit-rate expression for PCM that incorporates bandwidth, power, and noise. In Chap. 6 we shall discuss the celebrated Shannon capacity expression that provides a precise limit on the rate of data transmission over a channel obeying the assumptions in Shannon's model.

To further distinguish between symbols/s and bits/s the term *baud* is sometimes used to represent 1 symbol/s. If binary symbols are used, the number of bauds and the bit rate are identical. If multivalued symbols are used, however, the two differ. A device operating at 110 baud, for example, produces 110 symbols/s. If two such symbols only are used, the device outputs 110 bits/s. If eight such symbols are used, the bit rate is 330 bits/s. The term "baud" was introduced historically in the telegraph field and is still used extensively to describe the transmission rates of lower-speed data terminals. The asynchronous terminals of Sec. 3-8 fall in this category. They are often described by their baud characteristic. The higher-speed synchronous terminals are commonly rated in bits/s, however. (Some confusion occasionally arises, a 9,600-bits/s device, for example, being called, erroneously, a 9,600-baud device even though it may be outputting 2,400 symbols/s. The correct usage would be 2,400 baud, if baud usage were desired.)

Although the stress in this section thus far has been on waveshaping to reduce intersymbol interference, binary coding can be used as well to adjust the signal spectrum. One example, commonly used to reduce dc spectral components, is the *bipolar signal* discussed in Sec. 3-6 in connection with the Bell System T1 system. Recall that in this type of signal, successive 1's are represented by alternating polarity pulses. An example appears in Fig. 3-66a. (In practice, shaping as discussed up to this point in this section would be used in addition to get the desired pulse spectrum.) If a long sequence of 1's appears, the average of the signal is still 0, so that drift problems do not arise. This scheme is useful in cases where dc and low-frequency signal components cannot be tolerated. The spectrum of the bipolar pulse sequence has been calculated and appears in the literature. It does have the desired null at 0 frequency.[59]

---

[59] Bennett and Davey, *op. cit.*, pp. 27, 28, 338–341; J. J. Spilker, Jr., *Digital Communications by Satellite*, Prentice-Hall, Englewood Cliffs, N.J., 1977, pp. 475, 476.

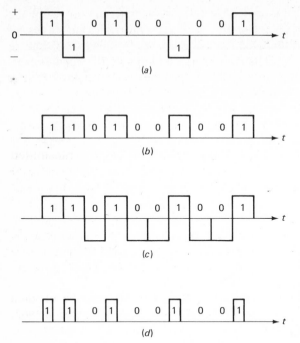

Figure 3-66 Examples of binary signals (waveshaping not shown). (*a*) Bipolar signal. (*b*) Nonreturn-to-zero (NRZ) signal: on-off. (*c*) Nonreturn-to-zero (NRZ) signal: polar. (*d*) Return-to-zero signal.

The most common mode of binary transmission is the one called *nonreturn-to-zero* (NRZ) transmission. Two examples appear in Fig. 3-66*b* and *c*, respectively. These are the binary codes (again shown without shaping) we have been using thus far in this book. The first is also referred to as an *on-off signal,* since the 1 is represented by a positive pulse, the 0 by the absence of a pulse. This signal has a nonzero average value at half its positive amplitude (if 0's and 1's are equally likely), corresponding to its dc value. The second signal is often called a *polar signal,* since 1's and 0's alternate in polarity. The return-to-zero signal of Fig. 3-66*d* is sometimes used as well. Here the 1 pulse is on only half the pulse interval. Other types of binary codes have been used also.[60]

In this section we have focused on waveshaping of digital pulses to reduce intersymbol interference. In the next section we discuss in detail a specific digital transmission technique, correlative coding or partial-response signaling, that has begun to be used in practice quite frequently to provide desirable spectral shaping, as well as to increase the bit rate that can be transmitted over a given channel. This *introduces intersymbol interference,* in a controlled way, to provide these features, rather than attempting to eliminate such interference.

[60] Bennett and Davey, *op. cit.,* pp. 27, 28; W. C. Lindsey and M. K. Simon, *Telecommunications System Engineering,* Prentice-Hall, Englewood Cliffs, N.J., 1973, fig. 1-5.

3-10 DUOBINARY TECHNIQUES AND CORRELATIVE CODING[61]

In the previous section we discussed the problem of intersymbol interference and the possibility of using signal pulse shaping to eliminate it. The result is that with a bandwidth of B hertz available for transmission, up to $2B$ symbols/s, or 2 symbols/Hz, may be transmitted over the channel. This maximum rate of symbol transmission is referred to as the Nyquist rate. In practice the transmission rate varies from B to somewhat less than $2B$ symbols/s, depending on the transmission channel and system used. It was also pointed out in the previous section that using M-ary signaling, with one of M possible symbols transmitted in a T-second signaling interval, the *bit rate* over a band-limited channel could be increased. Specifically, with $M = 2^n$, up to $2n = 2 \log_2 M$ bits/Hz can be transmitted. We noted some problems with M-ary signaling, however, some of which we will return to later in this book, in considering M-ary transmission in more detail. Included were problems with timing, pulse jitter, and pulse shaping that become more acute, and noise problems that manifest themselves with fixed power. An additional critical problem is that system complexity increases as the number of possible symbols increases.

An alternative procedure for increasing the bit rate that can be handled over a given channel of specified bandwidth was invented by A. Lender in 1962. The procedure, termed the *duobinary technique,* combines $n = 2$ successive binary pulses together to form a multilevel signal. The combining process *introduces* intersymbol interference in a *controlled* way, however, rather than attempting to eliminate it. The technique was later extended by Lender and others to higher-level transmission. The basic idea is to combine successive binary pulses together in a known fashion. The resultant signals, after passing through a band-limited channel, are *correlated* in time and the original signal may be readily reconstructed. The general case of combining $n \geq 2$ successive bits to provide signal correlation over a span of n output signals has been termed *correlative-level coding*[62] or *partial-response signaling.*[63] These extensions of duobinary signaling will be described briefly later. They require fewer levels for transmission than does M-ary signaling, and the complexity of implementation is less as well. As will be noted later, they allow the signal frequency spectrum to be shaped to fit specified transmission constraints.

---

[61] The author is indebted to Adam Lender of GTE Lenkurt for his help with this section. A good summary of the techniques, with detailed references, appears in S. Pasupathy, "Correlative Coding: A Bandwidth-Efficient Signaling Scheme," *IEEE Commun. Soc. Mag.,* vol. 15, no. 4, pp. 4–11, July 1977.

[62] A. Lender, "Correlative Level Coding for Binary Data Transmission," *IEEE Spectrum,* vol. 3, no. 2, pp. 104–115, February 1966.

[63] E. R. Kretzmer, "Binary Data Communication by Partial Response Transmission," *Conference Record, 1965 IEEE Annual Communication Convention,* pp. 451–455; and "Generalization of a Technique for Binary Data Communication," *IEEE Trans. Commun. Technol.,* vol. COM-14, pp. 67, 68, February 1966.

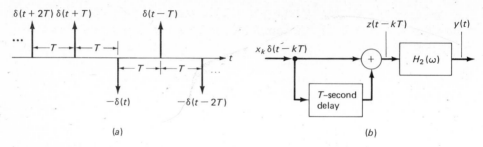

Figure 3-67 Duobinary signaling. (*a*) Binary input sequence, $x_k \delta(t - kT), x_k = \pm 1$. (*b*) Duobinary signal generation.

Duobinary techniques have been applied both to the transmission of higher-speed data over telephone channels,[64] and to the transmission of two T1 data streams over a channel normally handling one T1 stream at the 1.544-Mbits/s rate.[65] Note that in combining successive bits using the duobinary technique or other correlative coding techniques the bits do not have to be from the same bit stream. Two independent bit streams may be merged, as is normally done with time multiplexing, by taking successive bits from each stream in turn. The streams are then separated out at the receiving point.

The duobinary idea is quite simply described. Consider an input sequence of binary impulses (very narrow pulses) $x_k \delta(t - kT)$, $k = 0, \pm 1, \pm 2, \ldots$, spaced a binary interval T apart. The x_k's are taken to be ± 1 representing the sequence of 1's and 0's. A possible sequence is shown in Fig. 3-67*a*. These are the sampled binary outputs of a data terminal, computer, PCM system, or any other binary source to be transmitted over a channel of bandwidth $B \geq 1/2T$. In a normal binary system these would now be fed to a Nyquist-type shaping filter of bandwidth B as described in the last section, to produce the desired transmitted pulses. (Recall again that the case $B = 1/2T$ corresponds to the ideal low-pass filter deemed unrealizable in the last section.) In the duobinary scheme the sum of two successive binary pulses is first formed before shaping, however. Letting $z(t - kT)$ represent this sum, we have

$$z(t - kT) = x_k \delta(t - kT) + x_{k-1} \delta(t - [k - 1]T) \qquad k = 0, \pm 1, \pm 2, \ldots$$
$$(3\text{-}101)$$

The $z(t - kT)$ symbols are impulses that can now take on the values $+2$, 0, or -2. These multilevel impulses are in turn fed into a shaping filter with transfer function $H_2(\omega)$. This process of generation of the duobinary signals is shown schematically in Fig. 3-67*b*.

[64] GTE Lenkurt markets a 2,400-bits/s data set, the 261A, that uses the duobinary encoding technique combined with binary FM.

[65] This is accomplished by the GTE Lenkurt 9148A repeater. This system transmits 48 voice channels (in binary form) or any binary signal at rates of up to 3.152 Mbits/s over lines designed to handle 1.544 Mbits/s. Modified duobinary signaling, to be discussed later, is actually used for the implementation.

Consider, in particular, the special case of $H_2(\omega)$, an ideal low-pass filter of bandwidth $1/2T$. We choose its amplitude to be T for normalization purposes. We thus have

$$H_2(\omega) = T \qquad -\frac{1}{2T} \le f \le \frac{1}{2T}$$
$$= 0 \qquad |f| > \frac{1}{2T} \tag{3-102}$$

It is apparent that the output of the filter, $y(t)$, is the sum of a sequence of $(\sin x)/x$ pulses, peaking at intervals T seconds apart, whose amplitudes are

$$y_k = x_k + x_{k-1} = 0, +2, \text{ or } -2 \tag{3-103}$$

Thus
$$y(t) = \sum_{k=-\infty}^{\infty} y_k \frac{\sin \pi(t - kT)/T}{\pi(t - kT)/T} \tag{3-104}$$

Three levels are required to transmit these signals, in contrast to the four levels required for the M-ary case with $n = 2$ and $M = 4$.

At the receiver, assuming for the time being no transmission distortion or impairment occurs, $y(t)$ is sampled at T-second intervals and the various sample values $y_k = 0, +2,$ or -2 are obtained. An example of a typical input sequence x_k and the corresponding output sequence y_k appears in Fig. 3-68a. Since we had $y_k = x_k + x_{k-1}$, it is apparent that the way to decode the current binary digit

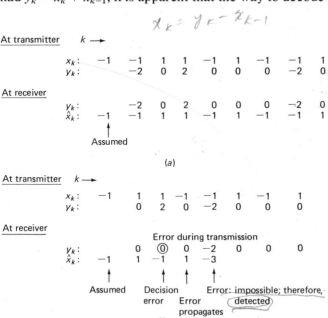

Figure 3-68 Duobinary signal patterns. (a) Perfect reception. (b) Error during transmission.

is to subtract the previous decoded digit from the current received sample y_k. Specifically, letting $\hat{x}_k$ represent the estimate at the receiver of the original binary digit x_k at time $t = kT$, we have

$$\hat{x}_k = y_k - \hat{x}_{k-1} \tag{3-105}$$

Obviously, if y_k is received correctly and $\hat{x}_{k-1}$ corresponds to a correct decision at the previous time interval $t = (k-1)T$, the current estimate $\hat{x}_k$ is correct as well. This is indicated in Fig. 3-68a. This technique of using the stored estimate of the previous symbol is called *decision feedback*.

Note, however, a potential problem here. Not only will a decision error at the receiver be made if y_k is received incorrectly, but this error may tend to propagate. An example appears in Fig. 3-68b. An error in one time slot results in an error in the following slot, and may continue into successive time slots. To eliminate this potential problem (a characteristic of decision feedback schemes) a precoding procedure is used. This will be discussed later.

The example of Fig. 3-68b points up a useful property of duobinary signaling to be discussed briefly later as well. This is the ability to detect errors (although not necessarily to correct them) through the occurrence of impossible error patterns. In the example of Fig. 3-68b an impossible estimate $\hat{x}_k = -3$ appears, and is detected as an error.

An additional potential problem with the duobinary scheme described here is the use of the ideal low-pass filter as the shaping filter $H_2(\omega)$ in Fig. 3-67. This is of course unrealizable and undesirable, as noted in the previous section. But actually it is not needed. For the combination of the pulse addition prior to the shaping filter plus the shaping operation give rise to an effective transfer function or filtering operation that *is* realizable to a good approximation. Specifically, call the transfer function of the delay and summing operations in Fig. 3-67 $H_1(\omega)$. The overall transfer function, from input to output, is then

$$H(\omega) = H_1(\omega)H_2(\omega) \tag{3-106}$$

This is shown schematically in Fig. 3-69a. To find $H_1(\omega)$ it is sufficient to find the impulse response of the delay summer and take its Fourier transform. In

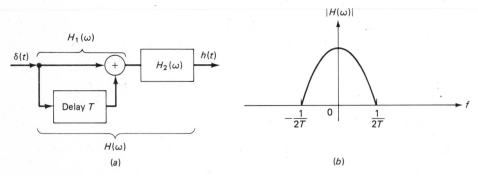

Figure 3-69 Spectrum, duobinary signaling. (a) Overall transfer function. (b) Amplitude spectrum, overall filter.

particular, assume that an impulse function $\delta(t)$ is applied at the system input. It is left to the reader to show, using the shifting theorem, that

$$H_1(\omega) = 1 + e^{-j\omega T} = 2 \cos \frac{\omega T}{2} e^{-j\omega T/2} \tag{3-107}$$

With $H_2(\omega)$ the ideal low-pass filter of amplitude T [Eq. (3-102)], we then have

$$H(\omega) = 2T \cos \frac{\omega T}{2} e^{-j\omega T/2} \quad -\frac{1}{2T} \leq f \leq \frac{1}{2T}$$

$$= 0 \qquad\qquad\qquad |f| > \frac{1}{2T} \tag{3-108}$$

The amplitude spectrum $|H(\omega)|$ is shown in Fig. 3-69b. This cosinusoidal spectral characteristic can be approximated in practice and can be used for generating a duobinary signal sequence from a binary input sequence. One can thus achieve Nyquist rates using this scheme.

Modified Duobinary Signaling

In many transmission systems (the telephone plant is an example), dc signals cannot be passed because of the presence of transformers. This was noted earlier. Alternatively, single-sideband (SSB) modulation techniques to be discussed in Chap. 4 may be used. The zero-frequency point for the baseband signals under discussion here is shifted up to the carrier frequency with SSB modulation. A small pilot carrier signal is often inserted at this frequency for use in SSB reception. It is thus desirable not to have any spectral components in the SSB signal in the vicinity of the carrier. This corresponds to having no spectral components in the vicinity of zero frequency for the baseband signals. The duobinary signal with its cosinusoidal spectrum about the origin (Fig. 3-69) is thus not appropriate for such applications. It was noted earlier that the bipolar signal was introduced in place of the NRZ signal to handle a similar problem. The modified duobinary scheme[66] overcomes the dc frequency problem as well. This scheme combines two binary pulses so as to eliminate the dc component, yet retains the desirable property of allowing signaling at the Nyquist rate.

In this scheme pulses spaced *two* time slots apart are *subtracted*. Correlation is thus introduced over the span of three binary intervals rather than the two of the duobinary scheme. Figure 3-70a shows how the modified duobinary signal is formed.

It is apparent, repeating the notation previously adopted for the duobinary technique, that the output signal $y(t)$ has sample values at time $t = kT$ given by

$$y_k = x_k - x_{k-2} \tag{3-109}$$

A three-level signal is again generated: if $x_k = \pm 1$, as assumed previously, y_k can again take on the values 2, 0, -2. If $x_k = 0$ or 1, y_k is ± 1 or 0.

[66] Lender, *op. cit.*, pp. 113, 114.

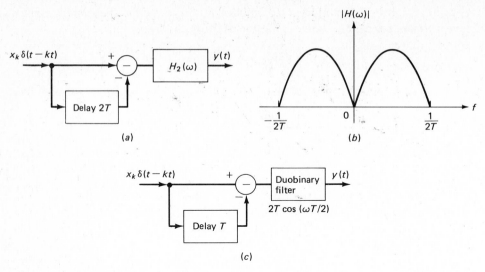

Figure 3-70 Modified duobinary technique. (*a*) Basic implementation. (*b*) Spectral characteristic. (*c*) Alternate implementation.

The transfer function of the delay-subtracter unit of Fig. 3-70*a* is given by

$$H_1(\omega) = 1 - e^{-j2\omega T}$$
$$= 2je^{-j\omega T} \sin \omega T \tag{3-110}$$

The amplitude characteristic is thus

$$\left| H_1(\omega) \right| = 2 \left| \sin \omega T \right| \tag{3-111}$$

and the overall transfer characteristic is

$$H(\omega) = 2je^{-j\omega T} \sin \omega T \, H_2(\omega) \tag{3-112}$$

In particular, if $H_2(\omega)$ is again the ideal low-pass filter with amplitude T, the overall amplitude spectrum is

$$|H(\omega)| = 2T \sin \omega T \qquad -\frac{1}{2T} \le f \le \frac{1}{2T} \tag{3-113}$$

$$= 0 \qquad |f| > \frac{1}{2T}$$

This characteristic is sketched in Fig. 3-70*b*. It has the desired null at $f = 0$ (because of the subtraction), as well as a null again at the Nyquist bandwidth $B = 1/2T$.

The $\sin \omega T$ filter appearing is not easily implementable. Instead, note that the transfer function for the delay-subtraction operation may be rewritten in different form:

$$H_1(\omega) = 1 - e^{-2j\omega T} = (1 - e^{-j\omega T})(1 + e^{-j\omega T}) \tag{3-114}$$

The second term in parentheses is precisely that obtained for the delay summer of the duobinary scheme [Eq. (3-107)]. Incorporating the ideal low-pass filter $H_2(\omega)$ into the overall transmission characteristic, it is apparent that $H(\omega)$ may be written in the following alternative form:

$$H(\omega) = (1 - e^{-j\omega T})2T \cos \frac{\omega T}{2} e^{-j\omega T/2} \qquad -\frac{1}{2T} \leq f \leq \frac{1}{2T} \qquad (3\text{-}115)$$

$$= 0 \qquad \text{elsewhere}$$

This form is implementable by first carrying out the differencing operation shown, and then following with the cosinusoidal duobinary filter. The differencing operation is implemented digitally by subtracting two successive pulses. This implementation is shown schematically in Fig. 3-70c.

To extract the original binary symbols from the modified duobinary samples y_k, recall that y_k was generated as $y_k = x_k - x_{k-2}$. Then $x_k = y_k + x_{k-2}$. Assume now that $x_k = 0$ or 1. Then $y_k = 0$ or ± 1. But if $y_k = 1$, x_k *must* be 1. Similarly, if $y_k = -1$, x_k *must* be 0. Finally, if $y_k = 0$, $x_k = x_{k-2}$: i.e., the current binary value is the same as the one estimated two time slots back. If we again let $\hat{x}_k$ represent the binary estimate at time $t = kT$, with y_k the modified duobinary sample measured at that time, we thus have as the decoding rule,

$$\hat{x}_k = 1 \qquad \text{if } y_k = 1$$

$$\hat{x}_k = 0 \qquad \text{if } y_k = -1 \qquad\qquad (3\text{-}116)$$

$$\hat{x}_k = \hat{x}_{k-2} \qquad \text{if } y_k = 0$$

An example of modified duobinary signaling, including the application of this decoding rule, appears in Fig. 3-71. In part (b) of that figure errors are assumed to occur in two time slots during the transmission of y_k. In the first there is no error propagation because y_k two slots later is -1. The decision is thus based on y_k alone. In the second case y_k two slots after the error is 0. The decision, being based on $\hat{x}_{k-2}$ at that point, is then in error also. Note, therefore, that error propagation can still occur if the y_k's received in alternate slots after an error is made happen to be 0's.

Precoding

It was noted earlier that precoding is used to eliminate the possibility of error propagation. The precoding operations for the duobinary and modified duobinary schemes are described separately.

1. *Duobinary precoding.* Consider the amplitude values x_k only. Let these now be 1 or 0 for simplicity. Form the modulo-2 (exclusive-or) sum,

$$a_k = x_k \oplus a_{k-1} \qquad (3\text{-}117)$$

This operation is shown schematically in Fig. 3-72. (Recall that the modulo-2 or exclusive-or logical operation was introduced earlier in Sec. 3-4, in discussing

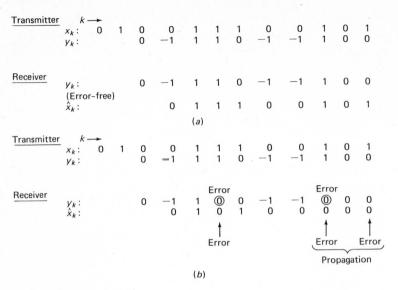

Transmitter	$k \longrightarrow$											
$x_k:$	0	1	0	0	1	1	1	0	0	1	0	1
$y_k:$		0	−1	1	1	0	−1	−1	1	0	0

Receiver										
$y_k:$	0	−1	1	1	0	−1	−1	1	0	0
(Error-free)										
$\hat{x}_k:$	0	1	1	1	0	0	1	0	1

(a)

Transmitter	$k \longrightarrow$											
$x_k:$	0	1	0	0	1	1	1	0	0	1	0	1
$y_k:$		0	=1	1	1	0	. −1	−1	1	0	0

Receiver					Error				Error		
$y_k:$	0	−1	1	⓪	0	−1	−1	⓪	0	0
$\hat{x}_k:$	0	1	0	1	0	0	0	0

Error ↑   Error ↑   Error ↑

Propagation

(b)

Figure 3-71 Modified duobinary signal patterns. (a) Error-free transmission. (b) Error during transmission.

the Gray code.) The a_k symbols are now fed into the duobinary signal generator, including the signal shaper, as shown in Fig. 3-72. The sampled output of the duobinary output stream is thus given by

$$y_k = a_k + a_{k-1}$$
$$= (x_k \oplus a_{k-1}) + a_{k-1} \tag{3-118}$$

(Note that the duobinary operation still calls for ordinary addition, in contrast to the precoded mod-2 logical addition.)

Since $a_k = 0$ or 1, $y_k = 0$, 1, or 2. A little thought will indicate, using (3-118), that $y_k = 0$ or 2 can only come from $x_k = 0$. Similarly, $y_k = 1$ must correspond to $x_k = 1$. Hence the decoding rule is very simple: the estimate of x_k must just be y_k, interpreted modulo-2. At the receiver, then, the reconstruction of the original binary information is given by

$$\hat{x}_k = y_k \quad \text{mod-2} \tag{3-119}$$

Since each binary decision depends only on the *current* received sample y_k, there can be no error propagation.

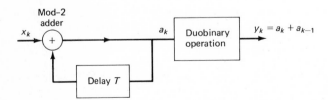

Figure 3-72 Precoding for duobinary signaling.

Transmitter $\quad k$

x_k:		0	0	1	1	0	1	0	0
a_k:	1	1	1	0	1	1	0	0	0

↑
Assumed

Receiver

y_k:	2	2	1	1	2	1	0	0
$\hat{x}_k = y_{k,\,\text{mod-2}}$:	0	0	1	1	0	1	0	0

(a)

Transmitter $\quad k$

x_k:		0	0	1	1	0	1	0
a_k:	1	1	1	0	1	1	0	0
y_k:	2	2	1	1	2	1	0	

Receiver

Error

y_k:	2	①	1	1	2	1	0

Error

$\hat{x}_k = y_{k,\,\text{mod-2}}$:	0	①	1	1	0	1	0

No propagation

(b)

Figure 3-73 Example signals, precoded duobinary. (*a*) Error-free transmission. (*b*) Error during transmission.

An example of the operation of the rule, in both the error-free and error-occurring cases, appears in Fig. 3-73.

2. *Modified duobinary precoding.* A similar precoding procedure is used to eliminate the possibility of error propagation in the modified duobinary case. A mod-2 logical addition is again used here, prior to the modified duobinary signal generation, but on signals two binary intervals ($2T$ seconds) apart. Thus the operation

$$a_k = x_k \oplus a_{k-2} \qquad (3\text{-}120)$$

is first carried out. (Note that mod-2 addition and subtraction are the same.) The successive a_k samples are now fed into the modified duobinary generator. The precoding operation in this case is shown schematically in Fig. 3-74. The output signal samples are now given by

$$
\begin{aligned}
y_k &= a_k - a_{k-2} \\
&= (x_k \oplus a_{k-2}) - a_{k-2}
\end{aligned}
\qquad (3\text{-}121)
$$

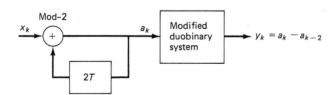

Figure 3-74 Precoding for modified duobinary signals.

It is apparent that $y_k = 0, +1,$ or -1 in this case. It is also apparent that x_k can be extracted from y_k by disregarding the polarity of y_k. Since plus and minus operations are the same in mod-2 logic, the modulo-2 value of y_k also provides the desired value of x_k. Thus at the receiver, one extracts x_k by writing

$$\hat{x}_k = |y_k| = y_k \text{ mod-2} \tag{3-122}$$

Again error propagation is eliminated since each binary decision depends only on the current received value. An example of precoded signaling for modified duobinary transmission appears in Fig. 3-75.

Error Detection

It has been noted earlier that the introduction of correlation into the transmitted bit patterns by the duobinary techniques enables certain error patterns to be detected. In the case of duobinary signaling the following rule must be observed: Let the top and bottom levels of the transmitted samples be called the extreme levels. (These are the two values, $y_k = 0$ or 2 shown occurring in Fig. 3-73.) Then the values of two successive bits at the extreme levels must be different if the number of intervening bits at the center level is odd. They must have the same value if the number of intervening bits at the center level is even. (This includes the case of no intervening bits at the center level.) This rule can be checked from the successive samples of y_k of Fig. 3-73a. Note that two successive values of 2 or 0 (the extreme levels) occur with either two or no values of 1 (the center level) in this short sample. The sample values 2 and 0 following one another have one 1 between them. An error is detected in the case of Fig. 3-73b, since three 1's occur between the two 2's in the y_k sequence of that figure. The *particular* digit in error cannot be determined, so error correction is not possible without added coding.

The modified duobinary rule for error detection is expressed as follows. Divide the pulse train (the sequence of y_k samples) into two trains of odd-numbered and even-numbered samples. In either train, then, successive samples at the extreme levels must always alternate in value. (If $y_k = -1, 0, 1,$ the extreme values are ± 1.) Thus, if the current extreme level in one of the two alternating trains is a 1, the previous extreme value must have been a -1. The next extreme value must be a -1 as well. This rule can be checked for the y_k

Transmitter

			x_k	1	1	1	1	0	1	0	0	1	0	0	0	1
a_k	1	0	0	1	1	0	1	1	1	1	0	1	0	1	1
y_k			-1	1	1	-1	0	1	0	0	-1	0	0	0	1

Receiver

$\hat{x}_k = |y_k|$	1	1	1	1	0	1	0	0	1	0	0	0	1
$= y_{k, \text{mod-2}}$													

Figure 3-75 Precoded modified duobinary signaling.

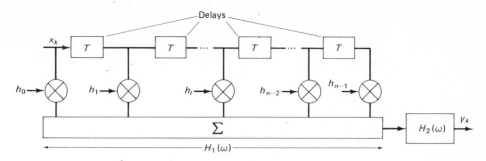

Figure 3-76 Generalized correlative or partial response signaling.

signal sequence of Fig. 3-75. Note that the odd-numbered train has the values $-1\ 1\ 0\ 0\ -1\ 0\ 1$. Extreme levels alternate. The even-numbered train has the values $1\ -1\ 1\ 0\ 0\ 0$. Again -1 and 1 alternate. This rule can also be used for error detection.

Correlative or Partial-Response Coding

As noted earlier, the duobinary schemes may be generalized to introduce *controlled* intersymbol interference over a span of several digits. Such schemes are variously called correlative coding, or partial-response signaling, schemes.[67] The binary impulses (narrow pulses), at T-second intervals, are fed into the digital filter of Fig. 3-76, prior to shaping with $H_2(\omega)$, as shown. The digital filter, called a *transversal* or *nonrecursive* digital filter,[68] combines delayed and weighted samples of the input. The h_l coefficients shown are integers and are chosen to produce the transfer function desired. It is apparent that the transfer function for this filter is given by

$$H_1(\omega) = \sum_{l=0}^{n-1} h_l e^{-jl\omega T} \tag{3-123}$$

As an example, if $n = 2$ and $h_0 = h_1 = 1$, the duobinary scheme is obtained. If $h_0 = 1$, $h_2 = -1$, and all the other coefficients are zero, the modified duobinary scheme is obtained. By appropriate adjustment of the h_l coefficients, in general, one may shape the transfer characteristic between 0 and B in any way desired. [If $H_2(\omega)$ is again chosen to have the ideal low-pass characteristic with bandwidth B, $H(\omega) = H_1(\omega)H_2(\omega)$ only extends out to B hertz.]

One of the advantages of correlative coding over straight M-ary transmission is that fewer levels are needed for the same rate of transmission. To show this, assume that duobinary or modified duobinary signaling is used, but now let

[67] Lender, *op. cit.*; Kretzmer, *op. cit.*

[68] M. Schwartz and L. Shaw, *Signal Processing: Discrete Spectral Analysis, Detection, and Estimation*, McGraw-Hill, New York, 1975, pp. 62–70.

the input sequence $\{x_k\}$ be multilevel. In particular, say that $2^{n/2}$ levels are used. It is apparent that by adding or subtracting successive input samples, as is the case with duobinary and modified duobinary, respectively, an output signal sequence $\{y_k\}$ having $2(2^{n/2}) - 1 = 2^{(n/2)+1} - 1$ possible levels is generated. But 2 symbols/Hz may be transmitted using the duobinary technique with Nyquist shaping such that the symbol spacing $T = 1/2B$ seconds. Since each symbol carries $n/2$ bits (one of $2^{n/2}$ levels is input every T seconds), the system trans-mits at the rate of n bits/Hz. As an example, if 2 bits/Hz are desired, only a binary input is required. This is the case already discussed. The output se-quence consists of 3-level symbols. If $n = 4$ bits/Hz is desired, 4 input levels must be used, and the output sequence consists of 7-level symbols. Note that M-ary transmission with 100 percent roll-off raised cosine transmission would require $M = 2^n = 16$-level transmission. For $n = 6$ bits/Hz, 8 input levels are used, and the output symbols have 15 levels. M-ary transmission would require $2^6 = 64$ levels for raised cosine shaping with 100 percent roll-off.

3-11 SUMMARY

In this chapter, the first on communication systems per se in this book, we focused on digital systems. These are assuming an ever-more-important role in all areas of communication technology as the use of digital technology and computer processing of data proliferate at a rapidly increasing rate.

We distinguished between two categories of digital signals in this chapter—those that are innately digital in their format (e.g., data terminal or computer outputs), and those that begin as analog signals. The latter class, whether voice signals, telemetry data, facsimile, TV, etc., must first be con-verted to digital format using sampling and A/D conversion techniques. Once converted to digital form, they are indistinguishable from originally generated digital data, and both classes of signals are intermingled as they wend their way through a communications network.

In discussing the sampling of analog waveshapes, we made use of the Fourier analysis of Chap. 2 to show that for signals band-limited to B hertz, at least $2B$ samples per second were necessary to retain all the information in the original (analog) signal. The sampling process is fundamental not only in prepar-ing signals for digital transmission, but plays a key role as well in data analysis and processing by computer.

For the transmission of *digital data,* i.e., discrete numbers rather than continuous (analog) waveshapes, the signal samples must further be quantized into a specified number of amplitude levels. This results in quantization noise—the difference between the actual sample amplitude and its quantized approximation. The use of an appropriate number of levels makes this effect tolerable. The resultant stream of signal digits may then be coded into any number system desired. The binary system is the one most commonly used, since the resultant stream of binary pulses is least susceptible to noise and intersymbol interference. Where bandwidth limitations pose a problem, however, multilevel signaling may be used to pack higher data rates over a given bandwidth channel. This concept is explored in more detail in Chap. 4 in discussing multisymbol modulation techniques for digital communications.

For a fixed number of quantization levels, it is found appropriate to use nonuniform level spacing to accommodate a wider dynamic range of analog signals. This is carried out in practice by compressing the input signals prior to the quantization procedure. The reverse process of expansion is carried out at the receiver. The composite process is referred to as companding. In a brief analysis focusing on a particular nonlinear companding law used in practice, we demonstrated the improvement obtained using the companding technique.

Although most analog signals are transmitted using A/D conversion and time multiplexing in the overall process called pulse code modulation (PCM), delta modulation has been adopted as an alternative procedure in a variety of existing systems. In the delta-modulation approach, the difference between the actual signal sample and its predicted estimate is quantized into two levels and the appropriate level transmitted. Delta-modulation systems generally require a higher sampling rate or some adaptive form to provide the same or better performance than the equivalent PCM system.

Time multiplexing of digital signals is commonly used to combine a multiplicity of signal sources over a common, shared channel. These signals can be either digital representations of analog signals or signals originally created in digital format. The time-multiplexing process serves as one important example of the efficient processing of signals that may be carried out if they are in digital form. Two types of multiplexing may be broadly distinguished, although boundaries are tending to disappear. One type combines digital streams bit by bit; the other combines blocks of bits, grouped into fixed-length characters. A common procedure is to periodically sample each of the input digital streams to be multiplexed, each at a rate commensurate with its own bit rate and that of the much higher output (multiplexed) rate. Known synchronization sequences must also be introduced periodically to enable demultiplexing to take place at a receiving destination.

Since the output bit stream, whether from a single digital source or a time-multiplexed output, must ultimately be transmitted over a channel to its destination point, the output pulses must be shaped appropriately to reduce intersymbol interference introduced due to signal distortion arising during transmission. A class of shaping referred to generically as Nyquist shaping was shown to eliminate intersymbol interference completely under ideal conditions. A special case, referred to as sinusoidal roll-off shaping, is used a great deal in practice, and was discussed in detail in this chapter. Reference will be made to this form of shaping in Chap. 4 and subsequent chapters. As a general conclusion, it is found that a channel with B hertz bandwidth will allow from somewhat more than B to $2B$ symbols per second to be transmitted over the channel. Alternately, if $1/T$ symbols (pulses) per second are to be transmitted, the bandwidth required ranges from an unattainable minimum of $B = 1/2T$ hertz to

$B = 1/T$ hertz, depending on the shaping used. We use the word *symbol* here since multilevel signaling will allow correspondingly higher bit rates to be transmitted. This point is explored in more detail in Chap. 4.

We concluded this chapter on digital communication systems by discussing some important techniques for transmitting at higher bit rates over a given channel, using precoding or signal shaping introduced for this purpose at the transmitter. These techniques are generically referred to as correlative coding or partial-response signaling. The special case of duobinary signaling enables two bit streams to be combined for transmission over a channel normally capable of handling the bit rate of only one of them. These techniques allow a variety of signal shapes to be introduced at the transmitter to accommodate special channel needs. For example, dc signal energy can be eliminated, as can energy in the vicinity of a specified frequency that can then be used for other required purposes.

The treatment in this chapter was devoted entirely to *baseband* signals, those that transmit the basic signals with no frequency translation carried out. In Chap. 4 we point out the common need for transmission of these digital signals at high frequencies, and discuss various ways of performing the requisite frequency translation up at the transmitter (*modulation*) and down at the receiver (*demodulation*). Again Fourier analysis and bandwidth concepts will be found to play a key role.

In Chap. 5 we return to baseband digital systems, calculating their performance in the presence of noise. We then discuss error-correction and error-detection coding in Chap. 6. These are methods for introducing additional noninformation bits, in a controlled way, at the transmitter, to detect and correct errors occurring due to noise during transmission.

PROBLEMS

3-1 A sinusoidal input at 1 Hz, $\sin 2\pi t$, is to be sampled periodically.

(*a*) Find the maximum allowable interval between samples.

(*b*) Samples are taken at intervals of $\frac{1}{3}$ s. Perform the sampling operation graphically, and show to your satisfaction that no other sine wave (or any other time function) of bandwidth less than 1.5 Hz can be represented by these samples.

(*c*) The samples are spaced $\frac{2}{3}$ s apart. Show graphically that these may represent another sine wave of frequency less than 1.5 Hz.

3-2 A function $f_1(t)$ is band-limited to 2,000 Hz, another function $f_2(t)$ to 4,000 Hz. Determine the maximum sampling interval if these two signals are to be time-multiplexed, using a single sampling rate.

3-3 Consider the system shown in Fig. P3-3.

(*a*) $P(t)$ is the periodic square pulse train shown in part (*a*) of the figure, with period $T = \pi/W$. Determine the parameters K and β of the ideal filter so that $g(t) = f(t)$.

(*b*) Because of a failure in the pulse generator, the pulses of $P(t)$, although still periodic, come

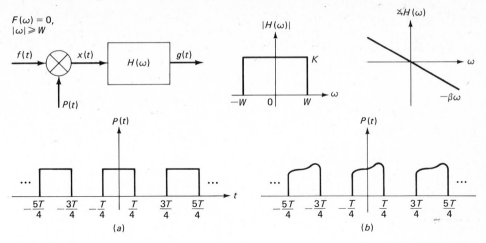

$F(\omega) = 0,$
$|\omega| \geq W$

Figure P3-3

out nonflat and arbitrarily shaped, as shown in part (b). Can the shape of $f(t)$ still be recovered, to within a constant multiplier, at the output of the system? For simplicity, take $\beta = 0$.

3-4 Twenty-four signal channels plus one synchronization (marker) channel, each band-limited to 3,300 Hz, are sampled and time-multiplexed at an 8-kHz rate. Calculate the minimum bandwidth needed to transmit this multiplexed signal in a PAM system.

3-5 Five signal channels are sampled at the same rate and time-multiplexed. The multiplexed signal is then passed through a low-pass filter. Three of the channels handle signals covering the frequency range 300 to 3,300 Hz. The other two carry signals covering the range 50 Hz to 10 kHz. Included is a synchronization signal.
 (a) What is the minimum sampling rate?
 (b) For this sampling rate what is the minimum bandwidth of the low-pass filter?

3-6 Temperature measurements covering the range -40 to $+40°C$, with $\frac{1}{2}°C$ accuracy, are taken at 1-s intervals. They are converted to binary format for PCM transmission. What is the bit rate required?

3-7 The readings from 100 pressure sensors, covering the range 10 to 50 psi, with 0.25-psi resolution, are taken every 10 s, and are then transmitted using time-multiplexed PCM techniques. What is the bit rate of the PCM output? Approximately what bandwidth is required to transmit the PCM signal?

3-8 Consider the PCM system shown in Fig. P3-8 (p. 204) which time-multiplexers 10 signal channels.
 (a) What is the minimum sampling rate?
 (b) 30,000 samples/s are taken.
  (1) What is the bit rate (bits/s) at the PCM output?
  (2) What are the *minimum* bandwidths required at points 1, 2, 3?

3-9 Ten 10-kHz signal channels are sampled and multiplexed at a rate of 25,000 samples/s per channel. Each sample is then coded into six binary digits.
 (a) Find the PCM output rate, in bits/s. Estimate the bandwidth needed to transmit the PCM stream.
 (b) Using the same number of quantization levels as above, each sample is now transmitted as a sequence of four-level pulses. What is the output rate, in bits/s? Estimate the transmission bandwidth required.
 (c) Repeat (b) for the case in which each quantized sample is transmitted as a multilevel pulse without any further coding.

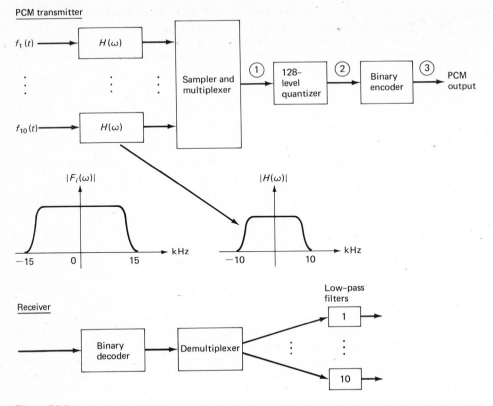

PCM transmitter

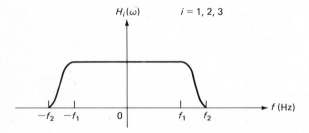

Figure P3-8

3-10 Three signal channels are sampled at a rate of 100 samples/s and then time-multiplexed. They have the following bandwidths:

Channel 1:   0–10 Hz.
Channel 2:   0–10 Hz.
Channel 3:   0–20 Hz.

(*a*) Sketch the multiplexed waveform if the signals in each of the channels is given, respectively, by $s_1(t) = 10 \cos 20\pi t$, $s_2(t) = 0$, $s_3(t) = 5 \sin 40\pi t$.

(*b*) Now assume that arbitrary signals of the appropriate bandwidth are present in each channel. After demultiplexing (separation) at the receiver, each signal is passed through a low-pass filter with transfer function $H_i(\omega)$ with characteristics as shown in Fig. P3-10 (f_1 and f_2 may differ

Figure P3-10

from channel to channel). Determine the minimum permissible value of f_1 and maximum permissible value of f_2 for perfect signal recovery in each channel.

(c) Show, by means of a block diagram and waveforms, how the three channels could be time-multiplexed without loss of information into a single PAM signal with a composite pulse rate of 80 pulses/s.

3-11 A single information channel carries voice frequencies in the range of 50 to 3,300 Hz. The channel is sampled at an 8-kHz rate, and the resulting pulses are transmitted over either a PAM system or a PCM system.

(a) Calculate the minimum bandwidth of the PAM system.

(b) In the PCM system the sampled pulses are quantized into eight levels and transmitted as binary digits. Find the transmission bandwidth of the PCM system, and compare with that of (a).

(c) Repeat (b) if 128 quantizing levels are used. Compare the rms quantization noise in the two cases if the peak-to-peak voltage swing at the quantizer is 2 V.

3-12 Repeat Prob. 3-11 for the case of 12 voice channels each carrying frequencies of 50 to 3,300 Hz, which are sampled and time-multiplexed. Include block diagrams of both the PAM and PCM systems.

3-13 The sinusoidal voltage $10 \sin 6,280t$ is sampled at $t = 0.33$ ms and thereafter periodically at a 3-kHz rate. The samples are then quantized into eight voltage levels and coded into binary digits. Draw the original voltage and below it to scale the outputs of the sampler, the quantizer, and the encoder. Calculate the rms quantization noise.

3-14 A signal voltage in the frequency range 100 to 4,000 Hz is limited to a peak-to-peak swing of 3 V. It is sampled at a uniform rate of 8 kHz, and the samples are quantized to 64 evenly spaced levels. Calculate and compare the bandwidths and ratios of peak signal to rms quantization noise if the quantized samples are transmitted either as binary digits or as four-level pulses.

3-15 Derive Eq. (3-47) for the mean-squared quantization noise using the μ-law compander.

3-16 Calculate the signal-to-quantization noise ratio for a 7-bit code and a $\mu = 255$ companding law for several values of input signal power, checking the appropriate entries in Table 3-3. Consider both laplacian and gaussian statistics.

3-17 Repeat Prob. 3-16 for a $\mu = 100$ companding law. Plot the SNR calculated vs. σ_x^2 for both values of μ and compare. How are the results affected if a 6-bit code is used? An 8-bit code?

3-18 A PCM system uses 7-bit quantization and a $\mu = 255$ companding law at its output. The maximum voltage at the input to the compressor is limited to 2 V. Compare the signal-to-quantization noise ratio for two signals, both modeled by laplacian statistics, one with a standard deviation of 0.4 V, the other with a standard deviation of 0.02 V.

3-19 *Carefully* draw a slowly varying analog function. This function is to be sampled 10 times, with samples spaced somewhat closer together than the usual Nyquist interval. Assume that the samples are fed to a two-level quantizer as part of a delta-modulation system. Start the system off by comparing the first sample with 0. Zero is also used as the first initial estimate. Previous sample estimation is to be used: $g_j = 0.9\,\hat{x}_{j-1}$, with $\hat{x}_{j-1}$ the estimate at sample $(j - 1)$.

(a) Show the sequence of delta-modulator outputs, $\pm k'$.

(b) Start off with a relatively large value of k'; track the original function with its estimates. Reduce k' until overload noise occurs.

3-20 Consider a sine-wave test signal $x(t) = A \sin \omega_m t$. Sketch the sine wave over the interval $-\pi/2 < \omega_m t < \pi/2$. The sine wave is to be sampled many times at a rate $f_s \gg f_m$.

(a) Demonstrate that if $k'f_s/2\pi Af_m < 1$, overload noise will result. (k' is the delta-modulator step size.)

(b) Let $\alpha \equiv k'f_s/2\pi Af_m = 0.707$. Show that the overload-noise region is entered at the point $\omega_m t = -\pi/4 \equiv -\theta_1$. Assuming that the quantization noise is negligible so that the delta-modulator receiver output is very nearly a straight line, draw the receiver output on top of the sine-wave curve and find the time $\omega_m t \equiv \theta_2$ at which the overload region ends.

(c) Verify Eq. (3-64) for the overload noise.

3-21 Refer to Prob. 3-19. Let the signal at the input to the delta-modulation system be $A \cos \omega_m t$. Sketch both the binary output sequence $\pm k'$ and the quantized estimate to the original signal, over several cycles of the cosine wave, for the following cases:

(a) $f_s = 10 f_m, A = 4k'$

(b) $f_s = 30 f_m, A = 4k'$

(c) $f_s = 30 f_m, A = 10k'$

Comment on the effect of each of these cases on quantization noise and slope overload.

3-22 Consider the M12 multiplexer described in the text. As pointed out there, it uses three 1's to denote a stuffed bit in one of the four T1 channels being multiplexed. Find the average time between losses of a frame in one of the channels due to a stuffing bit error if the probability of a bit error is 10^{-5} and bit errors are assumed to be independent. Repeat for a bit-error probability of 10^{-3}. What would the average times between stuffing errors for the same two error probabilities be if one bit only were used to denote stuffing? *Note:* Assume that a stuffing bit is needed only every three frames, on the average, for each T1 channel.

3-23 (a) The Bell System T1 short-haul PCM system multiplexes twenty-four 64-kbits/s channels. One framing bit is added at the beginning of each T1 frame for synchronization purposes. Show that the nominal bit rate of the T1 system is 1.544 Mbits/s.

(b) Four T1 channels are combined to form a T2 stream at 6.312 Mbits/s, as described in the text. Show these numbers provide a margin in the bit rate of each T1 channel of about 1.78 kbits/s. Show that the average bit stuffing rate per channel is 1 bit in three T2 frames.

3-24 A character-interleaved multiplexer accepts a variety of 10, 15, and 30 character/sec asynchronous terminals at its input and outputs 8-bit characters for each terminal over a 4,800-bits/s line. One 8-bit synchronous character must be introduced in every 20 characters transmitted. In addition, another 3 percent of the allowable output rate is to be used to accommodate possible changes in the terminal transmission speeds.

Indicate at least two combinations of such terminals that can be accommodated by this multiplexer, sketching a possible framing pattern for each. Assuming that several bits in the sync character can be set aside as a separate control field, to what uses involving the multiplexing operation could they be put? Explain.

3-25 A character-interleaved multiplexer is used to time-multiplex a variety of 110-, 150-, and 300-bits/s asynchronous terminals over a 2,400-bits/s line. The output characters are 8 bits in length. Three percent of the output rate is to be set aside for synchronization and possible variations in input terminal rates.

(a) 110-bits/s terminals only are multiplexed. How many may be combined by this multiplexer/line speed combination? Draw a timing diagram indicating which terminals will be allocated to which outgoing characters. Sketch an entire frame. (This is the interval between repeats of the full pattern.) A synchronization character is to be inserted every 100 characters. Special control characters may be transmitted to fill gaps in transmission or to accommodate possible variations in input terminal speed.

(b) Repeat (a) if a synchronous 1,200-bits/s terminal is to be multiplexed with the 110-bits/s terminals.

(c) Choose a particular combination of 110-, 150-, and 300-bits/s terminals that can be accommodated by the multiplexer with the 2,400-bits/s outbound line speed. Repeat (a). How do your results differ if the synchronization character is inserted once every 120 time slots?

3-26 Refer to the Codex 920 TDM described in the text. This multiplexer is to be used over a 9,600-bits/s trunk. It is to handle a 2,400-bits/s synchronous input channel, twelve 110-bits/s terminals, twelve 150-bits/s terminals, and a number of 300-bits/s terminals.

Find the number of 300-bits/s terminals that may be accommodated by this multiplexer, and then set up a scanning pattern that will handle all the inputs to be multiplexed together. Sketch the resultant output time slot sequence, each slot consisting of 9 bits.

3-27 Put together a number of synchronous channel and asynchronous terminal combinations that can be handled by the Codex 920 TDM, over 2,400-, 480-, and 7,200-bits/s trunk lines. For each combination design an appropriate scanning pattern and sketch the output time slot sequence.

3-28 A high-speed synchronous data channel transmits at a rate of 20 Mbits/s. A framing pattern n bits long is introduced into the data stream every L bits.

  (a) Find the average time, in seconds, between frame losses if the probability of a random bit error is $p = 10^{-5}, n = 8, L = 1,000$ bits, and three frames in a row must be declared in error before a frame loss is declared present.

  (b) How many frames are required to detect a frame loss and then reacquire frame if the framing pattern, once detected, must appear at the same position two times in a row? How long a time interval is this?

  (c) Repeat (a) and (b) if four frames in a row must be declared in error before a frame loss is declared as such.

3-29 A computer outputs binary symbols at a 56-kbits/s rate. Find the baseband bandwidths required for transmission for each of the following roll-off factors if sinusoidal roll-off spectral shaping is used: $r = f_x/f_c = 0.25, 0.5, 0.75, 1$.

3-30 Repeat Prob. 3-29 if three successive binary digits are coded into one of eight possible amplitude levels. These eight-level pulses, shaped using sinusoidal roll-off shaping, are then transmitted.

3-31 A baseband bandwidth of 3.6 kHz is available.

  (a) Find the bit rates possible if binary pulses with sinusoidal roll-off shaping are to be transmitted. Consider the cases $r = f_x/f_c = 0.25, 0.5,$ and 1.

  (b) Repeat if four successive binary digits are combined into one pulse of 16 possible values.

3-32 Find the baseband bandwidth needed to transmit the multiplexer output stream of Prob. 3-24 if binary pulses with raised cosine shaping are used. Repeat if 25 percent roll-off shaping is used instead.

3-33 Consider the sinusoidal roll-off spectrum of Fig. 3-64 in the text. Assume linear phase shift, $\theta(\omega) = -\omega t_0$.

  (a) Show that the impulse response is given by Eq. (3-98).

  (b) Sketch $|H|(\omega)$ and $h(t)$ for the roll-off factor $\omega_r/\omega_c = 0.5, 0.75,$ and 1.0, and compare. Pay specific attention to the bandwidth and pulse tails in each case. Sketch $h(t)$ for several symbol intervals.

  (c) The raised cosine spectrum is given by $\omega_r/\omega_c = 1.0$. Show that the impulse response in this case has an additional zero halfway between the usual zero crossings. Discuss the possibility in this case of detecting successive pulses by "slicing" the received signal at the half-amplitude levels.

3-34 An analog source is sampled, quantized, and encoded into binary PCM. Four levels of quantization are used. The PCM pulses are transmitted over a bandwidth of 4,800 Hz using Nyquist shaping with 50 percent roll-off.

  (a) Find the maximum possible PCM pulse rate.

  (b) Find the maximum permissible source bandwidth.

3-35 A source is sampled, quantized, and encoded into PCM. Each sample is encoded into a "word" consisting of three information (data) pulses plus a synchronizing pulse. The information pulses can take on four possible levels. Transmission is accomplished over a channel of bandwidth 6,000 Hz using Nyquist pulses with 50 percent roll-off. Find:

  (a) The maximum possible PCM pulse rate.

  (b) The corresponding information rate of the PCM signal.

  (c) The maximum permissible source bandwidth. It is desired to trade quantization levels for source bandwidth. If the number of quantization levels is reduced by a factor of 4, how large could the new source bandwidth be?

3-36 Refer to Prob. 3-9. Find the bandwidths required in all three parts of the problem if Nyquist shaping with 50 percent roll-off is used. How would these numbers change if 25% roll-off is used?

3-37 Verify Eqs. (3-108) and (3-112) for the transfer functions of duobinary and modified duobinary signaling, respectively. Show that $H(\omega)$ for the modified duobinary case can also be written as Eq. (3-115). Show that Fig. 3-70c does in fact serve as an alternative implementation.

3-38 Pick any arbitrary sequence of binary digits. Find the corresponding set of duobinary and modified duobinary signal samples y_k. Assume in each case that an error takes place in the reception of one of these samples. Show the effect of this error on the decoding of the signals. Are any subsequent errors detectable as such?

3-39 Using the same sequence of binary input digits chosen in Prob. 3-38, let precoding be carried out first, in the two cases of duobinary and modified duobinary transmission [see Eqs. (3-117) and (3-120), respectively]. Calculate the sequence of y_k's transmitted in the two cases. Reconstruct the original sequence of x_k's from the sequence of y_k's, using (3-119) and (3-122), respectively. Again let an error occur in the reception of one of the transmitted y_k samples. Show the effect on the reconstructed binary signal sequence.

3-40 Check that the duobinary and modified duobinary output signals in Prob. 3-39 obey the error-detection rules described in the paragraph on error detection, Sec. 3-10.

MODULATION TECHNIQUES

4-1 INTRODUCTION

In Chap. 3 we focused on digital communication systems because of their great technological significance: digital data transmission is increasing at an extremely rapid rate and analog signals are more and more frequently being converted to digital format before transmission. The stress in Chap. 3 was on the basic signals themselves, however. We described methods by which analog signals are converted to digital format (using, for example, normal A/D conversion or some form of delta modulation). We also discussed ways in which digital signals are combined, using time-multiplexing techniques or some form of partial-response signaling, to allow more efficient use of a transmission channel.

In all cases, however, the digital signals as well as the analog signals discussed were those originally generated. These are referred to generically as *baseband* signals. Although we discussed both the generation and reception of the baseband signals in Chap. 3 we had little to say about the actual medium over which the signals are ultimately transmitted. This channel separating the transmitter and receiver may in some cases be the air, in others, a set of wires, a hollow conducting tube (a waveguide), or a set of optical fibers. Efficiency of transmission requires that the information-bearing signals be processed in some manner before being transmitted over an intervening medium.

Most commonly, the baseband signals must be shifted to higher frequencies for efficient transmission. This is done by varying the amplitude, phase, or frequency (or combinations of these) of a high-frequency sine-wave carrier, in accordance with the information to be transmitted. This process of altering the characteristics of a sine-wave carrier is referred to as *sine-wave* or *continuous-wave* (c-w) modulation. The baseband signals constitute the modulating signal and the resultant signal is a high-frequency modulated carrier. The use of higher frequencies provides more efficient radiation of electric energy and makes available wider bandwidths for increased information transfer than is possible at the lower frequencies. It is a well-known phenomenon of electromagnetic theory that an efficient radiator of electric energy (the antenna) must be at least the order of magnitude of a wavelength in size. Since the wavelength of a 1-kHz tone is 300 km, radio transmission is hardly practicable at audio frequencies. But a 1-MHz carrier wave could be (and is) transmitted efficiently with an antenna 300 ft high. Similarly, the outputs of low-speed data sets must be converted to c-w modulated signals for transmission over telephone channels, and 1.5-Mbits/s PCM signals must be converted to microwave c-w modulation for proper launching over a microwave transmission medium.

In this chapter we thus focus on sine-wave modulation techniques.

Although the term modulation in this chapter refers specifically to sine-wave or c-w modulation, it has been used more generally to refer to the procedure of processing a signal for more efficient transmission. We have already seen some examples of this use of the term in Chap. 3. There we referred to pulse amplitude modulation (PAM) and to pulse code modulation (PCM). In the first case an analog signal is sampled periodically to produce a series of pulses whose height varies in accordance with the analog signal value at the time of sampling. One may visualize this transmission scheme as one in which the height of a periodic set of pulses (the carrier) is altered in a definite pattern corresponding to the information to be transmitted. This form of modulation enables one to time-multiplex many information channels for sequential transmission over a single channel. In the case of pulse code modulation the PAM pulses are further processed by using A/D conversion (quantization) and then encoding into binary format. This serves the purpose of coding the signals to better overcome noise and distortion ultimately encountered on the channel (medium), as well as to enable digital processing to be utilized throughout the transmission path.

We have indicated above that pulse-modulation systems generally require some form of c-w modulation to finally transmit the signals over the desired channel; this is also true of analog signals that are to be transmitted, without conversion to pulse or digital forms, to remote points. Examples, of course, include the usual radio and TV broadcasts, as well as many telemetered signals where it is more expedient to transmit the data directly, without digitizing them. Although the modulation techniques used for digital (pulse-coded) signals and analog signals are conceptually the same, we shall distinguish between the two cases in this chapter. There are several reasons for this:

1. The increasing importance of the transmission of digital data, and the consequent development of a sizable industry devoted to the production of specialized digital modulation equipment dictates emphasis on this area.
2. The resultant carrier systems are sufficiently distinct to warrant separate categorization in the literature. This is particularly true of multilevel or multisymbol digital carrier systems.

3. The study of digital carrier systems, particularly of the binary type, is often simpler than their analog counterparts. By focusing first on these we gain insight into the sine-wave modulation process that is then quite useful in understanding the operation of analog carrier systems. This is particularly true of the spectral properties of such systems. These can be obtained rather simply for digitally modulated sine-wave signals, particularly in the case of binary pulse modulation. The results can then readily be extended to more complex types of signals, involving analog modulating, or combinations of analog and digital modulating, signals.

We accordingly begin the chapter by focusing on binary communications, discussing the three basic ways of modulating sine-wave carriers, amplitude, phase, and frequency, using baseband binary signals as the modulating source. We then generalize this to include multisymbol modulation, with combinations of multiple phase and multiple amplitude modulation as the prime practical example. We then study various types of analog c-w modulation, with emphasis primarily on amplitude modulation (including both normal and single-sideband modulation) and frequency modulation.

Not only is c-w modulation necessary to launch signals effectively over a desired medium, but it also leads to the possibility of *frequency multiplexing,* or staggering of frequencies over the specified band. This is directly analogous to the *time-multiplexing* technique introduced in Chap. 3, in which many signal channels are sequentially sampled and transmitted serially in time. In the frequency-multiplexed case signal channels are transmitted over adjacent, nonoverlapping frequency bands. They are thus transmitted in parallel, simultaneously in time. This means that many telephone conversations can be transmitted over a single pair of wires (depending of course on the bandwidth allowed for each conversation and the total bandwidth of the system, including the wires). Very often a group of digital channels, each incorporating many time-multiplexed signals, may further be combined by frequency-multiplexing techniques.

Operation at higher carrier frequencies makes more bandwidth available and hence leads to the possibility of frequency multiplexing more signals or transmitting wider-band signals than is possible at lower frequencies. As an example, the amplitude-modulation (AM) broadcast band in this country is fixed at 550 to 1,600 kHz. This provides for 100 channels, spaced 10 kHz apart. This entire band, however, is just a fraction of the band, 6 MHz, needed for one TV channel. Operating TV broadcasts at the 60-MHz range and up (the VHF band) makes several channels available. Increasing the carrier frequency to 470 MHz and up (the UHF band) makes many more TV channels available. This is one reason for the current interest in optical frequencies for communication purposes.

In this chapter we shall concentrate, as in the previous chapter, on the spectral (bandwidth) characteristics of the various types of c-w systems, as well as on general systems aspects of their generation and reception. This will further solidify the Fourier analysis of Chap. 2. It parallels the approach used in Chap. 3 in discussing pulse-modulation systems. After introducing a discussion of noise effects in Chap. 5, we shall be in a position to further compare the various c-w systems, as well as the pulse systems already studied, on the basis of their performance in noise.

4-2 BINARY COMMUNICATIONS[1]

As noted earlier there are essentially three ways of modulating a sine-wave carrier: variation of its amplitude, phase, and frequency in accordance with the information being transmitted. In the binary case this corresponds to switching the three parameters between either of two possible values. Most commonly the amplitude switches between zero (*off* state) and some predetermined amplitude level (the *on* state). Such systems are then called *on-off-keyed* (OOK) systems. Similarly, in *phase-shift keying* (PSK), the phase of a carrier switches by π radians or 180°. Alternately, one may think of switching the polarity of the carrier in accordance with the binary information stream. In the *frequency-shift-keyed* (FSK) case, the carrier switches between two predetermined frequencies, either by modulating one sine-wave oscillator or by switching between two oscillators locked in phase. Although other types of binary sine-wave signaling schemes are in use as well, we shall concentrate here on the basic schemes.

On-Off Keying

Assume a sequence of binary pulses, as shown in Fig. 4-1a. The 1's turn on the carrier of amplitude A, the 0's turn it off (Fig. 4-1b). It is apparent that the spectrum of the OOK signal will depend on the particular binary sequence to be transmitted. Call a particular sequence of 1's and 0's $f(t)$. Then the amplitude-modulated or OOK signal is simply

$$f_c(t) = Af(t) \cos \omega_c t \qquad (4\text{-}1)^2$$

where $f(t) = 1$ or 0, over intervals T seconds long. But note that this is in exactly the form of the modulated signal discussed in Chap. 2. As shown there, upon

---

[1] W. R. Bennett and J. R. Davey, *Data Transmission,* McGraw-Hill, New York, 1968, discuss these systems in much more detail than is possible here. They include a thorough treatment of the spectral characteristics of these systems. M. Schwartz, W. R. Bennett, and S. Stein, *Communications Systems and Techniques,* McGraw-Hill, New York, 1966, also discuss these systems in more detail than is possible here, although their emphasis is primarily on the relative noise immunity of the systems. Some of this material will be discussed in Chaps. 5 and 6 of this text after the basic aspects of noise phenomena are introduced.

[2] The carrier frequency ω_c should not be confused with the same symbol used for low-pass cutoff frequency in Sec. 3-9.

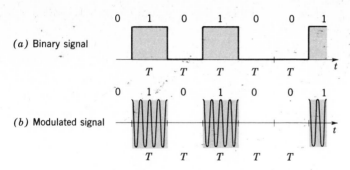

Figure 4-1 On-off-keyed signal.

taking the Fourier transform of the amplitude-modulated (OOK) signal $f_c(t)$, and using the frequency shifting theorem of (2-53), we have

$$F_c(\omega) = \frac{A}{2} \left[F(\omega - \omega_c) + F(\omega + \omega_c) \right] \qquad (4\text{-}2)$$

A similar example appeared in (3-3) in discussing sampling.

The effect of multiplication by $\cos \omega_c t$ is simply to shift the spectrum of the original binary signal (the baseband signal) up to frequency ω_c (Fig. 4-2). This is the general form of an AM signal. It contains upper and lower sidebands symmetrically distributed about the carrier or center frequency ω_c. Note the important fact that with an initial baseband bandwidth of $2\pi B$ rad/s (B hertz), the AM or transmission bandwidth is twice that, i.e., $\pm 2\pi B$ rad/s or $\pm B$ hertz about the carrier, for a total bandwidth of $2B$ hertz.

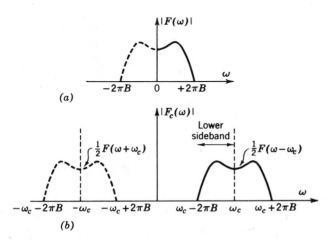

Figure 4-2 Amplitude spectrum—amplitude-modulated wave. (*a*) Spectrum of modulating signal. (*b*) Spectrum of amplitude-modulated wave.

Although the signals of Fig. 4-1 are shown sketched as rectangular in shape for simplicity, they could equally well have any shaping desired. The baseband time function $f(t)$ would then incorporate the specific shaping parameter used. Equation (4-2) and Fig. 4-2 then indicate that the shaped modulated signal of (4-1) would have the spectrum of the baseband signal shifted up to, and centered about, the carrier frequency. As an example, say that sinusoidal roll-off shaping is used, either by shaping the baseband pulses, or by shaping the high-frequency modulated pulses. The spectrum of the modulated signal looks like the baseband spectrum, shifted up to the carrier frequency f_c hertz and with a transmission bandwidth $B_T = 2B = (1/T)(1 + r)$, with r the roll-off factor [see (3-99)]. An example appears in Fig. 4-3.

Because of the form of (4-1), the frequency shift of a signal $f(t)$ due to multiplication by $\cos \omega_c t$ is a general result for AM signals. It is true for all modulating signals $f(t)$ and not just for the binary case we are in the process of considering. This will be stressed again in Sec. 4-4 in discussing AM signals in general. As an example, let $f(t) = \cos \omega_m t$, a single sine wave of frequency ω_m. Then, by trigonometry,

$$\cos \omega_m t \cos \omega_c t = \tfrac{1}{2} \cos (\omega_m + \omega_c)t + \tfrac{1}{2} \cos (\omega_m - \omega_c)t$$

The single-line spectral plot representing $\cos \omega_m t$ is thus replaced by *two* lines, symmetrically arrayed about ω_c. Similarly, if $f(t)$ is a finite sum of sine waves, each sine wave is translated up in frequency by ω_c.

As another special case, assume the signal $f(t)$ to be the single rectangular pulse of Chap. 2. (This is then the special case of a binary train in which all

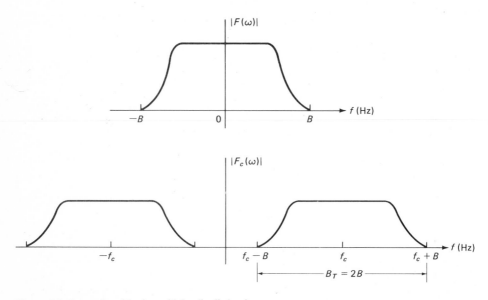

Figure 4-3 Example with sinusoidal roll-off shaping.

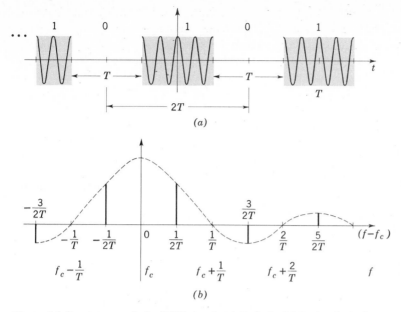

Figure 4-4 Spectrum, periodic OOK signal. (*a*) Periodic OOK signal. (*b*) Spectrum (positive frequencies only).

symbols are 0, except for one 1.) For a pulse of amplitude A and width T (the binary interval), the spectrum of the AM signal becomes simply

$$\frac{AT}{2}\left[\frac{\sin(\omega-\omega_c)T/2}{(\omega-\omega_c)T/2}+\frac{\sin(\omega+\omega_c)T/2}{(\omega+\omega_c)T/2}\right]$$

With an initial bandwidth of approximately $1/T$ hertz (from 0 frequency to the first zero crossing), we now have a transmission bandwidth of $2/T$ ($\pm 1/T$ about the carrier). Another special case is a binary train of alternating 1's and 0's, resulting in a periodically alternating OOK signal. The spectrum of this signal is just the $(\sin x)/x$ line spectrum of a pulse of width T, periodic with period $2T$, translated up to frequency f. This is shown in Fig. 4-4.

Frequency-Shift Keying

Here, if we first consider rectangular shaping for simplicity,

$$\left.\begin{array}{l} f_c(t) = A\cos\omega_1 t \\ f_c(t) = A\cos\omega_2 t \end{array}\right\} \quad -\frac{T}{2}\le t\le\frac{T}{2} \tag{4-3}$$

or

A 1 corresponds to frequency f_1, a zero to frequency f_2 (Fig. 4-5). (*Note:* Generally, f_1 and $f_2 \gg 1/T$. In some systems, particularly over telephone lines, f_1 and $f_2 \sim 1/T$, as shown here.) An alternative representation of the FSK wave

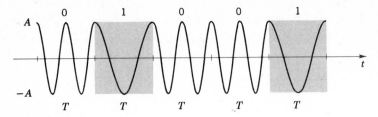

Figure 4-5 FSK wave.

consists of letting $f_1 = f_c - \Delta f, f_2 = f_c + \Delta f$. The two frequencies then differ by $2\Delta f$ hertz. Then

$$f_c(t) = A \cos (\omega_c \pm \Delta\omega)t \qquad -\frac{T}{2} \le t \le \frac{T}{2} \qquad (4\text{-}3a)$$

The frequency then deviates $\pm \Delta f$ about f_c. Δf is commonly called the *frequency deviation*. The frequency spectrum of the FSK wave $f_c(t)$ is in general difficult to obtain. We shall see this is a general characteristic of FM signals. However, one special case which provides insight into the spectral characteristics of more complex FM signals, and leads to a good rule of thumb regarding FM

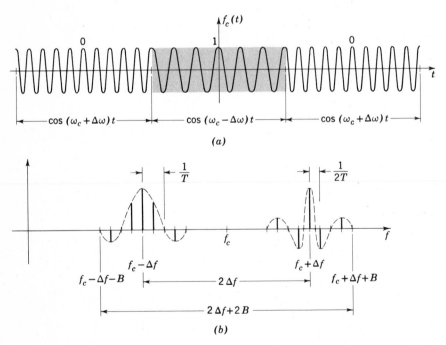

Figure 4-6 Spectrum, periodic FSK wave. (*a*) Periodic FSK signal. (*b*) Spectrum (positive frequencies only).

bandwidths, may be readily evaluated. Assume that the binary message consists of an alternating sequence of 1's and 0's. If the two frequencies are each multiples of the reciprocal of the binary period T (i.e., $f_1 = m/T$, $f_2 = n/T$, m and n integers), and are synchronized in phase, as assumed in Eq. (4-3), the FSK wave is the periodic function of Fig. 4-6. Note, however, that this may also be visualized as the linear superposition of two periodic OOK signals such as the one of Fig. 4-4, one delayed T seconds with respect to the other. The spectrum is then the linear superposition of two spectra such as the one in Fig. 4-4. Specifically, it is left for the reader to show that the positive frequency spectrum is of the form

$$\frac{\sin\left[(\omega_1 - \omega_n)T/2\right]}{(\omega_1 - \omega_n)T/2} + (-1)^n \frac{\sin\left[(\omega_2 - \omega_n)T/2\right]}{(\omega_2 - \omega_n)T/2}$$

with $\omega_n - \pi n/T$, $\omega_1 = \omega_c - \Delta\omega$, $\omega_2 = \omega_c + \Delta\omega$. This spectrum is shown sketched in Fig. 4-6 for the special case $\Delta f \gg 1/T$. The bandwidth of this periodic FSK signal is then $2\Delta f + 2B$, with B the bandwidth of the baseband signal. (The dashed lines shown in Fig. 4-6 are simply used to connect the discrete lines and have no significance.)

Two extreme cases are of interest:

1. If $\Delta f \gg B$, the bandwidth approaches $2\Delta f$. Thus, if one uses a wide separation of tones in an FSK system, the bandwidth is essentially just that separation. It is virtually independent of the bandwidth of the baseband binary signal. *This is distinctly different from the AM case.*
2. If $\Delta f \ll B$, the bandwidth approaches $2B$. In this case, even with the tones chosen very close together, the minimum bandwidth is still that required to transmit an OOK (AM) signal; here the bandwidth *is* determined by the baseband signal.

The first case is commonly called *wideband FM*, the second *narrowband FM*. We shall see shortly that the bandwidth $2\Delta f + 2B$ and its two extreme values are quite good approximations to FM bandwidths with complex modulating signals. This analysis through the use of simple binary signals provides, in addition, much more physical insight into FM bandwidth determination than is possible with complex signals. In particular, if the baseband signal is an arbitrary string of binary pulses, each shaped according to sinusoidal shaping with a roll-off factor, r, the approximate bandwidth of the corresponding FSK signal is given by $2\Delta f + 2B$, with $B = (1/2T)(1 + r)$, T being the baseband (or FSK) pulse width. The exact shape of the FSK spectrum is difficult to calculate, but its form would be roughly that shown in Fig. 4-7.[3]

Note that the FM transmission bandwidth is generally much greater than that for AM, which is always $2B$, that is, twice the baseband bandwidth. Then

[3] Bennett and Davey, *op. cit.*, chap. 5.

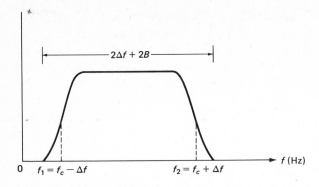

Figure 4-7 FSK spectrum, sinusoidal roll-off shaping (positive frequencies only).

why use FM? We shall show in a later chapter that it is just this wideband property of FM that makes its performance generally far superior to AM in a noisy environment. This is analogous to the pulse modulation results noted in Chap. 3. Encoding of pulse-amplitude-modulation (PAM) signals into binary pulse code modulation (PCM) results in an expansion of the system bandwidth but the noise immunity increases considerably. A general characteristic of communication systems to which we shall refer after discussing noise in systems is that one can generally improve system performance in the presence of noise by encoding or modulating signals into equivalent wideband forms. Binary PCM and FM are examples of such wideband signals.

It is common in FM analysis to denote the dependence of transmission bandwidth on the relative magnitudes of the frequency deviation Δf and baseband bandwidth B by defining a parameter β, the *modulation index*, as the ratio of the two. Thus

$$\beta \equiv \frac{\Delta f}{B} \tag{4-4}$$

In terms of β the FM transmission bandwidth is

$$\begin{aligned} B_T = \text{FM bandwidth} &= 2\Delta f + 2B \\ &= 2B(1 + \beta) \end{aligned} \tag{4-5}$$

Narrowband FM systems correspond to $\beta \ll 1$, wideband systems to $\beta \gg 1$. We shall find the modulation index β playing a significant role throughout our discussion of FM.

Phase-Shift Keying

In this case we have the phase-shift-keyed signal given as

$$f_c(t) = \pm\cos \omega_c t \qquad -\frac{T}{2} \leq t \leq \frac{T}{2} \tag{4-6}$$

if rectangular shaping is assumed. Here a 1 in the baseband binary stream corresponds to positive polarity, and a 0 to negative polarity. The PSK signal

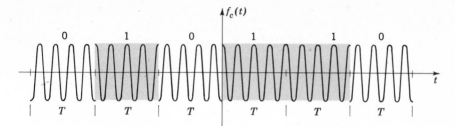

Figure 4-8 PSK signal.

thus corresponds essentially to a polar NRZ binary stream (Fig. 3-66c), translated up in frequency. An example is shown in Fig. 4-8. The discontinuous-phase transitions shown at the beginning and end of each bit interval, whenever a transition from 1 to 0 or 0 to 1 takes place, are actually smoothed out during transmission because of the shaping used. The information regarding polarity is, however, retained in the center of each interval, so that decoding at the receiver is normally timed to be carried out in the vicinity of the center. This is also true for OOK and FSK signals. The PSK signal has the same double-sideband characteristic as OOK transmission. Introducing roll-off shaping in the high-frequency pulses of (4-6) results in a spectrum centered at the carrier frequency f_c, with a bandwidth twice that of the shaped baseband spectrum.

Detection of Binary Signals

One may very well ask at this point which of the three binary signaling techniques discussed (and other possible ones as well) one would prefer using in practice. What are the relative advantages and disadvantages of the different techniques? We have already noted that FSK systems perform better in a noisy environment than do OOK systems. We shall show later in this book that PSK systems perform still better and, in fact, may be shown to be the optimum possible for binary signaling in the presence of additive noise. (This ignores intersymbol interference and other types of distortion; it assumes that additive noise is the sole form of disturbance during transmission.)

Then why not always use PSK techniques? The answer lies in the detection process at the receiver. Recall that we modulate a sine-wave carrier with the baseband binary stream of necessity to shift the resultant modulated signal to an appropriate frequency for transmission. At the receiver we must undo this process or *demodulate* the signal to recover the original binary stream. This process of demodulation is often also called *detection*. (The binary stream must then be further processed as in Chap. 3 to eventually retrieve the individual signals contained within it.) There are essentially two common methods of demodulation. One, called *synchronous* or *coherent detection,* simply consists of multiplying the incoming signal by the carrier frequency, as locally generated at the receiver, and then low-pass-filtering the resultant multiplied signal. The

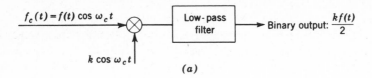

(a)

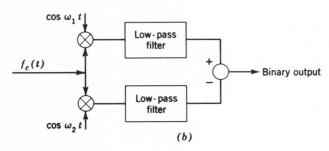

(b)

Figure 4-9 Synchronous detection. (a) OOK or PSK signals. (b) FSK signals.

other method is called *envelope detection*. The synchronous detection procedure is diagrammed in Fig. 4-9. Note that the FSK signals require *two* sine waves, one for each frequency transmitted. This is just the reverse of the original modulation process at the transmitter and serves to translate the binary signals back down to baseband.

To demonstrate the synchronous method, assume the high-frequency binary signal to have the AM form $f_c(t) = f(t) \cos \omega_c t$. [If $f(t) = \pm 1$, we have the PSK signal; if it is 1 or 0, we have the OOK case.] If we multiply this by $k \cos \omega_c t$ as indicated (k is an arbitrary constant of the multiplier), we get $kf(t) \cos^2 \omega_c t = (k/2)(1 + \cos 2\omega_c t)f(t)$.[4] But the term $f(t) \cos 2\omega_c t$ represents $f(t)$ translated up to frequency $2f_c$, the second harmonic of the carrier frequency f_c. This is rejected by the low-pass filter and the output is $(k/2)f(t)$, just the desired baseband binary sequence. (The constant factor has no significance since the output signal can always be amplified or attenuated by any derived amount.) So the synchronous detector does the desired job of reproducing the signal $f(t)$.

But we have assumed the local carrier $\cos \omega_c t$ is exactly at the same frequency as the incoming carrier term $\cos \omega_c t$, and in phase with, or synchronized to, it as well. If the locally generated sine wave were at a frequency $\cos (\omega_c + \Delta\omega)t$, the multiplication would generate $kf(t) \cos (\omega_c + \Delta\omega)t \cos \omega_c t = (k/2)[\cos (2\omega_c + \Delta\omega)t + \cos \Delta\omega t]f(t)$. The output of the low-pass filter would then be $[kf(t)/2] \cos \Delta\omega t$ if $\Delta\omega$ were within the filter passband, which is not at all the desired signal! (This technique for shifting carrier frequencies or superheterodyning AM signals will be discussed later.) Alternately,

---

[4] This discussion again assumes rectangular shaping for simplicity. Non-rectangular-shaped pulses are easily included by letting $f(t)$ provide the desired baseband shape.

if the local signal were at the right frequency ω_c, but θ radians out of phase with the incoming carrier, that is, cos $(\omega_c t + \theta)$, it is apparent that the low-pass filter output would be $[kf(t)/2]$ cos θ. This is the desired baseband output but is attenuated in amplitude. In particular, as θ increases, cos θ decreases. For θ close to $\pi/2$ the output is very nearly zero. As θ increases beyond $\pi/2$, the output signal reverses sign. If the baseband signal is a polar NRZ sequence, the entire signal reverses polarity and all 1's become 0's, all 0's become 1's! So not only must the locally generated carrier be at the same frequency, but also synchronized in phase as well. This is the reason for the term *synchronous* detection.

Phase synchronism is quite difficult to attain, particularly if transmission takes place over long distances. This means that a receiver clock which provides the synchronism must be synchronized or slaved to the transmitter clock to within a fraction of a carrier cycle, no mean task. As an example, if transmission is at $f_c = 3$ MHz with a period $1/f_c = 0.3$ μs (this is in the so-called h-f band, used for short-wave broadcasting), a phase difference $\theta \ll \pi/2$ radians implies clock synchronization to within much less than a quarter of a period, or much less than 0.07 μs! At a carrier frequency of 100 MHz, $\pi/2$ radians corresponds to 2.5 ns, while at a carrier frequency of 1,000 MHz, this is 0.25 ns. Phase synchronization is thus quite a difficult task. It is particularly difficult if either the transmitter or receiver is mounted on a rapidly moving vehicle introducing doppler phase shifts proportional to the relative velocity between transmitter and receiver. If the relative velocities change rapidly enough, the phase shifts in turn change rapidly and synchronism cannot be maintained. The same problem arises if the signal transmission takes place through a fading medium in which randomly moving scatterers also introduce random doppler phase shifts.

The reader may have recognized already that this problem of maintaining synchronism is similar to that encountered in baseband binary transmission in Chap. 3. The problem here, however, is a much more difficult one; the synchronism previously considered was, for example, from bit interval to bit interval. With high carrier frequencies, $1/f_c \ll T$, the bit interval, so that the problem is compounded. As with the baseband digital synchronization case, various methods are available for obtaining the required phasing information.

1. A pilot carrier may be transmitted, and superimposed on the high-frequency binary signal stream, which may be extracted at the receiver and used to synchronize the receiver local oscillator (this is similar to the marker or timing pulses transmitted with the baseband data stream to maintain bit interval, word, and frame synchronism).
2. A phase-locked loop, locking on either the data stream or a pilot tone, may be used at the receiver to drive the phase difference to zero,[5] etc.

[5] A. J. Viterbi, *Principles of Coherent Communication*, McGraw-Hill, New York, 1966, chap. 2.

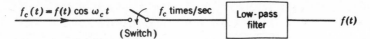

Figure 4-10 Another synchronous detector.

We shall have more to say about synchronous detection later in this chapter, as well as in the discussions on binary transmission in the presence of noise. It is interesting to note, however, that it is not really necessary to physically multiply by a pure sine wave to obtain the desired demodulated form $f(t) \cos^2 \omega_c t$. Switching or gating $f_c(t) \cos \omega_c t$ on and off at a rate of f_c times per second and then using low-pass filtering will accomplish the same job (see Fig. 4-10). This is apparent from a consideration of the switching function of Chap. 3. All-digital gating circuits are in fact often simpler to design and use for this purpose than is multiplication by a pure sine wave. [If the baseband binary stream contains time-multiplexed information, further switching (see Chap. 3) is then necessary to separate out the individual signals. But this switching, gating, or commutating function is of course carried out at the much slower, baseband, rate.]

The difficulty of maintaining phase synchronism notwithstanding, PSK transmission and phase-locked loop synchronous detection has been successfully used in space communications. The Pioneer V deep-space probe, as an example, used biphase modulation in its telemetry system. The Pioneer IV system also used phase-coherent techniques.[6]

The other common form of detection, *envelope detection,* avoids the timing and phasing problems of synchronous detection. Here the incoming high-frequency signal is passed through a nonlinear device and low-pass filter. Envelope detection will be considered later in this chapter in discussing more general AM signals and receivers. Suffice it to say at this point that one common form of envelope detector is a diode half-wave rectifier (the nonlinear device) followed by a RC low-pass filter (Fig. 4-11). As the name indicates the output of the detector represents the envelope of the incoming high-frequency wave. The RC time constant is long enough to hold the incoming amplitude over many

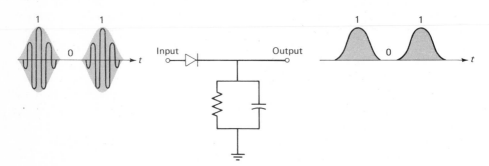

Figure 4-11 Envelope detector.

[6] See, e.g., A. V. Balakrishnan (ed.), *Space Communications,* McGraw-Hill, New York, 1963, chaps. 14 and 15.

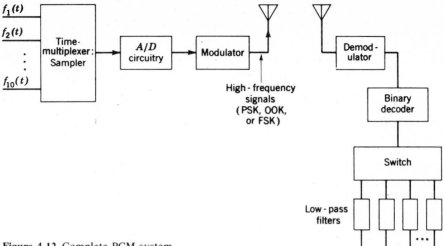

Figure 4-12 Complete PCM system.

carrier cycles, yet short enough compared to a binary period to discharge once the binary signal changes.

Notice one difficulty, however. The PSK signal has a *constant* envelope (Fig. 4-8) so that it cannot be used with an envelope detector. Thus the PSK system *requires* synchronous detection. As usual there is a trade-off in economics. One may avoid using synchronous detection by adopting envelope detection, but at the price of requiring FSK or OOK transmission, rather than the more optimum (in the presence of noise) PSK system.[7] We shall also show in discussing the effects of noise, in later chapters, that envelope detection of OOK or FSK signals is somewhat inferior to synchronous detection of these signals. An additional price must thus be paid for this much simpler circuit, although in this case it can be demonstrated that the cost will not be too great.

The envelope-detection process requires the presence of an unchanging carrier term in addition to the varying high-frequency binary signal. This is the reason envelope detection cannot be used with PSK signals. To demonstrate this point consider an OOK signal $f_c(t) = Af(t) \cos \omega_c t$. Recall that $f(t)$ is a random sequence of 1's and 0's. Squaring the expression (this is the simplest nonlinear operation that demonstrates the envelope-detection process), we get $A^2 f^2(t) \cos^2 \omega_c t$. With rectangular shaping $f(t) = 1$ or 0, so $f^2(t) = 1$ or 0 as well. The output of a low-pass filter is then $A^2/2$ or 0, reproducing the derived 1,0 sequence. In the case of the PSK signal, however, $f(t) = \pm 1$, $f^2(t) = 1$, and the output is *always* $A^2/2$. There are other ways of analyzing the envelope detector, and these are considered in discussions of AM systems, as well as in the problems at the end of this chapter.

To conclude the discussion of binary signaling at this point, we show in Fig. 4-12 a diagram of a complete time-multiplexed PCM system. This includes the

---

[7] This is why the space communication systems have used synchronous detection, the cost notwithstanding. The reduction possible in transmitter power is worth the price there.

initial time-multiplexing and A/D circuitry (the latter includes the quantization and binary encoding operations); the modulator which produces the high-frequency binary signals; then, at the receiver, the demodulator, which includes a synchronous or envelope detector, binary decoder, a switch, or commutator circuitry for unscrambling the time-multiplexed signals; and finally low-pass filters in each output channel for providing the final output signals. Note that this is essentially the same set of blocks discussed in Chap. 3, but with the addition of the high-frequency modulator and demodulator.

4-3 MODULATION TECHNIQUES FOR DIGITAL COMMUNICATIONS: MULTISYMBOL SIGNALING

In the previous section we have focused on the simplest forms of digital carrier systems, those involving binary amplitude, phase, or frequency modulation. We noted in Chap. 3, particularly in Secs. 3-4 and 3-9, that the bandwidth required to transmit a baseband digital sequence could be reduced by going to multilevel signaling: combining successive binary pulses to form a longer pulse requiring a correspondingly smaller bandwidth for transmission. This was generalized in Sec. 3-9 to include the concept of multisymbol signaling. Specifically, with ideal Nyquist shaping 2 symbols/s/Hz can be transmitted over the Nyquist bandwidth of B hertz. If a set of $M = 2^n$ symbols is used, with n the number of successive binary digits combined to form the appropriate symbol to be transmitted, $2n$ bits/s/Hz may be transmitted using the Nyquist band.

In this section we specifically discuss multiphase, multiamplitude, and combined multiphase/multiamplitude signaling schemes as examples of multisymbol systems. These are used quite commonly in both telephone and satellite data communications and we shall refer to examples from both of these application areas later in this section. We consider the simplest signaling schemes only in this book, leaving more complex schemes to the references.[8] Multifrequency schemes are also used in practice, but for a different purpose: they generally result in *larger,* rather than reduced bandwidths, because of the requirement to space multiple frequency carriers far enough apart. They provide improved noise immunity as a result, however. Multifrequency schemes will be discussed in Chap. 6. Multisymbol signals are often called M-ary signals.

As the first example of a multisymbol scheme, consider a system in which two successive binary pulses are combined, and the resultant set of four binary pairs, 00, 01, 10, 11, is used to trigger a high-frequency sine wave of four possible phases, one for each of the binary pairs. This is the obvious extension to four phases of binary PSK transmission, discussed in the previous section. The ith signal, of the four possible, can be written

---

[8] See, e.g., J. J. Spilker, Jr., *Digital Communications by Satellite,* Prentice-Hall, Englewood Cliffs, N.J., 1977, chap. 11.

$$s_i(t) = \cos(\omega_c t + \theta_i) \qquad i = 1, 2, 3, 4 \qquad -\frac{T}{2} \le t \le \frac{T}{2} \qquad (4\text{-}7)$$

with rectangular shaping assumed at this point for simplicity. This thus extends the binary PSK representation of (4-6).

Two possible choices for the four phase angles are

$$\theta_i = 0, \pm \frac{\pi}{2}, \pi \qquad (4\text{-}8a)$$

$$\theta_i = \pm \frac{\pi}{4}, \pm \frac{3\pi}{4} \qquad (4\text{-}8b)$$

In both cases the phases are spaced $\pi/2$ radians apart. Signals of this type are called *quaternary PSK* (QPSK) signals. They are a special case of multi-PSK (MPSK) signals. Binary PSK signals are sometimes labeled BPSK as well.

In general, as already noted, n successive binary pulses are stored up and one of $M = 2^n$ symbols is outputted. If the binary rate is R bits/s, each binary pulse interval is $1/R$ seconds long. The corresponding output symbol is then $T = n/R$ seconds long. The signals of (4-7) may be represented, by trigonometric expansion, in the following form:

$$s_i(t) = a_i \cos \omega_c t + b_i \sin \omega_c t \qquad -\frac{T}{2} \le t \le \frac{T}{2} \qquad (4\text{-}9)$$

For the case of (4-8a), the (a_i, b_i) pairs are given, corresponding respectively to the angles $\theta_i = 0, -\pi/2, \pi$, and $\pi/2$, by

$$(a_i, b_i) = (1, 0), (0, 1), (-1, 0), (0, -1) \qquad (4\text{-}10)$$

The corresponding sets of (a_i, b_i) for (4-8b) are given by

$$(\sqrt{2}\,a_i, \sqrt{2}\,b_i) = (1, 1), (-1, 1), (-1, -1), (1, -1) \qquad (4\text{-}11)$$

Transmission of this type is often called *quadrature transmission,* with two carriers in phase quadrature to one another ($\cos \omega_c t$ and $\sin \omega_c t$) transmitted simultaneously over the same channel.

It is useful to represent the signals of (4-9) in a two-dimensional diagram by locating the various points (a_i, b_i). The horizontal axis corresponding to the location of a_i is called the *inphase axis*. The vertical axis, along which b_i is located, is called the *quadrature axis*. The four signals of (4-10) then appear as shown in Fig. 4-13a, those of (4-11) in Fig. 4-13b. The signal points are said to represent a *signal constellation*.

The inphase (cosine) and quadrature (sine) representation of the QPSK signals $s_i(t)$ suggests one possible way of generating these signals: two successive binary input pulses are stored up and the pair of numbers (a_i, b_i), taken every $T = 2/R$ seconds, is used to modulate two quadrature carrier terms, $\cos \omega_c t$ and $\sin \omega_c t$, respectively. Where one of the numbers is zero, that carrier is of course disabled. A modulator of this type is shown in Fig. 4-14a. An example, using the (a_i, b_i) pairs of (4-10), is shown in Fig. 4-14b.

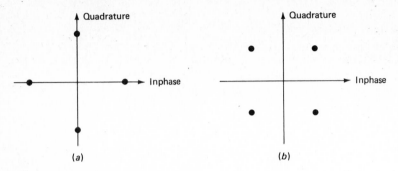

Figure 4-13 QPSK signal constellations.

It is apparent that demodulation is carried out by using two synchronous detectors in parallel, one in quadrature with the other. A comparison of the two detector outputs then determines the particular binary pair transmitted. A block diagram of such a demodulator appears in Fig. 4-15.

Quadrature Amplitude Modulation

More general types of multisymbol signaling schemes may be generated by letting a_i and b_i in (4-9) take on multiple values themselves. The resultant

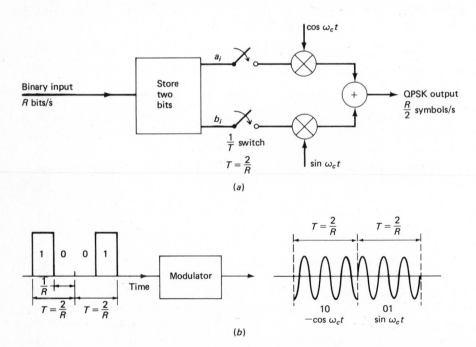

Figure 4-14 Generation of QPSK signals. (*a*) Modulator. (*b*) Example corresponding to Eq. (4-10).

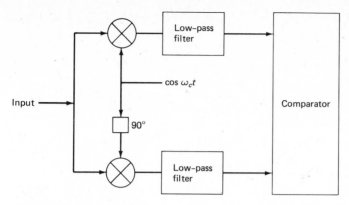

Figure 4-15 QPSK demodulator (a possible implementation).

signals are called *quadrature amplitude modulation* (QAM) signals. These signals may be interpreted as having multilevel amplitude modulation applied independently on each of the two quadrature carriers. The demodulator of Fig. 4-15, with a level detector applied at the output of each synchronous detector, could then be used to recover the desired digital information. The constellation for a 16-state QAM signal set appears in Fig. 4-16. Note that this signal can be considered as being generated by two amplitude-modulated signals in quadrature. Since four amplitude levels are used on each of the carriers, the signal is sometimes referred to as a four-level QAM signal. All points in the constellation are equally spaced.

It is apparent that the general QAM signal may also be written

$$s_i(t) = r_i \cos (\omega_c t + \theta_i) \tag{4-12}$$

with the amplitude r_i and phase angle θ_i given by the appropriate combinations of (a_i, b_i). A phase detector-amplitude level detector combination could then also be used to extract the digital information.

We have neglected signal shaping up to now, just as in the previous section. This is apparent from the rectangular input signals and the sharp phase discon-

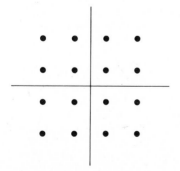

Figure 4-16 Four-level (16-symbol) QAM constellation.

tinuities shown in the output modulated signals of Fig. 4-14b. In practice, signal shaping, using, for example, the sinusoidal roll-off spectrum of Sec. 3-9 (Fig. 3-64) must be used to reduce intersymbol interference. This is particularly important for multilevel signaling, as already noted in Sec. 3-9. An actual modulator would thus have the input binary pulses shaped before modulating the carrier. Alternatively, the successive output signals of Fig. 4-14 would each be passed through an appropriate bandpass shaping filter before being transmitted.

As the result of shaping, an individual output symbol, nominally designed to fit into the interval T seconds long, may now span several T-second intervals. (The object is to shape the pulses, however, so that they go through zero at the decision points spaced T seconds apart in the other intervals. See Fig. 3-61, for example.) The transmitted signal in any given T-second interval thus has added to it contributions from the tails of signals on either side, depending on the actual shaping used. A typical signal at time t, in the current slot, may thus be written as

$$s(t) = \sum_n \left[a_n h\left(t - \frac{n}{T}\right) \cos \omega_c t + b_n h\left(t - \frac{n}{T}\right) \sin \omega_c t \right] \quad (4\text{-}13)$$

$h(t)$ represents the impulse response of the shaping filter. The extent of this time response determines the number of values of n, positive and negative, that must be included in the sum shown. $n = 0$ corresponds to the current T-second interval; n positive, intervals into the future (due to any "precursor" tails of symbols transmitted); n negative, intervals in the past. The (a_n, b_n) pair in any interval represents the one of the possible pair values actually transmitted in that interval.

Letting $t = 0$ (the center of the interval) be the signal sampling point, $h(-n/T), n \neq 0$, should be zero for intersymbol interference to be absent. If $h(-n/T) \neq 0$, (4-13) can be used to determine the extent of intersymbol interference.

From the form of (4-13) it is apparent that the general QAM signal must have a spectrum centered about the carrier frequency $f_c = \omega_c/2\pi$. There will be upper and lower sidebands extending a bandwidth B hertz, respectively, above and below the carrier frequency, corresponding to the baseband signal shifted up to frequency f_c. The shaping of the sidebands depends on the shaping filter $h(t)$.[9] An example appears in Fig. 4-17. The transmission bandwidth B_T is $2B$ hertz, as shown.

---

[9] With no intersymbol interference the spectrum of an individual pulsed carrier in any one T-second interval is given exactly by the transform $H(\omega)$ of $h(t)$, the Nyquist shaping filter characteristic, shifted up to f_c. More generally, one must take into account the contribution of all intervals in finding the spectrum of the transmitted signal $s(t)$. This depends on the statistics of the amplitude-level sequence $\{a_n, b_n\}$. Details appear in Bennett and Davey, *op. cit.*, chap. 11. Suffice it to say that the spectrum is determined primarily by $H(\omega)$, so that we may still talk of Nyquist shaping in Chap. 3, applied to single digital pulses, as appropriate to a complete sequence of such pulses.

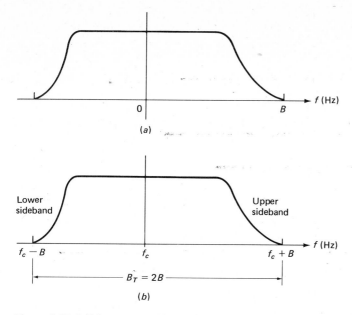

Figure 4-17 QAM spectrum. (*a*) Baseband spectrum. (*b*) QAM spectrum.

In practice, shaping is done at both the transmitter, as part of the modulation process, and at the receiver, in conjunction with the demodulation process. The Nyquist shaping characteristic to which we have referred many times, beginning in Chap. 3, represents the combination of these two filtering operations.

Using the discussion of Nyquist shaping of Sec. 3-9 and the discussion thus far in this section one may determine the particular form of QAM, and the type of shaping required, to transmit specified bit rates over various channels. Specifically, say the transmission bandwidth is B_T hertz. This then corresponds to a baseband bandwidth of $B = B_T/2$ hertz. We know from Sec. 3-9 that the symbol rate that may be transmitted over a channel with baseband bandwidth B hertz is $2B/(1 + r)$, where the roll-off factor r varies from an ideal value of 0 (ideal low-pass filtering) to 1, for raised cosine filtering. The symbol rate allowable over the equivalent transmission channel of bandwidth B_T hertz is thus $B_T/(1 + r)$ symbols/s. For a QAM signal with $M = 2^n$ possible symbols or states, the allowable bit rate is $nB_T/(1 + r)$ bits/s, or $n/(1 + r)$ bits/s/Hz of *transmission* bandwidth. Some examples of the allowable bit rate per hertz appear in Table 4-1.

As an example, a channel with a 2.4-kHz bandwidth will allow $2,400 \times 1.8 = 4,300$ bits/s transmission if four-state QAM (equivalent to four-phase PSK or QPSK; see Fig. 4-13) is used, and pulse shaping with a 10 percent roll-off factor is adopted. If 100 percent roll-off is used instead, 2,400 bits/s transmission using QPSK is possible. If a 20-MHz transmission bandwidth is available and

Table 4-1 Allowable bit rates, QAM transmission (bits/s/Hz of transmission bandwidth)

M (number of states)	Roll-off factor, r			
	0.1	0.25	0.5	1
2	0.9	0.8	0.67	0.5
4	1.8	1.6	1.33	1.0
8	2.7	2.4	2.0	1.5
16	3.6	3.2	2.67	2.0

QPSK is again used, the equivalent transmission rates become 36 Mbits/s and 20 Mbits/s for 10 percent and 100 percent roll-off factors, respectively.

Alternatively, say that the transmission bandwidth B_T is specified and a certain bit rate is desired. With a particular QAM constellation chosen the roll-off factor necessary to attain the desired bit rate may be found. The question then arises as to whether the shaping filter required may be designed economically for the application in question. As an example, the SPADE multiple-access satellite communications system used for PCM communications via the Intelsat satellite global network uses four-phase PSK to attain a bit rate of 64 kbits/s over a transmission bandwidth of 38 kHz.[10] It is left to the reader to show that the pulse shaping required corresponds to a Nyquist roll-off factor of about 19 percent. (The SPADE system will be described briefly in Sec. 4-11.)

Since the bit rate allowable over a given channel depends on the number of symbols or states chosen, why not go on indefinitely increasing the size of the QAM signal constellation? One answer already given in Chap. 3 was that as the number of amplitude levels used increases, the intersymbol interference problem becomes more severe. In addition, it is apparent that as the number of distinguishable phases increases, the phase spacing between signals reduces correspondingly. (Compare Figs. 4-13 and 4-16.) Phase jitter and timing problems make it more difficult to accurately determine phase as the number of distinguishable phases increases. Ultimately, ever-present noise added during transmission and in reception at the receiver makes it more difficult to distinguish individual points in a constellation as the number of points increases.

A little thought will indicate that the specific location of the points in the constellations of Figs. 4-13 and 4-16 depends on the actual amplitudes of the signals. As the amplitudes increase, the points move out. As they decrease, the points move in. For a fixed transmitter power level, the location of the points is restricted. The only way to add more states, or points in the constellation, is to

[10] J. G. Puente and A. M. Werth, "Demand-assigned Service for the Intelsat Global Network," *IEEE Spectrum*, vol. 8, no. 1, pp. 59–69, January 1971.

put points between those already existing. The resultant points are spaced closer together, and noise and phase jitter will produce errors in detection more often. (The effect of noise will be considered in detail in Chap. 6.) There is thus a limit on the number of QAM states that may be used. In practice, 16-state QAM, as in Fig. 4-16, is the maximum that has been used.

Modems: Application to Telephone Data Sets

A complete system for generating the modulated signals of (4-13), or the equivalent signal constellation as shown, for example, in Fig. 4-16, and then for reproducing the baseband binary signals after reception, consists of a transmitter and a receiver. A simplified block diagram of a QAM transmitter-receiver combination, based on the discussion thus far, appears in Fig. 4-18. The transmitter shown there is often referred to as a modulator, even though the modulator itself may only represent a portion of the transmitter. It must contain input buffer memory to store the n successive binary pulses required to generate the particular output signal corresponding to the binary sequence, shaping filters, and an oscillator for generating the inphase and quadrature carrier terms.

Two-way transmission (either simultaneously or one way at a time) is commonly carried out over transmission channels. Binary digits are thus accepted by the transmitter for transmission over the channel, while modulated signals, coming from the other direction, are processed by a receiver and con-

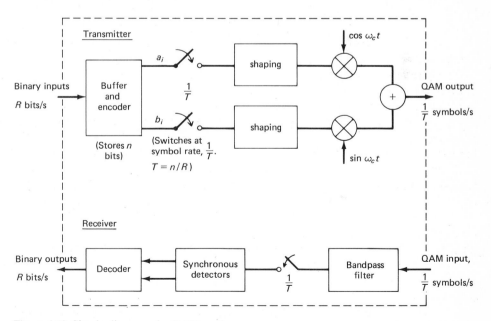

Figure 4-18 Simple diagram of a QAM modem.

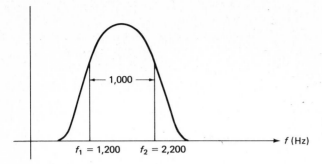

Figure 4-19 1,200-bits/s FSK transmission over telephone lines.

verted to a desired binary output. Figure 4-18 shows such a receiver, often called a demodulator. The two devices, *mo*dulator and *dem*odulator, are most often packaged in one unit called a *modem,* as shown in Fig. 4-18. (The word "modem" is taken from the initial letters of both, shown italicized above.) Left out in the basic diagram of Fig. 4-18 are such important implementation details as carrier tracking and synchronization circuitry; a binary scrambler-descrambler combination if a Gray code is used (Sec. 3-4); additional adaptive filters at the receiver, called *adaptive equalizers,*[11] which automatically adjust their characteristics to compensate for intersymbol interference that may appear; etc.

Modems have been widely adopted for the transmission of digital data over various transmission media. The example of a four-phase PSK modem for digital transmission over a 38-kHz channel in the SPADE satellite system has already been cited. The most widely used modems, however, are those designed to transmit data over the ever-present telephone lines that normally handle voice signals. The bandwidth properties of these lines were mentioned briefly in Secs. 3-2 to 3-4 in discussing the generation of PCM signals. They were referred to again in Sec. 3-9 when discussing Nyquist shaping. These lines accept signals in the range 300 to 3.3 kHz. They will normally allow the direct transmission, over dialed lines, of 600 to 1,200 bits/s data rates. FSK is used for this purpose. Figure 4-19 shows the spectrum of a 1,200-bits/s FSK signal chosen to fit into the transmission characteristics of a typical telephone line. The two carrier frequencies, one at 1,200 Hz, the other at 2,200 Hz, are spaced 1,000 Hz apart.

For higher-bit-rate transmission over the telephone line, multisymbol signaling must be used. Examples of three signal constellations and their corresponding transmission spectra, used in 2,400-, 4,800-, and 9,600-bits/s modems, respectively, appear in Fig. 4-20.[12] The amplitude spectra shown are scaled in

[11] R. W. Lucky, J. Salz, and E. J. Weldon, *Principles of Data Communication,* McGraw-Hill, New York, 1968.

[12] E. R. Kretzmer, "The Evolution of Techniques for Data Communication over Voiceband Channels," *IEEE Commun. Soc. Mag.,* vol. 16, no. 1, pp. 10–14, January 1978.

decibels, the 6-dB points corresponding to one-half the peak amplitude. These points thus correspond to the cutoff frequency of the low-pass Nyquist shaping filters discussed in Sec. 3-9. The 2,400-bits/s modem uses a QPSK or four-phase PSK constellation, with raised cosine signal shaping, as shown. The 4,800-bits/s modem uses eight-phase PSK (the signal points are spaced 45° apart on a circle) and a 50 percent roll-off filtering characteristic. A 16-point QAM signal constellation, with points equally spaced, as shown in Fig. 4-20c, is used for the 9,600-bits/s modem. The lower-bit-rate modems use the frequency range 600 to 3,000 Hz as the transmission bandwidth, with the carrier at the

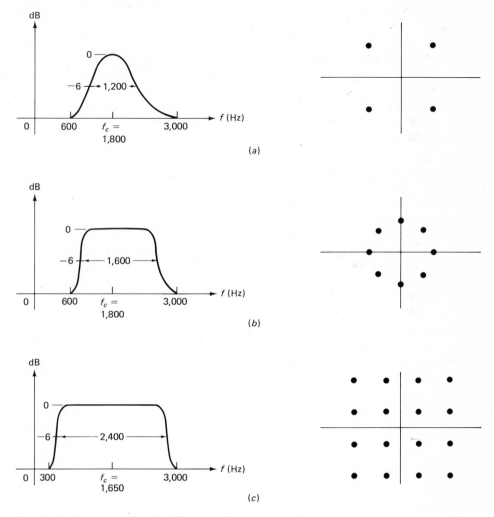

Figure 4-20 Spectra and constellations for higher-speed modems. (a) 2,400-bits/s, 4-phase PSK, raised cosine characteristic. (b) 4,800-bits/s, 8-phase PSK, 50 percent roll-off. (c) 16-state QAM, 9,600 bits/s, 10 percent roll-off.

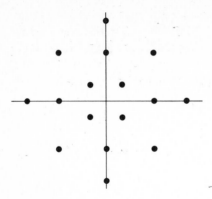

Figure 4-21 16-state constellation, 9,600-bits/s modem; low sensitivity to phase jitter.

center. For the 9,600-bits/s modem a wider frequency range, from 300 to 3,000 Hz, is needed. Special line conditioning (filtering) must be used to obtain these characteristics. This is available in lines leased privately from the telephone companies only, so that dialed transmission at 9,600 bits/s over the regular telephone is generally not possible. It is left to the reader to demonstrate that the signals of Fig. 4-20 do provide the transmission capabilities indicated.

Another 9,600-bits/s signal constellation, used by the independent modem manufacturers in the United States and adopted as an international standard by the CCITT, appears in Fig. 4-21. This constellation, called a modified four-phase/four-amplitude system, has its points located at the relative distances of ±1, ±3, and ±5 units from the origin. It provides particularly low sensitivity to phase jitter.

These modems have been implemented using LSI circuitry as well as microcomputer technology. The reader is referred to the literature for details of implementation.[13]

The time-division multiplexers discussed in Sec. 3-8 focused on baseband multiplexing of varying bit-rate data signals. In practice, the multiplexer outputs must be modulated for appropriate transmission to geographically distant points. Modems are commonly used in conjunction with multiplexers to carry out the appropriate signal modulation and demodulation. A typical example of a small data communication network involving communication between a geographically distributed group of 1,200-bits/s data sources and a central processing system appears in Fig. 4-22. Some of the data sources are shown located close to a multiplexer; hence, a modem is not needed in these cases. Other examples would involve data sources and/or small computers of varying bit-rate outputs communicating with one another, or with a large central computer.[14]

---

[13] P. J. Van Gerwen et al., "Microprocessor Implementation of High-Speed Data Modems," *IEEE Trans. Commun.*, vol. COM-25, no. 2, pp. 238–250, February 1977; K. Watanabe, K. Inoue, and T. Sato, "A 4800 Bit/s Microprocessor Data Modem," *IEEE Trans. Commun.*, vol. COM-26, no. 5, pp. 493–498, May 1978.

[14] M. Schwartz, *Computer-Communication Network Design and Analysis,* Prentice-Hall, Englewood Cliffs, N.J., 1977. See specifically the examples in chaps. 2 and 12 and in the appendix.

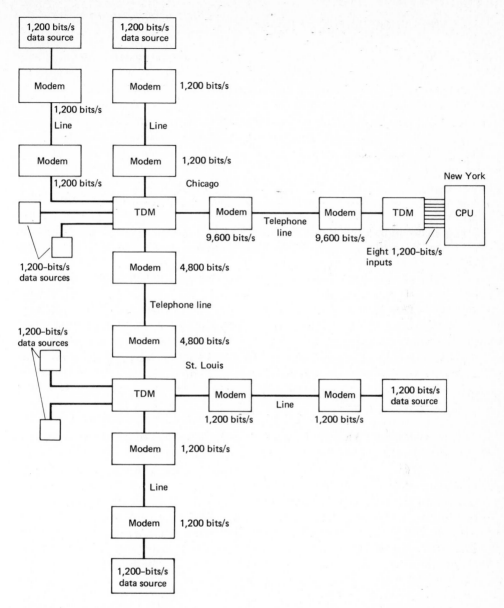

Figure 4-22 Example of a small data network: communication between terminals and central computer.

The use of multiplexers and the corresponding higher-speed modems, as shown in Fig. 4-22, results in reduced communications cost. Only one telephone line need be used, in this example, in linking terminals in the Midwest of the United States to the computer in New York City.

4-4 AMPLITUDE MODULATION

In Sec. 4-2 we discussed sine-wave or c-w modulation systems in which the modulating signal consisted of a digital pulse train. Such systems are important in their own right because of the continuously increasing use of digital systems. The discussion was also important because it enabled us to introduce the basic concept of sine-wave modulation and demodulation (or detection), as well as to introduce spectrum considerations for AM and FM systems in which the modulating signal takes a particularly simple form. We now generalize our approach somewhat, and consider AM and FM systems designed to handle more complex signals than a sequence of binary digits. The signals could be voice, TV, as encountered in standard broadcasting, or complex combinations of both digital and analog waveshapes, obtained by time multiplexing and frequency multiplexing many individual signal channels into one master signal. This composite signal would in turn amplitude modulate or frequency modulate a sine-wave carrier.

In this section we concentrate on AM systems. Much of the material is identical to that already discussed previously, so that our treatment will be rather brief. We shall then go on to single-sideband (SSB) systems, concluding with a discussion of FM systems. Here too the material, particularly that relating to spectral considerations, will be similar to that presented previously.

Recall that when we talk of sine-wave modulation we imply that we have available a source of sinusoidal energy with an output voltage or current of the form

$$v(t) = A \cos (\omega_c t + \theta) \qquad (4\text{-}14)$$

This sinusoidal time function is called a *carrier*. Any one of the three quantities A, ω_c, or θ may be varied in accordance with the information-carrying, or modulating, signal. We restrict ourselves in this section to AM systems in which A only is assumed to be varied. A basic assumption to be adhered to in all the work to follow will be that the modulating signal varies slowly compared with the carrier. This then means that we can talk of an envelope variation, or variation of the locus of the carrier peaks. Figure 4-23 shows a typical modulating signal $f(t)$ and the carrier envelope variation corresponding to it. The amplitude-modulated carrier of Fig. 4-23 can be described in the form

$$f_c(t) = K[1 + mf(t)] \cos \omega_c t \qquad (4\text{-}15)$$

(We arbitrarily choose our time reference so that θ, the carrier phase angle, is zero.) Obviously, $|mf(t)| < 1$ in order to retain an undistorted envelope, in which case the envelope is a replica of the modulating signal. Note that in both Chap. 3 and Sec. 4-2 we discussed modulated carrier terms of the form

$$f_d(t) = f(t) \cos \omega_c t \qquad (4\text{-}16)$$

The present expression, Eq. (4-15), differs by the addition of the carrier term K $\cos \omega_c t$. This term is necessary to ensure the existence of an envelope, as shown

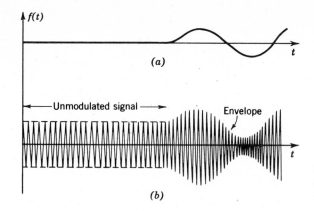

(a)

(b)

Figure 4-23 Amplitude modulation of a carrier. (*a*) Modulating signal. (*b*) Amplitude-modulated carrier.

in Fig. 4-23. This was no problem in the case of the OOK signals of Sec. 4-2, since the on-off sequence of pulses ensured the presence of an envelope. With more complex modulating signals $f(t)$, one must ensure this by adding the appropriate carrier term.

Some systems actually transmit signals of the form of Eq. (4-16). These are called *double-sideband* (DSB) or *suppressed-carrier* systems. As indicated earlier (Figs. 4-2 to 4-4), the effect of multiplying an arbitrary signal $f(t)$ by cos $\omega_c t$ is to translate or shift the spectrum up to the range of frequencies surrounding the carrier frequency f_c. The further addition of the carrier term, as in Eq. (4-15), provides a discrete spectral line at frequency f_c as well. The resultant spectrum of the AM signal of (4-15) is thus similar to those of Figs. 4-2 to 4-4 with the addition of a discrete-line component at the carrier frequency f_c. An example appears in Fig. 4-24. The only stipulation here is that the bandwidth B of the modulating signal be less than the carrier frequency f_c. Two sets of side frequencies appear, as noted earlier: an upper and a lower set. Each set contains a band of frequencies corresponding to the band of frequencies covered by the original signal and is called a *sideband*. Each sideband contains all the spectral components—both amplitude and phase—of the original signal and so

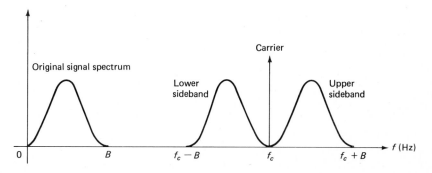

Figure 4-24 AM spectrum, including carrier (positive frequencies only are shown).

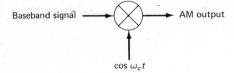

Baseband signal → ⊗ → AM output

cos $\omega_c t$

Figure 4-25 Product modulator for generation of AM signals.

presumably contains all the information carried by $f(t)$. It is left for the reader to show that the phase angles of the upper and lower sidebands differ in sign. [As a hint, recall from Chap. 2 that for real $f(t)$, the phase of the Fourier transform $F(\omega)$ is odd-symmetrical about the origin.]

We now ask ourselves the very pertinent question: How do we physically produce an AM signal? One method indicated in Chap. 3 and referred to again in Sec. 4-2 was that of gating or switching $f(t)$ on and off at the carrier rate. This translated $f(t)$ up to all harmonic multiples of frequency f_c. Bandpass filtering at one of these frequencies, we get the DSB form of the AM signal.

This technique is one of several that can be used to generate AM signals. From the form of Eqs. (4-15) and (4-16) it is apparent that we need some device that will provide at its output the product of the two input functions $f(t)$ and cos $\omega_c t$. Such a device is called a *product modulator*. This generic device required for the generation of an AM signal is shown schematically in Fig. 4-25. If the baseband signal shown being applied at the input has the form $K[1 + mf(t)]$, as in (4-15), normal AM with a carrier present is produced at the output. If the input is simply $f(t)$, a baseband signal without the addition of a fixed constant (or dc value), suppressed carrier DSB or AM results. Where DSB AM is specifically desired, two such modulators are often combined in a balanced configuration to ensure that no dc terms leak through to produce a carrier component. Details appear in the next section.

A simple example of a product modulator might be a galvanometer with separate windings for the magnet and the coil. The force on the coil is then proportional to the product of the currents in each winding. If

$$F \propto i_1 i_2 \tag{4-17}$$

$$i_i = a[1 + mf(t)] \tag{4-18}$$

(i.e., a dc bias plus the modulating signal) and

$$i_2 = b \cos \omega_c t \tag{4-19}$$

the force on the coil, as a function of time, is of the same form as Eq. (4-15). The coil motion would then appear as an amplitude-modulated signal.

A motor with stator and rotor separately wound would give the same results. A loudspeaker, with the modulating signal and an appropriate dc bias applied to the voice coil and the carrier to the field winding, is again a similar device. These are all examples of *product* modulators.

These product modulators and other types of modulators commonly used may be classified in one of two categories:

1. Their terminal characteristics are nonlinear.
2. The modulation device contains a switch, as just described, which changes the system from one linear condition to another.

A *linear* time-invariant device by itself can never provide the product function needed. We may recall from our discussion of linear systems in Chap. 2 that *a non-time-varying linear system can generate no new frequencies.* The response to all sine-wave inputs may be superposed at the output. Such a system can therefore never be used as a modulator.

If we now add a switch to a linear system which switches the system in a specified manner from one linear condition to another (a simple example would be a switch turning the system on and off) and which does this independently of the signal, the system is still basically linear; the two basic characteristics of a linear system still apply. Such a system is still governed by linear differential equations, but the coefficients of the equations now vary with time according to the prescribed switching action. The system is now a linear time-varying system. As we shall see shortly, such a system can be used for modulation and will generate new frequencies. This is then a generalization of our previous switching process.

If instead of a switch we now introduce a device whose static terminal characteristics are nonlinear, the differential equations governing the system become nonlinear differential equations. Again, as we shall see, modulation or generation of new frequencies becomes possible. Two typical kinds of nonlinear characteristics are shown in Fig. 4-26. Actually, all physical devices have some curvature or nonlinearity in their static terminal characteristics, but we usually assume the region of operation to be small enough so that the characteristics are very nearly linear over this region.

Figure 4-26a might be the characteristic of a typical square-law diode or transistor. Figure 4-26b might represent an approximation to a linear rectifier. The two curves differ in the type of nonlinearity. The linear rectifier of Fig. 4-26b has a strong nonlinearity at one point and is approximated by linear characteristics elsewhere. (The use of the term linear rectifier and the piecewise-linear characteristics assumed have sometimes led to the mistaken conclusion that the device is linear.) The characteristics of Fig. 4-26a have no one point of strong nonlinearity but are everywhere nonlinear, owing to the curvature of the characteristics.

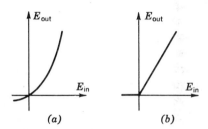

(a)   (b)

Figure 4-26 Nonlinear characteristics. (*a*) Nonlinearity due to curvature of characteristics. (*b*) "Strong" discontinuity; piecewise-linear characteristic.

Figure 4-26b can also be used as the representation of the characteristics of a switching device and, in fact, serves as a good example of the distinction between linear time-invariant, nonlinear, and switching characteristics. Thus, if the device is operated in one of the linear portions of the curve only, it behaves as a linear time-invariant device. Typically, the device would be biased to operate in a region far from the discontinuity, and the signals applied would be small enough to keep it in that region. The output is then a linear replica of the input, and if two signals are applied, the output is a superposition of the response to each. If the signal applied to the device is now increased to the point where the discontinuity is crossed and the two linear regions are involved, the output is no longer a simple linear function of the input. If one signal is applied, the output is a distorted version, and if two signals are applied, the output no longer can be found by superposing the two input responses separately. The output is then a nonlinear function of the input; as the input amplitude changes, the output does not change proportionally.

If the input-signal level is now dropped back to its original small value but is somehow switched in a predetermined fashion between the two linear regions of Fig. 4-26b, we have a linear switch with the corresponding linear time-varying equations.

Most modulators in use can thus be classified as having one of the two types of characteristics of Fig. 4-26 if switches are included as represented by Fig. 4-26b.

We shall first analyze simple modulators incorporating the nonlinear characteristics of Fig. 4-26a. The analysis of the piecewise-linear characteristic of Fig. 4-26b will be very similar to that describing the switching operation in Chap. 3.

Square-Law Modulator

The simplest form of this type of modulator appears in Fig. 4-27. e_o and e_i represent the incremental output and input variations, respectively, away from some fixed operating point (Fig. 4-27b). The nonlinear characteristics of this case are assumed continuous so that e_o may be expanded in a power series in e_i,

$$e_0 = a_1 e_i + a_2 e_i^2 + a_3 e_i^3 + \text{(higher-order terms)} \tag{4-20}$$

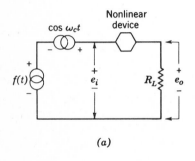
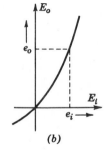

(a) (b)

Figure 4-27 Simple modulator. (a) Simple circuit. (b) Terminal characteristics.

The quadratic term in the power series denotes the presence of curvature in the characteristics and is all-important in the modulation process.

For the circuit in Fig. 4-27a

$$e_i = \underbrace{\cos \omega_c t}_{\text{Carrier}} + \underbrace{f(t)}_{\substack{\text{Modulating} \\ \text{signal}}} \tag{4-21}$$

Then, retaining just the first two terms of the power series of Eq. (4-20), we get

$$e_o = a_1 \cos \omega_c t + a_1 f(t) + a_2 [\cos^2 \omega_c t + 2f(t) \cos \omega_c t + f^2(t)]$$

$$= \underbrace{a_1 f(t) + a_2 \cos^2 \omega_c t + a_2 f^2(t)}_{\text{Unwanted terms}} + \underbrace{a_1 \cos \omega_c t \left[1 + \frac{2a_2}{a_1} f(t) \right]}_{\substack{\text{Amplitude-modulated} \\ \text{terms}}} \tag{4-22}$$

The second term thus contains the desired AM signal. (Let $m \equiv 2a_2/a_1$, $K \equiv a_1$.) The first term contains unwanted terms that can be filtered out. $\cos^2 \omega_c t$ gives $\frac{1}{2}(1 + \cos 2\omega_c t) = dc$ and twice the carrier frequency. $f^2(t)$ contains frequency components from dc to twice the maximum frequency component of $f(t)$, that is, $2B$. This is left as an exercise for the reader. It may be noted in passing that this device could also be used as a second-harmonic generator if $\cos \omega_c t$ only were introduced and its second harmonic retained.

Piecewise-Linear Modulator (Strong Nonlinearity)

A typical circuit (identical with the circuit of Fig. 4-27a) and the piecewise-linear characteristics for such a device (here chosen as a "linear" rectifier) are shown in Fig. 4-28. R_L is the load resistance; R_d, the diode forward resistance.

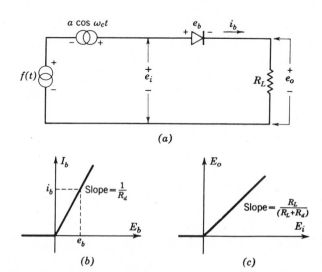

Figure 4-28 Piecewise-linear modulator. (a) Simple modulator. (b) Diode characteristic. (c) Circuit characteristic.

We should like to demonstrate that the piecewise-linear characteristics can also be used for simple AM. The discontinuity assumed in the terminal characteristics prevents the use of the power-series expansion as was done in the previous modulator.

We could, of course, approximate the characteristics by a polynomial. This would convert them to the continuous form of Fig. 4-27 with the discontinuity more pronounced at the origin. Both types of nonlinearity could thus be treated simultaneously. We prefer, however, to use a different approach similar to that used in discussing the switching operations in Chap. 3.

We shall assume that the carrier amplitude is much greater than the maximum value of $f(t)$,

$$|f(t)| \ll a$$

If the carrier alone were present, the rectifier would clip the negative halves of the carrier. Under the assumption of a strong carrier the clipping of the carrier plus modulating signal will occur approximately at the same point in the cycle as the carrier alone. We thus have

$$e_i = a \cos \omega_c t + f(t) \qquad |f(t)| \ll a$$

$$e_o \doteq \frac{R_L e_i}{R_L + R_d} = b e_i \qquad a \cos \omega_c t > 0 \tag{4-23}$$

$$e_o \doteq 0 \qquad a \cos \omega_c t < 0$$

Since the output now varies between two values (be_i and 0) periodically, at the frequency of the carrier, the input may be looked on as having been switched between two regions of the diode operation. By the simple expedient of assuming a weak signal compared with the carrier we have converted the nonlinear device into a linear switching device. Since the transitions from one region of operation to another are now variables of time independent of the signal $f(t)$, we have effectively replaced a nonlinear equation by a linear time-varying one. This approach is the one actually used in solving some types of nonlinear differential equations. It is to be emphasized, however, that this is an approximate technique, valid only for small signals. Equation (4-23) can be written mathematically as

$$e_o \doteq [a \cos \omega_c t + f(t)]S(t)$$

with
$$S(t) = b \qquad -\tfrac{1}{4}T < t < \tfrac{1}{4}T, \ T = \frac{1}{f_c} \tag{4-24}$$

$$S(t) = 0 \qquad t \text{ elsewhere}$$

and repeating at multiples of $T = 1/f_c$ seconds.

$S(t)$ is again our familiar periodic switching function as shown in Fig. 4-29.

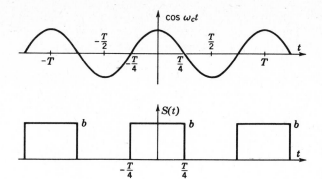

Figure 4-29 Switching function.

Again using the Fourier representation,

$$S(t) = b \left[\frac{1}{2} + \sum_{n=1}^{\infty} \frac{\sin (n\pi/2)}{n\pi/2} \cos n\omega_c t \right] \qquad (4\text{-}25)$$

But

$$e_o = [a \cos \omega_c t + f(t)]S(t)$$

By using the Fourier-series expansion for $S(t)$, performing the indicated multiplication, and collecting terms, e_o can be written in the form

$$
\begin{aligned}
e_o(t) = b \Bigg(\bigg\{ &\tfrac{1}{2}f(t) + \frac{2}{\pi} a \cos^2 \omega_c t \\
&+ \sum_{n=3}^{\infty} \frac{\sin (n\pi/2)}{n\pi/2} [f(t) + a \cos \omega_c t] \cos n\omega_c t \bigg\} \\
&+ \frac{a}{2} \cos \omega_c t \left[1 + \frac{4}{\pi a} f(t) \right] \Bigg)
\end{aligned} \qquad (4\text{-}26)
$$

Again, we can filter this output. The terms in the braces give direct-current, low frequencies up to B [the maximum frequency component of $f(t)$], and then higher frequencies: $2f_c$, $3f_c - B$, $3f_c$, $3f_c + B$, etc. After filtering,

$$e_o(t) \doteq K[1 + mf(t)] \cos \omega_c t$$

where

$$K = \frac{ab}{2} \qquad (4\text{-}27)$$

$$m = \frac{4}{\pi a}$$

The piecewise-linear modulator of Fig. 4-28 thus produces an amplitude-modulated output upon proper filtering.

We have actually demonstrated this modulation property for the device considered as a switch rather than as a nonlinear device, under the assumption of a small signal (relative to the carrier). The truly nonlinear solution leads to

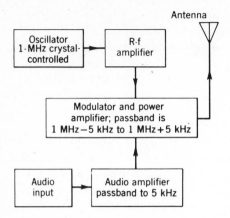

Figure 4-30 Standard high-level broadcast transmitter.

harmonics of the signal $f(t)$ and higher-order modulation products of the carrier and its harmonics. These additional terms would be filtered out in an actual modulator, leaving Eq. (4-27) as the solution in the more exact analysis also.

The two nonlinear devices just discussed are representative of many types that are used for low-level (i.e., low-power) modulation purposes. Examples include nonlinear-resistance modulators, semiconductor diode modulators, mechanical-contact modulators (also called "choppers," or *vibrators,* which are basically switching devices of the piecewise-linear type previously discussed), semiconductor modulators such as photocells and Hall-effect modulators using germanium crystals, magnetic modulators, nonlinear-capacitor modulators, vacuum-tube modulators, transistor modulators, etc.

For standard radio-broadcast work these low-level modulators are highly inefficient. (Their use in balanced modulator circuits for SSB transmission or as parts of servo systems is discussed later in this chapter.) Class C amplifier operation provides higher efficiency of power generation and is commonly used for standard broadcast transmitters. The modulation process here involves the gross nonlinearity due to class C operation. A tuned circuit at the amplified output then provides the necessary filtering to obtain the desired amplitude-modulated wave: a carrier plus sidebands.[15] A modulator-power amplifier combination provides the basic elements of the typical standard broadcast transmitter. A block diagram of the transmitter is shown in Fig. 4-30. This is a simplified version with only the essential elements shown.

Frequency Conversion

Multiplication by a sine wave, with its resultant shift in frequency, is used not only to generate AM signals, but to shift signals from one frequency band to

[15] Details of practical modulators, as well as other devices found in communication systems, may be found in the book *Communication Circuits: Analysis and Design,* by K. K. Clarke and D. T. Hess, Addison-Wesley, Reading, Mass., 1971.

another. Synchronous detection, described in Sec. 4-2 for the case of binary carrier transmission, and to be generalized in Sec. 4-7 to the case of any type of baseband signal, is an example in which high-frequency signals are shifted back down to baseband after multiplication by a sinewave at the carrier frequency.

More generally, the shift from one frequency band to another, using multiplication by a sine wave, is called *frequency conversion*. This process is often used to boost modulated carriers to a desired frequency range (one example would be the shifting of signals up to the microwave range for transmission via terrestrial microwave links or for transmission up to a stationary satellite), to shift carrier signals down to a specified intermediate frequency (i-f) range in which filtering and amplification are readily carried out (the superheterodyne radio receiver to be described in Sec. 4-7 includes precisely this function), or, in a series of conversion steps, to frequency multiplex a group of signals together. (Examples of frequency-multiplexing techniques appear in Sec. 4-11.) The term *mixer* is sometimes also used to designate a device in a receiver system that carries out frequency conversion.

The frequency conversion process is sketched schematically in Fig. 4-31. Assume that the signal to be multiplied by the sine wave is itself centered at a carrier frequency f_1, as shown in Fig. 4-31b. The input signal may then be written $g(t) = f(t) \cos \omega_1 t$, with $f(t)$ a baseband signal. Multiplication by $\cos \omega_2 t$

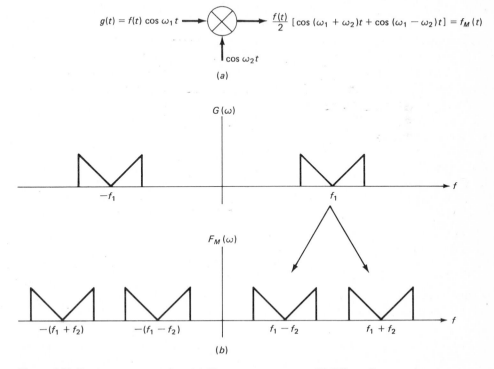

Figure 4-31 Frequency conversion. (a) Frequency converter. (b) Effect of conversion.

produces $f_M(t) = [f(t)/2][\cos{(\omega_1 + \omega_2)t} + \cos{(\omega_1 - \omega_2)t}]$, as shown in Fig. 4-31$a$. The spectrum of $f_M(t)$ is then shifted up and down by f_2 hertz, as shown in Fig. 4-31b. Either one of the two DSB signals, one centered at $(f_1 + f_2)$ hertz, the other at $(f_1 - f_2)$ hertz, may be selected by appropriate filtering. This process may be repeated if desired. As a special case, note that if $f_1 = 0$, we have just the product modulation discussed previously. If $f_2 = f_1$, we generate both the baseband signal $f(t)$ centered about 0 Hz and a signal at twice the original carrier frequency. If the carrier is to be doubled, we follow by a bandpass filter centered at $2f_1$. If we wish to select the baseband signal $f(t)$, we follow by a low-pass filter. The resultant process is the one we previously called synchronous detection. If $f_1 - f_2$ is required to be some specified intermediate frequency f_0, we must have $f_2 = f_1 + f_0$ or $f_1 - f_0$. (Note that f_2, the frequency of the sine wave multiplying the input signal, can thus be above f_1 or below f_1. Either case is possible.) If the output of the product modulator is followed by a bandpass filter centered at $f_0 = f_1 - f_2$, the difference frequency terms only are transmitted.

In modern systems the injected sine wave $\cos{\omega_2 t}$ is often synthesized digitally. Discrete samples of $\cos{\omega_2 t}$, enough to generate a full cycle of the sine wave, can be stored in tabular form in a random-access memory.

Specific examples of the frequency-conversion process appear in the sections following, as well as in problems at the end of the chapter.

4-5 SINGLE-SIDEBAND TRANSMISSION AND BALANCED MODULATORS

We note from the frequency plot of the amplitude-modulated carrier (Fig. 4-24) that the desired information to be transmitted [as originally given by variations of the modulating signal $f(t)$] is carried in one of the two sidebands. The carrier itself carries no information. Transmission of the entire spectrum thus represents a waste of frequency space since transmitting both sidebands requires double the bandwidth needed for transmitting one sideband. It also represents a waste of power (especially power in the carrier). This has resulted in the widespread use of SSB transmission for transoceanic radiotelephone circuits and wire communications. In this type of transmission the carrier and one sideband are suppressed and only the remaining sideband transmitted. More recently, the Bell System in the United States has developed a single-sideband microwave radio, the AR6A, that allows 6,000 voice circuits to be transmitted over a 28.6-MHz-wide channel.[16] This is three times the number previously possible using FM techniques. Details will be presented later in this chapter in discussing frequency division multiplexing.

---

[16] R. E. Markle, "Single Sideband Triples Microwave Radio Route Capacity," *Bell Lab. Rec.*, vol. 56, no. 4, pp. 104–110, April 1978.

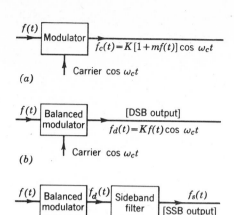

(a)

(b)

(c)

Figure 4-32 Amplitude-modulation systems. (*a*) Normal AM. (*b*) Suppressed carrier, or DSB. (*c*) SSB.

With the apparent advantages of SSB over AM, why is it that SSB is not used for standard radio broadcasting? The answer is that the circuitry *required at the receiver* is too complex *at present*. Compatible SSB techniques have been proposed to alleviate this problem.[17] To demonstrate these statements more quantitatively, we must discuss methods of producing and then detecting SSB signals.

A method commonly used to generate an SSB signal is first to suppress the carrier of the AM signal. The resulting transmission is, as noted earlier, *suppressed-carrier,* or *double-sideband* (DSB), transmission. One of the two sidebands is then filtered out. The carrier suppression is usually accomplished by means of a *balanced modulator,* as indicated in the block diagrams of Fig. 4-32.

A normal AM output has the form

$$f_c(t) = K[1 + mf(t)] \cos \omega_c t \qquad (4\text{-}28)$$

while the corresponding mathematical expression for a DSB output is

$$f_d(t) = K'f(t) \cos \omega_c t \qquad (4\text{-}29)$$

The DSB system differs from normal AM simply by suppressing the carrier term. This means that with no modulating signal applied the output should be zero. If we recall the nonlinear devices discussed previously, the carrier term arose because of a quiescent condition or dc bias somewhere in the modulator. If we can balance out this quiescent condition, we have a simple product modulator with the output given by Eq. (4-29). (In the case of the magnetic

[17] L. R. Kahn, "Compatible Single Sideband," *Proc. IRE,* vol. 49, pp. 1503–1527, October 1961. The article describes a compatible SSB system developed for use with normal AM radio sets. See also Schwartz, Bennett, and Stein, *op. cit.,* pp. 193–198, for a discussion of this and other compatible SSB systems.

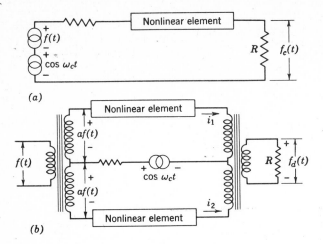

(a)

(b)

Figure 4-33 Balanced modulator. (a) Modulator. (b) Balanced modulator.

product modulators mentioned—the galvanometer, motor, loudspeaker—the assumption was that a dc term appeared in one of the currents. If we balance out this dc current, we have the desired suppressed-carrier output.)

One possible scheme for suppressing or balancing out the carrier term is shown in Fig. 4-33. This method uses two nonlinear elements of the types previously considered in a balanced arrangement. The resulting device is then an example of a *balanced modulator*.

To demonstrate the balancing effect, assume that the transformer between the load R and the modulator is ideal. The modulator then sees the resistor load. The upper and the lower sections of the balanced modulator are then each identical with Fig. 4-33a, and the previous analysis of such nonlinear circuits holds. In particular, the two currents flowing may be written in the form

$$i_1 = K[1 + mf(t)] \cos \omega_c t \qquad (4\text{-}30)$$

and

$$i_2 = K[1 - mf(t)] \cos \omega_c t \qquad (4\text{-}31)$$

(the other terms generated in the nonlinear process are assumed filtered out). But the load voltage $f_d(t)$ is proportional to $i_1 - i_2$, or

$$f_d(t) = K'f(t) \cos \omega_c t \qquad (4\text{-}32)$$

Since the switching operation discussed in Chap. 3 in connection with sampling was shown to produce terms of exactly the DSB form, these switches may also be considered balanced modulators. One such switch, the shunt-bridge diode modulator, is shown in Fig. 4-34. The diodes in the bridge mod-

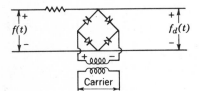

Figure 4-34 Balanced modulator: shunt-bridge diode modulator.

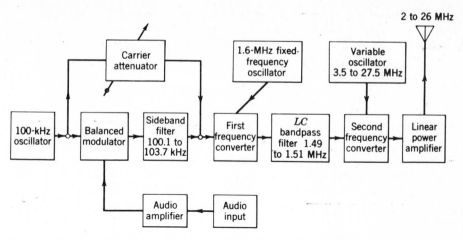

Figure 4-35 Single-sideband transmitter. (From B. Fisk and C. L. Spencer, "Synthesizer Stabilized Single-Sideband Systems," *Proc. IRE,* vol. 44, no. 12, p. 1680, December 1956, by permission.)

ulator may be treated as switches, switching on and off at the carrier rate. With the diodes assumed identical (i.e., balanced), the bridge is balanced when $f(t) = 0$ and there is no output; with the carrier input of the polarity indicated the diodes are essentially short-circuited out, and the output is again zero. When the carrier reverses polarity, however, the diodes open and $f_d(t)$ is equal to $f(t)$. The output is thus alternately $f(t)$ and 0, switching at the carrier frequency rate, just as required here, and for the sampling function in Chap. 3 as well.

The balanced modulator is essentially the heart of the SSB transmitters that are most commonly used at the present time. The balanced modulators are operated at low power levels, with linear amplifiers then following to reach the required transmitted power. This is in contrast to ordinary AM (Fig. 4-30), where the modulation is frequently carried out at high power levels, with class C amplifiers.

In addition to the balanced modulator, the SSB system requires a sideband filter with sharp cutoff characteristics at the edges of the passband. (Attenuation at the carrier frequency must be especially high.) Because of the stringent requirements on the fitter, the modulation is commonly performed at a relatively low fixed frequency at which it is possible to design the required sideband filter.

A typical SSB transmitter incorporating the two features mentioned—low-level balanced modulator and low-frequency (l-f) filtering—is shown in Fig. 4-35.[18] The modulation process in this transmitter is carried out at 100 kHz. Two levels of frequency conversion, or mixing, are employed, with the signal successively shifted up in frequency by the frequency of the two injected signals (1.6 MHz and the final variable frequency).

[18] B. Fisk and C. L. Spencer, "Synthesizer Stabilized Single-Sideband Systems," *Proc. IRE,* vol. 44, no. 12, p. 1680, December 1956.

In the discussion of SSB detection to follow, the need for reinjecting the suppressed carrier will be discussed. To ensure that the reinjected carrier is of precisely the right frequency a pilot carrier of known amplitude is transmitted (usually 10 to 20 dB below the carrier in normal AM). The carrier attenuator in Fig. 4-35 provides the pilot carrier.

4-6 SSB SIGNAL REPRESENTATION: HILBERT TRANSFORMS[19]

Note that although we have written mathematical expressions for both normal AM and DSB (suppressed-carrier) signals [Eqs. (4-28) and (4-29)], the SSB generation has been described only in terms of filtering out one of the sidebands of the DSB signal. It is of interest to develop an explicit expression for a SSB signal not only for its own sake, but because it also serves two other purposes: (1) as a by-product it indicates an alternative way of generating SSB signals (the *phase-shift method* of SSB generation); and (2) it enables us to discuss SSB detection in a quantitative way.

To obtain the desired expression for a single-sideband signal assume we have available a baseband or low-frequency time function $z(t)$ whose Fourier transform $Z(\omega)$ is nonzero for positive frequencies only. This is admittedly not a real or physical possible time function, since we showed in Chap. 2 that all real-time functions had Fourier transforms defined over both negative and positive frequency ranges. [In fact, recall that $|F(\omega)| = |F(-\omega)|$ for real $f(t)$, while the phase angle of $F(\omega)$ is odd symmetric about zero frequency.] By definition, then,

$$Z(\omega) = 0 \qquad \omega < 0 \qquad (4\text{-}33)$$

An example of such a Fourier transform is shown in Fig. 4-36a. Now assume $z(t)$ multiplied by $e^{j\omega_c t}$. A little thought will indicate that this corresponds to a translation up by frequency ω_c, so that the transform of $z(t)e^{j\omega_c t}$ is just $Z(\omega - \omega_c)$. This is shown in Fig. 4-36b. It is apparent from the figure that this is just a SSB signal with *upper* sideband only present. A function $z(t)$ with the peculiar one-sided spectral property is called an *analytic signal*.[20] Since $z(t)e^{j\omega_c t}$ is complex, we simply take its real part to obtain the SSB signal $f_s(t)$:

$$f_s(t) = \text{Re}\,[z(t)e^{j\omega_c t}] \qquad (4\text{-}34)$$

Now given a modulating or information-bearing signal $f(t)$, how does one generate the analytic signal $z(t)$ from it? Once we show this we have in essence closed the loop, indicating how one goes from $f(t)$ to its SSB version $f_s(t)$. Consider $f(t)$ passed through a $-90°$ phase shifter, a device which shifts the phase of all positive frequency components of $f(t)$ by $-90°$ and all negative

[19] Schwartz, Bennett, and Stein, *op. cit.*, pp. 29–35.
[20] *Ibid*.

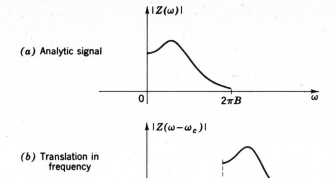

(a) Analytic signal

(b) Translation in
frequency

Figure 4-36 Generation of a SSB signal.

components by 90°. (The phase characteristic is always an *odd* function of frequency.) Call the output of the phase shifter $\hat{f}(t)$. We then have

$$\begin{aligned}\hat{F}(\omega) &= -jF(\omega) \qquad \omega \geq 0 \\ &= +jF(\omega) \qquad \omega < 0\end{aligned} \tag{4-35}$$

Note now that $j\hat{F}(\omega) = F(\omega)$, $\omega \geq 0$; $= -F(\omega)$, $\omega < 0$. This indicates that we can generate the analytic signal $z(t)$ by adding $j\hat{f}(t)$ to $f(t)$:

$$z(t) = f(t) + j\hat{f}(t) \tag{4-36}$$

For we then have

$$\begin{aligned}Z(\omega) &= 2F(\omega) \qquad \omega \geq 0 \\ &= 0 \qquad\qquad \omega < 0\end{aligned} \tag{4-37}$$

just as desired (Fig. 4-37). By this expedient of defining an analytic signal $z(t)$

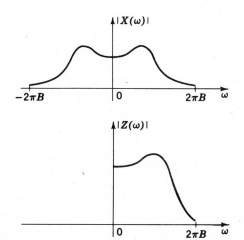

Figure 4-37 Modulating signal and its analytic signal.

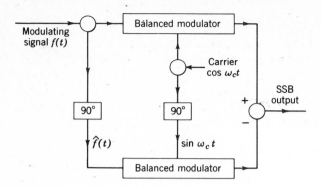

Figure 4-38 Phase-shift method of generating SSB (upper-sideband case).

and then showing how it may be generated from $f(t)$ and its 90° phase-shifted version $\hat{f}(t)$, we have finally the desired SSB representation in terms of $f(t)$:

$$f_s(t) = \text{Re}[z(t)e^{j\omega_c t}] = \text{Re}\{[f(t) + j\hat{f}(t)]e^{j\omega_c t}\}$$
$$= f(t) \cos \omega_c t - \hat{f}(t) \sin \omega_c t \qquad (4\text{-}38)$$

Equation (4-38) is not only the explicit expression for a SSB signal in terms of the baseband signal $f(t)$; it tells us in addition how to generate $f_s(t)$; that is, we generate a DSB signal $f(t) \cos \omega_c t$ by using a product or balanced modulator, generate another DSB signal $\hat{f}(t) \sin \omega_c t$ by phase shifting $f(t)$ by 90° and then multiplying by $\sin \omega_c t$ and then subtracting the two. This phase-shift method of generating SSB signals is diagrammed in Fig. 4-38.

It is left to the reader to show that the *lower* sideband only may be similarly generated by writing

$$f_s(t) = f(t) \cos \omega_c t + \hat{f}(t) \sin \omega_c t \qquad (4\text{-}39)$$

As a simple example of these results, and to develop somewhat more insight into the significance of the operations indicated by Eqs. (4-38) and (4-39), consider $f(t)$ to be just the single-frequency term $\cos \omega_m t$. The DSB signal is then

$$f_{d1}(t) = \cos \omega_m t \cos \omega_c t = \tfrac{1}{2}[\cos (\omega_m + \omega_c)t + \cos (\omega_c - \omega_m)t] \qquad (4\text{-}40)$$

containing the expected two sideband frequencies. How do we now cancel one of these sideband terms? If it is desired to retain the upper-sideband term only, we must subtract from Eq. (4-40) an expression of the form

$$f_{d2}(t) = \tfrac{1}{2}[\cos (\omega_c - \omega_m)t - \cos (\omega_m + \omega_c)t] = \sin \omega_m t \sin \omega_c t \qquad (4\text{-}41)$$

Upon subtracting Eqs. (4-40) and (4-41), the lower sideband cancels, leaving the desired upper sideband,

$$\cos (\omega_c + \omega_m)t$$

Similarly, upon *adding* Eq. (4-40) and (4-41), the upper sideband cancels, leaving the lower sideband

$$\cos (\omega_c - \omega_m)t$$

These are, of course, the operations indicated by Eqs. (4-38) and (4-39).

The 90° phase shift of all frequency components of $f(t)$ is commonly called a Hilbert transformation, and $\hat{f}(t)$ is called the *Hilbert transform* of $f(t)$. We shall use this transform in the next section in discussing the detection of SSB signals.

One major problem in the design of phase-shift SSB systems is the practical realization of the wideband 90° phase-shift network, for *all* the frequency components of the modulating signal $f(t)$ must be shifted by 90°.

4-7 DEMODULATION, OR DETECTION

The process of separating a baseband or modulating signal from a modulated carrier is called demodulation, or detection. Detection is, of course, necessary in all radio receivers where the information to be received is carried by the modulating signal.

We have discussed the demodulation process previously in Sec. 4-2, in connection with high-frequency digital signal detection. We considered there two basic methods of demodulating high-frequency signals: envelope detection and synchronous (coherent) detection. Synchronous detection was also mentioned briefly in Sec. 4-4 in connection with frequency conversion. Both methods again appear here in discussing AM and SSB demodulation. In particular, we shall find that synchronous detection is generally required for SSB (or DSB) systems, while normal AM may use either detection method.[21] Since, as noted earlier, envelope detection is much more simply instrumented, this is the preferred method of AM detection.

Demodulation is basically the inverse of modulation and requires nonlinear or linear time-varying (switching) devices also. Since the nonlinear circuits used are essentially the same, the details of detector operation are left as exercises for the reader. (Again, two types of nonlinear detector may be considered: the square-law detector with the current-voltage characteristics represented by a power series, and the piecewise-linear detector with a nonlinearity concentrated at one point.)

Recall from Sec. 4-2 that envelope detection consisted of passing the amplitude-modulated carrier through a nonlinear device and then low-pass filtering the nonlinear output. One common configuration, already noted in Sec. 4-2, consists of a half-wave diode rectifier followed by a parallel RC circuit. The circuit is diagrammed in Fig. 4-39. The rectified output is shown in the two cases of filtering and no filtering.

The analysis of the detector of Fig. 4-39a is carried out very simply by

---

[21] Some thought will convince the reader that amplitude modulation is the generalized form of on-off keying, while double-sideband modulation corresponds to phase-shift keying in the binary case.

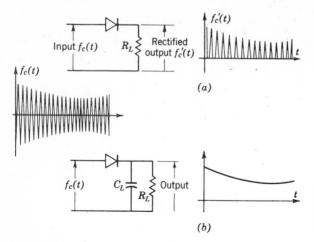

Figure 4-39 Amplitude-modulation detector. (*a*) Detection, no filtering. (*b*) Detection with filtering.

either treating the diode as a square-law device (as in Sec. 4-2),[22] or as a piecewise device. In the former case, it is apparent that with the input given by

$$f_c(t) = K[1 + mf(t)] \cos \omega_c t \qquad (4\text{-}42)$$

the output may be written

$$f'_c(t) = [f_c(t)]^2 = K^2[1 + mf(t)]^2 \cos {}^2\omega_c t \qquad (4\text{-}43)$$

Expanding the expression and selecting the low-pass terms, it is easily shown that these latter contain the desired $f(t)$ plus distortion $[f^2(t)]$ components. Similarly, if the diode is approximated as a piecewise-linear device, switching at the carrier-frequency rate, the output may be written

$$f'_c(t) = K[1 + mf(t)] \cos \omega_c t \, S(t) \qquad (4\text{-}44)$$

with $S(t)$ the switching function previously defined. By again expanding $S(t)$ in its Fourier series, it can be shown very simply that the output contains a component proportional to $f(t)$ plus higher-frequency terms (sum and difference frequencies of carrier and modulating signal).

The capacitor of Fig. 4-39*b* serves to filter out these higher-frequency terms. Looked at another way, C_L is chosen so as to respond to envelope variations, but the circuit time constant (including the diode forward resistance) does not allow the circuit to follow the high-frequency (h-f) carrier variations.[23] A block diagram of a typical superheterodyne radio receiver incorporating such a detector is shown in Fig. 4-40.

The incoming signal passes first through a tuned radio-frequency (r-f) amplifier which can be tuned variably over the radio band 550 to 1,600 kHz.

---

[22] Other nonlinear representations lead to essentially the same results, but with much more algebraic complexity and manipulation.

[23] Practical envelope detectors are discussed in Clarke and Hess, *op. cit.*, chap. 10.

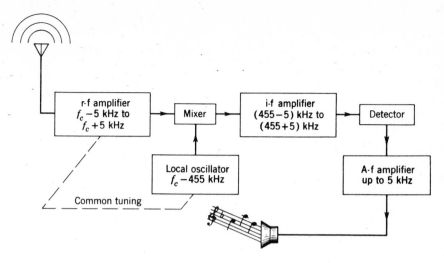

Figure 4-40 Superheterodyne AM receiver.

This signal is then mixed with a locally generated signal. The sum and difference frequencies generated contain a term centered about 455 kHz. (The local oscillator and r-f amplifier are tuned together so that there is always a difference frequency of 455 kHz between them.) The mixer acts as a frequency converter, shifting the incoming signal down to the fixed intermediate frequency of 455 kHz. Several stages of amplification are ordinarily used, with double-tuned circuits providing the coupling between stages. The intermediate-frequency (i-f) signal is then detected as described above, amplified further in the audio-frequency (a-f) amplifiers and applied to the loudspeaker. The superheterodyning operation refers to the use of a frequency converter and fixed, tuned i-f amplifier before detection.

The synchronous detection operation as indicated in Secs. 4-2 and 4-4 consists simply of multiplying the incoming carrier signal by a locally generated carrier $\cos \omega_c t$ and low-pass-filtering the resultant. This is shown again in Fig. 4-41. Based on our discussion in the previous sections, it is apparent that the multiplication by $\cos \omega_c t$ serves to translate the frequency components in the incoming-modulated carrier up and down by the carrier signal. The low-pass filter rejects the terms centered about $2f_c$, and passes those at baseband frequencies. The synchronous detection process can thus equally well be considered to consist of a frequency converter plus filter. This conversion process is

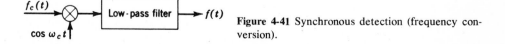

Figure 4-41 Synchronous detection (frequency conversion).

diagrammed in Fig. 4-42, for the special case of SSB demodulation. The multiplication by $\cos \omega_c t$ may be accomplished in innumerable ways as already noted previously: by the use of nonlinear devices, switching, digital synthesis, etc. The key factor, however, is the necessity of having available at the receiver the carrier frequency term $\cos \omega_c t$.

In the case of normal AM it is apparent that multiplication by $\cos \omega_c t$ results in the prefiltered expression

$$K[1 + mf(t)] \cos^2 \omega_c t$$

For a DSB signal the equivalent expression is

$$f(t) \cos^2 \omega_c t$$

This latter expression is of course a special case of the one discussed in Sec. 4-4 in connection with frequency conversion (Fig. 4-31). For the SSB signal the prefiltered term is, using Eq. (4-38),

$$f(t) \cos^2 \omega_c t - \hat{f}(t) \sin \omega_c t \cos \omega_c t$$

It is apparent that the desired term $f(t)$ appears at the filter output in all three cases.

For most types of baseband signals the local carrier must not only be of the right frequency but must be synchronized in phase with the carrier as well. This was emphasized in Sec. 4-2. If the carrier shifts in phase the resultant output signal may be a considerably distorted version of the baseband signal $f(t)$. To demonstrate this for the SSB case consider multiplication by a carrier term $\cos (\omega_c t + \theta)$. It is left for the reader to show that in this case the filtered output signal is given by

$$2f_o(t) = f(t) \cos \theta + \hat{f}(t) \sin \theta \tag{4-45}$$

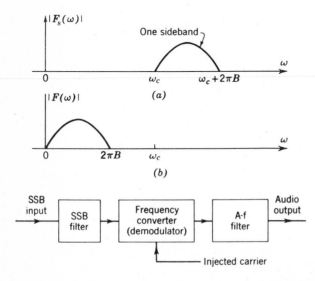

Figure 4-42 Single-sideband demodulation. (*a*) Before frequency conversion. (*b*) After frequency conversion.

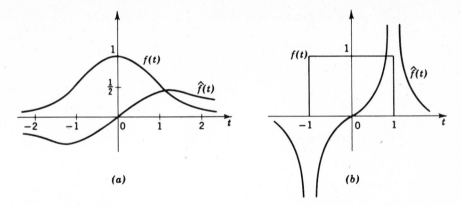

Figure 4-43 Examples of Hilbert transforms. (a) $f(t) = 1/(1 + t^2)$. (b) $f(t)$, a unit pulse.

Note that not only is the desired output reduced as θ increases from 0, but a distortion term $\hat{f}(t) \sin \theta$ appears. If $\theta = \pi/2$, in particular, only $\hat{f}(t)$ appears at the output.

Interestingly, it turns out that the human ear is relatively insensitive to phase changes in signals. It cannot distinguish the Hilbert transform $\hat{f}(t)$ from $f(t)$, so that there is no noticeable distortion in received signals that are destined to be *heard*. For all other signals—digital, TV, etc.—this distortion term limits the operation of the SSB system. SSB receivers must thus be synchronized in phase, which again presents the problem noted in Sec. 4-2.

The kind of distortion to be expected with lack of phase synchronization may be determined by evaluating the Hilbert transform $\hat{f}(t)$. This is readily accomplished by using the 90° phase-shift definition. As an example, assume the signal $f(t)$ is an *even* function of time. (This simplifies the discussion.) Test pulses of the kind considered in Chap. 2 are examples. Then $F(\omega)$ is an even function of frequency. The Hilbert transform $\hat{f}(t)$ is then given by

$$\hat{f}(t) = \frac{1}{2\pi} \int_{-\infty}^{\infty} \hat{F}(\omega) e^{j\omega t}\, d\omega$$

$$= \frac{1}{\pi} \int_{0}^{\infty} F(\omega) \sin \omega t\, d\omega \qquad (4\text{-}46)$$

Notice that $\hat{f}(t)$ is then an odd function of time. Using Eq. (4-46) we may calculate $\hat{f}(t)$ quite readily. Two simple examples are tabulated below, and sketched in Fig. 4-43.

1. $f(t) = 1/(1 + t^2)$. Then

$$F(\omega) = \pi e^{-|\omega|} \qquad \hat{f}(t) = \frac{t}{1 + t^2}$$

2. $f(t)$ = the unit pulse of Fig. 4-43b. Then

$$F(\omega) = 2\,\frac{\sin \omega}{\omega} \qquad \hat{f}(t) = \frac{1}{\pi} \ln \frac{|t + 1|}{|t - 1|}$$

Note that in the case of the sharply varying unit pulse the Hilbert transform becomes infinite at the points of sharp variation. In the case of the more realistic pulse of Fig. 4-43a, these peaks in the Hilbert transform are less pronounced but still quite high. This distortion becomes quite noticeable in TV and other signal receivers when the phase shift θ approaches $\pi/2$.

Various methods of synchronizing the local carrier exist, as noted in Sec. 4-2. A small amount of unmodulated carrier energy is commonly sent as a pilot signal, and after extraction at the receiver, is used to provide the necessary synchronization.

A typical SSB-receiver block diagram is shown in Fig. 4-44.

The carrier filter and sideband filter are used to separate the sideband and the pilot carrier. Double-frequency conversion (the first and the second mixers) is used, just as in the case of the SSB transmitter (Fig. 4-35), to obtain a convenient range of frequency for the filters. The output of the carrier filter is

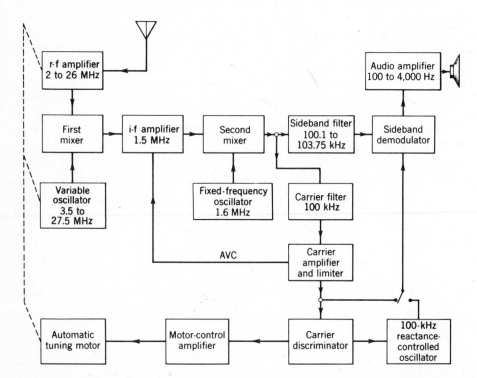

Figure 4-44 Single-sideband receiver. (From B. Fisk and C. L. Spencer, "Synthesizer Stabilized Single-Sideband Systems," *Proc. IRE*, vol. 44, no. 12, p. 1680, December 1956, by permission.)

amplified and limited and used for automatic tuning controls, either as a reinsertion carrier directly or to lock the frequency of a local oscillator that furnishes a suitable insertion carrier.

4-8 FREQUENCY MODULATION

We have investigated thus far in the past few sections the effect of slowly varying the amplitude of a sinusoidal carrier in accordance with some information to be transmitted. The desired information is then found to be concentrated in sidebands about the carrier frequency. By choosing the carrier frequency high enough, information transmission by means of radio (through the air) becomes practicable. Alternatively, with AM the signal-frequency spectrum may be shifted to a frequency range where circuit design becomes more feasible, where circuit components are more readily obtained and more economically built, or where equipment size and weight can be reduced.

In addition, many information channels may be transmitted simultaneously by means of frequency-multiplexing techniques.

For these reasons and others, AM systems of the standard, SSB, or DSB type are commonly used in both the communication and the control field.

Amplitude modulation is, however, not the only means of modulating a sine-wave carrier. We could just as well modulate the phase and frequency of a sine wave in accordance with some information-bearing signal. And such frequency-modulation (FM) systems are of course also utilized quite commonly.

We first discussed all three methods of modulating a sine-wave carrier in Sec. 4-2 in the special case of binary modulating signals. Just as we then extended the AM analysis to more general forms of information-bearing signals in the sections that followed, we propose to investigate FM in more generality in this section and in those that follow.

We noted previously, in discussing binary FM or FSK systems, that FM transmission requires wider transmission bandwidths than the corresponding AM systems. We shall find this to be generally true for all types of modulating signals. Why then use FM? Again, as noted earlier, FM provides better discrimination against noise and interfering signals. This fact will be demonstrated in Chap. 5 after we have studied noise in systems.

Interestingly, the expression found for FM transmission bandwidth in the binary case, $2\Delta f + 2B$, with Δf the frequency deviation away from the carrier and B the baseband bandwidth, will be found to serve as a quite useful rule of thumb for more complex signals as well.

Although we shall briefly consider phase modulation (PM), the stress will be placed on FM. Actually, for complex modulating signals, one form is easily derived from the other; and both are encompassed in the class of *angle-modulation* systems, as will be demonstrated below.

Before going further we need first to specify what we mean by *frequency modulation*. We might start intuitively by saying that we shall consider a frequency-modulation system one in which the frequency of the carrier is caused to vary in accordance with some specified information-carrying signal. Thus we might write the frequency of the carrier as $\omega_c + Kf(t)$, where $f(t)$ represents the signal and K is a constant of the system. This is, of course, analogous to the AM case, and was possible in the binary FM case. We run into some difficulty, however, when we attempt to express the more general frequency-modulated carrier mathematically. For we can talk about the frequency of a sine wave only when the frequency is constant and the sine wave persists for all time. Yet here we are attempting to discuss a variable frequency!

The difficulty lies in the fact that, strictly speaking, we can talk only of the sine (or cosine) of an *angle*. If this angle varies linearly with time, we can specifically interpret the frequency as the derivative of the angle. Thus, if

$$f_c(t) = \cos \theta(t) = \cos (\omega_c t + \theta_o) \tag{4-47}$$

the usual expression for a sine wave of frequency ω_c, we are implicitly assuming $\theta(t)$ to be linear with time, with ω_c its derivative.

When $\theta(t)$ does not vary linearly with time, we can no longer write Eq. (4-47) in the standard form shown, with a specified frequency term. To obviate this difficulty, we shall define an *instantaneous radian frequency* ω_i to be the derivative of the angle as a function of time. Thus with

$$f_c(t) = \cos \theta(t) \tag{4-48}$$

we have

$$\omega_i \equiv \frac{d\theta}{dt} \tag{4-49}$$

(This then agrees, of course, with the usual use of the word *frequency* if $\theta = \omega_c t + \theta_0$.)

If $\theta(t)$ in Eq. (4-48) is now made to vary in some manner with a modulating signal $f(t)$, we call the resulting form of modulation *angle modulation*. In particular, if

$$\theta(t) = \omega_c t + \theta_0 + K_1 f(t) \tag{4-50}$$

with K_1 a constant of the system, we say we are dealing with a *phase-modulation* system. Here the phase of the carrier wave varies linearly with the modulating signal. Binary signaling is a special case here.

Now let the *instantaneous frequency,* as defined by Eq. (4-49), vary linearly with the modulating signal,

$$\omega_i = \omega_c + K_2 f(t) \tag{4-49a}$$

Then
$$\theta(t) = \int \omega_i dt = \omega_c t + \theta_0 + K_2 \int f(t) \, dt \tag{4-51}$$

This, of course, gives rise to an FM system. As an example, if we again consider binary FM (FSK), we have the baseband signal $f(t)$ switching between either of two states. Then $\omega_i = \omega_c \pm \Delta\omega$, and the instantaneous frequency switches correspondingly between two frequencies. The phase angle $\theta(t)$ increases linearly with time in any one binary interval T, switching back to its initial value θ_0 as a new interval begins.

Both phase modulation and frequency modulation are seen to be special cases of angle modulation. In the phase-modulation case the phase of the carrier varies with the modulating signal, and in the frequency-modulation case the phase of the carrier varies with the integral of the modulating signal. If we first integrate our modulating signal $f(t)$ and then allow it to phase modulate a carrier, this gives rise to a frequency-modulated wave. This is the method used for producing a frequency-modulated carrier in the Armstrong indirect FM system.

A frequency-modulated carrier is shown sketched in Fig. 4-45c. The modulating signal is assumed to be a repetitive sawtooth of period T $(2\pi/T \ll \omega_c)$. Compare this with Fig. 4-6a, in which the modulating signal is a periodic square wave. As the sawtooth modulating signal increases in magnitude, the FM oscillates more rapidly. Its amplitude remains unchanged, however.

Frequency modulation is a nonlinear process, and so, as pointed out in previous sections, we would expect to see new frequencies generated by the modulation process. As indicated by Eqs. (4-49a) and (4-51), the FM signal oscillates more rapidly with increasing amplitude of the modulating signal. We would, therefore, expect the frequency spectrum of the FM wave, or its bandwidth, to widen correspondingly.

The analysis of the FM process is inherently much more complicated than that for AM, particularly in dealing with a general modulating signal. This is due to the nonlinearity of the FM process. Superposition cannot be used, so the

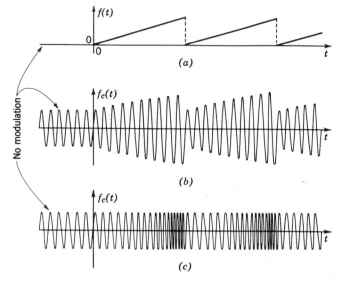

Figure 4-45 Frequency modulation. (a) Modulating wave. (b) AM carrier. (c) FM carrier.

analysis for one particular type of modulating signal cannot be readily applied to another. However, at the two extremes of small-amplitude modulating signal (narrowband FM) and large-amplitude modulating signal (wideband FM), it turns out that spectral analyses for different modulating signals produce the transmission bandwidth requirements: $2B$ hertz and $2\Delta f$ hertz, respectively, found in the binary FSK case, while at intermediate amplitudes of modulating signal the bandwidth requirement obtained for the binary FSK signal serves as a good rule of thumb for all types of signals.

Because of the difficulty of analyzing general FM signals, we shall consider here only one nonbinary modulating signal; namely, we shall assume $f(t)$ a single sine wave. Although this is a poor "approximation" to information-carrying signals, which may vary randomly and unpredictably with time, the results obtained are in substantial agreement with the much simpler binary FSK approach of Sec. 4-2, and will serve to further solidify the discussion there. (One would expect agreement between the two approaches, since a sine wave is similar in appearance to a square wave of the same frequency. The sine wave is in fact the first term in the square-wave Fourier series.)

Spectral analyses of other classes of modulating signals may be found in the literature.[24] The results obtained are, as noted, not substantially different than those obtained here for square-wave (FSK) and sine-wave modulating signals. FM spectral analysis with a random or noiselike modulating signal (often used as a model for real signals) leads to similar results.[25] Assume then a sinusoidal modulating signal at frequency f_m:

$$f(t) = a \cos \omega_m t \tag{4-52}$$

The instantaneous radian frequency ω_i is

$$\omega_i = \omega_c + \Delta\omega \cos \omega_m t \qquad \Delta\omega \ll \omega_c \tag{4-53}$$

where $\Delta\omega$ is a constant depending on the amplitude a of the modulating signal and on the circuitry converting variations in signal amplitude to corresponding variations in carrier frequency.

The instantaneous radian frequency thus varies about the unmodulated carrier frequency ω_c, at the rate ω_m of the modulating signal and with a maximum deviation of $\Delta\omega$ radians. Just as in the binary case $\Delta f = \Delta\omega/2\pi$ gives the maximum frequency deviation away from the carrier frequency and is called the *frequency deviation*.

The phase variation $\theta(t)$ for this special case is given by

$$\theta(t) = \int \omega_i \, dt = \omega_c t + \frac{\Delta\omega}{\omega_m} \sin \omega_m t + \theta_0 \tag{4-54}$$

[24] H. S. Black, *Modulation Theory*, Van Nostrand, Princeton, N.J., 1955; S. Goldman, *Frequency Analysis, Modulation, and Noise*, McGraw-Hill, New York, 1948; D. Middleton, *An Introduction to Statistical Communication Theory*, McGraw-Hill, New York, 1960, chap. 14.

[25] Schwartz, Bennett, and Stein, *op. cit.*, sec. 3.10; Middleton, *op. cit.*

θ_0 may be taken as zero by referring to an appropriate phase reference, so that the frequency-modulated carrier is given by

$$f_c(t) = \cos(\omega_c t + \beta \sin \omega_m t) \qquad (4\text{-}55)$$

with

$$\beta \equiv \frac{\Delta\omega}{\omega_m} = \frac{\Delta f}{f_m} \qquad (4\text{-}56)$$

Again, as in Sec. 4-2, for binary FSK, β is called the *modulation index* and is by definition the ratio of the frequency deviation to the baseband bandwidth. For a single sine wave of frequency f_m, the baseband bandwidth B is just f_m.

We noted previously that increasing the amplitude of the modulating signal should increase the bandwidth occupied by the FM signal. Increasing the modulating-signal amplitude corresponds to increasing the frequency deviation Δf or the modulation index β. We would thus expect the bandwidth of the FM wave to depend on β. This will be demonstrated in the sections to follow. The average power associated with the frequency-modulated carrier is independent of the modulating signal, however, and is in fact the same as the average power of the unmodulated carrier. This is again in contrast to the AM case, where the average power of the modulated carrier varies with the modulating-signal amplitude.

That this statement is true may be demonstrated by using Eq. (4-55) for a sinusoidal modulating signal. Assuming that $f_c(t)$ represents the instantaneous voltage impressed across a 1-Ω resistor, the average power over a cycle of the modulating frequency is given by

$$\frac{1}{T}\int_0^T f_c^2(t)\,dt = \frac{1}{T}\int_0^T \cos^2(\omega_c t + \beta \sin \omega_m t)\,dt$$

where $T = 1/f_m$. This expression may be rewritten as

$$\frac{1}{T}\int_0^T \frac{1 + \cos(2\omega_c t + 2\beta \sin \omega_m t)}{2}\,dt$$

The second term of the integral goes to zero since it is periodic in T. This assumes $\omega_m = 2\pi/T \ll \omega_c$. The first term gives $\frac{1}{2}$ watt. If the amplitude of the carrier had been written as A_c volts, the average power would have been found to be $A_c^2/2$ watts for a 1-Ω resistor.

Although shown only for a sinusoidal modulating signal, the foregoing result is true for any modulating signal whose highest-frequency component B hertz is small compared with the carrier frequency f_c. This is left as an exercise for the reader.

Narrowband FM

To simplify the analysis of FM, we shall treat it in two parts. We shall first consider the sinusoidally modulated carrier with $\beta \ll \pi/2$, and then with $\beta >$

$\pi/2$. Small β corresponds to narrow bandwidths, and FM systems with $\beta \ll \pi/2$ are thus called *narrowband FM systems*. The equations for *narrowband FM* appear in the form of the equations of the product modulator of the previous sections on AM and so give rise to sideband frequencies equally displaced about the carrier, just as in the case of AM.

To demonstrate this point, consider the sinusoidally modulated carrier of Eq. (4-55), and assume $\beta \ll \pi/2$. (This implies that the maximum phase shift of the carrier is much less than $\pi/2$ radians. This is ordinarily taken to mean $\beta < 0.2$ rad, although $\beta < 0.5$ rad is sometimes used as a criterion.) We have

$$f_c(t) = \cos (\omega_c t + \beta \sin \omega_m t)$$
$$= \cos \omega_c t \cos (\beta \sin \omega_m t) - \sin \omega_c t \sin (\beta \sin \omega_m t) \qquad (4\text{-}57)$$

But, for $\beta \ll \pi/2$, $\cos (\beta \sin \omega_m t) \doteq 1$, and $\sin (\beta \sin \omega_m t) \doteq \beta \sin \omega_m t$.

The frequency-modulated wave for small modulation index thus appears in the form

$$f_c(t) \doteq \cos \omega_c t - \beta \sin \omega_m t \sin \omega_c t \qquad \beta \ll \frac{\pi}{2} \qquad (4\text{-}58)$$

Note that this expression for $f_c(t)$, the frequency-modulated carrier, has a form similar to that of the output of a product modulator; it contains the original unmodulated carrier term plus a term given by the product of the modulating signal and carrier. For the sinusoidal modulating signal, $\beta \sin \omega_m t$, this product term provides sideband frequencies displaced $\pm \omega_m$ radians from ω_c. The bandwidth of this narrowband FM signal is thus $2f_m = 2B$ hertz, agreeing with the FSK result.

If we had assumed a general modulating signal $f(t)$ instead of the sinusoidal modulating signal used here, we would have obtained similar results. For, as shown by Eqs. (4-49a) and (4-51), we then have

$$\omega_i = \omega_c + K_2 f(t)$$
$$\theta(t) = \int \omega_i dt = \omega_c t + \theta_0 + K_2 \int f(t) \, dt$$

Suppressing the arbitrary phase angle θ_0, and defining a new time function $g(t)$ to be the integral of $f(t)$, $g(t) \equiv \int f(t) \, dt$, we have, as the frequency-modulated carrier,

$$f_c(t) = \cos [\omega_c t + K_2 \, g(t)] \qquad (4\text{-}59)$$

If K_2 and the maximum amplitude of $g(t)$ are now chosen small enough so that $|K_2 g(t)| \ll \pi/2$,

$$f_c(t) = \cos \omega_c t - K_2 \, g(t) \sin \omega_c t \qquad (4\text{-}60)$$

Our previous discussion of product modulators indicates that the spectrum of this general case of narrowband FM consists of the carrier plus two sidebands, one on each side of the carrier and each having the form of the spectrum of the function $g(t)$. Narrowband FM is thus equivalent, in this sense, to AM. The bandwidth of a narrowband FM signal, in general, is $2B$ hertz, where B is the highest-frequency component of either $g(t)$ or its derivative $f(t)$, the original modulating signal. (Remember that the linear process of integration adds no new frequency components. The lower-frequency components, however, are accentuated in comparison with the higher-frequency components.) The FSK result is, of course, a special case here.

Although AM and narrowband FM have similar frequency spectra and their mathematical representations both appear in the product-modulator form, they are distinctively different methods of modulation. In the AM case we were interested in variations of the carrier envelope, its frequency remaining unchanged; in the FM case, we assumed the carrier amplitude constant, its phase (and effectively the instantaneous frequency also) varying with the signal. This distinction between the two types of modulation must be retained in the narrowband FM case also (here maximum phase shift of the carrier is assumed less than 0.2 rad).

To emphasize this distinction in the two types of modulation, note that the product modulator or sideband term in either Eq. (4-58) or Eq. (4-60) appears in phase quadrature with the carrier term ($\sin \omega_c t$ as compared with $\cos \omega_c t$). In the AM case, we had both carrier and sideband terms in phase,

$$f_c(t) = \cos \omega_c t + mf(t) \cos \omega_c t$$

For the narrowband FM case, we have

$$f_c(t) = \cos \omega_c t - K_2 g(t) \sin \omega_c t$$

That the inphase, or phase-quadrature, representation is fundamental in distinguishing between AM and narrowband FM (or, alternatively, small-angle phase modulation) is demonstrated very simply by the use of rotating vectors.[26]

We rewrite Eq. (4-58) in its sideband-frequency form,

$$f_c(t) = \cos \omega_c t - \beta \sin \omega_m t \sin \omega_c t$$

$$= \cos \omega_c t - \frac{\beta}{2} [\cos (\omega_c - \omega_m)t - \cos (\omega_c + \omega_m)t] \qquad (4\text{-}61)$$

This can also be written in the form

$$f_c(t) = \text{Re} \left[e^{j\omega_c t} \left(1 - \frac{\beta}{2} e^{-j\omega_m t} + \frac{\beta}{2} e^{j\omega_m t} \right) \right] \qquad (4\text{-}62)$$

where Re [ ] represents the real part of the expression in the square brackets.

$e^{j\omega_c t}$ may be represented as a unit vector rotating counterclockwise at the rate of ω_c rad/s. Superimposed on this rotation are changes in the vector due to the three terms in parentheses. We then take the real part of the resultant vector, or its projection on the real axis. If we suppress the continuous ω_c rotation and concentrate solely on the three terms in parentheses, we may plot these as the three vectors of Fig. 4-46a. Note that the resulting vector deviates

[26] Black, *op. cit.*, pp. 186–187.

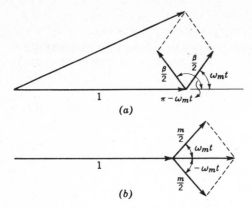

Figure 4-46 Vector representation. (a) Narrowband FM. (b) AM.

in phase from the unmodulated vector, while its amplitude is very nearly unchanged. (This is for the case $\beta \ll \pi/2$. Here β is shown larger for the sake of clarity.)

We can use the same vector approach for the AM case and get

$$f_c(t) = \cos \omega_c t + m \cos \omega_m t \cos \omega_c t$$

$$= \cos \omega_c t + \frac{m}{2} \left[\cos (\omega_c + \omega_m)t + \cos (\omega_c - \omega_m)t \right]$$

$$= \operatorname{Re} \left[e^{j\omega_c t} \left(1 + \frac{m}{2} e^{j\omega_m t} + \frac{m}{2} e^{-j\omega_m t} \right) \right] \qquad (4\text{-}63)$$

The three quantities in parentheses here are likewise shown plotted in Fig. 4-46b. Note that the two sideband vectors have been rotated by $\pi/2$ radians as compared with the FM case. The resultant vector here varies in amplitude but remains in phase with the unmodulated carrier term.

We can summarize by saying that the resultant of the two sideband vectors in the FM case will always be perpendicular to (or in phase quadrature with) the unmodulated carrier, while the same resultant in the AM case is colinear with the carrier term. The FM case thus gives rise to phase variations with very little amplitude change ($\beta \ll \pi/2$), while the AM case gives amplitude variations with no phase deviation.

The distinction and similarity between AM and narrowband FM (or phase modulation as well) leads us to a commonly used method of generating a frequency-modulated wave with small modulation index ($\beta < 0.2$ rad).

We demonstrated in our discussion of AM systems that the output of a balanced modulator provides just the product or sideband term required by Eq. (4-60). For an *amplitude-modulated* output we would then add to this output the *inphase* carrier term. This is shown in block diagram form in Fig. 4-47a.

To obtain a *phase-modulated* output, according to our previous discussion, we must add a *phase-quadrature* carrier term to the balanced-modulator output.

Such a system is shown in Fig. 4-47*b* in block-diagram form. (Remember that this is restricted only to small phase variations of 0.2 rad or less.)

As we demonstrated previously, however, phase modulation and FM differ only by a possible integration of the input modulating signal [Eq. (4-51)]. To obtain a narrowband frequency-modulated output, then, we need only integrate our input signal and then apply it to the phase-modulation system of Fig. 4-47*b*. The resultant narrowband FM system ($\beta < 0.2$ rad) is shown in Fig. 4-47*c*.

The narrowband FM system of Fig. 4-47*c* is used in the Armstrong indirect FM system.

Wideband FM

We have just shown that a frequency-modulated signal with small modulation index ($\beta \ll \pi/2$) has a frequency spectrum similar to that of an amplitude-modulated signal. The significant distinction between the two cases arises from the fact that in the FM case the sidebands are in phase quadrature with the carrier, while in the AM case they are in phase with the carrier. The bandwidth of the narrowband FM signal, just like the AM signal, is thus $2B$, with B the maximum-frequency component of the modulating signal.

The noise and interference reduction advantages of FM over AM, mentioned previously, become significant, however, only for large modulation index ($\beta > \pi/2$). The bandwidths required to pass this signal become correspondingly large, as first noted in Sec. 4-2. Most FM systems in use are of this wideband type.

Here the comparison between FM and AM that was valid for small modulation index ($\beta \ll \pi/2$) ends. We shall show that the previous results for narrowband FM can be extended to the wideband case, however.

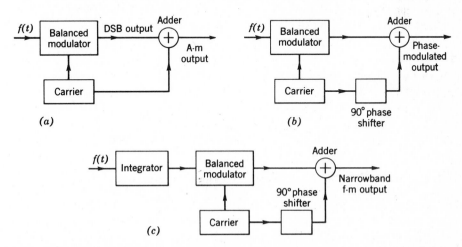

Figure 4-47 Evolution of a narrowband FM system. (*a*) Possible AM system. (*b*) Phase-modulation system (small phase deviations < 0.2 rad). (*c*) Narrowband FM system ($\beta < 0.2$ rad).

We demonstrate the increase in signal bandwidth with increasing β here for the idealized model of a sinusoidal modulating signal. The results, as noted earlier, are similar to those obtained in the binary FM case. The frequency-modulated carrier is again written in the expanded form

$$f_c(t) = \cos \omega_c t \cos (\beta \sin \omega_m t) - \sin \omega_c t \sin (\beta \sin \omega_m t) \qquad (4\text{-}64)$$

For $\beta \ll \pi/2$ we would, of course, get our previous result of a single carrier and two sideband frequencies. But now let β be somewhat larger at first. We can expand $\cos (\beta \sin \omega_m t)$ in a power series to give us

$$\cos (\beta \sin \omega_m t) \doteq 1 - \frac{\beta^2}{2} \sin^2 \omega_m t \qquad \beta^2 \ll 6 \qquad (4\text{-}65)$$

If we assume $\beta \ll \sqrt{6}$ and retain just the first two terms in the power series for the cosine, we get the additional term $\sin^2 \omega_m t \cos \omega_c t$ in the expression for $f_c(t)$. This term gives, upon trigonometric expansion, additional sideband frequencies spaced $\pm 2\omega_m$ radians from the carrier and also contributes a term $-\beta^2/4$ to the carrier. $\sin (\beta \sin \omega_m t)$ can still be represented by $\beta \sin \omega_m t$, the first term in its power-series expansion, for $\beta^2 \ll 6$, so that $f_c(t)$ becomes

$$f_c(t) \doteq \left(1 - \frac{\beta^2}{4}\right) \cos \omega_c t - \frac{\beta}{2} [\cos (\omega_m - \omega_c)t - \cos (\omega_m + \omega_c)t]$$

$$+ \frac{\beta^2}{8} [\cos (\omega_c + 2\omega_m)t + \cos (\omega_c - 2\omega_m)t] \qquad \beta^2 \ll 6 \qquad (4\text{-}66)$$

Note that the carrier term has now begun to decrease somewhat with increasing β, the first-order sidebands at $\omega_c \pm \omega_m$ increase with β, and a new set of sidebands (the second-order sidebands) appear at $\omega_c \pm 2\omega_m$. The spectrum of this signal is shown in Fig. 4-48b, while the narrowband case is plotted in Fig. 4-48a. This appearance of new sidebands with increasing modulation index corresponds to the widening of the spectrum in the binary FM case as the two

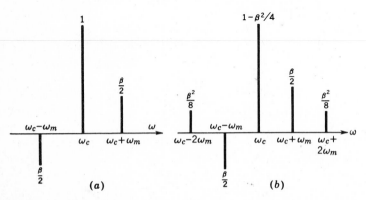

Figure 4-48 Effect of increasing β on FM spectrum ($\beta^2 \ll 6$). Sine-wave modulation. (a) Narrowband case, $\beta \ll \pi/2$. (b) Increasing β.

frequencies there are deviated further away from the carrier. This is distinctly different from the AM case, where the number of sideband frequencies was dependent solely on the number of modulating frequencies, and not on the amplitude of the modulating signal. Here new sets of significant sidebands appear as the modulation index increases. For a fixed modulating frequency, β is proportional to the amplitude of the modulating signal, so that increases in signal amplitude generate new sidebands. This produces a corresponding increase in bandwidth. (The bandwidth has been doubled in this case, going from $2f_m$ to $4f_m$.)

Since the average power in the frequency-modulated wave is independent of the modulating signal, increasing power in the sideband frequencies must be accompanied by a corresponding decrease in the power associated with the carrier. This accounts for the decrease in carrier amplitude that we have already noted.

As β increases further, we require more terms in the power-series expansion for both $\cos (\beta \sin \omega_m t)$ and $\sin (\beta \sin \omega_m t)$. This gives rise to increasingly more significant sideband components, and the bandwidth begins to increase with β, or with increasing amplitude of the original modulating signal. This power-series approach can be used to explore the characteristics of wideband FM. It becomes a tedious job of trigonometric manipulation to determine the significant sideband frequencies and their associated amplitudes, however, so that we shall resort to a somewhat different approach.

We are basically interested in determining the frequency components of the frequency-modulated carrier given by

$$f_c(t) = \cos (\omega_c t + \beta \sin \omega_m t)$$
$$= \cos \omega_c t \cos (\beta \sin \omega_m t) - \sin (\beta \sin \omega_m t) \sin \omega_c t$$

But we note that both $\cos (\beta \sin \omega_m t)$ and $\sin (\beta \sin \omega_m t)$ are periodic functions of ω_m. As such, each may be expanded in a Fourier series of period $2\pi/\omega_m$. Each series will contain terms in ω_m and all its harmonic frequencies. Each harmonic term multiplied by either $\cos \omega_c t$ or $\sin \omega_c t$, as the case may be, will give rise to two sideband frequencies symmetrically situated about ω_c. We thus get a picture of a large set of sideband frequencies in general, all displaced from the carrier ω_c by integral multiples of the modulating signal ω_m. The sidebands corresponding to the $\sin \omega_c t$ term will be quadrature sidebands, while those corresponding to the $\cos \omega_c t$ term will be inphase sidebands. For small values of β, $\cos (\beta \sin \omega_m t)$ and $\sin (\beta \sin \omega_m t)$ vary slowly, and so only a small number of the sidebands about ω_c will be significant in amplitude. As β increases, these two terms vary more rapidly, and the amplitudes of the higher-frequency terms become more significant. This picture of course agrees with our previous conclusion that increasing β produces a wider-bandwidth signal and agrees as well with the binary FM analysis. It is again in contrast to the AM case, where only the carrier and a single set of sidebands appear.

These remarks are given more significance by considering some plots of

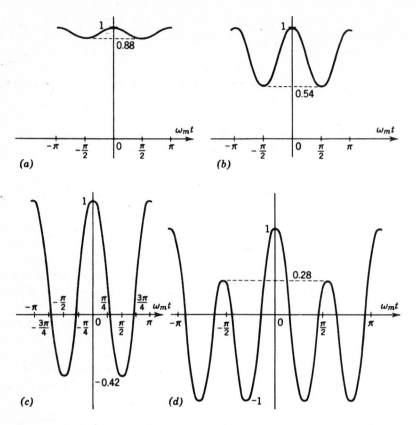

Figure 4-49 Plots of cos (β sin $\omega_m t$) for various β. (a) $\beta = 0.5$, cos (0.5 sin $\omega_m t$). (b) $\beta = 1$, cos (sin $\omega_m t$). (c) $\beta = 2$, cos (2 sin $\omega_m t$). (d) $\beta = 5$, cos (5 sin $\omega_m t$).

cos (β sin $\omega_m t$). These are shown in Fig. 4-49 for various values of β. The four curves drawn demonstrate some interesting points:

1. For $\beta < 0.5$ the curve can be represented approximately by a dc component plus a small component at twice the fundamental frequency ω_m. But these terms multiplied by cos $\omega_c t$ give just the carrier and the second-order sideband terms that we obtained previously for $\beta \ll \sqrt{6}$.

2. For $\beta < \pi/2 = 1.57$ the function remains positive and appears as a dc component with some ripple superimposed. One would thus expect a Fourier analysis to give a large dc component plus successively decreasing harmonics. This dc component, or average value of the function, decreases with β, however. Again, if we multiply this by cos $\omega_c t$, we obtain a picture of a carrier term decreasing with β, while the sidebands increase with β. These sidebands are all displaced from the carrier by even integral values of ω_m ($\pm 2\omega_m$, $\pm 4\omega_m$, etc.).

3. For $\beta > \pi/2 = 1.57$ the function takes on negative values. As β increases, the positive and negative excursions become more rapid and one would expect a greatly increased harmonic content, with much less energy at direct current. Again converting these results to variations about the carrier frequency, we would expect a rapid decrease in the energy content of the carrier for $\beta > \pi/2$ and increased energy in the sideband components.

The value $\beta = \pi/2$ thus represents the transition from a more or less slowly varying periodic time function with most of the spectral energy appearing in the carrier to a rapidly varying function with the spectral energy spread out over a wide range of frequencies.

We would have obtained similar results had we plotted the quadrature term $\sin (\beta \sin \omega_m t)$, with the one difference that the average value of this function is always zero. Its frequency components are all odd integral multiples of ω_m, so that they give rise to odd-order sidebands about the carrier when multiplied by $\sin \omega_c t$.

We can sum up these qualitative results by stating that for $\beta < \pi/2$ the frequency-modulated wave consists of a large carrier term plus smaller sideband terms. The carrier and even-order sidebands (those at even integral multiples of ω_m away from ω_c: $\pm 2\omega_m$, $\pm 4\omega_m$, $\pm 6\omega_m$, etc.) are contributed by the inphase term of the expression for the frequency-modulated carrier. The odd-order sidebands ($\pm \omega_m$, $\pm 3\omega_m$, etc.) are contributed by the quadrature-phase term.

For $\beta > \pi/2$, we get a picture of a wave with only a small carrier term plus increased energy in the sidebands. The bandwidth of the FM signal increases rapidly with $\beta > \pi/2$. This implies wide bandwidths and is, of course, as pointed out previously, the desirable situation in most FM systems used for their good noise- and interference-rejection properties.

By wideband FM, we shall thus mean a frequency-modulated signal with $\beta > \pi/2$.

To demonstrate these conclusions more quantitatively, we must actually determine the frequency components and their amplitudes for the frequency-modulated signal of arbitrary β and sinusoidal modulating signal. As was noted previously, this may be done by expanding both $\cos (\beta \sin \omega_m t)$ and $\sin (\beta \sin \omega_m t)$ in their respective Fourier series. Multiplying the cosine term by $\cos \omega_c t$ and the sine term by $\sin \omega_c t$ then gives us our FM signal.

Both series may be found simultaneously by considering the periodic complex exponential

$$v(t) = e^{j\beta \sin \omega_m t} \qquad -\frac{T}{2} < t < \frac{T}{2} \qquad (4\text{-}67)$$

The real part of this function gives us our cosine function; the imaginary part, the sine function. If we expand this exponential in its Fourier series, we can expect to get a real part consisting of even harmonics of ω_m and an imaginary part consisting of the odd harmonics. [This is deduced from our previous discussion of cos (β sin $\omega_m t$) and the associated Fig. 4-49.] By equating reals and imaginaries, we shall then obtain the desired Fourier expansions for cos (β sin $\omega_m t$) and sin (β sin $\omega_m t$), respectively.

The Fourier coefficient of the complex Fourier series for the exponential of Eq. (4-67) is given by

$$c_n = \int_{-T/2}^{T/2} e^{j(\beta \sin \omega_m t - \omega_n t)} \, dt \qquad \omega_m = \frac{2\pi}{T}, \, \omega_n = \frac{2\pi n}{T} = n\omega_m \qquad (4\text{-}68)$$

Normalizing this integral by letting $x = \omega_m t$, we get

$$\frac{c_n}{T} = \frac{1}{2\pi} \int_{-\pi}^{\pi} e^{j(\beta \sin x - nx)} \, dx \qquad (4\text{-}69)$$

This integral can be evaluated only as an infinite series (as was the case in Chap. 2 for the sine integral Si x). It occurs very commonly in many physical problems, however, and so has been tabulated in many books.[27] It is called the *Bessel function of the first kind* and is denoted by the symbol $J_n(\beta)$. [Note from Eq. (4-69) that c_n is a function of both β and n.]

In particular,

$$J_n(\beta) \equiv \frac{1}{2\pi} \int_{-\pi}^{\pi} e^{j(\beta \sin x - nx)} \, dx \qquad (4\text{-}70)$$

so that
$$c_n = T J_n(\beta) \qquad (4\text{-}71)$$

For $n = 0$, we get $c_0 = T J_0(\beta)$, the dc component of the Fourier-series representation of the periodic complex exponential of Eq. (4-67). Increasing values of n give the corresponding Fourier coefficients for the higher-frequency terms of the Fourier series. The spectrum of the complex exponential of Eq. (4-67) (and ultimately that of the frequency-modulated signal) will thus be given by the value of the Bessel function and will depend on the parameter β. Thus

$$e^{j\beta \sin \omega_m t} = \frac{1}{T} \sum_{n=-\infty}^{\infty} c_n e^{j\omega_n t} = \sum_{n=-\infty}^{\infty} J_n(\beta) e^{j\omega_n t} \qquad \omega_n = n\omega_m \qquad (4\text{-}72)$$

As an example, let $n = 0$. Then

$$J_0(\beta) \equiv \frac{1}{2\pi} \int_{-\pi}^{\pi} e^{j\beta \sin x} \, dx \qquad (4\text{-}73)$$

is the dc component of the Fourier series of Eq. (4-72). But

$$e^{j\beta \sin x} = 1 + (j\beta \sin x) + \frac{(j\beta \sin x)^2}{2!} + \frac{(j\beta \sin x)^3}{3!} + \cdots \qquad (4\text{-}74)$$

using the series expansion for the exponential.

The power terms in sin x may be rewritten as sines and cosines of integral multiples of x, so that Eq. (4-74) can also be written

[27] See, for example, E. Jahnke and F. Emde, *Tables of Functions*, Dover, New York, 1945.

$$e^{j\beta \sin x} = 1 + j\beta \sin x - \frac{\beta^2}{2} \frac{1 - \cos 2x}{2} + \cdots \tag{4-75}$$

If we now integrate over a complete period of 2π radians [as called for by Eq. (4-73)], the terms in $\sin x$, $\cos 2x$, etc., vanish, leaving us with

$$J_0(\beta) = 1 - \frac{\beta^2}{4} + \cdots \tag{4-76}$$

an infinite series in β. [This is, of course, one method of evaluating the integral of Eq. (4-70).] But a comparison with Eq. (4-66) shows that this is exactly the coefficient of the carrier term of our FM signal, obtained there by a power-series expansion of $\cos (\beta \sin \omega_m t)$. The advantage of the present approach, using a Fourier-series expansion, is that the Bessel function is already tabulated so that we do not have to repeat the evaluation of Eq. (4-70) for different values of n by means of an infinite series.

From Eq. (4-72), the Fourier-series expansion of the complex exponential, we can obtain our desired Fourier series for $\cos (\beta \sin \omega_m t)$ and $\sin (\beta \sin \omega_m t)$. It can be shown, either from the integral definition of $J_n(\beta)$ [Eq. (4-70)] or from the power-series expansion of $J_n(\beta)$, that

$$J_n(\beta) = J_{-n}(\beta) \qquad n \text{ even}$$

and $\qquad J_n(\beta) = -J_{-n}(\beta) \qquad n \text{ odd}$ $\qquad (4\text{-}77)$

Writing out the Fourier series term by term, and using Eq. (4-77) to combine the positive and negative terms of equal magnitude of n, we get

$$e^{j\beta \sin \omega_m t} = J_0(\beta) + 2[J_2(\beta) \cos 2\omega_m t + J_4(\beta) \cos 4\omega_m t + \cdots] \\ + 2j[J_1(\beta) \sin \omega_m t + J_3(\beta) \sin 3\omega_m t + \cdots] \tag{4-78}$$

But $\qquad e^{j\beta \sin \omega_m t} = \cos (\beta \sin \omega_m t) + j \sin (\beta \sin \omega_m t) \tag{4-79}$

Equating real and imaginary terms, we get

$$\cos (\beta \sin \omega_m t) = J_0(\beta) + 2J_2(\beta) \cos 2\omega_m t + 2J_4(\beta) \cos 4\omega_m t + \cdots \tag{4-80}$$

and

$$\sin (\beta \sin \omega_m t) = 2J_1(\beta) \sin \omega_m t + 2J_3(\beta) \sin 3\omega_m t + \cdots \tag{4-81}$$

Equations (4-80) and (4-81) are the desired Fourier-series expansions for the cosine and sine terms. As noted previously, the cosine term has only the even harmonics of ω_m in its series, the sine term containing the odd harmonics.

The spectral distribution of the frequency-modulated carrier is now readily obtained. As previously written,

$$f_c(t) = \cos \omega_c t \cos (\beta \sin \omega_m t) - \sin \omega_c t \sin (\beta \sin \omega_m t) \tag{4-82}$$

Using the Fourier-series expansions for the cosine and sine terms, and then utilizing the trigonometric sum and difference formulas (as in the AM analysis), we get

$$f_c(t) = J_0(\beta) \cos \omega_c t - J_1(\beta)[\cos (\omega_c - \omega_m)t - \cos (\omega_c + \omega_m)t]$$
$$+ J_2(\beta)[\cos (\omega_c - 2\omega_m)t + \cos (\omega_c + 2\omega_m)t]$$
$$- J_3(\beta)[\cos (\omega_c - 3\omega_m)t - \cos (\omega_c + 3\omega_m)t]$$
$$+ \cdots \tag{4-83}$$

We thus have a time function consisting of a carrier and an infinite number of sidebands, spaced at frequencies $\pm f_m$, $\pm 2f_m$, etc., away from the carrier. This is in contrast to the AM case, where the carrier and only a single set of sidebands existed. The odd sideband frequencies arise from the quadrature term of Eq. (4-82); the even sideband frequencies arise from the inphase (cos $\omega_c t$) term. This, of course, agrees with our previous qualitative discussion based on Fig. 4-49.

The magnitudes of the carrier and sideband terms depend on β, the modulation index, this dependence being expressed by the appropriate Bessel function. Again, this is at variance with the AM case, where the carrier magnitude was fixed and the two sidebands varied only with the modulation factor.

We showed previously, from qualitative considerations of the time variation of the function $f_c(t)$, that for $\beta \ll \pi/2$, we should have primarily a carrier and one or two sideband pairs. For $\beta > \pi/2$, we should have increasingly more significant sideband pairs as β increases. The magnitude of the carrier should also decrease rapidly.

We can now verify these qualitative conclusions by referring to plots or tabulations of the Bessel function.[28] As an example, a graph of the functions $J_0(\beta)$, $J_1(\beta)$, $J_2(\beta)$, $J_8(\beta)$, and $J_{16}(\beta)$ is shown in Fig. 4-50. Note that for $\beta > \pi/2$ the value of $J_0(\beta)$ decreases sharply. $J_0(\beta)$ represents the magnitude of the carrier term, so that this result agrees with that obtained from the curves of Fig. 4-49.

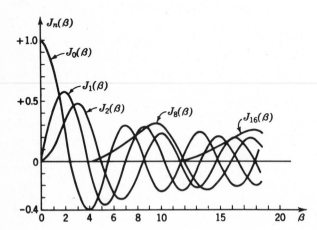

Figure 4-50 Examples of Bessel functions of the first kind.

[28] See, for example, the various curves and tables in A. Hund, *Frequency Modulation*, McGraw-Hill, New York, 1942, chap. 1.

For β small ($\beta \ll \pi/2$), the only Bessel functions of significant magnitude are $J_0(\beta)$ and $J_1(\beta)$. The FM wave thus consists essentially of the carrier and the two first-order sideband frequencies. This is, of course, the narrowband FM case considered previously. As β increases, however, the magnitudes of the higher-order sidebands begin to increase, the carrier magnitude begins to decrease, and the bandwidth required increases. The significant number of sideband terms depends on β, the modulation index. Note that $J_8(\beta)$ is essentially zero up to $\beta = 4$ and then begins to increase to a peak value at $\beta = 9.5$. $J_{16}(\beta)$ begins to increase significantly for $\beta > 12$ and peaks at about $\beta = 18$.

It is apparent from the curves of Fig. 4-50, and can be shown in general, that the higher-order Bessel functions, $J_n(\beta)$ with $n \gg 1$, are essentially zero to the point $\beta = n$. They then increase to a peak, decrease again, and eventually oscillate like damped sinusoids. This means that for $\beta \gg 1$ the number of significant sideband frequencies in the frequency-modulated wave is approximately equal to β. [$J_n(\beta) \doteq 0$, $n > \beta$, so that the corresponding sideband terms are negligible.] Since the sidebands are all f_m hertz apart, and since there are two sets of them, on either side of the carrier, the bandwidth B_T of the FM signal is approximately

$$B_T \doteq 2\beta f_m = 2\frac{\Delta f}{f_m} f_m = 2\Delta f \qquad \beta \gg 1 \qquad (4\text{-}84)$$

This agrees, of course, with the results of the binary FM case. This assumes a sinusoidal modulating signal of frequency f_m hertz. Δf represents the maximum-frequency deviation away from f_c, the unmodulated carrier frequency, and depends on the amplitude of the modulating signal. So for large β the bandwidth is directly proportional to the amplitude of the modulating signal. This is again to be compared with the AM or narrowband FM case where the bandwidth is $2B$, B being the baseband bandwidth (f_m in the case of a single sine wave at that frequency).

The bandwidth is equal to $2\Delta f$ only for very large modulation index. For smaller values of β, we can determine the bandwidth by counting the significant number of sidebands. The word significant is usually taken to mean those sidebands which have a magnitude of at least 1 percent of the magnitude of the unmodulated carrier. We have been assuming an unmodulated carrier of unit amplitude ($\cos \omega_c t$), so that the significant sidebands will be those for which $J_n(\beta) > 0.01$. The number will vary with β and can be determined readily from tabulated values of the Bessel function.

An example of such a determination is shown by Table 4-2, giving values of β up to 2. Using such a table, we can plot bandwidth versus the modulation index β. Such a curve has been drawn in Fig. 4-51. The transmission bandwidth B_T is shown normalized to both Δf, the frequency deviation, and to the baseband bandwidth $B(f_m$ in the sine-wave case). Also shown for reference is the rule-of-thumb relation

$$B_T = 2\Delta f + 2B = 2B(1 + \beta) \qquad (4\text{-}84a)$$

Table 4-2 Significant sidebands

β	$J_0(\beta)$	$J_1(\beta)$	$J_2(\beta)$	$J_3(\beta)$	$J_4(\beta)$	Number of sidebands	Bandwidth
0.01	1.00	0.005	—	—	—	1	$2f_m$
0.20	0.99	0.100	—	—	—	1	$2f_m$
0.50	0.94	0.24	0.03	—	—	2	$4f_m$
1.00	0.77	0.44	0.11	0.02	—	3	$6f_m$
2.00	0.22	0.58	0.35	0.13	0.03	4	$8f_m$

as well as a sine-wave analysis curve with sideband frequencies having amplitudes 10 percent or greater of the carrier included. The bandwidth in this case is, of course, less than that for the 1 percent amplitude case, and agrees closely with the rule-of-thumb relation. All curves show B_T equal to $2B$ for small β, and approaching $2\Delta f$ for large β.

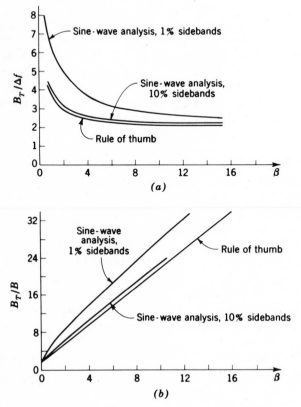

Figure 4-51 FM bandwidth versus modulation index. (*a*) Normalized to frequency deviation. (*b*) Normalized to baseband bandwidth.

We are now in a position to make some calculations of the required bandwidth for typical sinusoidal signals. Since the bandwidth ultimately varies with Δf, the frequency deviation, or the amplitude of the modulating signal, some limit must be put on this amplitude to avoid excessive bandwidth. The FCC has fixed the maximum value of Δf at 75 kHz for commercial FM broadcasting stations. What does this imply in the way of required bandwidth? If we take the modulating frequency f_m to be 15 kHz (typically the maximum audio frequency in FM transmission), $\beta = 5$, and the required bandwidth is 240 kHz, from Fig. 4-51. (Alternatively, for $\beta = 5$, there are eight significant sideband frequencies, or $2 \times 8 \times 15 = 240$ kHz is the bandwidth required.) For $f_m < 15$ kHz (the lower audio frequencies) β increases above 5, and the bandwidth eventually approaches $2\Delta f = 150$ kHz. So it is the *highest* modulating frequency that determines the required bandwidth. (These are the extreme cases, since $\Delta f = 75$ kHz corresponds to the maximum possible amplitude of the modulating signal.) The corresponding rule-of-thumb result is 180 kHz for the $\beta = 5$ case. The difference is, of course, due to the difference in definition of bandwidth. For the 10 percent sideband level case, the bandwidth is 190 kHz, which is in close agreement with the rule of thumb.

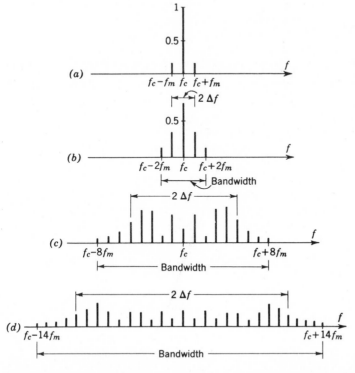

Figure 4-52 Amplitude-frequency spectrum, FM signal (sinusoidal modulating signal, f_m fixed, amplitude varying). (*a*) $\beta = 0.2$. (*b*) $\beta = 1$. (*c*) $\beta = 5$. (*d*) $\beta = 10$.

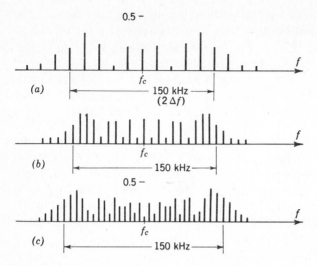

Figure 4-53 Amplitude-frequency spectrum, FM signal (amplitude of Δf fixed, f_m decreasing). $(a) f_m = 15$ kHz $(\beta = 5)$. $(b) f_m = 7.5$ kHz $(\beta = 10)$. $(c) f_m = 5$ kHz $(\beta = 15)$.

Frequency modulation is used for transmission of the sound channel in commercial TV, and the maximum-frequency deviation there (corresponding to the maximum amplitude of the modulating signal) has been fixed at 25 kHz by the FCC. For a 15-kHz audio signal this gives $\beta = 1.7$, or a bandwidth of 110 kHz. The lower audio frequencies require bandwidths well within this limit, for with f_m decreasing and $\beta > 15$, the bandwidth required approaches $2\Delta f = 50$ kHz.

The relatively large bandwidths required for commercial FM, as compared with AM sound transmission, are the price one has to pay to obtain significant improvement in noise and interference rejection. This question will be discussed in detail in Chap. 5.

The amplitude spectra of a frequency-modulated signal are shown plotted in Fig. 4-52 for $\beta = 0.2, 1, 5, 10$. The sinusoidal modulating signal is assumed to be of constant modulation frequency f_m, so that β is proportional to the signal amplitude. The amplitude of the spectral line at frequency $f_c \pm n f_m$ is given by $J_n(\beta)$. If the amplitude of the modulating signal is fixed (for example, $\Delta f = 75$ kHz for all signals) and different audio frequencies are considered, β increases as f_m decreases and we get spectrum plots similar to those of Fig. 4-52. For these plots Δf is fixed, however, so that we get a picture of more and more spectral lines crowding into a fixed-frequency interval. An example of such a plot is shown in Fig. 4-53. Δf has been chosen as 75 kHz, with f_m varying from 15 kHz down to 5 kHz. This plot shows clearly the lines concentrating within the $2\Delta f$ or 150-kHz points as β increases. (Remember again that the bandwidth approaches $2\Delta f$ for large β.)

4-9 GENERATION OF FREQUENCY-MODULATED SIGNALS

The methods of generating the wideband FM signals discussed in the last section can be grouped essentially into two types:

1. *Indirect FM* Here integration and phase modulation are first used to produce a narrowband FM signal. Frequency multiplication is then utilized to increase the modulation index to the desired range of values.
2. *Direct FM* Here the carrier frequency is directly modulated or varied in accordance with the input modulating signal (hence, the classification *direct*). This is the technique most commonly used and will be the one discussed here. Indirect techniques are discussed in the reference.[29]

At the heart of the direct FM transmitter is the device employed to vary the carrier frequency. The carrier frequency is normally generated by an oscillator whose frequency-determining circuit is a high-Q resonant circuit or crystal. Variations in the inductance or capacitance of this resonant circuit will then change the resonant frequency or the oscillator frequency. Thus, assume the capacitance of the resonant (tank) circuit is made proportional to the baseband modulating signal $f(t)$. Then

$$C = C_0 + \Delta C = C_o + Kf(t) \tag{4-85}$$

Here C_0 is the zero-signal capacitance. If the change ΔC in capacitance is small compared with C_0, it is simple to show that the instantaneous frequency ω_i of the tuned circuit becomes linearly proportional to $f(t)$, as desired. Specifically,

$$\omega_i = \frac{1}{\sqrt{LC}} = \frac{1}{\sqrt{LC_0}} \frac{1}{\sqrt{1 + \Delta C/C_0}}$$

$$\doteq \omega_c \left(1 - \frac{\Delta C}{2C_0} \right)$$

$$= \omega_c \left[1 - \frac{Kf(t)}{2C_0} \right]$$

$$= \omega_c - K_2 f(t) \qquad \omega_c^2 \equiv \frac{1}{LC_0}, \ \Delta C \ll C_0 \tag{4-86}$$

and

$$\Delta f = \frac{\Delta C}{2C_0} f_c = \frac{Kf_c}{2C_0} f(t) \tag{4-87}$$

It is apparent that small variations in the inductance L produce the same result.

How small ΔC should be compared with C_0 depends on how accurate the linear approximation to the square-root term in Eq. (4-86) must be. With $\Delta C/C_0 < 0.013$, the distortion due to this approximation is less than 1 percent. (The reader can check this for himself by evaluating the next, nonlinear term, in the power-series expansion of the square root.) Although the change in capacitance is of necessity small, the frequency deviation Δf may be quite large if the zero-signal resonant frequency f_c is large enough. As an example, if $\Delta C/2C_0 = 0.005$, and $f_c = 15$ MHz, $\Delta f = 75$ kHz, which is the maximum deviation specified for standard radio broadcasting. Generally, this method of *directly*

---

[29] Clarke and Hess, *op. cit.*, sec. 11.8.

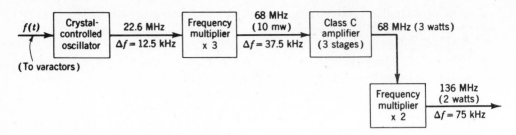

Figure 4-54 Typical direct FM transmitter, space telemetry. [From A. V. Balakrishnan (ed.), *Space Communications,* McGraw-Hill, New York, 1963, p. 190.]

varying the instantaneous frequency requires substantially less additional frequency multiplication and conversion than does the indirect method.

There are various ways of obtaining a capacitance (or inductance) variation proportional to the signal intelligence $f(t)$. A reverse-biased *PN* junction provides one example. Other examples appear in the reference.[30]

FM transmitters have been used quite commonly in space communications and telemetry. A typical block diagram of a direct-type FM transmitter for space applications is shown in Fig. 4-54.[31] This all-solid state system provides 2 W output power at 136 MHz. A series reactive network consisting of varactor diodes and an inductor provides the required variable capacitance in the 22.6-MHz crystal-controlled oscillator. The FM signal is then tripled in frequency and amplified to provide 3 W output at 68 MHz. A frequency doubler then provides the desired output. The initial frequency deviation is 12.5 kHz. The final Δf, after frequency multiplication by 6, is 75 kHz.

Frequency multiplexing of several data channels is quite commonly used in space telemetry.[32] In this case the individual data channels each frequency-modulate a carrier, the resultant FM signals being arranged to occupy adjacent frequency bands. These FM signals are then summed and the composite complex signal used to frequency modulate a very high sine-wave carrier for final transmission. An example of such an *FM-FM system* is shown in Fig. 4-55. (The notation FM-FM is commonly used to indicate two steps of FM. AM-FM similarly implies initial frequency multiplexing of multiple data channels using AM techniques, with the composite set of AM signals then used to frequency modulate a high-frequency carrier.) In this example, the ratio of frequency deviation to subcarrier frequency in each subchannel is held fixed at 7.5 percent. For a modulation index of 5 for each subchannel, the corresponding signal bandwidths increase with the channel number. Low-bandwidth signals would

[30] *Ibid.,* secs. 11.5–11.7.

[31] W. B. Allen, in A. V. Balakrishnan (ed.), *Space Communications,* McGraw-Hill, New York, 1963, pp. 190–192.

[32] See FM-FM Telemetry Systems, in E. L. Gruenberg (ed.), *Handbook of Telemetry and Remote Control,* McGraw-Hill, New York, 1967, chap. 6. Also see R. Stampfl, "The Nimbus Satellite Communication System," in Balakrishnan, *op. cit.,* chap. 18.

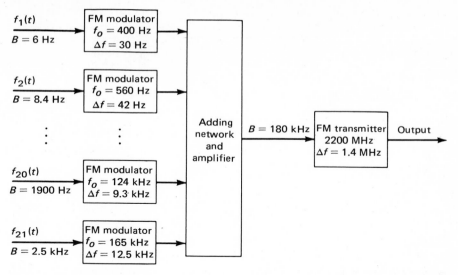

Figure 4-55 FM-FM system for space telemetry (proportional-bandwidth subcarrier channels).

thus be used in the low-numbered channels, wider-bandwidth signals in the higher-number channels.

Very often one or more of the subchannels may be used for the transmission of time-multiplexed digital data, or several subchannels may be combined to accommodate wider-band data signals.

Frequency multiplexing will be discussed in more detail in Sec. 4-11.

4-10 FREQUENCY DEMODULATION

Frequency demodulation, the process of converting a frequency-modulated signal back to the original modulating signal, can be carried out in a variety of ways. Ultimately, however, the process used must provide an output voltage (or current) whose *amplitude* is *linearly* proportional to the *frequency* of the input FM signal. The term *frequency discriminator* is commonly used to characterize a device providing this frequency-amplitude conversion.

Various schemes have been proposed to accomplish the task of frequency demodulation. We shall discuss two types of frequency discriminators in this section: a balanced discriminator involving the use of tuned circuits and so-called zero-crossing detectors. Others appear in the reference.[33]

Consider again the FM signal given by

$$f_c(t) = A \cos \theta(t) = A \cos \left[\omega_c t + K \int f(t) \, dt \right] \tag{4-88}$$

Here $f(t)$ is the baseband or modulating signal. It is apparent that to extract $f(t)$

[33] Clarke and Hess, *op. cit.*, chap. 12.

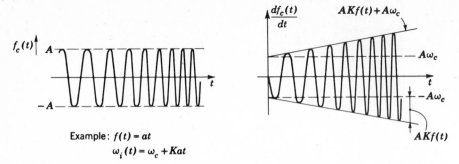

Figure 4-56 Typical FM signal $f_c(t)$ and its derivative $df_c(t)/dt$ [example: $f(t) = at$].

from this expression, we must somehow generate the instantaneous frequency

$$\omega_i(t) = \frac{d\theta(t)}{dt} = \omega_c + Kf(t) \tag{4-89}$$

If we then balance out the constant term ω_c, we have the desired output $f(t)$.

We shall show later that zero-crossing detectors do provide a measure of ω_i and, hence, $f(t)$. The approach we consider now is that of studying Eq. (4-88) for the FM signal. Some thought indicates that ω_i may be generated from that equation by differentiating $f_c(t)$ with respect to time. Specifically, if we assume the amplitude A constant (we shall have to include a limiter to ensure this later), the derivative of $f_c(t)$ is just

$$\frac{df_c(t)}{dt} = -A \sin \theta(t) \left[\frac{d\theta}{dt}\right]$$

$$= -A[\omega_c + Kf(t)] \sin \theta(t) \tag{4-90}$$

Note that this expression is precisely in the form of an AM signal, whose *envelope* is given by

$$A[\omega_c + Kf(t)] = A\omega_c \left[1 + \frac{Kf(t)}{\omega_c}\right]$$

Since we have been assuming the frequency deviation $\Delta\omega = Kf(t) \ll \omega_c$, the envelope never goes to zero, and, in fact, varies only slightly about the average quantity $A\omega_c$. It is then apparent that we may now obtain $f(t)$ by envelope detecting $df_c(t)/dt$ (see Fig. 4-56).

The FM detector thus consists of a differentiating circuit followed by an envelope detector (Fig. 4-57). To ensure that the amplitude A is truly constant (otherwise additional, distortion, terms involving dA/dt appear at the output),

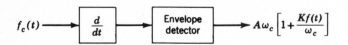

Figure 4-57 FM detector.

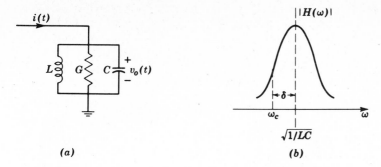

(a)

(b)

Figure 4-58 Single-tuned circuit. (*a*) Circuit. (*b*) Amplitude response.

one normally inserts a *limiter* prior to differentiation. The limiter serves to keep amplitude variations from appearing in the output. If one uses hard clipping to keep the amplitude invariant, square waves at the varying instantaneous frequency $\omega_i(t)$ are produced. It is then necessary to follow the limiter with a bandpass filter centered about ω_c to convert the square waves back to the cosinusoidal form of Eq. (4-88); i.e., terms centered about $2\omega_c$ and the other higher harmonics of ω_c are filtered out. This is commonly incorporated in the differentiator, as will be shown in the paragraphs that follow.

There are various ways of performing the necessary differentiation. We consider here only one, that involving the single-tuned circuit of Fig. 4-58.

To demonstrate that this circuit does provide the required differentiation and envelope detection, we first note that if the circuit is detuned so that the unmodulated carrier frequency ω_c lies on the sloping part of the amplitude-frequency characteristic and the frequency variations occur within a small region about the unmodulated carrier, the amplitude of the output wave will follow the instantaneous frequency of the input. This is shown in both Figs. 4-58 and 4-59. The region over which the characteristics are very nearly linear must be wide enough to cover the maximum-frequency deviation.

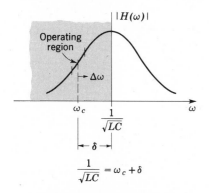

Operation along left-hand slope   Operation along right-hand slope

Figure 4-59 FM demodulation.

We demonstrate this quantitatively by calculating the frequency response of the circuit. It is apparent that the magnitude of the transfer impedance is given by

$$|H(\omega)| = \left| \frac{V_o(\omega)}{I(\omega)} \right| = \frac{1/G}{\sqrt{1 + (\omega C - 1/\omega L)^2/G^2}} \qquad (4\text{-}91)$$

We now define $\Delta\omega = \omega - \omega_c$ as the frequency deviation away from the carrier frequency ω_c, and let $1/\sqrt{LC} = \omega_c \pm \delta$, as in Fig. 4-59. (The sign depends on the slope along which we operate.) To ensure linearity of the operating region, assume that

$$\Delta\omega \ll \omega_c$$

(This is of course normally true anyway.) Also, let $\delta \ll \omega_c$. (The carrier frequency ω_c is then close to the center frequency $1/\sqrt{LC}$.) It is then easily shown that Eq. (4-91) reduces to

$$|H| \doteq \frac{1/G}{\sqrt{1 + 4(C/G)^2(\delta - \Delta\omega)^2}} \qquad (4\text{-}92)$$

in the case of operation along the left-hand slope.

Now assume further that $\delta \gg \Delta\omega$. This implies, of course, that although ω_c is close to $1/\sqrt{LC}$, we never deviate far enough away from ω_c to approach $1/\sqrt{LC}$. This is necessary to ensure linearity of the operating region. Further, assume that $\delta \ll G/2C \equiv \alpha$. Equation (4-92) then reduces to

$$|H| \doteq \frac{1}{G} \left[\left(1 - \frac{\delta^2}{2\alpha^2} \right) + \frac{\delta}{\alpha^2} \Delta\omega \right] \qquad (4\text{-}93)$$

Note that with the assumptions made $|H|$ does have a component linearly proportional to $\Delta\omega$, as desired. This shows that the single-tuned circuit *has* provided the desired differentiation, while the envelope of the output voltage *does* provide the desired output $f(t)$.

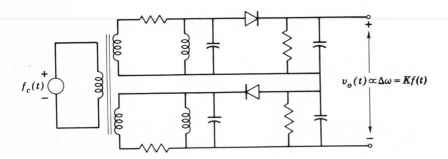

Figure 4-60 Balanced FM demodulator.

Collecting the assumptions made, we have

$$\Delta\omega \ll \delta \ll 1/\sqrt{LC}$$
$$\Delta\omega \ll \omega_c$$
$$\delta \ll \alpha$$

As an example, assume that we are to demodulate an FM signal with a carrier frequency $f_c = 10$ MHz. The maximum-frequency deviation $\Delta f = 75$ kHz. We then use as the differentiating circuit a single-tuned circuit centered at $1/2\pi\sqrt{LC} = 10$ MHz $+ 200$ kHz, with $\alpha/2\pi = 10^6$ per second. (Then $\delta/2\pi = 200$ kHz.)

The requirements on the inequalities to obtain linearity may be relaxed somewhat and the $\delta^2/2\alpha^2$ term in Eq. (4-93) (which is large compared with the $\Delta\omega$ term) eliminated by using a *balanced demodulator*, that is, two single-tuned circuits, one tuned to $\omega_c + \delta$ and the other to $\omega_c - \delta$. If the individual outputs are envelope detected and subtracted, the final output is

$$\left. |H| \right|_{\omega_c+\delta} - \left. |H| \right|_{\omega_c-\delta} = \frac{2\delta}{G\alpha^2}\Delta\omega \tag{4-94}$$

as desired. An example of such a balanced demodulator is shown in Fig. 4-60.

A block diagram of a typical FM receiver, covering the broadcast range of 88 to 108 MHz, is shown in Fig. 4-61. Note that except for the limiter and discriminator circuits, the form of the receiver is similar to that of a conventional AM receiver. All h-f circuits prior to the discriminator must be designed for the FM bandwidth of 225 kHz, however, while the audio amplifier which amplifies the recovered modulating signal need cover only the 50- to 15-kHz range. The i-f amplifiers are tuned to a center frequency of 10.7 MHz. The audio amplifier normally includes a deemphasis circuit. This circuit, in conjunction with a preemphasis circuit in the transmitter, provides additional discrimination against noise and interference. It is discussed in detail in Chap. 5.

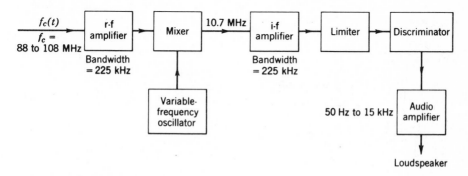

Figure 4-61 Typical FM receiver.

Zero-crossing Detectors

It was noted above that FM detectors normally include hard limiters to eliminate any amplitude fluctuations that could then be converted (erroneously) to a
detected FM output. It is thus apparent that the FM information must be
contained in the points at which the FM signal $f_c(t)$ crosses the origin, the *zero
crossings*. We shall demonstrate this statement below, providing another means
of detecting FM signals.

Consider then again the FM signal given by

$$f_c(t) = A \cos \theta(t) = A \cos [\omega_c t + K\textstyle\int f(t)\, dt] \qquad (4\text{-}88)$$

Again, as an example, let $f(t) = at$, $0 \leq t \leq T$, a repetitive ramp. Then $\omega_i = \omega_c + Kat$, $\theta(t) = \omega_c t + Kat^2/2$. Let t_1 be a zero crossing, as shown in Fig. 4-62,
with $t_2 = t_1 + \Delta t$ the next zero crossing. Then $\theta(t_2) - \theta(t_1) = \pi$. Assume that
the bandwidth B of $f(t)$ is much less than f_c, the carrier frequency. The
information-bearing signal $f(t)$ changes much more slowly than f_c. In the interval $(t_2 - t_1)$, then, $f(t)$ may be assumed effectively constant, so that

$$\theta(t_2) - \theta(t_1) = \pi = \omega_c(t_2 - t_1) + K \int_{t_1}^{t_2} f(t)\, dt$$

$$\doteq \omega_c(t_2 - t_1) + Kf(t_1)(t_2 - t_1)$$

$$= \underbrace{[\omega_c + Kf(t_1)]}_{\dfrac{d\theta}{dt} \equiv \omega_i}(t_2 - t_1) \qquad (4\text{-}95)$$

From Eq. (4-95) we thus have

$$\omega_i = \omega_c + Kf(t) \doteq \frac{\pi}{t_2 - t_1}$$

and

$$f_i = f_c + \frac{K}{2\pi} f(t) \doteq \frac{1}{2(t_2 - t_1)} \qquad (4\text{-}96)$$

The desired output $f(t)$ may thus be found by measuring the spacing between
zero crossings.

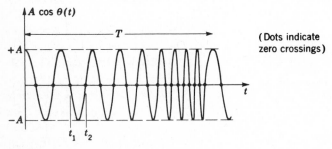

Figure 4-62 Zero-crossing determination.

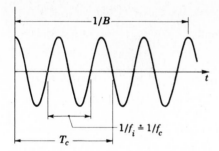

Figure 4-63 Counting intervals.

If positive-going zero crossings only are considered [those at which the slope of $f_c(t)$ is positive], we get

$$f_i \doteq \frac{1}{t_2 - t_1} \qquad (4\text{-}96a)$$

A simple way of measuring the spacing of zero crossings is to actually *count* the number of zero crossings in a given time interval.

Thus, consider a counting interval T_c long enough so that it counts a significant number of zero crossings, yet short enough compared with $1/B$ so that $f(t)$ still does not change too much in this interval (see Fig. 4-63). Then

$$\frac{1}{f_c} < T_c \ll \frac{1}{B} \qquad (4\text{-}97)$$

(As an example, let $f_c = 10$ MHz, $T_c = 1\ \mu$s, $B = 20$ kHz.) Let n_c be the number of positive zero crossings in T_c seconds. We then have

$$n_c \doteq \frac{T_c}{t_2 - t_1} \qquad (4\text{-}98)$$

or

$$f_i = \frac{n_c}{T_c} \qquad (4\text{-}99)$$

(n_c is approximately 10 in the example noted.) The instantaneous frequency f_c, and from this the derived $f(t)$, are thus found directly in terms of the measured count n_c. [In practice, one would half-wave rectify $f_c(t)$, differentiate to accentuate the zero-crossing points, again rectify to eliminate the negative pulses due to the negative-going zero crossings, and pass the resultant sequence of positive pulses into a low-pass device with time constant T_c. The output would then be a direct measure of $f_i = f_c + Kf(t)/2\pi$. By using a balanced device one may again obtain $f(t)$ directly.]

Other types of FM detectors in addition to those described above have been developed and used in practice. One particular class, which is of particular importance in deep-space communications where power is at a premium, uses feedback techniques to improve signal detectability in the presence of noise. Among the different types of detectors in this group one may consider the phase-locked loop, the FM demodulator with feedback (FMFB),[34] and the frequency-locked loop FM demodulator.[35]

4-11 FREQUENCY-DIVISION MULTIPLEXING: TELEPHONE HIERARCHY

In Secs. 3-3 and 3-8 we introduced time-division multiplexing or combining of signals as a means of transmitting multiple signals over one common signal channel. Another common technique, particularly with analog signals, for combining many independent signals so that they can be transmitted over a common channel is that of *frequency-division multiplexing* (FDM). In this procedure individual baseband signal channels are each multiplexed up in frequency, the carrier for each channel being chosen so that the resultant modulated signals occupy adjacent, nonoverlapping frequency bands. The composite signal, made up of the sum of the individual modulated signals, is then transmitted as one wider-band analog signal.

The technique is thus analogous to time-division multiplexing. There samples of individual signal channels occupy adjacent, nonoverlapping time slots. In the case of FDM the signals occupy adjacent frequency bands. In both cases the transmission bandwidth is increased in proportion to the number of signal channels multiplexed together. The advantage of multiplexing—either in time or in frequency—is that transmission facilities are *shared* among the various channels multiplexed together.

The various modulation techniques discussed in this chapter may be used to carry out the FDM process. An example of an FM-FM frequency multiplexed scheme has already been presented briefly in Sec. 4-9 (see Fig. 4-55). AM, DSB, and SSB techniques have all been used to multiplex or combine independent signal channels. Combinations of these have been used as well. The sound signal and the video (picture) signal for TV are multiplexed together for common transmission over a given frequency band. In this case the voice signal is transmitted using FM, while the picture signal is sent using *vestigial sideband* transmission: the upper sideband plus a fraction of the lower sideband are transmitted.[36] Frequency multiplexing is used as well in FM stereo transmission to transmit two independent signals. (Details appear in Prob. 4-57, at the end of the chapter.)

[34] Schwartz, Bennett, and Stein, *op. cit.*, sec. 3-9, pp. 157–163. (The phase-locked loop is treated in detail, both as a tracking and acquisition device and as a demodulator, in A. J. Viterbi, *Principles of Coherent Communication,* McGraw-Hill, New York, 1966, chap. 2.)

[35] K. K. Clarke and D. T. Hess, "Frequency Locked Loop Demodulator," *IEEE Trans. Commun. Technol.,* pp. 518–524, August 1967. See also Clarke and Hess, *Communication Circuits, op. cit.,* sec. 12.7.

[36] H. Stark and F. B. Tuteur, *Modern Electrical Communications Theory and Systems,* Prentice-Hall, Englewood Cliffs, N.J., 1979, pp. 274–297.

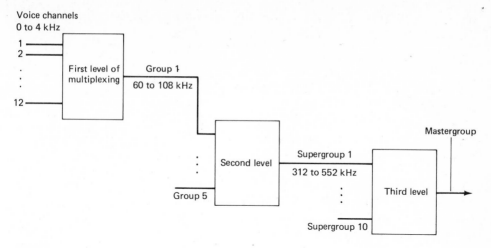

Figure 4-64 North American FDM hierarchy.

To demonstrate the concept of frequency-division multiplexing in more detail, we focus in this section on the North American FDM telephone hierarchy that uses SSB techniques to transmit up to 600 voice channels over a common transmission channel.[37] [FM FDM techniques are commonly used as well to multiplex up to 1,800 voice channels over the microwave links used for long-haul (long-distance) transmission.[38]] Three successive levels of multiplexing, using SSB modulation at each level, are used to attain the final combination of 600 voice channels. First, 12 voice channels, of 4-kHz bandwidth, are multiplexed together to form a *group* covering the range 60 to 108 kHz. Five groups in turn are further combined to form a *supergroup* covering the range 312 to 552 kHz. Finally, 10 supergroups are combined to form a *mastergroup*. (No standard arrangement is specified for the mastergroup generation. We shall instead mention two mastergroup examples.) The FDM hierarchy is shown schematically in Fig. 4-64.

The first level of multiplexing uses 12 carriers spaced 4 kHz apart to generate 12 SSB signals that are then summed to form the group. The lower sideband is selected in each case, resulting in *inverted* spectra compared to the baseband originals. There are various ways of generating the desired SSB signals, as noted earlier in this chapter. The Bell System embodiment of this FDM hierarchy first generates DSB signals, using product modulators, and then filters out the upper sideband of each DSB signal. Details are shown in Fig. 4-65. The sideband inversion noted above is indicated by following two points *a* and *b* on the spectral curve. Although developed specifically for voice (telephone) transmission, any signal fitting into the voice channel band, with a spectrum

[37] Bell Telephone Laboratories, *Transmission Systems for Communications,* 4th ed., 1970, pp. 128–139.

[38] *Ibid.,* pp. 523–547.

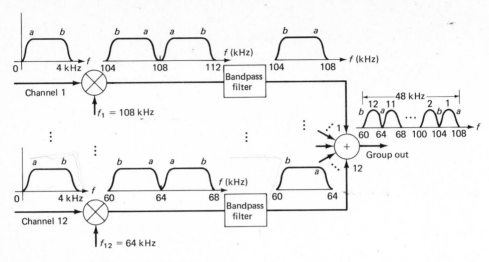

Figure 4-65 Group generation, Bell System.

from 200 to 3,400 Hz, may be transmitted in a voice channel. This enables data modem outputs to be multiplexed as well.

An alternative method of group generation, arriving at the same output structure of Fig. 4-65, is employed by Northern Telecom of Canada. A two-step modulation process is used. An upper sideband SSB signal is generated for each of the 12 voice channels by first shifting all channels individually up to a common center frequency of 8.14 MHz and then filtering out the lower sidebands. Each upper sideband SSB signal is then shifted down to its appropriate frequency band in the group band by beating with one of 12 equally spaced car-

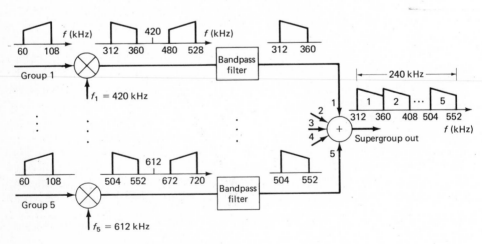

Figure 4-66 Bell System supergroup generation.

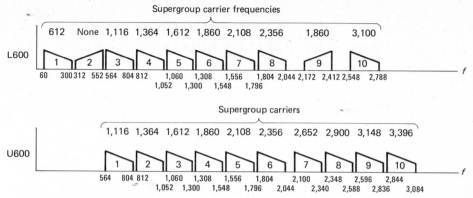

Figure 4-67 L600 and U600 mastergroups. (From *Transmission Systems for Communications,* 4th ed., Bell Laboratories, 1970, fig. 6-8, p. 135. Copyright (1970) Bell Telephone Laboratories; reprinted by permission.)

riers, ranging from 8.204 MHz to 8.248 MHz. It is left to the reader to demonstrate that this two-step modulation process results in the same group spectral occupancy as that of Fig. 4-65.

The second level of this FDM hierarchy combines five groups, each in the range 60 to 108 kHz, by again using SSB modulation. The resultant supergroup (Fig. 4-64) occupies the frequency range 312 to 552 kHz. Group 1, as an example, is shifted to a center frequency of 420 kHz and the lower (inverted) sideband, ranging from 312 to 360 kHz, is selected by again using bandpass filtering. The five carrier frequencies used, at spacings of 48 kHz, are 420, 468, 516, 564, and 612 kHz. Details appear in Fig. 4-66.

As in the case of the original voice channels combined to form a group, *any* 48 kHz-bandwidth signal may be inserted in place of a group signal. This allows higher-bit-rate data signals, as an example, to be transmitted over the telephone system, side by side with voice and/or lower-bit-data signals.

The highest level of the FDM hierarchy of Fig. 4-64 consists of a mastergroup made up of 10 supergroups, or the equivalent of six hundred 4-kHz voice channels. SSB modulation techniques are again used to generate the mastergroup. Various frequency allocations are used in practice. Two examples appear in Fig. 4-67. Both are used by the Bell system.[39] Note that the L600 system has eight inverted (lower) sidebands, while all sidebands in the U600 mastergroup are inverted. The L600 system is used for wideband transmission over coaxial cable and microwave radio relay systems. Shown in Fig. 4-67 are the carrier frequencies used in the SSB modulation of the various supergroups. It is left to the reader to show that the L600 and U600 frequency characteristics shown in Fig. 4-67 are in fact obtained by SSB modulation with these carrier

[39] *Ibid.,* p. 135.

frequencies. Any 240-kHz signal, data or analog, can be used in place of a supergroup. One example would be a 250-kbits/s digital signal.

The U600 mastergroup is used as a building block for larger channel groupings. One example is the Bell System AR6A, which packs 6,000 voice channels into one wideband channel for transmission over microwave long-haul (long-distance) radio relay links.[40] It does this by combining 10 U600 mastergroups. The resultant 6,000-channel signal covers the frequency range 59.844 MHz to 88.840 Mhz. This very wideband signal is then modulated up to microwave frequencies (4 or 6 GHz) for transmission over microwave radio.

FDM using FM techniques has long been commonly used for microwave long-haul transmission. This is because of the distortion encountered in power-amplifier tubes and frequency-converting circuits when SSB is used. The wide variations in amplitude encountered with SSB transmission encompassing thousands of channels result in distortion unless devices which are linear over wide amplitude ranges are available. Since the amplitude of an FM signal is effectively constant (or can be made so with limiting), this problem does not arise with FM transmission. The development of ultralinear tubes (traveling-wave tubes are used in these microwave systems for power amplification) and a circuit to reduce transmitter distortion made SSB possible for long-distance microwave radio relay transmission. The AR6A system with 6,000 voice signals over one channel represents a 3-to-1 improvement over the FM systems, the channels for which carry 1,800 voice signals.[41]

4-12 SATELLITE SYSTEMS APPLICATIONS: FDMA, SPADE SYSTEMS[42]

The use of satellites for both worldwide and domestic communications has expanded tremendously over the past years. Telephone (voice) traffic is now carried routinely between member nations of the Intelsat organization, augmenting international telephone communications possible previously only by submarine cable strung under the various oceans of the world. The transmission of data traffic via satellite has also expanded throughout the world.

The communication systems used for satellite communications provide an important and technologically significant application of both the baseband and modulation techniques discussed in this chapter and the previous one. In addi-

---

[40] R. E. Markle, "Single Sideband Triples Microwave Radio Route Capacity," *Bell Lab. Rec.*, vol. 56, no. 4, pp. 105–110, April 1978.

[41] Bell Telephone Laboratories, *op. cit.*, pp. 523–547.

[42] A good summary paper is the one by J. G. Puente and A. M. Werth, "Demand-assigned Service for the Intelsat Global Network," *IEEE Spectrum*, vol. 8, no. 1, pp. 59–69, January 1971. See J. J. Spilker, Jr., *Digital Communications by Satellite*, Prentice-Hall, Englewood Cliffs, N.J., 1977, part II, for more details. The Intelsat IV System is described in detail in "The Intelsat IV Communications System," P. L. Bargellini (ed.), *Comsat Tech. Rev.*, vol. 2, no. 2, pp. 437–570, Fall 1972.

tion, because of the geographic remoteness of satellites, new methods of gaining entree to the satellite system—called *access methods*—had to be developed, and it is useful to describe these in some detail.

Although we shall be referring primarily to voice traffic in discussing existing satellite communication systems, for simplicity's sake, digital transmission plays a significant role as well. This is particularly true for the U.S. domestic satellites which are used for data and record transmission, as well as commercial TV broadcasts. In fact, just as is the case with terrestial communications, PCM techniques converting analog voice signals into binary pulses are beginning to be used on satellite systems. As the drive toward all-digital signal transmission continues, the distinctions between analog and digital signal transmission tend to blur.

We focus here on synchronous satellites. These are satellites located at a height of 35,860 km above the earth's surface, in orbit over the equator, and rotating in synchronism with the earth. They thus appear to be in a stationary position, as seen from the earth. These satellites are provided with receiving antennas to receive transmissions from earth stations, and with transmitting antennas to relay the transmissions to other stations located geographically distant on the earth. By adjusting the antenna beam patterns one can generate global beams covering all portions of the earth in sight of a satellite. (One Intelsat satellite over the Atlantic Ocean covers portions of Africa, western Europe, South America, and the eastern part of the United States.) One can generate beams shaped to cover one country only (as in the case of domestic satellites), or multiple narrow beams that can be switched between a variety of locations.

Various frequency bands have been allocated to commercial satellite use, worldwide. The most common one consists of a 500-MHz wide band centered at 6 GHz in the *up-link* direction (toward the satellite), and centered at 4 GHz in the *down-link* direction (toward the earth). These directions are indicated in the schematic picture of Fig. 4-68.

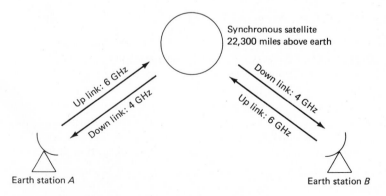

Figure 4-68 Commercial satellite communications.

The 500-MHz band at each frequency is typically subdivided into 12 bands of 36 MHz each, each covered by a separate transmitter/receiver called a transponder. (Spacing between the bands accounts for the missing bandwidth.) Each transponder band is in turn divided into a number of frequency channels, depending on the type of application or signal to be transmitted.

Various modes of accessing a satellite are either in use or planned. The most common procedure, called *frequency-division multiple access* (FDMA), is similar to frequency-division multiplexing discussed in the last section. Earth stations are allocated specific channel assignments and must use these frequencies in transmitting, or in receiving signals. An earth station will transmit a frequency-modulated signal, using a specific carrier in the 6-GHz range. The modulating or baseband signal for the transmitted FM signal consists of the frequency-division multiplexed (FDM) sum of a number of 4-kHz channels. All earth stations accessing a particular transponder have their FM carriers spaced so that the FM signals for each occupy adjacent bands, the composite FDMA signal at the satellite covering the entire 36-MHz transponder band. The overall access procedure is referred to as FDM/FM/FDMA.

A more recent technique introduced to handle light traffic situations is that of *demand assignment multiple access* (DAMA). Here frequency bands are assigned, on demand, from a pool of bands available, to those stations requesting them. Unlike FDMA, where idle channels go unused, all channels in the demand assignment technique are made available to all stations, so that there is a greater likelihood that a channel will be used. (In a heavy traffic situation this is not necessary, since an assigned channel will rarely be idle.) The third access technique used in satellite communication is that of *time-division multiple access,* analogous to TDM, with different stations assigned time slots during which to transmit. It is apparent that global synchronization of station clocks is a necessity here. In the paragraphs that follow we discuss the FDMA and demand assignment techniques in more detail. As noted earlier, this provides additional applications for our earlier discussions of multiplexing, modulation, and bandwidth allocations.

Frequency-Division Multiple Access (FDMA)

In this scheme, the most common one used in the Intelsat satellite system, fixed "connections" are maintained between pairs of earth stations in various countries. These connections consist of groups of dedicated 4-kHz frequency slots or voice channels. The number of such channels assigned to a particular pair of stations depends on the traffic needs of the pair. We describe the FDMA technique in terms of an example, portrayed in detail in Fig. 4-69.[43] In this example a particular country, A, is assigned sixty 4-kHz channels, divided on an equal basis among five other countries, B to F. The 12 channels assigned to each country would be frequency-multiplexed to form the 60-channel 252-kHz

---

[43] Puente and Werth, *op. cit.*

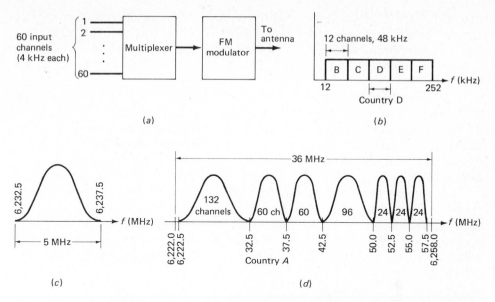

Figure 4-69 Typical FDMA generation, one transponder. (*a*) FDM/FM generation, earth station, country A. (*b*) Multiplexer output, frequency allocations. (*c*) FDM/FM output, country A. (*d*) Composite FDMA spectrum, at transponder.

baseband signal shown in Fig. 4-69*b*.[44] This baseband signal frequency-modulates a carrier at the assigned carrier frequency of 6.235 GHz to produce the FM signal of 5-MHz bandwidth of Fig. 4-69*c*. In this example, six other countries are assumed to access the same transponder. The composite FDMA spectrum of the seven accessing signals has the form shown in Fig. 4-69*d*,[45] with country *A*'s FM signal indicated among the seven. Each of the six other countries accessing this transponder is assumed to have the channel allocation and band location shown in Fig. 4-69*d*.

In this typical case one of the countries has been allocated 132 channels, with an FM bandwidth of 10 MHz; another country has a 96-channel allocation with a bandwidth of 7.5 MHz; one, besides country *A*, has a 60-channel allocation; others have 24-channel allocations with a bandwidth of 2.5 MHz each.

Bandwidth allocations in the Intelsat system are normally varied in increments of 2.5 MHz, as indicated in this example. To accommodate as wide a variety of traffic choices as possible, Intelsat has standardized on a specified number of 4-kHz channels per accessing carriers, with the bandwidth specified for each. The example of Fig. 4-69 indicates some of these assignments: 24 channels with a bandwidth of 2.5 MHz, 60 channels with a bandwidth of 5 MHz, 96 channels with a bandwidth of 7.5 MHz, and 132 channels with a

[44] *Ibid.*, fig. 2.
[45] *Ibid.*, fig. 12.

bandwidth of 10 MHz. Note that the system becomes less efficient as the channel assignment per carrier is reduced. Thus the smaller channel assignments require proportionately more bandwidth and the total number of channels per transponder is correspondingly reduced as the channel assignments are reduced. The example of Fig. 4-69d shows 420 channels accessing the transponder. Were the 36-MHz transponder bandwidth to be allocated exclusively to 24-channel carriers, with a 2.5-MHz bandwidth for each, 14 accesses, for a total of 336 channels, would be accommodated. At the other extreme, the entire transponder could be covered by one carrier with 900 channels multiplexed. The number of channels accommodated by one transponder thus ranges from a low of 336 to a high of 900, depending on the traffic usage and requirements.

To complete the transmission, the 6-GHz up-link signals are rebroadcast at specified frequencies in the 4-GHz down-link range. Receiving stations select the frequency band (or bands) containing the channels directed to their subscribers, use FM receivers to demodulate the appropriate multichannel baseband signal, and then further demultiplex the individual 4-kHz channels.

Demand-assigned Multiple Access (DAMA)

The FDMA scheme described in the previous paragraph establishes permanently connected or fixed channel assignments between accessing countries. This provides an efficient way of maintaining communications between earth stations that are often in communication with one another. The scheme becomes less efficient for those cases where the traffic between earth stations may be classed as light or moderate. It means tying up a relatively scarce commodity for lightly loaded traffic situations. In such situations it becomes more cost-effective to provide channels (connections between stations) on a demand-assigned basis. Demand assignment becomes in fact a necessity as the number of possible connections increases. Consider as a single example 50 earth stations wanting to communicate with one another using the same satellite. There are $50 \times 49/2 = 1,225$ possible station connections that can be made. With the 12 transponders used in a relatively efficient manner, 5,000 to 10,000 voice channels are available to serve these connections. This thus means that only 4 to 8 channels can be made available between pairs of earth stations. As the number of possible connections increases (for 100 earth stations this becomes 4,900 possible connections) the problem becomes exacerbated. One is forced to go to some form of demand-assigned connection scheme.

This problem of providing appropriate connections between any pair of subscribers is obviously not new. This is exactly the problem faced daily by all users of public communication facilities.

Rather than provide permanent connections between all possible users, connections are made only when needed. The result is a far less costly system. The price paid by the user is an occasional busy signal (assuming the communication network is appropriately designed!). Since most satellite connections may only be occasionally used anyway, it is apparent that demand assignment techniques should be a very effective way of solving the connection problem. They allow many more users to access the system than would be the case with fixed connections only.

Just as with the usual multiplexing techniques two ways of demand assignment are possible: an FDMA approach using frequency assignments, on demand, from a pool of frequencies available, or a TDMA approach, in which time slots are made available from those not currently occupied. We shall discuss the FDMA approach only here.[46]

With a demand-assigned technique it becomes necessary to provide a signaling channel through which the connection is made. This channel is also used to signal completion of a connection, the frequency allocated to this particular call then being returned to the pool. We discuss FDMA demand assignment in terms of the particular scheme, the SPADE system, used in the Intelsat global network.[47]

SPADE System

The SPADE demand assignment system uses 800 independently assignable frequency channels covering the 36-MHz bandwidth of a satellite transponder. In the Intelsat IV satellite series one transponder of the 12 available, covering the frequency ranges 6.302 to 6.338 GHz for reception (up link) and 4.077 to 4.113 GHz for transmission (down link), is used for the SPADE implementation. These channels are each 38 kHz wide and are spaced 45 kHz apart over the 36-MHz transponder band. Digital transmission at 64 kbits/s, using QPSK (four-phase PSK), is used to transmit information over each channel. The system can thus be used for either single-channel voice communication (telephony), using PCM techniques to convert a single 4-kHz voice channel to 64 kbits/s binary data, or, directly, for the transmission of digital data at the 64-kbits/s rate. (The primary purpose of the system is to provide telephone communication, however.) Note that 8-bit PCM samples are used in this scheme.

The setup of a call, involving assignment of a channel from those currently free among the 800 potentially available, is done through a separate signaling channel called the *common signaling channel*. Signaling is carried out digitally using time-division multiple-access (TDMA) techniques. 50-ms time frames, each divided into 1-ms time slots, are set up for this purpose. Forty-nine of the 50 time slots in a frame are allocated to stations, one per station. The remaining slot per frame is used as a reference.

[46] Spilker, *op. cit.*, chap. 10, discusses the TDMA concept.
[47] Puente and Werth, *op. cit.* The term SPADE is derived from the first letters of the words *s*ingle-channel-per carrier, *p*ulse-code-modulation, multiple-*a*ccess, *d*emand-assignment *e*quipment.

A station wishing to initiate a call to another station selects a frequency randomly from those available, as noted in its own allocation table. It then transmits this frequency, and the address of the station with which it would like to establish a connection, in the time slot allocated to it. The satellite continually rebroadcasts all signaling frames down-link to all stations associated with this transponder. All stations monitor the common signaling channel. Call requests and call disconnects are used to continually update the frequency allocation tables kept at each station. If another station had selected the same frequency earlier, as noted by the appearance of that frequency in a rebroadcast slot prior to its appearance in the slot of the station in question, both the calling and called stations would register a busy signal. The calling station would then select another frequency and try again. (Since the round-trip propagation delay up to the satellite and back is 240 ms, a station will hear its own request 240 ms after making it.)

If the frequency chosen by the calling station has not been selected by any other station prior to its broadcast on the signaling channel, the called station, noting its address, transmits an acknowledgment message back to the calling station in the time slot assigned to it. It then proceeds to set up its equipment to establish a connection at the frequency channel specified. The calling station will initiate communication over that channel on receipt of the acknowledgment message. The setup interval takes about 600 ms because of the double round-trip delay involved, time-slot locations, and other system delays.

A typical frame for the TDMA signaling scheme is shown in Fig. 4-70. Station B is shown setting up a call to station D. Both the frequency selected and the label of D would appear in the B slot shown. If the frequency has not previously been selected D sends an acknowledgment in its own slot. This is shown appearing about 600 ms later. Disconnect signals, releasing the connection and returning the frequency used to the common pool, are transmitted over the same signaling channel. Each slot in the signaling channel carries 128 bits. The signaling bit rate is thus 128 kbits/s, carried over a 160-kHz-wide frequency channel dedicated to signaling, using ordinary PSK.

The SPADE scheme is limited to only 49 stations per transponder. The stations selected for demand assignment are those with relatively light loading. A combination of both preassigned (fixed) and demand-assigned frequency allocations are used to handle the worldwide Intelsat communications.

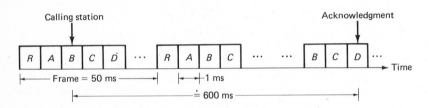

Figure 4-70 TDMA signaling in SPADE system.

4-13 SUMMARY

As noted at the beginning of this chapter, high-frequency sine-wave transmission is necessary for effective radiation of signals through space, water, or other transmission media. It is commonly used for transmitting signals over cables and microwave links in the telephone plant. The digital signals discussed in Chap. 3 must thus modulate sine-wave carriers for transmission over a desired channel.

Both digital and analog signals will generally modulate carriers in practice. For simplicity's sake and to maintain continuity with the digital communication systems introduced in Chap. 3, we first discussed in this chapter the types of sine-wave modulation peculiar to binary signal transmission. These include, among the most common types, PSK, OOK, and FSK transmission. Discussions of implementation, both at the transmitter and at the receiver, as well as bandwidth considerations, then carried over directly to the analog signal case as well. For example, AM transmission, with OOK as a special case, was shown to require a transmission bandwidth twice that of the baseband signal, although the use of SSB transmission reduces this to just that of the baseband signal. FM transmission, with FSK as a special case, requires, in general, wider bandwidths, with the transmission bandwidth approximately given by $2\Delta f + 2B$, with Δf the frequency deviation and B the modulating signal bandwidth.

Digital modulation techniques extend the binary modulation case to that of multisymbol signaling. Here a group of n consecutive binary digits are used to generate one of $M = 2^n$ different symbols. Examples studied in this chapter included multiple PSK (QPSK, with $M = 4$ possible signals, is one prominent example) and the combined amplitude/phase shift variation of a sine-wave carrier used in QAM transmission. These classes of signals are used extensively in high-bit-rate transmission over bandlimited telephone channels. The devices that convert the digital signals to the higher-frequency analog signals used in telephone practice are called modems.

AM systems were found to require some type of product modulator for the modulation process. Both nonlinear and piecewise linear devices were shown to provide the necessary product term. Product modulators are in general used to carry out frequency conversion, the translation of signals from one frequency band to another.

We concluded the chapter with a brief discussion of frequency-division multiplexing. Both telephone and satellite systems were used as examples. Frequency-division multiplexing, with groups of signals shifted to adjacent frequency bands, the combined wider band signal then used to modulate a higher-frequency carrier, is the analog of time-division multiplexing. In both cases a multiplicity of signals is combined for more economical transmission over a common, wider-band, channel.

In these last two chapters, we have explored the basic aspects of communication system design. We focused first on digital baseband systems, discussing PCM and delta modulation systems that convert analog signals to digital format, and then showing how digital signals, whether generated as such or derived from an analog form, may be combined using time-multiplexing techniques. We then studied the next step in the information transmission process, that of the modulation of a high-frequency carrier for long-range transmission. The concept of bandwidth was shown to play a fundamental role throughout. We have now seen how communication signals are handled, from generation at the transmitter to final processing at the receiver. Combining the transmitter and receiver we have the complete communication system first noted in our introduction to Chap. 1.

In the remaining two chapters of this book we study the performance of communication systems. As noted in Chap. 1, the need for a performance evaluation arises because of possibly deleterious effects introduced by the transmission channel. To simplify the analysis in this first approach to the subject, we focus on limitations due to noise only. (The effect of intersymbol interference on signal transmission was mentioned in Chap. 3.) We model the effect of noise in digital transmission in Chap. 5, obtaining useful equations and curves for the probability of error due to noise. Both baseband and high-frequency binary transmission are considered, and PSK, FSK, and OOK techniques are compared. We then study the effect of noise on AM and FM signal reception, obtaining the well-known FM improvement factor in the presence of noise.

In Chap. 6 we specialize to digital communications, and show how the error probability due to noise may be reduced over channels that are limited in their transmission by more complex signaling schemes and the use of error-correction coding. Shannon's capacity theorem, referred to previously in this book, is again introduced to provide a limit on how much improvement is possible through the use of coding techniques. Applications are drawn from both satellite and space communications systems.

The elements of probability theory are assumed known throughout the discussion in Chaps. 5 and 6. Those readers unfamiliar with the fundamentals of probability, or those wishing a review of the subject, are referred to the Appendix for a detailed introduction.

PROBLEMS

4-1 The output of a PCM system consists of a binary sequence of pulses, occurring at the rate of 2×10^6 bits/s. With raised cosine shaping used for the baseband pulses, compare the transmission bandwidths required in the following two cases:

(a) OOK transmission, amplitude modulation of a sine-wave carrier.

(b) FSK, switching between two sine waves of frequencies 100 and 104 MHz. Repeat the FSK calculation if the two frequencies are 100 and 120 MHz. Sketch the spectra in all cases and indicate assumptions made.

4-2 A binary message consists of an alternating sequence of 1's and 0's. FSK transmission is used, with the two frequencies each multiples of the binary period T and synchronized in phase (see Fig. 4-6a). Find the spectrum of the FSK wave and sketch, verifying the expression given in the text and sketched in Fig. 4-6b.

4-3 Refer to Prob. 3-29. Find the transmission bandwidths required if the binary pulses phase-modulate a high-frequency carrier. (PSK transmission is used.)

4-4 FSK transmission is used to transmit 1,200-bits/s digital signals over a telephone channel. The FSK signals are to fit into the range 500 to 2,900 Hz. A modulation index $\beta \doteq 0.7$ is desired, so that the two carrier frequencies are taken to be 1,200 Hz and 2,200 Hz. Find the baseband bandwidth required of the binary signals. Assuming sinusoidal roll-off shaping, what roll-off factor is required?

4-5 A telephone channel allows signal transmission in the range 600 to 3,000 Hz. The carrier frequency is taken to be 1,800 Hz.

(a) Show that 2,400-bits/s, four-phase PSK transmission with raised cosine shaping is possible. Show that the 6-dB bandwidth about the carrier is 1,200 Hz.

(b) 4,800 bits/s is to be transmitted over the same channel. Show that eight-phase PSK, with 50 percent sinusoidal roll-off, will accommodate the desired data rate. Show that the 6-dB bandwidth about the carrier is now 1,600 Hz.

4-6 9,600-bits/s transmission over a telephone line is desired. For this purpose the line must be specially conditioned to allow signal transmission over the range 300 to 3,000 Hz. Show that 16-state QAM, with 12.5 percent sinusoidal roll-off shaping, will provide the desired bit rate. The carrier is chosen in the center of the band at 1,650 Hz. Show that the 6-dB bandwidth about the carrier is 2,400 Hz.

4-7 A single voice channel is to be transmitted via PCM techniques using satellite communications. 8000 samples/s are taken, and 7-bit quantization (128 levels) is used. 32 synchronization bits are inserted into the binary stream for every 224 data bits transmitted. The resultant binary stream is then transmitted using sinusoidal roll-off shaping for each binary pulse. The roll-off factor is 20 percent.

(a) What is the PCM bit rate in bits/s?

(b) What is the baseband (PCM signal) bandwidth?

(c) PSK is used for transmission. What is the transmission bandwidth required?

(d) 4-Φ PSK is used instead: successive *pairs* of bits are used to phase-modulate a carrier. What is the transmission bandwidth required in this case?

(e) Repeat (c) if OOK transmission is used. Repeat if FSK is used with the frequency deviation chosen as 38 kHz. What is the modulation index β in this case? What is the frequency spacing of the two carriers?

4-8 A telephone line having an effective usable bandwidth of 2,400 Hz from 600 to 3,000 Hz is to be used to transmit data from a computer data terminal. A modem is used to connect the terminal to the line.

(a) A 1,200-bits/s terminal is connected. A modem with FSK transmission is to be used. Pick two appropriate frequencies for transmission if the modulation index β is to be as large as possible. What are the two frequencies? What is the resultant β? What shaping is to be used? Sketch a typical output wave pattern.

(b) The same line is to be used to provide 4,800 bits/s transmission capability for a time-division multiplexer outputting bits at that rate. Specify an appropriate modulation technique, the type of shaping used, and select an appropriate carrier frequency. Sketch the output wave pattern corresponding to a particular sequence of binary digits.

4-9 (a) Consider the synchronous detector implementation shown in Fig. P4-9. A high-frequency signal $f(t) \cos \omega_0 t$, with $f_0 \gg B$, the bandwidth of the baseband signal $f(t)$, is sampled synchronously

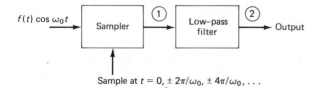

Sample at $t = 0, \pm 2\pi/\omega_0, \pm 4\pi/\omega_0, \ldots$ **Figure P4-9**

at the following intervals of time: $t = 0, \pm 2\pi/\omega_0, \pm 4\pi/\omega_0, \ldots$. The samples are then passed through a low-pass filter as shown. Demonstrate that $f(t)$ (this could be either a binary sequence or an analog signal) is reproduced at the filter output. *Hint:* Show this by representing the sampling operation either as a multiplication by a set of periodic impulses or as multiplication by a sampling function $S(t)$. Proceed as in Chap. 3, in discussing Nyquist sampling. Find and sketch the resultant spectra at points (1) and (2). What should the frequency characteristics of the filter be?

(*b*) What is the output like if the sampling times are displaced by $\pi/4\omega_0$ seconds? $\pi/2\omega_0$ seconds? What does this imply about the accuracy of sampling required if $f_0 = 100$ MHz?

4-10 The output of a 2,400-bits/s time-division multiplexer is fed into a modem. Compare the transmission bandwidths required at the modem output for the following modulation schemes (raised cosine shaping is used):

(*a*) FSK, with a frequency deviation of $\pm 2,400$ Hz about the carrier.

(*b*) OOK transmission.

(*c*) eight-phase PSK.

4-11 A 4,800-bits/s data terminal is connected to a modem. Calculate the transmission bandwidth B_T required at the modem output for each of the following schemes (50 percent roll-off shaping is used in all cases):

(*a*) OOK transmission.

(*b*) FSK transmission. The frequency deviates $\pm 3,600$ Hz about the carrier.

(*c*) 16-level QAM.

4-12 A transmission channel of 1 MHz bandwidth, centered at 100 MHz, is available for signal transmission.

(*a*) The PCM system and 240-kbits/s data source of Fig. P4-12 are time-multiplexed together, and the output then fed into a PSK modulator, as shown (25 percent roll-off shaping is used). Find the maximum number of quantization levels in the PCM system that may be used. (Neglect framing, signaling, and stuffing bits in this example.)

(*b*) The PCM system in (*a*) is required to have 256-level quantization. *Two* 240-kbits/s data sources are to be time-multiplexed with the PCM output. The same transmission channel must again be used. Indicate how the modulator might be modified to accomplish this.

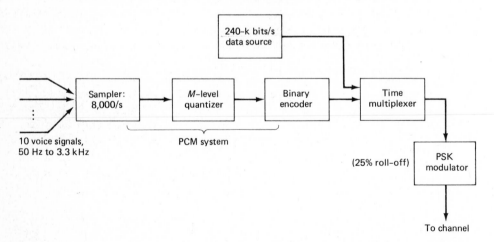

Figure P4-12

4-13 Consider the frequency conversion system sketched in Fig. P4-13. The input signal $s_1(t)$ is a double-sideband signal with 1-MHz sidebands on either side of 10 MHz, as shown in Fig. P4-13.

Specify the frequency $f_0 = \omega_0/2\pi$ of the locally generated input to the product modulator, $\cos \omega_0 t$, and sketch the filter characteristics required, for the following cases:

(*a*) The output is a baseband signal, 0 to 1 MHz.

(*b*) The output is a double-sideband signal, centered about 20 MHz.

(*c*) The output is a single-sideband signal with the spectral characteristic shown in Fig. P4-13.

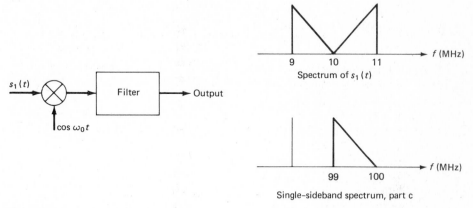

Spectrum of $s_1(t)$

Single-sideband spectrum, part c

Figure P4-13

4-14 An amplifier has a slight nonlinearity, so that its output-input characteristic is given by

$$e_o = A_1 e_i + A_2 e_i^2$$

(*a*) $e_i = \cos \omega_m t$. Find all the frequencies present in the output. How could this device be used as a second-harmonic generator?

(*b*) $A_1 = 10$, $A_2/A_1 = 0.05$. Determine the ratio of second-harmonic amplitude to fundamental amplitude. (This is frequently called the *second-harmonic distortion*.)

4-15 (*a*) Let $e_i = 0.1 \cos 500t$ in the amplifier of Prob. 4-14. Determine the frequency terms present in the output and the amplitude of each. Repeat for $e_i = 0.2 \cos 3,000t$.

(*b*) $e_i = 0.1 \cos 500t + 0.2 \cos 3,000t$. Determine the different frequency terms present in the output and the amplitude of each. What new frequency terms appear that were not present in the two cases of (*a*)? These are known as the *intermodulation* frequencies and are due to the nonlinear mixing of the two frequency terms.

4-16 A voltage $e_t = 0.1 \cos 2,500t + 0.2 \cos 3,000t$ is applied at the input to the amplifier of Prob. 4-14. The output voltage e_o is in turn applied to an ideal rectangular filter passing all frequencies between 50 Hz and $4,000/2\pi$ hertz with unity gain, rejecting all others. Calculate the power at the output of the filter in the frequency terms generated by the amplifier nonlinearity. Determine the ratio of power in these terms to the power in the 2,500 and 3,000 rad/s terms at the filter output.

4-17 A semiconductor diode has a current-voltage characteristic at room temperature given by

$$i = I_0 (e^{40v} - 1)$$

(*a*) What is the maximum variation in the voltage v to ensure i linearly proportional to v, with less than 1 percent quadratic distortion?

(*b*) Let $v = 0.01 \cos \omega_1 t + 0.01 \cos \omega_2 t$. Expand i in a power series in v, retaining terms in v, v^2, and v^3 only. Tabulate the different frequencies appearing in i, and the relative magnitudes of each.

4-18 A band-limited signal $f(t)$ may be expressed in terms of the finite Fourier series

$$f(t) = \frac{c_0}{T} + \frac{2}{T} \sum_{n=1}^{M} |c_n| \cos(\omega_n t + \theta_n) \qquad \omega_n = \frac{2\pi n}{T}.$$

The maximum-frequency component of $f(t)$ is $f_M = M/T$.

(a) Sketch a typical amplitude spectrum for $f(t)$. Then indicate the frequency components present in $f^2(t)$.

(b) What is the highest-frequency component of $f^2(t)$?

(c) Find the spectrum of $f(t) \cos \omega_c t$ ($\omega_c \gg \omega_M$), and compare with that of $f(t)$.

4-19 For each network shown in Fig. P4-19, calculate the maximum-frequency component in the output voltage.

(a) e_{in} has a single component of frequency f hertz.

(b) e_{in} has a single component of frequency $2f$ hertz.

(c) e_{in} has two frequency components, f and $2f$.

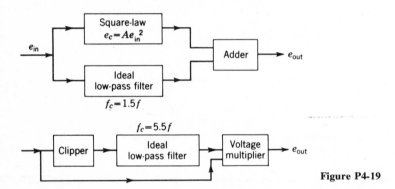

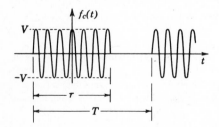

Figure P4-19

4-20 Spectrum of coherent radar pulses. These pulses may be generated by turning a sine-wave generator of frequency f_c on and off periodically at the same phase points. For the example shown in Fig. P4-20,

$$f_c(t) = V \cos \omega_c t \qquad \frac{-\tau}{2} < t < \frac{\tau}{2}$$

$$f_c(t) = 0 \qquad \text{elsewhere in a period}$$

(a) Find the spectrum of $f_c(t)$ by expanding the pulsed carrier in a Fourier series. (Use the exponential form of $\cos \omega_c t$.)

(b) Write $f_c(t) = V \cos \omega_c t \cdot S(t)$, with $S(t)$ a rectangular pulse train. Expand $S(t)$ in a Fourier series. Find the spectrum of $f_c(t)$, and compare with the result of (a).

Figure P4-20

4-21 A periodic function $f(t)$ is band-limited to 10 kHz and has a uniform or flat amplitude spectrum from 0 to 10 kHz. It is given by

$$f(t) = \frac{1}{2} + \sum_{n=1}^{10} \cos n\omega t \qquad \omega = 2\pi \times 1,000$$

Show that the function $f^2(t)$ has an amplitude spectrum decreasing linearly from 0 to 20 kHz. The envelope of the spectrum of $f^2(t)$ is thus triangular in shape.

4-22 A function $f(t)$ has as its Fourier transform

$$F(\omega) = K \qquad |\omega| < \omega_M \qquad F(\omega) = 0 \qquad |\omega| > \omega_M$$

If $f(t)$ is the input voltage to a square-law device with the characteristics $e_o = Ae_i^2$, show that the output has the triangular spectrum of Fig. P4-22 (compare with Prob. 4-21).

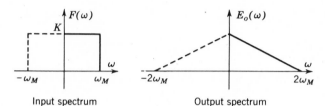

| Input spectrum | Output spectrum | **Figure P4-22** |

4-23 An AM signal consists of a carrier voltage $100 \sin (2\pi \times 10^6 t)$ plus the voltage $(20 \sin 6,280 t + 50 \sin 12,560 t) \sin (2\pi \times 10^6 t)$.

(*a*) Draw the amplitude-versus-frequency plot of the modulated signal.

(*b*) How much peak power will this signal deliver to a 100-Ω load?

(*c*) What is the average power (over a modulation cycle) delivered to a 100-Ω load?

(*d*) Sketch the envelope over a modulation cycle. (Indicate numerical values at key points.)

4-24 Assume that the carrier in Prob. 4-23 is suppressed after modulation. Repeat Prob. 4-23 under this condition.

4-25 *Frequency conversion.* An amplitude-modulated carrier

$$f_c(t) = A[1 + mf(t)] \cos \omega_c t$$

is to be raised or lowered in frequency to a new carrier frequency $\omega_c' = \omega_c \pm \Delta\omega$. The resultant time function will appear in the form $f_c'(t) = A'[1 + mf(t)] \cos \omega_c' t$.

(*a*) Show that a product modulator producing an output $f_c(t) \cos \Delta\omega t$ will accomplish this frequency conversion. Draw a simple block diagram for this frequency converter. Specify the filtering required in the two cases of frequency raised by $\Delta\omega$ and lowered by $\Delta\omega$ radians/s.

(*b*) Show that a nonlinear device with terminal characteristics $e_o = \alpha_1 e_i + \alpha_2 e_i^2$ will perform this frequency conversion if $f_c(t) + \cos \Delta\omega t$ is applied at the input. Sketch a typical amplitude spectrum of the output signal, and specify the filtering necessary to produce the frequency conversion.

4-26 A general nonlinear device has output-input characteristics given by $e_o = \alpha_1 e_i + \alpha_2 e_i^2 + \cdots$ (this could represent a diode, transistor, vacuum triode, etc.).

(*a*) Show that all combinations of sum and difference frequencies appear in the output if e_i is a sum of cosine waves.

(*b*) Show that this device can be used as a modulator if $e_i = f(t) + \cos \omega_c t$.

(*c*) The input is a normal AM wave given by $e_i = K[1 + mf(t)] \cos \omega_c t$. Show that this nonlinear device will demodulate the wave, i.e., reproduce $f(t)$.

(*d*) Show that this device can act as a frequency converter if

$$e_i = K[1 + mf(t)] \cos \omega_c t + \cos \Delta\omega t$$

In particular, show that with proper filtering the AM wave spectrum can be shifted up or down by $\Delta\omega$ radians/s.

4-27 A function appears in the form

$$f_c(t) = \cos \omega_c t - K \sin \omega_m t \sin \omega_c t$$

Using block diagrams, show how to generate this, given $\cos \omega_c t$ and $\sin \omega_m t$ as inputs.

4-28 Amplitude-modulation detector. An amplitude-modulated voltage

$$f_c(t) = K[1 + mf(t)] \cos \omega_c t$$

is applied to the diode-resistor combination of Fig. P4-28. The diode has the piecewise-linear characteristics shown. (It is then called a linear detector.) Show that for $|mf(t)| < 1$ the l-f components of the load voltage e_L provide a perfect replica of $f(t)$.

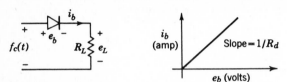

Figure P4-28 Linear detector.

4-29 Repeat Prob. 4-28 for the case where the diode-resistor combination of Fig. P4-28 has the square-law characteristic $e_L = \alpha_1 e_i + \alpha_2 e_i^2$. Show that e_L contains a component proportional to $f(t)$.

4-30 (*a*) A DSB (suppressed-carrier) signal of the form $f(t) \cos \omega_c t$ is to be demodulated. Show that multiplying this signal by $\cos \omega_c t$ in a product modulator reproduces $f(t)$.

(*b*) "The DSB demodulator is phase-sensitive." Demonstrate the validity of this statement by multiplying $f(t) \cos \omega_c t$ by $\cos (\omega_c t + \theta)$ and letting θ vary from 0 to π radians.

4-31 An AM signal has the form

$$s(t) = [1 + mf(t)] \cos \omega_0 t \qquad |mf(t)| \le 1$$

The bandwidth of $f(t)$ is $B \ll f_0$. Consider the receiver shown in Fig. P4-31.

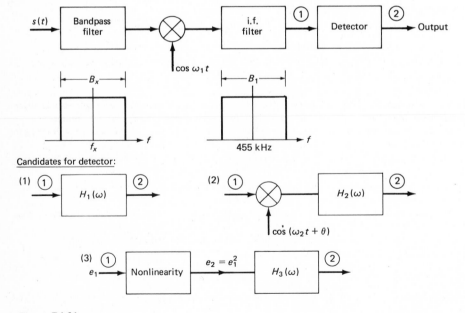

Figure P4-31

(a) Specify f_x, B_x, f_1, and B_1, if $f(t)$ is to be reproduced at the output.

(b) Show which of the three detectors shown could reproduce $f(t)$ at the output. Sketch carefully the corresponding filter characteristics. If the answer for device (2) is affirmative, specify what f_2 and θ should be.

4-32 The input signal $s(t)$ in Prob. 4-31 is applied to a somewhat modified receiver, as shown in Fig. P4-32.

(a) $f(t)$ is again to be reproduced at (2). What are B_y and f_1? Sketch carefully the i-f characteristic.

(b) Repeat (b) of Prob. 4-31.

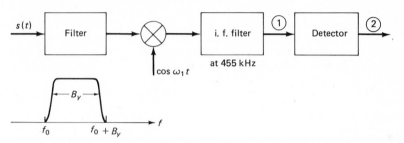

Figure P4-32

4-33 Prove the following properties of Hilbert transforms:

(a) If $f(t) = f(-t)$, then $\hat{f}(t) = -\hat{f}(-t)$.

(b) If $f(t) = -f(-t)$, then $\hat{f}(t) = \hat{f}(-t)$.

(c) The Hilbert transform of $\hat{f}(t)$ is $-f(t)$.

Hint: These are easily proven using the 90° phase-shift description of the transform process.

4-34 Define a function $y(t) = f(t) - j\hat{f}(t)$, with $\hat{f}(t)$ the Hilbert transform of $f(t)$.

(a) Show

$$Y(\omega) = 2F(\omega) \qquad \omega \leq 0$$
$$= 0 \qquad \omega > 0$$

(b) Show Re $[y(t)e^{j\omega_c t}]$ is a *lower-sideband* SSB signal and is given by Eq. (4-39).

4-35 Find the Hilbert transforms of the two pulses shown in Fig. 4-43 in the text, verifying the results shown there. *Hint:* Find the Fourier transforms of the two pulses, introduce the necessary 90° phase shift, and then take *inverse* transforms.

4-36 Show that detection of a SSB signal requires multiplication by a carrier term precisely in phase with the incoming signal. Show that a phase difference of θ radians between the SSB carrier and the carrier term at the receiver results in the output signal given by Eq. (4-45). Sketch the output function of Eq. (4-45) for the two pulses of Fig. 4-43 for $\theta = 0$, $\pi/4$, $\pi/2$.

4-37 *Single-sideband detection.* An audio tone $\cos \omega_m t$ and a carrier $\cos \omega_c t$ are combined to produce an SSB signal given by $\cos (\omega_c - \omega_m)t$.

(a) Show that $\cos \omega_m t$ can be reproduced from the SSB signal by mixing with $\cos \omega_c t$ as a local carrier.

(b) The local carrier drifts in frequency by $\Delta\omega$ radians to $\omega_c + \Delta\omega$. What is the demodulated signal now?

(c) The local carrier shifts in phase to $\cos (\omega_c t + \theta)$. Find the audio output signal.

4-38 (a) An amplitude-modulated voltage given by

$$v_c(t) = V(1 + m \cos \omega_m t) \cos \omega_c t \qquad \text{volts}$$

is applied across a resistor R. Calculate the percentage average power in the carrier and in each of the two sideband frequencies. Calculate the peak power.

(*b*) The voltage $v_c(t)$ is now a suppressed-carrier voltage $v_c(t) = V \cos \omega_m t \cos \omega_c t$ applied across R ohms. What is the fraction of the total average power in each of the sideband frequencies?

4-39 An SSB radio transmitter radiates 1 kW of average power summed over the entire sideband. What total average power would the transmitter have to radiate if it were operating as a DSB (suppressed-carrier) system and the same distance coverage were required?

4-40 An AM transmitter is tested by using the dummy load and linear narrowband receiver of Fig. P4-40. The r-f amplifier portion of the receiver is swept successively and continuously over the range 100 kHz to 10 MHz. Its frequency characteristic at any frequency f_c in this range is shown in the figure.

With no audio input the wattmeter reads 100 W average power. The peak-reading voltmeter (VM) and r-f amplifier tuning characteristic indicate an output of 10 V peaks at 1 MHz.

(*a*) With an audio input signal of 10 V peak at 1 kHz the wattmeter reads 150 W. At what frequencies are there receiver outputs? What are the various amplitudes, as read on the VM?

(*b*) The 1-kHz modulating signal is replaced by the composite signal $2 \cos 12,560t + 3 \cos 18,840t$. What is the new wattmeter reading? At what frequencies will there be receiver outputs? What are the various amplitudes?

(*c*) The transmitter modulator is modified so that the carrier is suppressed but the other characteristics are unchanged. An audio signal of 10 V peak at 1 kHz is applied. What is the wattmeter reading? What are the receiver output frequencies and amplitudes? Sketch the pattern that would be seen on an oscilloscope connected across the load. (The left-to-right sweep period is 2,000 μs; the return takes negligible time.)

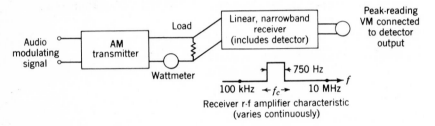

Figure P4-40

4-41 The radio receiver of Fig. P4-41 has the r-f amplifier and i-f amplifier characteristics shown. The mixer characteristic is given by $e_3 = e_2(\alpha_0 + \alpha_1 e_{01})$, with α_0 and α_1 constants, and $e_{01} = A \sin \omega_c' t$. The product-detector output is $e_5 = e_4 e_{02}$, with $e_{02} = B \cos (\omega_c - \omega_c')t$.

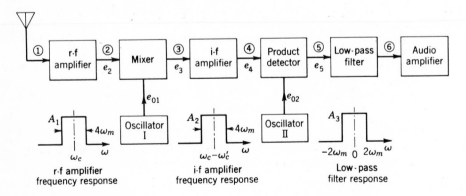

Figure P4-41

Find the signals at points 4 and 6 for each of the following signals at point 1:

(*a*) Normal AM, $(1 + \sin \omega_m t) \sin \omega_c t$

(*b*) Double sideband, suppressed carrier, $\sin \omega_m t \sin \omega_c t$

(*c*) Single sideband, $\cos (\omega_c - \omega_m)t$

4-42 The product detector and oscillator II in Fig. P4-41 are replaced by an envelope detector with the characteristic $e_5 = be_4{}^2$. The r-f amplifier can be continuously tuned over the range $f_c = \omega_c/2\pi = 535$ to $1,605$ kHz. The tuning capacitor of oscillator I is ganged to the tuning capacitor of the r-f amplifier so that $f'_c = \omega'_c/2\pi = f_c + 455$ kHz at all times. The circuit is then that of a superheterodyne radio receiver. The passband of the amplifiers is ± 5 kHz about the center frequency.

(*a*) Determine all frequencies present at points 3 and 4 if the signal at 1 is a 1-MHz unmodulated carrier.

(*b*) A 1-kHz audio signal amplitude modulates the 1-MHz carrier. Determine the frequencies present at points 2 and 4.

(*c*) The r-f amplifier is tuned to 650 kHz. To your amazement, however, a radio station at 1,560 kHz (WQXR in the New York City area) is heard quite clearly. Can you explain this phenomenon?

(*d*) The input at point 1 is the DSB signal of Prob. 4-41, with $f_c = 1$ MHz and $f_m = \omega_m/2\pi = 1$ kHz. Find the frequencies present at points 5 and 6. Compare with the product-detector output of Prob. 4-41.

4-43 Consider the system shown in Fig. P4-43. Show that the output $g(t)$ is a SSB signal. Do this by assuming $F(\omega)$ as shown ($B \ll f_c$) and carefully sketching the transforms at the output of each device. Preserve the distinction between the shaded and unshaded halves of $F(\omega)$. Is the lower or upper half of a conventional DSB signal retained?

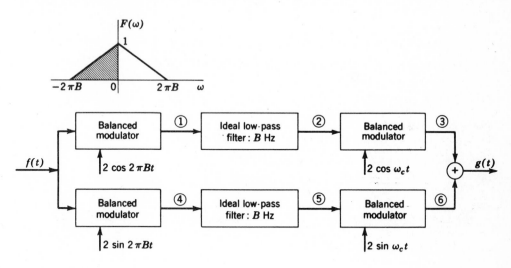

Figure P4-43

4-44 A DSB signal $(\cos \omega_m t + a \cos 2 \omega_m t) \cos \omega_c t$ is applied to the receiver shown in Fig. P4-44. Find the signals shown at points 1, 2, and 3 if

(*a*) $\omega_1 = \omega_c - 3\omega_m$

(*b*) $\omega_1 = \omega_c$

Discuss.

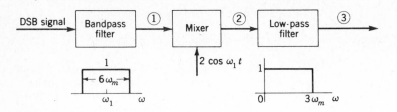

Figure P4-44

4-45 An AM signal $r(t) = a[1 + f(t)] \cos (\omega_0 t + \theta)$, where θ is a constant and $f(t)$ has a Fourier transform which is zero for $|\omega| > \Omega$. Assume that $\Omega \ll \omega_0$ and $|f(t)| \leq 1$. Consider the receiver shown in Fig. P4-45. The low-pass filters have transfer functions

$$H(\omega) = 1 \qquad |\omega| < \Omega$$
$$\qquad\quad 0 \qquad |\omega| > \Omega$$

By calculating the signals at points 1 through 6, show that the receiver can be used to demodulate $f(t)$.

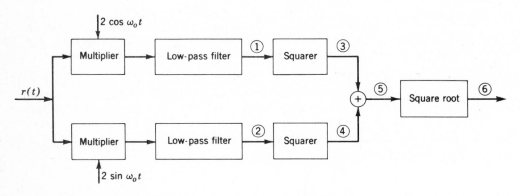

Figure P4-45

4-46 Consider a stereo broadcasting station that desires to transmit the two signals:

$$x(t) \to \text{left channel}$$
$$y(t) \to \text{right channel}$$

Both $x(t)$ and $y(t)$ have frequency components only in the range $(2\pi)30 \leq \omega \leq (2\pi)10,000$. The station actually broadcasts the signal $[A + x(t) + y(t)] \cos \omega_0 t + [A + x(t) - y(t)] \sin \omega_0 t$, where $\omega_0 = 2\pi(10^6)$ and $A \gg |x(t)| + |y(t)|$.

(*a*) Give and discuss a block diagram of a receiving system which provides as separate outputs $x(t)$ and $y(t)$. Assume the phase and frequency of the transmitted wave is known exactly at the receiver. You may use mixers, multipliers, adders, filters, etc.

(*b*) What would be the output of a receiver which merely consisted of an envelope detector and a low-pass filter? (Would it be intelligible?) *Hint:* Approximate for large A.

4-47 (*a*) A 1-kHz sine-wave frequency modulates a 10-MHz carrier. The amplitude of this audio signal is such as to produce a maximum-frequency deviation of 2 kHz. Find the bandwidth required to pass the frequency-modulated signal.

(*b*) Repeat (*a*), if the modulating signal is a 2-kHz sine wave.

(*c*) The 2-kHz sine wave of (*b*) is doubled in amplitude, so that $\Delta f = 4$ kHz. Find the required bandwidth.

4-48 A 100-MHz sine-wave carrier is to be frequency modulated by a 10-kHz sine wave. An engineer designing the system reasons that he can minimize bandwidth by decreasing the audio amplitude. He arranges for a maximum-frequency deviation Δf of only ± 10 Hz away from the 100-MHz carrier frequency and assumes that he requires only 20-Hz bandwidth.

(*a*) Find the fallacy in his reasoning, and specify the actual bandwidth required.

(*b*) Would he have been right to assume a bandwidth of 2 MHz if the audio amplitude had been chosen to produce a Δf of 1 MHz? Explain the difference between the results of (*a*) and (*b*).

4-49 An FM receiver similar to that of Fig. 4-61 is tuned to a carrier frequency of 100 MHz.

(*a*) A 10-kHz audio-signal frequency modulates a 100-MHz carrier, producing a β of 0.1. Find the bandwidths required for the r-f and i-f amplifiers and of the audio amplifier.

(*b*) Repeat (*a*) if $\beta = 5$.

(*c*) Two signals at 100 MHz are tuned in alternately. The carriers are of equal intensity. One is modulated with a 10-kHz signal and has $\beta = 5$; the other is modulated with a 2-kHz signal and has $\beta = 25$. Which requires the larger bandwidth? Explain. Compare the audio-amplifier outputs in the two cases.

(*d*) Two other signals are tuned in alternately. The carriers are again of equal intensity. One has a frequency deviation of 10 kHz with $\beta = 5$, the other a deviation of 2 kHz with $\beta = 25$. Which requires the larger bandwidth? Which gives the larger audio output?

4-50 A general frequency-modulated voltage has the form $f_c(t) = A_c \cos [\omega_c t + K \int f(t) \, dt]$. Show that the average power dissipated in a 1-Ω resistor is $A_c^2/2$ averaged over a modulating cycle.

4-51 (*a*) A modulating signal $f(t) = 0.1 \sin 2\pi \times 10^3 t$ is used to modulate a 1-MHz carrier in both an AM and an FM system. The 0.1-V amplitude produces a 100-Hz frequency deviation in the FM case. Compare the receiver r-f amplifier and audio-amplifier bandwidths required in the two systems.

(*b*) Repeat (*a*) if $f(t) = 20 \sin 2\pi \times 10^3 t$.

4-52 A sine wave is switched periodically from 10 to 11 MHz at a 5-kHz rate. Sketch the resultant waveshape. What is the form of the modulating signal if this frequency-shifted wave is considered an FSK signal? What is the approximate transmission bandwidth? Compare with the bandwidth required if the modulating signal is approximated by a 5-kHz sine wave of the same amplitude.

4-53 The tuned circuit of an oscillator is shown in Fig. P4-53. The current source $g_m e_g$ is controlled by the voltage $e_g(t)$ across R as shown.

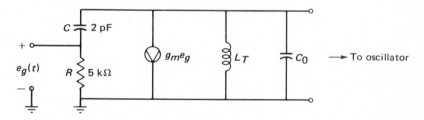

Figure P4-53

(*a*) Show that with $1/\omega_c \gg RC$, the total effective capacitance of the tuned circuit is given by

$$C_T = C_0 + C + g_m RC$$

(*b*) What is the oscillator frequency if $g_m = 4,000 \ \mu S$?

(*c*) What is the modulation index β if g_m varies from 3,000 to 5,000 μS at a 1-kHz rate?

4-54 Consider a circuit such as the one in Fig. P4-53, but with R and C interchanged. Calculate the effective inductance introduced by the controlled source. Repeat Prob. 4-53.

4-55 A controlled source circuit similar to that of Fig. P4-53 provides an effective capacitance $C' = [-2e_g(t) + 6]10^{-12}$ F appearing in parallel with C_0 and L_T. For this case $C_0 = 14 \ \mu\mu F$, $L_1 = 50 \ \mu H$, and $e_g(t) = 0.5 \sin 2\pi \times 500t$.

(a) What is the carrier frequency?

(b) What is the frequency deviation Δf?

(c) Determine the modulation index β. Show a rough sketch of the frequency spectrum of the FM wave indicating the approximate bandwidth.

(d) Repeat (c) if the frequency deviation is reduced to 0.01 of its value in (b), with the modulating frequency unchanged.

4-56. The capacitance of a PN junction is given in terms of its reverse bias voltage V by

$$C = \frac{C_0}{(1 + V/\psi)^K}$$

with ψ the contact potential, and K and C_0 known constants. Such a reversed-bias diode is to be used as a variable capacitor.

(a) $K = 0.5$, $\psi = 0.5$, $C_0 = 300$ pF. Plot C versus the reversed-bias voltage V.

(b) The diode is to be used as the sole source of capacitance in a tuned circuit. Assume the diode is biased at -6 V and the center frequency is to be adjusted to 10 MHz. Plot frequency deviation away from 10 MHz versus voltage deviation away from -6 V. Calculate the modulation sensitivity in hertz per volt in the vicinity of -6 V (this is the initial slope of the curve plotted), and the maximum voltage excursion in either direction for a maximum of 1 percent deviation away from a linear frequency-voltage characteristic.

(c) A fixed capacitance of 100 pF is added in series with the diode. Repeat part (b) and compare.

4-57 *FM Stereo.* Call the left-speaker output L, and the right-speaker output R. For stereo broadcasting the two signals are first combined (*matrixed*) to form composite signals $L + R$, and $L - R$. (As in monaural FM the nominal bandwidth of each signal is 15 kHz.) The $L - R$ signal is multiplied by a 38-kHz carrier in a balanced modulator to form a DSB signal, as shown in Fig. P4-57. The DSB signal, the baseband $L + R$ signal, and a 19-kHz tone from which the 38-kHz carrier is obtained are then summed to provide a frequency-multiplexed signal that in turn modulates the FM carrier. (The 19-kHz tone provides the pilot carrier for receiver synchronization.)

(a) Sketch a typical spectrum for the frequency-multiplexed signal.

(b) Estimate the frequency deviation and the modulation index if the transmission bandwidth is to be kept to 240 kHz. Compare with the monaural case ($L + R$ above).

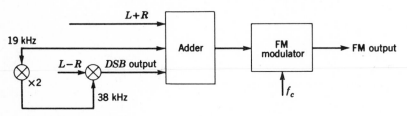

Figure P4-57

4-58 Consider the transmission system shown in Fig. P4-58.

(a) Find frequencies f_1 and f_2 ($f_2 > f_1$) such that the composite (summed) signal at (1) occupies as narrow a frequency range as possible. Show that this range is 12 kHz.

(b) The FM transmitter has a center frequency of 1 MHz. A modulation index of $\beta = 5$ is desired. Find the frequency deviation and bandwidth of the signal at (2).

(c) With everything else unchanged the *amplitude* of the signal at (1) is increased by a factor of *two*. Find the bandwidth and modulation index of the signal at (2).

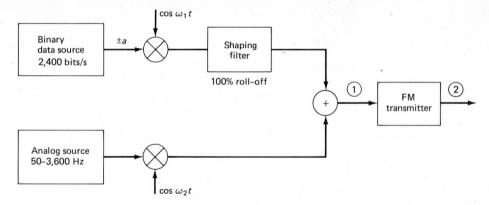

Figure P4-58

4-59 Two audio signals, each covering the range 50 Hz to 15 kHz, are to be combined and transmitted via FM. Two techniques shown in Fig. P4-59 are suggested.

(*a*) Find the frequency deviation and transmission bandwidth of the FM output in the two cases, and compare.

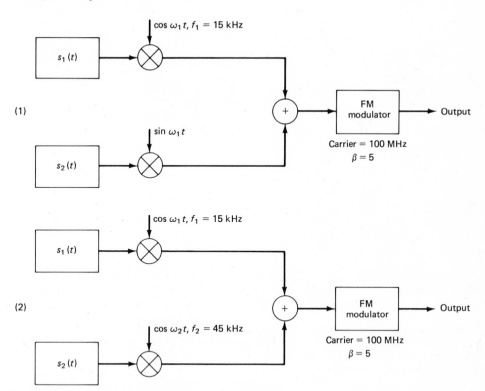

Figure P4-59

(b) Show a block diagram of the receiver in each case, indicating how $s_1(t)$ and $s_2(t)$ are reproduced.

(c) In (b), indicate whether envelope detection can be used, after FM detection, to reproduce $s_1(t)$ and $s_2(t)$ in each case, with the two transmitters as shown. If envelope detection is *not* possible in a particular case, can the transmitter be modified (prior to the FM modulator) to allow envelope detection to be used? How? Indicate with a block diagram where appropriate.

4-60 The outputs of ten 2,400-bits/s data sources are to be frequency-multiplexed using two levels of multiplexing, as shown in Fig. P4-60. Raised cosine shaping (100 percent roll-off) is used for the baseband binary outputs.

(a) Compare the transmission bandwidths of the multiplexed signals in the following cases:

(1) AM-AM (OOK transmission followed by AM).
(2) AM-SSB (OOK transmission followed by SSB modulation).
(3) FM-FM (FSK followed by FM).

Choose $\Delta f = 5$ kHz in the FSK modulators; $\Delta f = 740$ kHz in the FM modulator. The carrier frequencies $f_1, \ldots, f_{10}$, are to be chosen to minimize the bandwidth in each case.

(b) Sketch typical spectra at points (1), (2), (3) for the AM-AM and AM-SSB cases.

(c) Compare the bandwidths obtained in (a) with those required if the first stage of modulation is replaced by time multiplexing. Make any reasonable assumption about synchronizing or framing information needed.

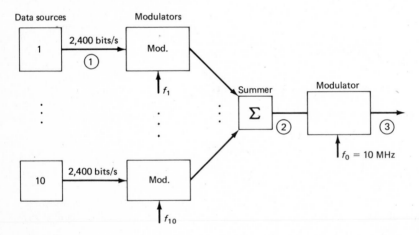

Figure P4-60

4-61 A 57-kbps binary signal is shaped so that it requires a 35-kHz bandwidth. Ten of these signals are then frequency-multiplexed using PSK modulation for each, as shown in Fig. P4-61a, with carriers selected to pack the signals as closely together as possible. ($f_1 = \omega_1/2\pi$ is selected as small as possible.) Ten such multiplexed outputs are now shifted in turn to the 100-MHz range using single-sideband transmission as shown in Fig. P4-61b: the first is multiplexed up to 100 MHz and its upper sideband only retained, the second is shifted up and placed as close to the first as possible, with its upper sideband retained, etc.

(a) What are the initial carrier frequencies f_1 and f_{10}?
(b) What is the final carrier frequency f_c?
(c) Specify the transmission characteristics of filters 1 and 10 shown in Fig. P4-61b.
(d) Sketch the output spectrum in Fig. P4-61b.
(e) Specify a block diagram of a receiver structure for extracting the original binary signals.

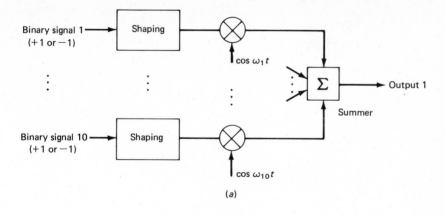

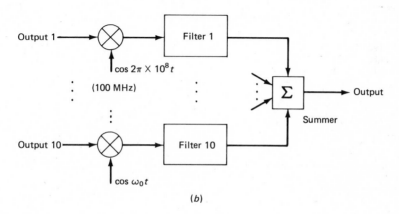

Figure P4-61

Be as quantitative as possible, indicating transmission characteristics of all filters used. Assume that perfect synchronization can be maintained between transmitter and receiver.

4-62 (*a*) Compare the final (100-MHz) transmission bandwidth of Prob. 4-61 with that required if each group of ten 57-kbits/s binary signals is time-multiplexed rather than frequency-multiplexed. Use the same percent of roll-off shaping.

   (*b*) Repeat (*a*) and compare, if two levels of time multiplexing are used, the final time-multiplexed output then fed into a 100-MHz SSB modulator.

4-63 Ten video signals, each of 4-MHz bandwidth, are to be transmitted using one carrier. Several methods of transmission are to be compared.

   (*a*) Each signal individually amplitude-modulates a carrier. The ten AM signals are then summed, and in turn amplitude-modulate a single carrier at a high frequency. Calculate the bandwidth of this final transmitted (composite) signal, if the ten AM signals are packed together as closely as possible.

   (*b*) PCM techniques are used instead: The ten video signals are sampled and then time-multiplexed, quantized to eight levels, coded into binary digits, and then transmitted using OOK transmission. Find the minimum possible bandwidth required for transmission in this case. (Neglect framing pulses.)

   (*c*) Repeat (*b*) if eight-phase PSK is used. Find the minimum possible bandwidth again.

4-64 (*a*) Consider the L600 Bell System Master group discussed in the text. Show, using a block diagram, how the mastergroup is formed from 10 supergroups.

(*b*) Repeat for the U600 Mastergroup.

4-65 Show how one may generate a 12-channel group from twelve 4-kHz voice channels in the FDM hierarchy of Fig. 4-64 by the two-step modulation process used by Northern Telecom of Canada as mentioned in the text: first modulate with an 8.14-MHz carrier; select the upper sideband; then use 12 carriers, spaced equally from 8.204 MHz to 8.248 MHz, selecting the lower sidebands. Provide a block diagram of the process. Specify the transmission characteristics of all filters used.

4-66 Computer output data at a 56-kbits/s rate is to be transmitted in one of the group channels of the FDM hierarchy of Fig. 4-64. Bipolar transmission is used to reduce low-frequency components. Sinusoidal roll-off shaping and SSB transmission are used to fit the signal in the 48-kHz bandwidth. Show a block diagram of the modulation process. What roll-off factor is needed?

PERFORMANCE OF COMMUNICATION SYSTEMS: LIMITATIONS DUE TO NOISE

In previous chapters we have focused on modern techniques for transmitting information. We first discussed digital communications, since it has become clear that the trend is moving more and more to the conversion of analog signals to digital format as soon as possible after the generation of the signals to be transmitted. We then described in detail various modulation techniques used in the transmission of communication signals over relatively long distances. We attempted where possible to compare alternative methods of handling signals. In doing this we focused on the transmitter, at the site where the signals are first generated, and then at the receiver, where the signals must either be reproduced, in the case of analog information, or correctly detected, if in digital format. One parameter of interest stressed throughout the book thus far has been the transmission bandwidth required to pass the communication signals relatively undistorted from transmitter to receiver. More generally, we stressed the spectral or frequency occupancy of the signals transmitted.

One important reason for this is that bandwidth is a resource that must be conserved. The electromagnetic spectrum used for the transmission of the overwhelming majority of communication signals is limited in its bandwidth allocation at the various frequency bands in use. It is thus important to know the spectral occupancy, as measured by the bandwidth of various signals to be transmitted in a given communication system. In addition, it is important to know the bandwidth requirements of the transmission medium or channel over which the signals are to be transmitted in order to have relatively undistorted signals arriving at the receiver. Where distortion is introduced, by transmission over band-limiting channels, its effect on the signals transmitted must be determined.

Bandwidth considerations represent one element in determining the performance of a communication system. As noted briefly in Sec. 1-1, in discussing the transmission of signals through a system, noise is added as the signal moves from transmitter to receiver, in some cases signal fading is encountered, interference from other signals may be introduced, and a variety of other adverse effects may appear. All these effects must ultimately be assessed as to their relative importance, and the significant ones incorporated into a model of the overall system from which the performance may be evaluated. In this chapter we stress the evaluation of the performance of communication systems using the simplest model of a transmission channel that limits system performance—one that adds noise to the signal as it propagates over the channel, as well as providing the usual bandwidth limitation. We do this for several reasons—first, because additive noise, as we shall note later in describing its effect on performance, represents the basic limitation on performance. The noise is always there whether we like it or not and so puts a limit on any performance measure. Second, it enables us to focus on simple modeling without any complications due to a variety of other performance-limiting factors. Third, it enables us to introduce useful measures of performance that can be extended to include other limiting factors, if so desired.

In summary, the performance of any communication system is ultimately limited by the nonzero response time of the system, or, equivalently, by the bandwidth, and by presence of noise in the system. This was made clear, in a somewhat qualitative way, in Chap. 1. These two fundamental limitations will again be stressed throughout this chapter. They will come up again in Chap. 6, in introducing and discussing Shannon's celebrated expression that describes fundamental limits on the rate of transmission in digital communication systems.

The two basic measures of communication system performance to be introduced in this chapter are, first, the probability of error in digital communication systems; second, the signal-to-noise ratio in analog systems, specifically AM and FM systems. We have already introduced signal-to-noise ratio as a measure of performance in discussing PCM and delta-modulation systems in Chap. 3. The discussion there focused on quantization and overload noise, both of which can be controlled by varying the quantization levels and sampling rate (albeit at the price of increases in the transmission bandwidth). The discussion here will concentrate on the unavoidable additive noise.

We start first here, as in the previous chapters, by considering baseband binary signals, then high-frequency binary modulated signals (PSK, OOK, FSK), and, finally, AM and FM systems. Topics on the representation of noise are introduced where needed.

5-1 ERROR RATES IN BINARY TRANSMISSION

The simplest example of the effect of noise on the performance of a communication system involves the transmission of baseband binary signals. To be more precise, assume that during transmission, noise has been added to the signals

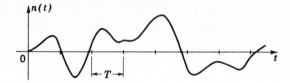

Figure 5-1 Typical oscillogram, noise voltage.

transmitted so that at the receiver the voltage measured consists of the sum of the two.

We shall explore the sources of this noise briefly at the end of this chapter, but suffice it to say at this point that noise is introduced in a variety of ways: where an antenna is used at the receiver, the antenna picks up noise radiation from the sky. Any dissipative elements generate noise as well. The dissipation in a transmission cable thus generates noise. Resistive elements in electrical circuits generate noise. All of these noise sources are temperature-dependent, and the noise generated is called generically *thermal noise*. In addition, active elements generate noise due to the random motion of current carriers (electrons and holes, for example). These two types of spontaneous fluctuation noise are always present in electrical circuits, although they may be reduced by lowering the temperature or changing the circuit design. This contrasts with "man-made" noise (electromagnetic pickup, mechanical vibrations converted to electrical disturbances, etc.), which can be eliminated or minimized with proper design, and erratic noise (effects of electrical storms, sudden and unexpected voltage change, etc.), which by its very description does not occur continuously. We focus here only on the spontaneous fluctuation noise.

A typical oscillogram or pen recording of the noise voltage $n(t)$ might appear as in Fig. 5-1. Although the noise is assumed random so that we cannot specify in advance particular voltage values as a function of time, we assume we know the noise statistics. In particular, we assume first that the noise is zero-mean gaussian; i.e., it has a gaussian probability-density function, with $E(n) = 0.$[1] Specifically, if we sample the noise at any arbitrary time t_1 the probability that the measured sample $n(t_1)$ will fall in the range n to $n + dn$ is given by $f(n)\, dn$, with

$$f(n) = \frac{e^{-n^2/2\sigma^2}}{\sqrt{2\pi\sigma^2}} \tag{5-1}$$

This is the most commonly used statistical model for additive noise in communications, and is in most applications a valid representation for actual noise present. We assume the noise variance σ^2 is known. (As will be demonstrated later in this chapter σ^2 may be either measured digitally or by a long time constant true power meter.) A brief discussion of the gaussian (or normal) density function appears in the Appendix. The function is shown sketched in Fig. 5-2. It has the typical bell-shaped curve, peaking at $n = 0$ (the most

---

[1] As in Chap. 3, the notation $E(n)$ stands for expectation, or mean value, of n.

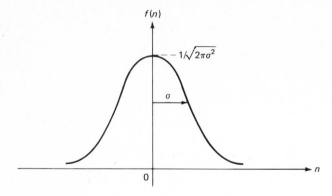

Figure 5-2 Gaussian probability-density function.

probable value of the random value). Its width is measured by the standard deviation σ. The probability that the random noise voltage n will be less than σ volts in magnitude at any time sampled is 0.68. The probability that values of n higher than several times σ will be attained falls off exponentially with n^2. The noise is equally likely to have positive and negative values. The driving force behind this gaussian model for the noise statistics (another model, that of *laplacian noise*, appears among the problems at the end of this chapter) is the *central limit theorem* of probability: the statistics of the sum of a number of random variables tends to gaussian under some rather broad conditions. (This theorem is discussed in the Appendix.)

Now assume we are receiving binary pulses in a digital communications system. The noise $n(t)$ is added to the incoming group of pulses in the receiver, and there is a possibility that the noise will cause an error in the decoding of the signal. Specifically, if the system is of the NRZ on-off type, in which pulses represent 1's (or marks) in the binary code, the absence of pulses representing 0's or spaces (see Chap. 3), the error will occur if noise in the absence of a signal happens to have an instantaneous amplitude comparable with that of a pulse when present or if noise in the presence of a signal happens to have a large enough negative amplitude to destroy the pulse. In the first case the noise alone will be mistaken for a pulse signal, and a 0 will be converted to a 1; in the second case the 1 actually transmitted will appear as a 0 at the decoder output.

How often will such errors occur on the average? Is it possible to decrease the rate of errors below a tolerable number by increasing the pulse amplitude? If so, how much increase is necessary? What is the effect on the error rate of reducing the noise? All these questions are readily solved if we know the noise statistics or have a reasonably good model for these. We shall demonstrate this using the gaussian noise model of Eq. (5-1).

Assume that the pulse amplitudes are all A volts, as in Chap. 3. The composite sequence of binary symbols plus noise is sampled once every binary interval and a decision is made as to whether a 1 or a 0 is present. One particularly simple way of making the decision is to decide on a 1 if the composite

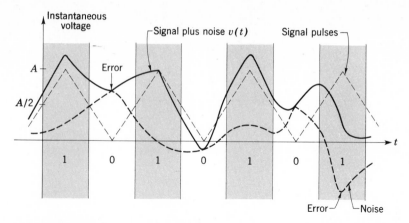

Figure 5-3 Effect of noise in binary pulse transmission.

voltage sample exceeds $A/2$ volts, and a 0 if the sample is less than $A/2$ volts. Errors will then occur if, with a pulse present, the composite voltage sample is less than $A/2$, or, with a pulse absent, if the noise alone exceeds $A/2$.

An example of a possible signal sequence, indicating the two possible types of error, is shown in Fig. 5-3. The signal pulses are shown triangular for simplicity's sake. The pulses and noise are shown dashed, while the resultant or composite voltage $v(t)$ is represented by the solid line. In this case samples are taken at the *center* of each binary interval, the system decoder then responding to the amplitude of these samples.

To determine the probability of error *quantitatively* we consider the two possible types of error separately. Assume first that a *zero* is sent, so that no pulse is present at the time of decoding. The probability of error in this case is just the probability that noise will exceed $A/2$ volts in amplitude and be mistaken for a pulse or a 1 in the binary code. Alternately, since $v(t) = n(t)$ if a 0 is present, the sampled value v is a random variable with the same statistics as the noise. The probability of error is then just the probability that v will appear somewhere between $A/2$ and ∞. Thus the density function for v, assuming a zero present, is just

$$f_0(v) = \frac{1}{\sqrt{2\pi\sigma^2}} e^{-v^2/2\sigma^2} \tag{5-2}$$

The subscript 0 denotes the presence of a 0 symbol and the probability of error P_{e0} in this case is just the area under the $f_0(v)$ curve from $A/2$ to ∞.

$$P_{e0} = \text{Prob}\left(v > \frac{A}{2}\right) = \int_{A/2}^{\infty} f_0(v)\, dv \tag{5-3}$$

The density function $f_0(v)$ is shown sketched in Fig. 5-4a, with the probability of error indicated by the shaded area.

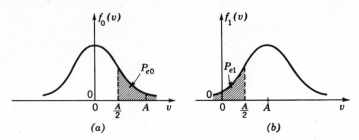

Figure 5-4 Probability densities in binary pulse transmission. (*a*) Noise only (0 transmitted). (*b*) Pulse plus noise (1 transmitted).

Assume now that a 1 is transmitted by the system encoder. This appears at the decoder as a pulse of amplitude A volts plus superimposed noise. A sample $v(t)$ of the composite voltage taken at time t is now the random variable $A + n(t)$. The fixed quantity A serves to shift the noise level from an average of zero volts to an average of A volts. The random variable v has the same statistics as n, fluctuating about A, however, rather than 0. Its density function is the same gaussian function, with the same variance, but with an average value of A. We have

$$f_1(v) = \frac{1}{\sqrt{2\pi\sigma^2}}\, e^{-(v-A)^2/2\sigma^2} \tag{5-4}$$

This is shown sketched in Fig. 5-4*b*. The probability of error now corresponds to the chance that the sample v of signal plus noise will drop below $A/2$ volts and be mistaken for noise only (or be judged, incorrectly, a 0). This is just the area under the $f_1(v)$ curve from $-\infty$ to $A/2$ and is given by

$$P_{e1} = \text{Prob}\left(v < \frac{A}{2}\right) = \int_{-\infty}^{A/2} f_1(v)\, dv \tag{5-5}$$

This probability of error is indicated by the shaded area in Fig. 5-4*b*.

How do we now find the probability of error of the *system*? Note that the two possible types of error considered belong to *mutually exclusive* events; the 0 precludes a 1 appearing, and vice versa. Probabilities can thus be added. However, in this case, it is apparent that P_{e0} and P_{e1} are both *conditional* probabilities, the first assuming a 0 present, the second a 1 present. To remove the conditioning we must multiply each by its appropriate a priori probability of occurrence. Thus, assuming the probability of transmitting a 0 is *known* to be P_0, while the probability of transmitting a 1 is *known* to be P_1, $(P_0 + P_1 = 1)$, we have as the total error of the system

$$P_e = P_0 P_{e0} + P_1 P_{e1} \tag{5-6}$$

It is apparent from Fig. 5-4 and from the symmetry of the gaussian curves that the two conditional probabilities P_{e0} and P_{e1} are equal in this example. (This may also be shown mathematically by a linear translation of coordinates, letting $x = v - A$ in Eq. (5-5), or by noting that the probability that noise alone will be less than $-A/2$ is the same as the probability that noise will exceed $A/2$.) If we also make the rather reasonable assumption that the two binary signals 0 and 1 are equally likely to occur, $P_0 = P_1 = \frac{1}{2}$, and we are left with the result that the total probability of error P_e is the same as P_{e0} or P_{e1}. It is left to the reader as an exercise to show that the probability of error is then simply given by

$$P_e = \frac{1}{2} \left(1 - \text{erf} \frac{A}{2\sqrt{2}\,\sigma} \right) \qquad \text{erf}\, x \equiv \frac{2}{\sqrt{\pi}} \int_0^x e^{-y^2}\, dy \qquad (5\text{-}7)$$

The error function erf x defined in (5-7) is tabulated in books on statistics or in various mathematical tables. With the 1's and 0's assumed equally likely in a long message, Eq. (5-7) gives the probability of an error in the decoding of any digit. Note that P_e depends solely on A/σ, the ratio of signal amplitude to the noise standard deviation. This latter quantity σ is commonly referred to as the *rms noise*. The ratio A/σ is then the peak *signal-to-rms-noise ratio*.

The probability of error is shown plotted versus A/σ, in decibels, in Fig. 5-5. Note that for $A/\sigma = 7.4$ (17.4 dB), P_e is 10^{-4}. This means that on the average 1 bit in 10^4 transmitted will be judged incorrectly. If 10^5 bits/s are being transmitted, this means a mistake every 0.1 s, on the average, which may not be satisfactory. However, if the signal is increased to $A/\sigma = 11.2$ (21 dB), a change of 3.6 dB, P_e decreases to 10^{-8}. For 10^5 bits/s this means a mistake every 1,000 s or 15 min on the average, which is much more likely to be

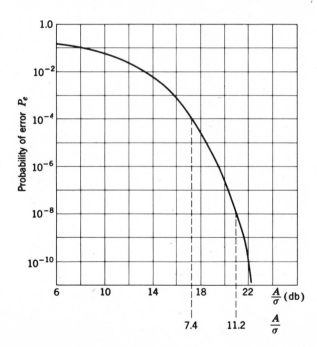

Figure 5-5 Probability of error for binary detection in gaussian note.

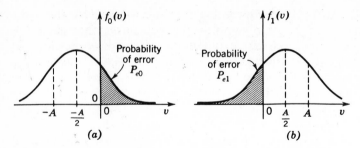

Figure 5-6 Probability densities in the transmission of NRZ-polar binary pulses. (*a*) Negative pulse transmitted. (*b*) Positive pulse.

tolerable. (In practice designers often use $P_e = 10^{-5}$ as a design goal for binary communication systems.)

Above $A/\sigma = 6$ or 16 dB (approximately), the probability of error decreases very rapidly with small changes in signal. In the example just cited, increasing the signal by a factor of 3.6 dB ($A/\sigma = 7.4$ to $A/\sigma = 11.2$) reduces the error rate by 10^4. The existence of a narrow range of signal-to-noise ratios above which the error rate is tolerable, below which errors occur quite frequently, is termed a *threshold effect*. The signal-to-noise ratio at which this effect takes place is called the threshold level. For the transmission of binary digits the threshold level is chosen somewhere between $A/\sigma = 6$ to $A/\sigma = 8$ (16 to 18 dB). Note that this does not imply complete noise suppression above the threshold level. It merely indicates that for pulse amplitudes greater than 10 times the rms noise, say, errors in the transmission of binary digits will occur at a tolerable rate.

That the foregoing error analysis for the transmission of on-off binary pulses holds true for NRZ-polar pulses is shown by Fig. 5-6. We recall from Chap. 3 that the positive and negative pulses need only be transmitted at $A/2$ or $-A/2$ volts, respectively. The decoder then determines which pulse is present from the polarity of the total instantaneous voltage (signal plus noise). Figure 5-6*a* shows the probability-density function for the negative pulse of $-A/2$ volts plus noise. Figure 5-6*b* shows the corresponding curve for the positive pulse plus noise. The error probability is in each case the same, and, comparing with Fig. 5-4, the same as for the case of on-off pulses. The error curve of Fig. 5-5 thus applies to either type of binary-digit transmission. The polar signal requires only half the signal amplitude of the on-off signal, however, or one-fourth of the peak power. (The *average* power is one-half that of the on-off signal, since that signal sequence is zero half the time, on the average.) It is thus apparent that a polar signal is preferable where possible.

Optimum Decision Levels

In this discussion of the probability of error in binary transmission we have so far relied on more-or-less intuitive judgments in developing the expression for the probability of error. Thus we have assumed the signals to be equally likely to be transmitted, we have arbitrarily chosen as our decision level the value $A/2$ for the on-off pulse sequence, or 0 in the case of the polar sequence, etc. One may readily ask how significant are these assumptions? Is it possible to *decrease* the probability of error by another choice of decision level? Is there a *minimum* P_e one can attain? We shall consider these questions and others related to them ("best" choice of binary waveshapes, the effect of multiple sampling, extension to M-ary transmission, etc.) in a quite general way in Chap. 6.

At this point, however, we can say something about the possibility of decreasing P_e for this particular problem of binary transmission by appropriate choice of decision level. In particular, assume a polar sequence of pulses transmitted with gaussian noise added, so that we have a composite sample $v(t)$ of signal plus noise, as previously. How shall we make the decision about whether a 1 or a 0 was transmitted? In this communications problem it is apparent that a reasonable design criterion is that of minimizing the probability of error P_e. A system with minimum P_e is then optimum from our point of view.

Since the decoder can only base its decision on the voltage amplitude of the sample $v(t)$ taken, it is apparent that the only possible way to adjust P_e or to optimize the system is to vary the amplitude level at which the decision is made. Call this decision level d. It is apparent from Fig. 5-7, in which this level is shown superimposed on the probability-density plots for polar transmission, that increasing d positively decreases the chance of mistaking a 0 for a 1 (P_{e0}) but at the cost of increasing P_{e1}. It is also apparent from the symmetry of the figure and the form of the gaussian functions shown that an optimum solution is $d = 0$, just our previous intuitive guess, *assuming 0's and 1's equally likely*. It is equally apparent that if for some reason 0's occur more often on the average ($P_0 > P_1$), one would tend to shift d positively. If on the other hand 1's tend to occur more often ($P_1 > P_0$), one would shift d negatively. The optimum choice of d thus depends on the a priori probabilities P_0 and P_1.

To make this discussion more quantitative we must return to our original formulation of the probability of error. From Eqs. (5-3), (5-5), and (5-6), we have, with d as an arbitrary decision level,

$$P_e = P_0 \underbrace{\int_d^\infty f_0(v)\, dv}_{P_{e0}} + P_1 \underbrace{\int_{-\infty}^d f_1(v)\, dv}_{P_{e1}} \tag{5-8}$$

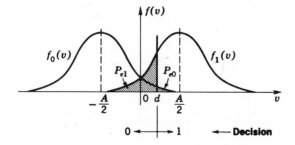

Figure 5-7 Choice of decision level in binary transmission.

(Recall that the previous equations were written for the on-off signal case. Had they been written for the polar case, 0 would have been used in place of $A/2$.)

An optimum choice of d corresponds to minimum P_e, according to our criterion. Since P_e is a function of d in Eq. (5-8), we simply differentiate with respect to d to find the optimum level. In particular, we then have

$$\frac{\partial P_e}{\partial d} = 0 = - P_0 f_0(d) + P_1 f_1(d)$$

or

$$\frac{f_1(d)}{f_0(d)} = \frac{P_0}{P_1} \tag{5-9}$$

invoking the usual rules of differentiation with respect to integrals.

It is thus apparent that the optimum d (in the sense of minimum error probability) depends on the form of the two conditional density functions [$f_0(v)$ and $f_1(v)$], as well as the a priori probabilities P_0 and P_1. If $P_0 = P_1 = \frac{1}{2}$, the optimum value of d is given by the point at which the two density functions intersect. For polar signals in gaussian noise this is just the point $d = 0$ (Fig. 5-7). For $P_0 \neq P_1$, the level shifts, as expected. Specifically, it is left for the reader to show that for the case of polar signals in additive gaussian noise (Fig. 5-7), the solution to Eq. (5-9) is given by

$$d_{\text{opt}} = \frac{\sigma^2}{A} \ln \frac{P_0}{P_1} \tag{5-10}$$

As expected, d increases positively if $P_0 > P_1$, and negatively, if $P_1 > P_0$. The actual shift depends on the signal amplitude, noise variance, and P_0/P_1.

In practice this choice of optimum d is not very critical, and in the case of polar signals in additive gaussian noise one would normally pick $d = 0$ as the decision level. For one generally does not know the a priori probabilities accurately, and even if one did, the signal-to-noise ratio required for effective binary communication makes d_{opt} from Eq. (5-10) very close to zero. Specifically, if $A/\sigma = 8$, and $P_0/P_1 = 3$ (highly unlikely, since this requires $P_0 = \frac{3}{4}$, $P_1 = \frac{1}{4}$), then $d_{\text{opt}} = \sigma/8$. The optimum shift away from a $d = 0$ decision level is thus a fraction of the rms noise, an insignificant change. (Some thought would indicate that this corresponds approximately to changing the signal amplitude by the same amount, with the level held fixed. With the signal-to-noise ratio initially at 18 dB, this represents a shift of approximately 0.13 dB away! From Fig. 5-5 the change in P_e is not noticeable.)

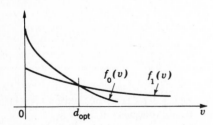

Figure 5-8 Example of determination of optimum decision level ($P_0 = P_1 = \frac{1}{2}$).

If the optimization is inconsequential in this case, why discuss it? There are several reasons.

1. We at least know that our first, intuitive, guess was a valid one. The optimization procedure tells us that *in this case* there is no sense looking for an improvement in system performance by varying the threshold level. This is very often exactly the reason for attempting to carry out system optimizations in more sophisticated situations.
2. If the received signal amplitude (or noise variance for that matter) is not accurately known, or varies due to disturbances along the transmission channel, the probability of error will change as well. The sensitivity of P_e to changes in the amplitude A is obviously related to the sensitivity due to changes in d. (Alternatively, one would approach the problem of sensitivity to changes in A in a way similar to that done here.)
3. If the statistics of the received signals plus noise are *not* gaussian, and do not have the nice symmetry of the density functions of Fig. 5-7, intuition fails in determining the desired decision level. Examples of such situations will be encountered later in this book in dealing with the noncoherent detection of binary signals.[2] One must thus resort to the solution of Eq. (5-9) or to the solution of equations like it. One simple pictorial example of two conditional density functions obtained under certain conditions in noncoherent detection is shown in Fig. 5-8. Here $P_0 = P_1 = \frac{1}{2}$ is assumed, so that the optimum decision level occurs at the intersection of the two conditional density functions $f_1(v)$ and $f_0(v)$.
4. This first discussion of optimum decision levels serves as a good introduction to the more general optimization problems to be considered in Chap. 6.

5-2 INFORMATION CAPACITY OF PCM SYSTEMS: RELATIONS AMONG SIGNAL POWER, NOISE, AND BANDWIDTH

The probability of error calculations for binary transmission in the previous section enable us to develop an interesting relation for the capacity, in bits/s, of a PCM system with output power S transmitting digital signals over a channel of bandwidth B and that introduces additive gaussian noise of variance σ^2. The capacity in this case represents the bit transmission rate possible for a prescribed probability of error. It turns out that this capacity expression is in precisely the same form as the Shannon capacity expression to be discussed in Chap. 6 that provides the maximum bit rate (or capacity) for error-free probability of a channel with the same constraints. Specifically, for an error probability of 10^{-5}, we shall see that the PCM system requires seven times the power

---

[2] See M. Schwartz, W. R. Bennett, and S. Stein, *Communication Systems and Techniques*, McGraw-Hill, New York, 1966, figs. 7-4-2 and 7-4-3, for the effect of threshold variation in on-off-keyed (OOK) signals with noncoherent detection.

(8.5 dB) of the theoretically optimum one with error-free transmission. The expression we shall obtain is also useful because it shows, in a manner similar to the Shannon expression, how signal power may be exchanged for bandwidth.

This exchange of signal power for bandwidth was also discussed earlier in Chap. 3 in describing the possibility of reducing bandwidth by combining several binary pulses and transmitting instead one M-level signal. This brief analysis also applies to digital transmission with successive symbols coded into one of several amplitude levels.

Specifically, assume the input analog signal has been quantized to M possible amplitude levels. Assume further that the signal of B hertz bandwidth has been sampled at the minimum Nyquist rate of $2B$ samples/s. With binary transmission, $M = 2^n$, $n = \log_2 M$ binary pulses would be transmitted for each sample, and the rate of information transmission in bits/s is then

$$C = 2B \log_2 M = 2nB \text{ bits/s} \tag{5-11}$$

The transmission channel must provide at least this transmission capability or capacity. (We have implicitly assumed here that all amplitude levels are equally likely. If this is not the case, the more likely levels would be represented by fewer bits than the less likely ones. See Sec. 1-4 for discussion.)

Rather than code into binary digits, however, say that the M levels are coded into n pulses of m amplitude levels each. Thus set $M = m^n$ in general. The information rate is still the same, so we now have

$$C = 2nB \log_2 m \text{ bits/s} \tag{5-12}$$

Let the bandwidth of the transmission channel be W hertz. With ideal Nyquist shaping we can transmit $2W$ symbols/s over this channel. Setting $2W = 2nB$, the desired symbol rate, we have

$$C = W \log_2 m^2 \text{ bits/s} \tag{5-13}$$

How is the number of amplitude levels m chosen? Obviously, one would like to make m as large as possible consistent with the power available for the system. Say that S watts average power is available. With m possible levels equally likely to be transmitted and spaced a units apart (equal spacing), the average signal power is found by averaging over all possible levels. Assuming that NRZ-polar transmission is used (which we have already seen provides a power improvement over the on-off case), the levels transmitted are actually $\pm a/2$, $+3a/2, \ldots, \pm(m-1)a/2$. The average power is then simply given by

$$S = \frac{2}{m} \left\{ \left(\frac{a}{2}\right)^2 + \left(\frac{3a}{2}\right)^2 + \cdots + \left[\frac{(m-1)a}{2}\right]^2 \right\}$$

$$= (a)^2 \frac{m^2 - 1}{12} \tag{5-14}$$

(Readers should check this relation for themselves.)

Solving for m^2 in terms of the average power S and substituting into Eq. (5-13) for the capacity, we have finally

$$C = W \log_2 \left(1 + \frac{12S}{a^2} \right) \tag{5-15}$$

with W the transmission bandwidth.

For a given capacity, then, one may reduce the transmission bandwidth W by increasing the average signal power S. This is just the procedure noted previously, where successive pulses are combined into one wider pulse with increased amplitude levels. But note how inefficient this bandwidth-power exchange is. One must increase the power *exponentially* to obtain a corresponding linear decrease in bandwidth. As an example, assume that $12S/a^2 \gg 1$. Then if the power is increased eightfold, the bandwidth may be reduced by a factor of 3. Similarly, by a linear *increase* in bandwidth, the required signal power may be *reduced* exponentially.

How does one now determine the level spacing a? This ultimately depends on the noise encountered while attempting to decode the received signals at the receiver. For the case of binary transmission this spacing between levels is precisely the signal amplitude A in the example of on-off transmission or the separation between signals in the example of polar transmission (see Figs. 5-4 and 5-6). The specific choice of A in that case, or a in general, depends on the noise variance σ^2 and the probability of error P_e deemed tolerable. This is made clear by Fig. 5-9, comparing m-level transmission with the two types of binary transmission. For the m-level case it is apparent that the spacing must be some constant K times the rms noise σ. In fact, if the error probability is small enough and the spacing a large enough, the chance that a given level being transmitted will be converted by noise to a level other than one of the adjacent ones is extremely small. To a good approximation the error analysis of the previous section is valid in this multilevel case also if the parameter A there is replaced by the spacing a.

In general, then, letting $a = K\sigma$, and writing N for σ^2 to emphasize *noise*

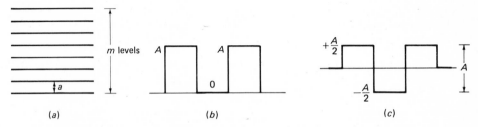

Figure 5-9 Digital pulse transmission. (*a*) m levels. (*b*) On-off pulses. (*c*) Polar pulses.

power (the connection between the noise variance and the noise power will be made clearer in Sec. 5-4), (5-15) can be rewritten as[3]

$$C = W \log_2 \left(1 + \frac{12}{K^2} \frac{S}{N} \right) \tag{5-16}$$

The quantity S/N, average signal power to average noise power ratio, or SNR as it is commonly abbreviated, will appear quite often in later discussions of the effect of noise on systems. At this point we simply note, as first pointed out in Chap. 1, that noise plays a role in determining system capacity through its effect on limiting the number of amplitude levels that may be used. This is exactly the reason it appears in the capacity expression of Eq. (5-16). Equation (5-16) thus emphasizes the point made in Chap. 1 that the capacity is limited by bandwidth and noise.

The capacity expression of Eq. (5-16) is interesting for another reason as well. Shannon[4] has shown that there is a maximum possible rate of transmission of binary digits over a channel limited in bandwidth to W hertz, with mean noise power N and mean signal power S. This rate of transmission, or capacity C, is maximum in the sense that if one tries to transmit information at a higher rate, the number of errors made in decoding the signals at the receiver begins to mount rapidly. In fact, the chance of an error in a code word of block length n can be shown to approach certainty as $n \to \infty$. On the other hand, if the information rate in bits per second is *less* than C, the chance of an error goes rapidly to zero. This maximum rate of transmission was found by Shannon to be given by

$$C = W \log_2 \left(1 + \frac{S}{N} \right) \tag{5-17}$$

Note that this is in precisely the same form as the PCM capacity expression of Eq. (5-16)! Shannon's maximum-capacity expression provides an upper bound on the rate at which one can communicate over a channel of bandwidth W, and signal-to-noise ratio SNR.[5] His original derivation indicated that it was theoret-

---

[3] We are assuming throughout this discussion that the power level remains fixed at all points from transmitter to receiver. In actual practice, of course, it will vary from point to point. For example, it will normally be much less at the receiver antenna, or receiver input, than at the transmitter output. But amplifiers may of course be used and normally *are* used to bring the signal up to the desired power level. The *relative* signal amplitude compared to the level spacing a remains unchanged throughout transmission and reception, however, so that the ratio S/a^2 of Eq. (5-15) remains fixed. One may, therefore, just as well consider *relative* signal power, and assume it unchanging throughout the transmission path. Alternatively, the power level S refers to the signal power measured at the same point as the noise N. In a real system, as will become apparent from examples discussed later, S is directly proportional to the power made available at the transmitter output. Although S and N are, strictly speaking, mean-squared voltages, we use the common term power to represent both. They would represent the power dissipated in a 1-Ω resistor.

[4] C. E. Shannon, "Communication in the Presence of Noise," *Proc. IRE*, vol. 37, pp. 10–21, January 1949.

[5] A detailed discussion of this equation appears in Chap. 6.

ically possible to transmit at almost this rate with the error rate made to approach zero as closely as desired, but he did not specify any particular system for so communicating. He showed only that complex encoding and indefinitely large time delays in transmission would be needed. Comparing Eqs. (5-16) and (5-17) we are now in a position to compare the PCM system with this hypothetical one transmitting at a maximum possible rate over a given channel.[6]

In particular note that the PCM system requires $K^2/12$ times the signal power of the optimum Shannon system for the same capacity, bandwidth, and noise power. For an error probability of 10^{-5}, we found in the previous section that $A/\sigma = 9.2$ for binary transmission. Using the argument described above, this should to a good approximation be the ratio of level spacing a to the rms noise σ for an m-level system. This is just the value of K required in (5-16). For this value of K, $K^2/12 = 7$, and the PCM system requires seven times as much power (8.5 dB) as the theoretically optimum one for the same channel capacity. (Note, however, that the optimum system can theoretically be made to transmit error free at the cost of large time delays in transmission, while the PCM system, in this example, has the nonzero error rate of 10^{-5}.) Both Eqs. (5-16) and (5-17) have the form of the capacity expression (1-3) introduced in Chap. 1. They make explicit the qualitative remarks made there that both noise and bandwidth play fundamental roles in limiting the performance of communication systems. A specific transmission channel will have a bandwidth W available for communication and will introduce some known additive noise power N. The Shannon expression (5-17) then describes the maximum rate of error-free digital transmission over that channel as a function of the signal power S. Equation (5-16) says that normal digital transmission over the same channel is limited in a similar manner: the transmission bandwidth W limits the number of symbols that may be transmitted over the channel to $2W$ per second at most. The number of bits carried by a symbol, with amplitude-level variation, is then limited by the average signal power S, and the noise N introduced on the channel.

5-3 ERROR ANALYSIS OF PCM REPEATERS

In this section we extend the binary error-probability analysis to the very practical problem of PCM repeaters. In discussing PCM systems in Chap. 3 we noted several times in passing that one significant reason for using all-digital signal transmission in communications is that the digital signals lend themselves so nicely to periodic conditioning and reshaping. Thus all communication channels serve to attenuate and distort signals passing through them. To ensure satisfactory reception at the final destination repeaters are commonly provided at appropriate spacings along the signal transmission route. These provide the

---

[6] An M-ary system to be described in Chap. 6 has a performance approaching the Shannon optimum as $M \to \infty$.

necessary amplification and correction of distortion (by appropriately designed filters) to enable signal recognition to be satisfactorily carried out at the receiver.

Such repeaters are commonly used in cable and wire transmission, underwater cable transmission, microwave communication links, optical communications, etc.[7] The number of repeaters and the spacing between them depends of course on the type of transmission medium used, its attenuation and phase distortion per unit length, the total transmission path length, carrier frequency used, etc. Typical attenuations range from a few tenths of a decibel per mile for wire lines and loaded cables in the voice frequency range to several decibels per mile at higher frequencies.[8] Phase shifts encountered also depend on the particular medium and frequency of transmission and vary roughly from a few tenths of a radian per mile to several radians per mile.

Although repeaters are used with all types of signals—analog and digital—we shall concentrate here on binary transmission only. We shall also ignore the modulation and demodulation processes necessary to get the signals on line (see Chap. 4), and will assume baseband transmission throughout.

Consider then a sequence of binary pulses transmitted along a line. The effect of the line may be characterized as equivalent to passing the signals through a linear network with frequency transfer function $H(\omega)$. The signals thus undergo amplitude and phase distortion, and filter networks, commonly called *equalizers,* must be introduced to compensate for the spectral distortion. In addition, noise and interference are introduced during transmission.

It is the introduction of the noise that poses a significant problem. For although the signal distortion may, at least in theory, be equalized to as fine a degree as required, the noise ultimately provides a limit on signal detectability. For over and above the relative amplitude-frequency distortion introduced by the line there is an overall attenuation of the signal level. If this level is allowed to drop too low, errors due to noise begin to limit the system transmission capability.

One may of course provide the proper amplification to bring the signal amplitude back up to the desired level, but this results in noise amplification as well. The only solution is to keep the signal level from dropping too low with respect to the noise. This therefore implies repeatedly conditioning the signal, at intervals along the line small enough to keep the attenuated signal well above the noise level. This provides the justification for the repeaters. In addition, the frequency distortion introduced by a short section of line is obviously less than that produced by a long line, so that amplitude and phase equalization is more readily carried out at each repeater, rather than at the ultimate destination.

Although repeaters are used to equalize and retime digital pulses, as well as

---

[7] See, e.g., the references to repeaters in D. H. Hamsher (ed.), *Communication System Engineering Handbook,* McGraw-Hill, New York, 1967. See also *Transmission Systems for Communications,* Bell Telephone Laboratories, 4th ed., 1970.

[8] Hamsher, *op. cit.*

to keep noise from accumulating during transmission, the discussion here will focus on the effects of noise only. This is admittedly an artificial situation, but the emphasis here is on the simplest model possible.

A little thought will indicate that two different classes of repeaters may be used. One employs straight amplification plus associated filtering to recoup the signal attenuation and compensate for the distortion. This is obviously the type of repeater that would be used for analog signal transmission. The second type, particularly appropriate for digital signal transmission, in essence makes a decision as to the binary symbol being transmitted (or decides on the appropriate level in the case of multilevel digital transmission), and then sends out a new, clean, noise-free binary symbol for further transmission down the line.

Which scheme is to be preferred for digital transmission? In the first scheme, involving amplifiers regularly spaced along the complete circuit, noise as well as signal is transmitted along the line, and in fact additional noise is introduced on each line section. A multilink line incorporating amplifiers is sketched in Fig. 5-10a. In the case of repeater signal conditioners, or *regenerative* repeaters as they are called (Fig. 5-10b), noise is effectively eliminated at each repeater, and is not carried along from repeater to repeater. But in the process of making a decision at each repeater, errors may be made, and these may be propagated along the entire line. We shall now show with some simple calculations that for relatively low noise (small probability of error) systems, this latter scheme is nonetheless to be preferred in detecting binary signals in noise.

Assume the binary signals are NRZ-polar, with transmitted amplitude $\pm a$ or transmitted power a^2. The individual lines, between repeaters, are each

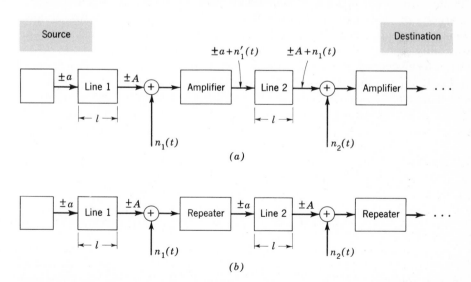

Figure 5-10 Repeaters for PCM transmission: $A^2 = a^2\epsilon^{-al}$. (a) Repeater amplifiers. (b) Repeater conditioners.

assumed to introduce a power attenuation α nepers/unit length. For a link length l, then, the power at the link output is $a^2 e^{-\alpha l}$. The corresponding (attenuated) signal amplitude is A.

Assume further that gaussian noise, with variance σ^2, is introduced on each line link, the different noises being independent of one another. At the input to the first repeater, then, the composite signal plus noise[9] is

$$\pm A + n_1(t)$$

In the case of the amplifier scheme, both signal and noise power are amplified by $e^{\alpha l}$ to recoup the desired signal level. The signal and noise now propagate down the second link, the power again being attenuated by $e^{-\alpha l}$. An additional noise term $n_2(t)$ is now added, so that the input to the second amplifier is

$$\pm A + n_1(t) + n_2(t)$$

Ultimately, after m line links, the detector at the final destination receives

$$v(t) = \pm A + n_1(t) + \cdots + n_m(t) \tag{5-18}$$

From the results of probability theory we know that the sum of m independent gaussian variables is again gaussian, with a variance equal to the sum of the variances, and the means summed as well. In this case the individual noise terms are assumed to have zero mean. Since all the variances are assumed equal, the composite signal plus noise $v(t)$ at the destination may be written simply as

$$v(t) = \pm A + n(t) \tag{5-19}$$

with the equivalent noise $n(t)$ gaussian of variance $m\sigma^2$. As in Sec. 5-1, the probability of an error is the probability that $v(t)$, when sampled, will be mistaken for the wrong binary symbol. Since the statistics are again gaussian, the probability of error is exactly the same as that calculated in Sec. 5-1, Eq. (5-7), with σ^2 replaced by $m\sigma^2$. (As in that section A remains the *received* signal amplitude.) The effect of m amplifiers in tandem is thus to increase the rms noise level by $\sqrt{m}$. Specifically, the probability of error in the m-link amplifier scheme is now given by

$$P_{amp}(e) = \frac{1}{\sqrt{2\pi}} \int_{A/\sqrt{m}\sigma}^{\infty} e^{-y^2/2} \, dy$$

$$= \frac{1}{2} \left[1 - \text{erf} \frac{A}{\sqrt{2m}\,\sigma} \right] \equiv \frac{1}{2} \text{erfc} \frac{A}{\sqrt{2m}\,\sigma} \tag{5-20}$$

again using the previous definition of the error function, and defining erfc x, the *complementary* error function, as $1 - \text{erf } x$.

Now consider a sequence of $(m - 1)$ repeaters which condition the signal and emit a clean noise-free pulse, as noted earlier. There are m sequential decisions that will be made on the one signal, including the final decision

---

[9] The noise is assumed primarily introduced at the input to the repeater.

at the ultimate destination. At the first repeater input the composite signal consists of $\pm A + n_1(t)$. The output of this repeater is simply $\pm a$ (the amplified and conditioned signal with the noise missing), or, if a mistake has been made in deciding on the correct signal, $\mp a$. Again the signal is attenuated to an amplitude level A. At the input to the second repeater the composite signal consists of $\pm A + n_2(t)$ if the first repeater decides correctly. This process is repeated on down the line.

It is apparent that errors initially made somewhere along the line may be corrected if an additional repeater further down the line makes a compensating error, or, in general, if there are an *even* number of incorrect decisions. A final error is made if an *odd* number of incorrect decisions is made along the line. If we assume as above that all the noises are independent and gaussian, with the same variance σ^2, it is apparent that the probability of an error at any one of the repeaters is

$$p = \frac{1}{\sqrt{2\pi}} \int_{A/\sigma}^{\infty} e^{-y^2/2} \, dy = \frac{1}{2} \operatorname{erfc} \frac{A}{\sqrt{2}\,\sigma} \tag{5-21}$$

The probability of making k errors at the m decision points is then

$$P(k \text{ errors}) = \binom{m}{k} p^k (1 - p)^{m-k} \tag{5-22}$$

with $\binom{m}{k} = m!/(m - k)!k!$ the number of possible ways in which a sequence of k incorrect and $(m - k)$ correct decisions can be made. This is just the binomial distribution. The probability of error, $P_{\text{rep}}(e)$, on the entire m-link system, is then obtained by summing over all odd values of k, and is given by

$$P_{\text{rep}}(e) = mp\,(1 - p)^{m-1} + \frac{m(m - 1)(m - 2)}{3!} p^3(1 - p)^{m-3} + \cdots$$

$$= \sum_{\substack{k=1 \\ (k \text{ odd})}}^{m} \binom{m}{k} p^k (1 - p)^{m-k} \tag{5-23}$$

In particular, if $mp \ll 1$,

$$P_{\text{rep}}(e) \doteq mp \qquad (mp \ll 1) \tag{5-24}$$

In this case the probability of a single incorrect decision is small enough that the probability of multiple errors is negligible. But the use of m decision points makes an individual error m times as likely.

The relative effectiveness of regenerative repeaters over amplifiers may now be found by comparing Eqs. (5-20) and (5-24.) Three sample calculations follow[10]:

---

[10] Figure 5-5 may be used in determining the necessary probabilities. Since that figure refers to on-off binary transmission, while we are assuming polar transmission here, 6 dB must be subtracted from the abscissa scale.

1. $$m = 100 \qquad \frac{A}{\sigma} = 6.36 \ (16 \ \text{dB}) \qquad p = 10^{-10}$$

Then $\qquad P_{\text{rep}}(e) \doteq 10^{-8} \qquad P_{\text{amp}}(e) = \dfrac{1}{2} \ \text{erfc} \ \dfrac{6.36}{10\sqrt{2}} > 0.1$

Note that here, with many repeaters and relatively high signal-to-noise ratio, the repeaters with conditioning circuitry far outperform the multiple amplifier scheme.

2. $$m = 5 \qquad \frac{A}{\sigma} = \sqrt{10} \ (10 \ \text{dB}) \qquad p = 8 \times 10^{-4}$$

$$P_{\text{rep}}(e) \doteq 4 \times 10^{-3} \qquad P_{\text{amp}}(e) = \frac{1}{2} \ \text{erfc} \ (1) = 0.08$$

Again the effect of the repeaters is to increase the overall probability of error, but with the conditioning scheme far outperforming the amplifier scheme. This is again due to the fact that the probability of more than one error is negligible in the repeater scheme, while the rms noise level is effectively increased by $\sqrt{5}$ in the amplifier scheme.

3. $$m = 3 \qquad \frac{A}{\sigma} = 1 \ (0 \ \text{dB}) \qquad p = 0.159$$

$$P_{\text{rep}}(e) \doteq 0.34 \qquad P_{\text{amp}}(e) \doteq 0.28$$

Here with low signal-to-noise ratio and small numbers of repeaters, the two techniques are roughly comparable.

An interesting problem is to find the optimum number of repeaters for a fixed total line length and fixed *transmitter* power. As the number of repeaters is decreased, the overall probability of an error decreases as well, since there are fewer chances of making errors. The length of line between repeaters increases, however, so that the *received* signal decreases, leading to a higher probability of error at any one repeater. Specifically, let the *total* length of line be L, and let a^2 be the transmitter power. Then for small probability of error we have, from Eq. (5-24),

$$P_{\text{rep}}(e) = \frac{m}{2} \ \text{erfc} \left(\frac{a}{\sqrt{2} \ \sigma} \ e^{-\alpha L/2m} \right) \qquad (5\text{-}25)$$

A trial-and-error minimization of this expression by varying m is left as an exercise for the reader.

The T1 carrier system operated by the Bell System and described in Chap. 3 has regenerative repeaters spaced 6,000 ft apart. The transmission medium in this case consists of 19- or 22-gauge cable pairs.[11]

---

[11] H. Cravis and T. V. Crater, "Engineering of T1 Carrier System Repeatered Lines," *Bell System Tech. J.*, vol. 42, pp. 431–486, March 1963.

5-4 NOISE POWER AND SPECTRAL REPRESENTATION OF NOISE

In the sections thus far in this chapter we have assumed the noise added to the signal to have known statistics (most commonly, gaussian), with a specified variance σ^2. This enabled us to calculate the probability of error in baseband binary signal transmission due to additive noise, to assess the performance of a PCM system, and to study the very practical problem of the effect of repeaters on PCM performance. To proceed further we must indicate ways of measuring the variance σ^2, as well as other statistical parameters of the noise. More important, since the noise function $n(t)$ is a time-varying wave, just as any signal it must be affected by the system through which it passes. How does one quantitatively determine the effect of systems on noise? What happens when noise, picked up, for example, by an antenna at a high frequency, is converted down to a lower frequency or to baseband, together with the signal to which it is added? How do linear filters and nonlinear devices act on the noise? All of these questions must be answered before proceeding further with our performance analysis of communication systems.

We thus digress somewhat, in this section and in the one following, to study ways of measuring and representing the noise. In particular, we shall find it useful, as with the signals in previous chapters, to develop a spectral representation for the noise. This will provide a means of studying the effect of systems—both linear and nonlinear—on noise. Interestingly, if we return to the premise of Chap. 1 that real *signals* must be time-varying and unpredictable (otherwise why transmit them?), it is apparent that the analysis we shall outline in this section and the one following is not only of importance in studying noise, but is exactly the analysis one must use in studying real time-varying signals in systems.

The frequency or spectral analysis of random signals and noise will be found to differ somewhat from the spectral analysis for deterministic signals studied in Chap. 2 and used in the chapters following. It will be found to be based on the spectral distribution of the *power* in the random wave. Nonetheless, many of the results obtained in our earlier study of deterministic signals will still be found to be valid. These include such things as bandwidths, the filtering effects of linear systems, the inverse time-bandwidth relationship, etc. This indicates the usefulness of the approach we have adopted of studying deterministic signals first.

Thus, although we shall stress the spectral representation of *noise,* as well as the effect of systems on noise in these sections, the ideas and concepts introduced are applicable as well to random signals. References given later will guide the reader interested in pursuing these topics further.

As noted in Sec. 5-1, a typical oscillogram of noise would appear as in Fig. 5-11. Random-signal waveshapes would also have a similar irregular, unpredictable appearance. We call this random time-varying function a *random process* $n(t)$. [Although we shall for simplicity's sake use the notation $n(t)$ and generally refer to the random wave as noise, random signals are included as

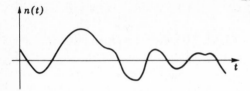

Figure 5-11 Random process.

well, as noted above.] A sample of $n(t)$ taken at an arbitrary time t is a random variable with some probability-density function $f_n(n)$. In the specific case discussed in Sec. 5-1, the noise was assumed to be gaussian, with $f_n(n)$ described by the gaussian function of (5-1). We shall most often assume gaussian noise in this book, but in general a random process will be described by any time function whose samples are random variables with some specified $f_n(n)$. (Consider as a simple, nongaussian, example, a two-level noise process: the random wave is equally likely to be either $+a$ or $-a$ when sampled. A special case is a binary pulse train, with 1's and 0's equally likely to occur. This is of course precisely the model of a binary signal randomized by the unpredictable occurrence of 1's and 0's.)

To determine $f_n(n)$ experimentally, if one did not know the underlying statistics, one could perform a histogram analysis, sampling $n(t)$ at intervals "far enough apart" to ensure the statistical independence of the samples, setting levels at n and $n + \Delta n$, and counting the number of times the samples fell in this range of n.[12]

But what is meant by "independent samples"? What constitutes "far enough apart"? It is questions like this that we shall discuss in this section and the next, in dealing with noise spectral analysis. To motivate the discussion of the spectral analysis of the random process $n(t)$, we first focus attention on the measurement of some simple statistical parameters such as $E(n)$, the variance σ^2, etc. In considering how one measures these quantities, we shall be led into the more general discussion of spectral analysis.

Although the average value and variance of n can be obtained from the same histogram analysis of $n(t)$ used to find $f_n(n)$, simply by averaging measured samples appropriately, it is apparent intuitively that one should be able to make the same measurements much more simply using time-averaging meters. For example, if we were to feed the wave $n(t)$ into a dc meter, we would intuitively expect to get a measure of the expected value $E(n)$. In this case we are implicitly comparing an *expected* or *statistical* average with a *time* average, as carried out by a dc meter. Specifically, if the meter has an effective time constant T, its reading should give a number[13]

---

[12] See M. Schwartz and L. Shaw, *Signal Processing,* McGraw-Hill, New York, 1975, chap. 3, for a brief treatment of this subject.

[13] A digital meter would replace the integral by the sum of samples, as noted above. We use the integral representation throughout for simplicity. See Schwartz and Shaw, *ibid.,* for the digital representation.

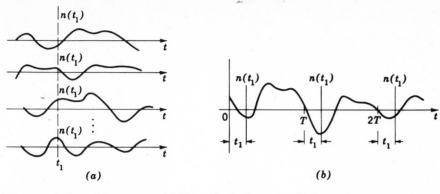

Figure 5-12 Ensemble averages. (*a*) Many identical sources. (*b*) By cutting one record.

$$\bar{n} = \frac{1}{T}\int_0^T n(t)\,dt \tag{5-26}$$

It is apparent that as $n(t)$ varies randomly, so will $\bar{n}$. Depending on *when* we perform the indicated average, we will get different numbers $\bar{n}$. So $\bar{n}$ is a random variable, with its own expected value, variance, etc. But we would still expect to find $\bar{n}$ some measure of $E(n)$, the statistical average of $n(t)$. To indicate the connection, we take the expected value of $\bar{n}$ itself; i.e., we visualize many meter readings over different sections of $n(t)$, each T seconds long, the expected value of $\bar{n}$ then being the statistical average of these. Then

$$E(\bar{n}) = E\left[\frac{1}{T}\int_0^T n(t)\,dt\right] \tag{5-27}$$

It may be shown that the operations of expectation and integration are interchangeable.[14] We then write

$$E(\bar{n}) = \frac{1}{T}\int_0^T E[n(t)]\,dt \tag{5-28}$$

We now assume that the expected value $E[n(t)]$ is independent of time t. This is reasonable, for if the expected value *were* varying with time, one would not expect the dc meter reading to be a fixed number anyway. Alternatively, if we were to visualize many such strips of $n(t)$ each T seconds long, placed one above the other [either obtained from the same record by cutting the one strip at T-second intervals, or by visualizing many identical sources providing independent outputs $n(t)$] (see Fig. 5-12), we could perform an ensemble average of the random variable $n(t_1)$ to actually find a close approximation to $E[n(t_1)]$. One would then assume that $E[n(t_2)]$, or $E[n(t)]$ at any *other*

---

[14] A. Papoulis, *Probability, Random Variables, and Stochastic Processes*, McGraw-Hill, New York, 1965, chap. 9.

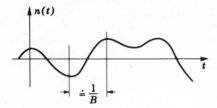

Figure 5-13 Significance of bandwidth B.

value of t, was the same. (This would be verified by averaging at each time interval, if so desired.)

With this assumption that $E[n(t)]$ is a *constant,* independent of time, we find, from Eq. (5-28), that

$$E(\bar{n}) = E(n) \tag{5-29}$$

So our intuition is justified here, indicating that at least in an *average* sense the time average $\bar{n}$ does provide a measure of $E(n)$. But we would also expect that the time interval T should play a role here. By making T longer, or by averaging over longer sections of $n(t)$, we would expect to find $\bar{n}$ approaching $E(n)$ more closely. In fact, this is easily demonstrated by calculating the variance var($\bar{n}$) of the random variable $\bar{n}$. Thus, by definition of the variance,

$$\text{var}\,(\bar{n}) = E[\bar{n} - E(\bar{n})]^2 = E\left\{ \frac{1}{T} \int_0^T [n(t) - E(n)]^2 \, dt \right\} \tag{5-30}$$

Although we shall not perform the calculation here, it is readily shown[15] that

$$\frac{\text{var}\,(\bar{n})}{E^2(n)} \to \frac{C}{BT} \qquad BT \gg 1 \tag{5-31}$$

where C is a fixed constant the order of 1, and B is the bandwidth of the process $n(t)$. [We shall define B precisely in the material to follow, but roughly speaking it is a measure of the rapidity of variation of $n(t)$, just as with the deterministic signals encountered previously. As indicated in Fig. 5-13 the time spread between successive dips and peaks in the wave is approximately the reciprocal of the bandwidth.] As an example, if $B = 100$ kHz and $T = 1$ ms (recall that this is the meter averaging time), $1/BT = 0.01$. The spread about the expected value is thus 0.1 of the expected value.

As the dc meter averages over longer and longer time intervals, its reading $\bar{n}$ approaches more and more closely the parameter $E(n)$. (In practical situations it is usually sufficient to have $T \gg 1/B$.) For the variance of the reading $\bar{n}$ goes to zero as $1/T$, indicating that the variations about $E(n)$ decrease in the same manner. Thus, in the previous example if the meter time constant is increased to 100 ms, $1/BT = 10^{-4}$. The spread about the expected value is the square root of this, or 10^{-2}. Increasing the integration time by a factor of

---

[15] *Ibid.,* pp. 324–328. This assumes of course that $E(n) \neq 0$.

100 has narrowed the deviation about $E(n)$ by a factor of 10, on the average. In the limit, as $T \to \infty$, we must have

$$\lim_{T\to\infty} \bar{n} = \lim_{T\to\infty} \frac{1}{T} \int_0^T n(t) \, dt = E(n) \qquad (5\text{-}32)$$

A random waveshape or process $n(t)$ for which Eq. (5-32) is true is said to be an *ergodic process;* i.e., time and ensemble (statistical) averages may be equated. Although there are processes for which this is not true,[16] we shall henceforth assume that $\bar{n}$ and $E(n)$ may be equated.

Since we have shown that one may use a dc meter to measure $E(n)$, one may reasonably ask if it is similarly possible to measure σ^2, the variance of the noise. The answer is of course "yes," and in fact one uses a *power meter* for this purpose. Specifically, if one defines the average power P_{av} over an interval T seconds long just as in previous chapters:

$$P_{av} \equiv \frac{1}{T} \int_0^T n^2(t) \, dt \qquad (5\text{-}33)$$

one shows, again by interchanging the order of ensemble averaging and integration, that

$$E(P_{av}) = E(n^2) = \sigma^2 + E^2(n) \qquad (5\text{-}34)$$

$$\sigma^2 = E(P_{av}) - E^2(n) \qquad (5\text{-}35)$$

if $E(n)^2$ is invariant with time. [Recall that $\sigma^2 = m_2 - m_1{}^2 = E(n^2) - E^2(n)$ for a random variable.]

Again P_{av}, as read by the power meter of time constant T, is a random variable, but *on the average* the readings will provide a measure of the second moment $E(n^2)$. Since $E^2(n)$ is very nearly the square of the dc value, it is apparent that the variance σ^2 must provide a measure of the fluctuating or non-dc power. To emphasize the fact that the variance on an ensemble average basis is the same as the time-averaged fluctuation power, we shall henceforth use the symbol N for the latter. One often calls this the ac power, as measured by true rms meters. N would then be the *square* of the rms meter reading.

One may again show[17] that

$$\frac{\text{var} \, (P_{av})}{E^2(P_{av})} = \frac{C'}{BT} \qquad BT \gg 1, \, C' \text{ a constant} \qquad (5\text{-}36)$$

---

[16] A trivial example of a nonergodic process is that consisting of an ensemble of constant-voltage sources. Each source maintains its output absolutely constant with time. Each source provides, however, a different output. Choosing one source at random and averaging its output with time, we of course measure the particular source voltage. Measuring all sources simultaneously, however, and averaging these (the *ensemble* average), we get a different result than the time average, the ensemble average depending on the distribution of the source output.

[17] Papoulis, *op. cit.* Here one must generally assume, however, that the statistics of $n(t)$ are *gaussian;* i.e., at any instant of time $f(n)$ is gaussian.

Thus, as $T \to \infty$ (in practice $T \gg 1/B$ again suffices), the reading P_{av} approaches $E(n^2)$ with probability of 1, and we have

$$\lim_{T \to \infty} (P_{av} - \bar{n}^2) = \lim_{T \to \infty} \frac{1}{T} \int_0^T [n(t) - \bar{n}]^2 \, dt = N \tag{5-37}$$

A process $n(t)$ for which time and ensemble averaging are equal, in the sense of Eq. (5-37), is again spoken of as an ergodic process. We assume henceforth that one may interchange these two averages, although we shall encounter at least one example later in which this is not valid.

We have just shown how one relates time and ensemble averages for a random wave or process $n(t)$. In particular, one may use a dc meter to measure $E(n)$, and, if the dc term is blocked or absent [$E(n) = 0$], one may use a true rms meter to measure N, the noise power or variance. These assume, however, that the meter time constant $T \gg 1/B$, with B the "bandwidth" of the process. How does one determine B? How is B related to the actual variations in time of $n(t)$? Is it possible to calculate N, and, if so, how does this depend on the physical systems through which $n(t)$ propagates? It is also apparent, extrapolating from our discussions of deterministic signals in previous chapters, that the bandwidth B must somehow relate to the physical system in which $n(t)$ is generated or through which it propagates. For example, one would not expect to find noise with a bandwidth of 1 MHz at the output of a system whose bandwidth is 100 Hz. Any noise terms varying this rapidly just would not appear at the system output.

Basically, we require some measure of how the noise process may vary in a given time interval. Specifically, if we consider the noise wave $n(t)$ of Fig. 5-14 we note that as $t_2 \to t_1$, $n(t_2)$ as a random variable becomes more "closely related to" (or "predictable by") $n(t_1)$. As $(t_2 - t_1)$ increases, we expect less dependence of one upon the other. We make this concept more precise by defining the autocorrelation function $R_n(t_1, t_2)$:

$$R_n(t_1, t_2) \equiv E[n(t_1)n(t_2)] \tag{5-38}$$

Although many different definitions of "dependence" of one random variable on one another are possible, the autocorrelation function is probably the simplest and has many desirable properties that we shall explore later. Note that it is the extension to the *same* random variable (hence the prefix *auto*) of the definition for the *covariance* of two random variables appearing in probability theory. It is apparent that if $t_2 \to t_1$, $R_n \to E(n^2)$, or just the statistical second

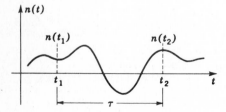

Figure 5-14 Autocorrelation definition.

moment. If at some spacing $(t_2 - t_1)$, $n(t_2)$ and $n(t_1)$ tend to become statistically independent (one would expect this to occur at intervals greater than $1/B$), R_n $\rightarrow E^2(n)$, or 0, if $E(n) = 0$. $R_n(t_1, t_2)$ thus provides one possible measure of the dependence of $n(t_2)$ and $n(t_1)$.

To simplify the discussion we shall assume that $R_n(t_1, t_2)$ depends only on the interval $(t_2 - t_1) \equiv \tau$, and not on the time origin t_1. [This is similar to our assumption, made previously, that $E(n)$ is independent of time.] Then we can write

$$R_n(\tau) \equiv E[n(t)n(t + \tau)] \qquad (5\text{-}38a)$$

A process for which this is true, and for which $E(n)$ is independent of time, is called a *stationary process*.[18]

Now how would we actually measure $R_n(\tau)$? As we did previously we set up a time integral with the property that its expected value equals $R_n(\tau)$. Consider the integral

$$\bar{R}_n(\tau) \equiv \frac{1}{T} \int_0^T n(t)n(t + \tau)\, dt \qquad (5\text{-}39)$$

Note that this provides some measure of the dependence with time τ of $n(t)$ and $n(t + \tau)$. For as $\tau \rightarrow 0$, $\bar{R}_n(0) = P_{av}$; as τ increases and $n(t)$ and $n(t + \tau)$ vary relatively independently of one another, one would expect that the product of the two would be negative as often as positive, approaching zero if $E(n) = 0$.

Again $\bar{R}_n(\tau)$ is a random variable, depending on the interval T seconds long over which evaluated. If we now ensemble average over all possible values of this variable, we get, from Eqs. (5-39) and (5-38a),

$$E[\bar{R}_n(\tau)] = R_n(\tau) \qquad (5\text{-}40)$$

(Again ensemble averaging and integration are interchanged.) So in an average sense the time average of Eq. (5-39) and the ensemble average of Eq. (5-38a) are the same.

As previously, one may calculate the variance of $\bar{R}_n(\tau)$ and show it goes to zero as $1/BT$, for large T.[19] We then have, as in the previous cases,

$$\lim_{T \rightarrow \infty} \bar{R}_n(\tau) = R_n(\tau) \qquad (5\text{-}41)$$

[Note from Eqs. (5-38a) and (5-39) that included as a special case here is the result $\lim_{T \rightarrow \infty} P_{av} = N + E^2(n) = E(n^2)$, previously shown as Eq. (5-37).]

We are now in a position to actually relate $R_n(\tau)$ to a spectral analysis of $n(t)$, and thus to a defined bandwidth. We could proceed by assuming

---

[18] Strictly speaking, such a process is usually called a *wide-sense* stationary process. The term stationary process is then reserved for a more general case in which distribution functions are invariant with time. See Papoulis, *op. cit.*, chap. 9.

[19] One must assume $n(t)$ is gaussian to actually carry out the averaging necessary.

temporarily that $n(t)$ is deterministic. We take a section of $n(t)$ T seconds long and expand it in a Fourier series. This will obviously consist of harmonics of the fundamental frequency $1/T$ (see Fig. 5-15). Thus,

$$n(t) = \frac{1}{T} \sum_{m=-\infty}^{\infty} c_m e^{j\omega_m t} \qquad \omega_m = \frac{2\pi m}{T} \qquad (5\text{-}42)$$

Here
$$c_m = \int_{-T/2}^{T/2} n(t) e^{-j\omega_m t}\, dt \qquad (5\text{-}43)$$

as in previous chapters. Note that c_m is a random variable because $n(t)$ is random. In fact, if $E(n) = 0$, it is easily shown that $E[c_m] = 0$. So we are in the peculiar situation of having a Fourier-series representation of $n(t)$ which is valid for the *particular time interval T* over which determined, but which varies statistically each time we take another strip of $n(t)$ T seconds long.

To obviate this difficulty one can calculate $|c_m|^2$, which can never go to zero, and use this to carry out a noise spectral analysis. This is in fact the approach commonly used in digital spectral analysis, as we shall see shortly. Instead, we adopt a more formal procedure at this time. Recall that the autocorrelation function $R_n(\tau)$ was introduced to provide a measure of the time variation of the random process $n(t)$. $R_n(\tau)$ thus plays the role here, in dealing with *random* processes, that the time function itself did in the deterministic case. In the deterministic case (Chap. 2) we found the spectral representation of a time function $f(t)$ by taking its Fourier transform. In the random case we formally *define* the spectral representation of $n(t)$ to be the *Fourier transform* of $R_n(\tau)$. Calling this quantity $G_n(f)$, we formally have

$$G_n(f) = \int_{-\infty}^{\infty} R_n(\tau) e^{-j\omega\tau}\, d\tau \qquad (5\text{-}44)$$

This function must have the property that as $R_n(\tau)$ takes on narrower and narrower values about zero [implying that $n(t)$ and $n(t + \tau)$ become less dependent for a given τ, or that $n(t)$ varies more rapidly], it becomes wider in frequency. This property is of course consistent with the desired inverse time-bandwidth relation.

From the Fourier transform relation connecting $G_n(f)$ and $R_n(\tau)$, it is apparent that the inverse Fourier transform exists, and that $R_n(\tau)$ may be formally found from $G_n(f)$ by writing

$$R_n(\tau) = \int_{-\infty}^{\infty} G_n(f) e^{j\omega\tau}\, \frac{d\omega}{2\pi}$$

$$= \int_{-\infty}^{\infty} G_n(f) e^{j\omega\tau}\, df \qquad (5\text{-}45)$$

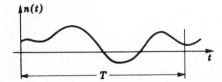

Figure 5-15 Fourier analysis of $n(t)$.

But recall that $R_n(0) = E(n^2)$ is the second moment of $n(t)$, and that $R_n(0) = P_{av}$, the total power in the noise wave, as measured by a true power meter. From (5-45), we have

$$R_n(0) = E(n^2) = \int_{-\infty}^{\infty} G_n(f) \, df \tag{5-46}$$

The function $G_n(f)$ defined formally as the Fourier transform of the autocorrelation function thus appears to have the dimensions of a power density: its integral over all frequencies is just the total power in the noise. For this reason $G_n(f)$ is termed specifically the *power spectral density* or frequently the *power spectrum*. It measures the distribution of noise power with frequency. A power meter tuned to a frequency f_0 and measuring the power in a narrow range Δf about f_0 would provide a good approximation to $2G_n(f)\Delta f$. [Since negative frequencies are just an artifice, one can equally well double $G_n(f)$ in (5-46) and integrate over positive frequencies only.] In the special case where the noise $n(t)$ is zero mean, $E(n) = 0$, and

$$N = \int_{-\infty}^{\infty} G_n(f) \, df \qquad E(n) = 0 \tag{5-47}$$

The noise power or variance N, the parameter on which the probability of error in the detection of pulses in noise was found to depend, is thus directly related to the spectral density $G_n(f)$. It is the sum of the noise-power contributions at all frequencies.

Some examples of autocorrelation function/spectral density pairs appear in Fig. 5-16. In each case the average value of the noise has been taken as zero, and the autocorrelation function thus approaches zero for τ large enough. Note also that in each case $R_n(0)$ has been set equal to N, and that the area under the spectral density curve is correspondingly equal to N as well.

Comparing the autocorrelation function/spectral density pairs in each of the three cases of Fig. 5-16, the inverse correlation time/bandwidth relation mentioned earlier becomes apparent. In particular, we define the bandwidth B of a noise wave in terms of the width of its spectral density function. The corresponding autocorrelation function then goes to zero for the spacing τ greater than $1/B$. Since $R_n(\tau) = E[n(t)n(t + \tau)]$ and $E(n) = 0$ by assumption, it is apparent that $n(t)$ and $n(t + \tau)$ (Fig. 5-17) become uncorrelated for $\tau > 1/B$. The reciprocal of the bandwidth thus plays an important role in determining the measure of correlation between a sample of $n(t)$ at time t, and a sample τ seconds later. Thus, in example 1 of Fig. 5-16, with f_0 the 3-dB bandwidth, $R_n(\pm 1/2\pi f_0) = e^{-1}R_n(0)$. [$G_n(f_0) = \frac{1}{2}G_n(0)$; this is *not* the same as half-power bandwidth, in which case $\int_{-B_{1/2}}^{B_{1/2}} G_n(f) \, df = N/2$.] In example 2, the noise $n(t)$ is truly *band-limited* to B hertz. It is apparent that at $\tau = \pm 1/2B$, and integral multiples thereof, $n(t)$ and $n(t + \tau)$ are *always* uncorrelated. This is also the case in example 3 for $|\tau| \geq T_n$, and we note that $1/T_n$, the first zero crossing of $G_n(f)$, is a measure of the bandwidth of $n(t)$.

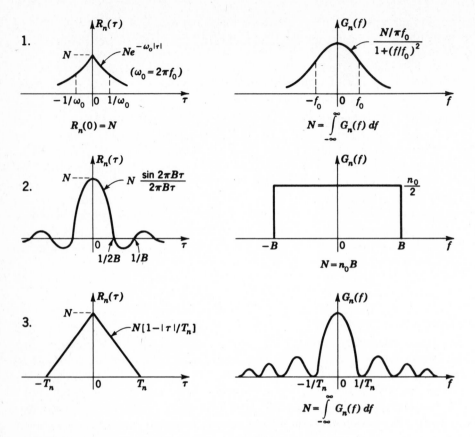

Figure 5-16 Correlation function and spectral-density pairs.

Specifically, if the bandwidth B is 1 MHz, samples spaced more than 1 μs apart are essentially uncorrelated in all three examples of Fig. 5-16. (In the case of example 3, if $T_n = 1$ μs, the samples *are* uncorrelated for all $\tau \geq 1$ μs.) Recall from probability theory that two uncorrelated gaussian variables are *independent* as well. If the random wave in Fig. 5-17 is then gaussian, $n(t)$ and $n(t + \tau)$ are essentially independent if $\tau > 1/B$ s.

Note that in examples 1 and 3 of Fig. 5-16 most of the noise power appears concentrated about the origin (dc), the bandwidths f_0 and $1/T_n$, respectively,

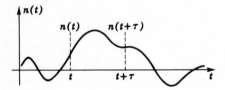

Figure 5-17 Spacing between noise samples.

then serving as measures of this concentration. The fact that relatively little noise power appears at the high frequencies thus indicates that the noise $n(t)$ rarely fluctuates at these rates, justifying our intuitive statements previously that 1/bandwidth is, roughly speaking, a measure of the time between significant changes in $n(t)$. This is of course also shown by the plots of the autocorrelation function $R_n(\tau)$; the value of τ for which $R_n(\tau)$ begins to decrease significantly is also a measure of the time between significant changes in $n(t)$ (this is exactly one possible interpretation of correlation), and is of course just 1/bandwidth, defined in some arbitrary sense. In example 2 *all* the power is assumed concentrated in the range 0 to B hertz. This case is of course the random signal equivalent of the band-limited deterministic signals of Chap. 2. $R_n(\tau)$ here is equivalent to $f(t)$ there; $G_n(f)$ is equivalent to $F(\omega)$.

One particular example of a spectral density that plays an extremely important role in communications and signal-processing analyses is that in which the spectral density $G_n(f)$ is flat or constant, say equal to $n_0/2$, over *all* frequencies:

$$G_n(f) = \frac{n_0}{2} \qquad (\text{all } f) \tag{5-48}$$

Although this is, strictly speaking, physically inadmissible, since it implies infinite noise power $[N = \int_{-\infty}^{\infty} G_n(f)\, df]$, it is a good model for many typical situations in which the noise bandwidth is so large as to be out of the range of our measuring instruments (or frequencies of interest to us). We shall describe a specific example of noise of this type later in this chapter in discussing thermal noise appearing at the input to a receiving antenna.

Noise $n(t)$ with a flat spectral density $n_0/2$, as in Eq. (5-48), is called *white noise* because of its "equal jumbling" of all frequencies (compare with the common appellation "white light").

Note that in the band-limited noise case of example 2 of Fig. 5-16, one may obtain *white noise* by letting $B \to \infty$. The noise of example 2 is therefore often called *band-limited white noise*. It is apparent from the transform-pair relations that the autocorrelation function for white noise is just an impulse or delta function centered at the origin:

$$G_n(f) = \frac{n_0}{2} \qquad R_n(\tau) = \frac{n_0}{2}\, \delta(\tau) \tag{5-48a}$$

The spectral-density and autocorrelation functions for white noise, as well as band-limited white noise of spectral density $G_n(f) = n_0/2$, $|f| \le B$, $G_n(f) = 0$,

Since the autocorrelation function is an impulse in the case of white noise, this indicates that $n(t)$ is *always* uncorrelated with $n(t + \tau)$, no matter how small τ may be. The implication then is that $n(t)$ may vary infinitely rapidly, since it contains power at all frequencies. In practice, of course, as just noted, this simply means that the high-frequency variations are beyond the capabilities of our instruments in a particular measurement we may be making. So although the white-noise model may appear to be physically inadmissible, we could

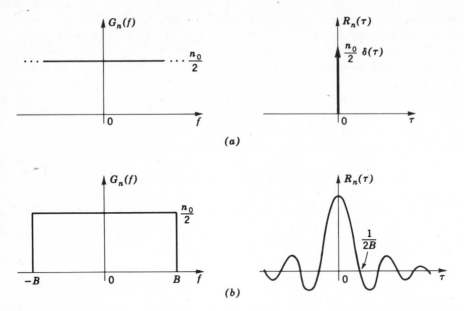

Figure 5-18 White-noise power spectrum and autocorrelation function. (*a*) White noise. (b) Band-limited white noise.

never measure the rapid variations anyway. As an example, if our measuring devices have a time response $\gg 1/B$, with B the noise bandwidth of an actual physical noise process, the noise looks to us for all practical purposes like white noise. Band-limited white noise, with $B \gg$ significant frequencies in the frequency response of our measuring devices, thus appears to us as white noise. To an oscilloscope of bandwidth 50 MHz, input noise with bandwidth 500 MHz would obviously appear like white noise.

How does one measure the spectral density $G_n(f)$ in practice? One possibility is to feed the noise wave (or random signal) under investigation into a parallel bank of narrow filters, spaced Δf hertz apart, with $\Delta f \ll B$, and measure the power output of each filter. Alternatively, one may use a scanning narrowband filter, whose center frequency is shifted sequentially in steps of Δf hertz, and measure its mean-squared output as the entire frequency range is scanned. The most common approach, however, involves the Fourier analysis mentioned earlier in connection with (5-42) and (5-43). It turns out that a possible estimate (approximation) of $G_n(f_l)$, the spectral density at frequency f_l, is given by

$$\hat{G}_n(f_l) = \frac{1}{T} |c_l|^2 = \frac{1}{T} \left| \int_{-T/2}^{T/2} n(t) e^{-j\omega_l t} \, dt \right|^2 \qquad (5\text{-}49)$$

with c_l just the Fourier coefficient of (5-43), using a strip of $n(t)$ T seconds long. Actually, one carries out the calculation of c_l digitally, taking samples of $n(t)$ spaced $< 1/2B$ seconds apart. The total number of samples processed (of the order of $2BT$) must be large to obtain good results. Fast Fourier-transform techniques developed to speed up the calculation task have made this approach to the determination of $G_n(f)$ the one most widely used at present.[20] One caveat however: It turns out that although the average of $G_n(f_l)$ in (5-49) is the desired spectral density $G_n(f_l)$ at frequency f_l, the *variance* of this estimate does not decrease with increasing T, as was the case in the discussion of $E(n)$ and σ^2 earlier. Instead one has to repeat the calculation for a number of sets of data samples (or strips T seconds long), and average them.[21]

5-5 RANDOM SIGNALS AND NOISE THROUGH LINEAR SYSTEMS

Using the concept of power spectral density, as developed for random signals and noise, we are now in a position to consider the effect of linear filtering on these nondeterministic signals. This then parallels and extends our discussion of deterministic signals through linear systems in Chap. 2.

Consider then noise $n_i(t)$ with prescribed spectral density $G_{n_i}(f)$ and hence noise power N_i, as well as autocorrelation function $R_{n_i}(\tau)$, passed through a linear system with frequency transfer function $H(\omega)$, as shown in Fig. 5-19. What are the properties of the noise $n_o(t)$, at the output? That is, what are its spectral density $G_{n_o}(f)$, autocorrelation function $R_{n_o}(\tau)$, and output noise power N_o?

One would expect to find results similar to those obtained in Chap. 2. If the input noise $n_i(t)$ is varying roughly at a rate defined by its bandwidth B which is *slow* compared to the system bandwidth B_{sys}, the output noise $n_o(t)$ differs very little from $n_i(t)$. If, on the other hand, $B \gg B_{\text{sys}}$, the rapid fluctuations of $n_i(t)$ cannot get through (the system will not respond rapidly enough), and one would expect to find $n_o(t)$ varying at roughly the rate B_{sys}. We can show that these intuitive arguments are of course valid, and that the effect of the system on the noise is given, quite simply, by the following relation between input and output spectral densities:

$$G_{n_o}(f) = |H(\omega)|^2 G_{n_i}(f) \tag{5-50}$$

The two extreme cases noted above ($B \ll B_{\text{sys}}$, $B \gg B_{\text{sys}}$) are summarized qualitatively in Fig. 5-20.

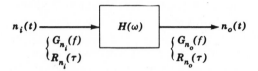

Figure 5-19 Noise through a linear system.

[20] See Schwartz and Shaw, *op. cit.*, chaps. 3 and 4, for a discussion of discrete spectral estimation.
[21] *Ibid.*

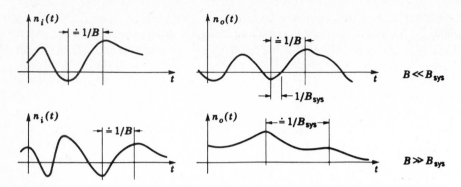

Figure 5-20 System response to input noise.

Equation (5-50) is demonstrated very simply by referring back to the defini-tion of the spectral density as the Fourier transform of the autocorrelation function. Recall from (5-38a) that $R_n(\tau)$ is the expectation of the product of $n(t)$ and $n(t + \tau)$. This is the definition of the autocorrelation function of the noise wave at both the input and the output of any linear system, with $n_i(t)$ used in place of $n(t)$ at the input, and $n_o(t)$ as the output noise function (Fig. 5-19). We know from linear system analysis that $n_i(t)$ and $n_o(t)$ are related by the convolu-tion integral:

$$n_o(t) = \int_{-\infty}^{\infty} h(t - \tau)n_i(\tau) \, d\tau \tag{5-51}$$

Here $h(t)$, the inpulse response at the system, is the inverse Fourier transform of $H(\omega)$, the system transfer function. Using (5-51) to evaluate $n_o(t)$ in the expression for $R_{n_o}(\tau)$, the output autocorrelation function, rewriting in terms of $R_{n_i}(\tau)$, the input autocorrelation function, and $h(t)$, the impulse response, and then taking Fourier transforms, one obtains (5-50). Details are left to the reader.

Equation (5-50) is an extremely important relation, basic to the understand-ing of random signals and noise passing through linear systems. It further substantiates the remark made in the previous section that in dealing with random signals, $R_n(\tau)$ and $G_n(f)$ play the roles, respectively, that a time func-tion $f(t)$ and its transform $F(\omega)$ do in dealing with deterministic signals. The spectral analysis of random signals focuses on the *power* distribution. This is related to the integrated square of signals and accounts for the $|H(\omega)|^2$ term appearing in the transfer relation of (5-50). The band-limiting effect of a linear system on an input random signal appears in the multiplication of the input spectral density by $|H(\omega)|^2$. This is directly analogous to the multiplication of the input Fourier transform by $H(\omega)$ for deterministic signals. These analogous relations account for the preservation of the concepts of bandwidth and inverse time-frequency relations in studying the passage of deterministic *or* random signals through linear systems. This is precisely why deterministic models of

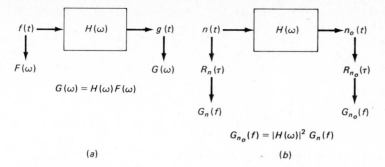

Figure 5-21 Handling of signals in linear systems. (*a*) Deterministic signals. (*b*) Random signals and noise.

signals can be used to represent random signals in systems, as has been done implicitly in the work thus far in this book.

The comparison of the handling of deterministic and random signals is summarized schematically in Fig. 5-21. The step in Fig. 5-21*b* showing the transition from a random wave to its autocorrelation function $R_n(\tau)$ is included to emphasize the role $R_n(\tau)$ plays as an equivalent to $f(t)$ in the deterministic case. One can bypass $R_n(\tau)$ completely and go directly from $n(t)$ to $G_n(f)$, if desired.

Now consider some examples of the application of (5-50). Assume first that we have white noise of spectral density $n_0/2$ applied at the input of an ideal low-pass filter of bandwidth B hertz (Fig. 5-22). The output noise is exactly the band-limited white noise of spectral density $G_n(f) = n_0/2$, $|f| \leq B$, $G_n(f) = 0$, $|f| > B$, mentioned previously. The output noise power is then just

$$N_o = n_0 B \tag{5-52}$$

By *increasing* the filter bandwidth B, we *increase* the output *noise* power. We also increase the rate of variation of noise at the output, or decrease the correlation between $n(t)$ and $n(t + \tau)$, for a fixed τ.

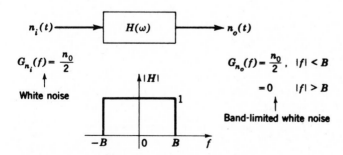

Figure 5-22 White noise through an ideal filter.

As a second simple example, consider the case of white noise applied to the input of an RC filter, as shown in Fig. 5-23. Since

$$|H(\omega)|^2 = \frac{1}{1 + (\omega RC)^2} = \frac{1}{1 + (2\pi RCf)^2}$$

for this filter, we have

$$G_{n_o}(f) = \frac{n_0/2}{1 + (2\pi RCf)^2} = \frac{n_0/2}{1 + (f/f_0)^2} \tag{5-53}$$

with $f_0 \equiv 1/2\pi RC$, just as in the first example of Fig. 5-16. The bandwidth of the output noise is thus inversely proportional to the filter time constant RC, just as in the case of the deterministic signals of Chap. 2. The correlation time is also of the order of RC. For we have, as in Fig. 5-16,

$$R_{n_o}(\tau) = N_o e^{-|\tau|/RC} \tag{5-54}$$

as the Fourier transform of this spectral density. The average noise power N_o may be readily found by integrating $G_{n_o}(f)$:

$$N_o = \int_{-\infty}^{\infty} G_{n_o}(f) \, df = \int_{-\infty}^{\infty} \frac{n_0/2}{1 + (f/f_0)^2} \, df = \frac{n_0 \pi}{2} f_0 \tag{5-55}$$

We now note an interesting fact shown in both of these examples—the low-pass filter and RC filter. The output *noise power* N_o is in both cases *proportional* to the system *bandwidth*. Thus in the first case we had $N_o = n_0 B$; in the second case we have $N_o = (n_0\pi/2)f_0$, with f_0 a measure of the system bandwidth. This thus indicates that as we *widen* the system bandwidth, we *increase* the *noise power* at the output. Similarly, we may *reduce* the noise at the output by *narrowing* the bandwidth.

We can generalize this statement to include all kinds of low-pass systems by defining a *noise equivalent bandwidth*. Assume white noise of spectral density $n_0/2$ applied at the input of an arbitrary linear system transfer function $H(\omega)$. The output noise power is then given by

$$N_o = \frac{n_0}{2} \int_{-\infty}^{\infty} |H(\omega)|^2 \, df \tag{5-56}$$

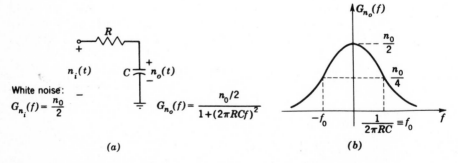

Figure 5-23 *RC* filtering of white noise. (*a*) Filter. (*b*) Output spectral density.

since $G_{n_o}(f) = (n_0/2)|H(\omega)|^2$. Assume that this same noise comes from an ideal low-pass filter of bandwidth B and amplitude $H(0)$, that is, the magnitude of the arbitrary filter transfer function at zero frequency. To have the same noise output we must then have

$$N_o = n_0 H^2(0)B = n_0 \int_0^\infty |H(\omega)|^2 \, df \qquad (5\text{-}57)$$

since $|H(\omega)|$ has even symmetry about the origin. Then we have the *noise equivalent bandwidth B* defined as

$$B \equiv \frac{1}{H^2(0)} \int_0^\infty |H(\omega)|^2 \, df \qquad (5\text{-}58)$$

This procedure essentially corresponds to replacing the arbitrary filter $H(\omega)$ by an equivalent ideal low-pass filter of bandwidth B, as shown in Fig. 5-24. One may also do this for bandpass filters, using in place of $H(0)$ the value of $H(\omega)$ at the center frequency. As an example, for the RC filter, we have, from Eq. (5-55), $B = (\pi/2)f_o = 1/4RC$.

To show the numerical quantities involved, assume that the white-noise spectral density at the input to a high-gain/low-pass amplifier is $n_0/2 = 10^{-14}$ V^2/Hz. The amplifier has a voltage gain of 10^3 and a high-frequency cutoff of 10 MHz. From Eq. (5-57), then, the mean-squared noise voltage at the amplifier output is $N_o = 2 \times 10^{-14} \times 10^6 \times 10^7 = 0.2 \ V^2$.

In terms of the noise equivalent bandwidth we have then, as in Eq. (5-57), the general relation

$$N_o = n_0 H^2(0)B \propto B \qquad (5\text{-}59)$$

This indicates that the output noise power N_o is *proportional* to the *bandwidth*. As an interesting rule of thumb, then, one reduces noise in systems by *decreasing* the bandwidth, and increases the noise by widening the bandwidth. We include here, in the word "systems," measuring instruments, receivers, signal processors, etc.

If one attempts to detect or otherwise measure signals in the presence of noise, one thus tries to use as narrow a system bandwidth as possible—assuming that the input noise is originally much wider in bandwidth than the system bandwidth. Of course one then has a limitation on the rate of variation of the signals themselves. For given classes of signals, one cannot narrow the bandwidth too far down to the point where the signal itself is adversely affected or disturbed in an undesirable manner. We shall have more to say about this

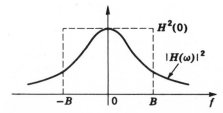

Figure 5-24 Noise equivalent bandwidth.

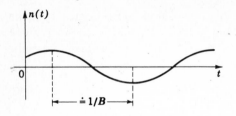
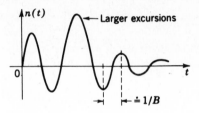

Figure 5-25 Effect of increased noise bandwidth.

trade-off between signal detectability and noise power increase with increasing bandwidth shortly.

We can now summarize our study of noise thus far with two simple statements that serve as useful rules of thumb:

1. Increasing the system bandwidth increases the rate of fluctuation of the output noise (assuming the system bandwidth is initially much less than the input noise bandwidth).
2. Increasing the system bandwidth increases the output noise power. Since the output noise power N_o is a measure of the mean-squared statistical fluctuations about the average value (assumed zero here) [recall that N was the noise variance and hence a measure of the width of the noise probability-density function $f(n)$ about $E(n)$], larger instantaneous values of $n(t)$ thus become more probable as well.

These two rules appear expressed pictorially in the curves of Fig. 5-25. As the bandwidth B increases, the random process $n(t)$ is expected to deviate more often and more violently (in terms of peak excursions) away from its expected or average value.

5-6 MATCHED FILTER DETECTION: APPLICATION TO BASEBAND DIGITAL COMMUNICATIONS

In Sec. 5-1 we calculated the probability of detection in a binary communication system subject to additive gaussian noise. We implicitly assumed there that the intersymbol interference discussed in Chap. 3 (Sec. 3-9) was no problem. (In data transmission over telephone channels the reverse is usually true, as noted earlier; additive noise generally poses no problem, intersymbol interference does.)

We found there that the probability of error ultimately depended on the peak signal-amplitude-to-rms-noise ratio, A/σ or $A/\sqrt{N}$, using the present noise terminology. An interesting and very practical question that we might pose would be: Is it possible to design the system up to the detector in order to maximize the ratio of peak signal amplitude to rms noise when sampled at the detector? Thus we envision white noise added to the sequence of binary pulses of known shape at the receiver input, with the composite sum passed through a

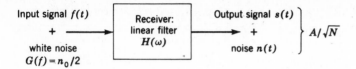

Figure 5-26 Matched-filter problem.

linear filter (representing the system), then sampled and a decision made at the detector. Is it then possible to design the linear filter prior to detection to maximize the ratio $A/\sigma = A/\sqrt{N}$ at its output? This is shown pictorially in Fig. 5-26.

(Although we assume throughout this section that the signal is a baseband pulse, the results obtained apply also to the case of binary carrier transmission. As discussed in Chap. 4 the high-frequency pulses may be synchronously detected, reducing them to the pulse problem under consideration. Alternatively, they may be envelope detected, if of the OOK or FSK type. This is shown in the sections that follow.)

We shall find that this problem is easily handled as an application of the power-spectrum concepts just introduced.

The results of the analysis to follow are useful in many other applications aside from binary or pulse-code-modulation (PCM) transmission. They apply as well to the design of pulse-amplitude-modulation (PAM) (nonquantized) systems. There the question posed would be: Given pulsed samples of the signal to be transmitted, how should these be filtered so as to have maximum signal-to-rms-noise ratio at the output of the system? We might alternatively state this: How should we shape our pulses (by filtering them) so that the signal-to-noise ratio will be maximized?

A similar problem is encountered in radar systems, where it is required to detect the presence of a signal echo embedded in fluctuation noise.[22] The amplitude of the signal relative to the noise should thus be maximized if possible.

In all these examples we are not specifically interested in maintaining fidelity of pulse shape. We are primarily interested in improving our ability to "see" or recognize a pulse signal in the presence of noise. This ability to see the pulse is assumed to be related to the ratio of peak signal to rms noise. An example of such a pulse signal embedded in noise is shown in Fig. 5-27.

How do we know that the peak signal-to-noise ratio can be maximized by properly choosing the filter characteristic? This is simply answered from our discussions of Chap. 2 and the previous section.

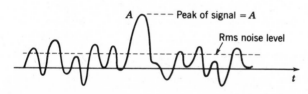

Figure 5-27 Pulse signal embedded in noise.

[22] *Ibid.*, chap. 5.

Assume, for simplicity's sake, that the input signal $f(t)$ is a rectangular pulse of width τ seconds. The system is assumed to have the idealized filter characteristics of Sec. 2-4, with a variable bandwidth B hertz.

For very small bandwidth ($B \ll 1/\tau$) the peak of the output signal is small and increases with bandwidth. This is shown by the curves of Fig. 2-35. It is also indicated by the curve of Fig. 5-28a, showing the familiar $(\sin x)/x$ spectrum with the filter cutoff frequency (B) superimposed. The output-signal amplitude is proportional to the area under the curve. For $B \ll 1/\tau$ the frequency spectrum is flat, so that the output signal increases linearly with B. The rms noise is proportional to $\sqrt{B}$ and so increases at a smaller rate than the signal for small bandwidth.

As the bandwidth increases, approaching $B = 1/\tau$, the signal amplitude begins to increase less rapidly with B. For $B > 1/\tau$ the signal remains approximately at the same amplitude as at $B = 1/\tau$ (Fig. 2-35). (Recall from Chap. 2 that increasing the bandwidth beyond $B = 1/\tau$ just served to fill out the fine details of the pulse. For a recognizable pulse a bandwidth $B = 1/\tau$ was all that was necessary.) The rms noise keeps increasing with bandwidth, however, so that the ratio of peak signal to rms noise begins to decrease inversely as $\sqrt{B}$. We would thus expect an optimum ratio at about $B = 1/\tau$. This will be borne out in the analysis to follow.

In general, not only the filter bandwidth but the shape of the filter characteristic can be adjusted to optimize the peak signal-to-noise ratio. To show this, consider $f(t)$ impressed across a linear filter with frequency transfer function $H(\omega)$. Defining $F(\omega)$ to be the Fourier transform of $f(t)$,

$$F(\omega) = \int_{-\infty}^{\infty} f(t)e^{-j\omega t}\, dt \tag{5-60}$$

The output signal $s(t)$ is given by

$$s(t) = \frac{1}{2\pi}\int_{-\infty}^{\infty} F(\omega)H(\omega)e^{j\omega t}\, d\omega$$

$$= \int_{-\infty}^{\infty} F(\omega)H(\omega)e^{j\omega t}\, df \qquad \omega = 2\pi f \tag{5-61}$$

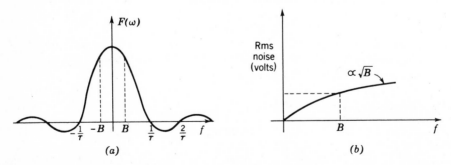

Figure 5-28 Rectangular pulse and noise passed through ideal filter. (a) Signal pulse spectrum, with filter cutoff superimposed. (b) rms noise output.

The magnitude of $s(t)$ at the sampling time t_0 is just the desired output-signal amplitude A. Thus,

$$A = |s(t_0)| = \left| \int_{-\infty}^{\infty} F(\omega)H(\omega)e^{j\omega t_0} \, df \right|$$

The power spectrum of the white noise at the filter input is taken as

$$G(f) = \frac{n_0}{2} \tag{5-62}$$

as in the previous section. The power spectrum at the filter output is then

$$G_n(f) = \frac{n_0}{2} |H(\omega)|^2 \tag{5-63}$$

and the average output noise power (or mean-squared noise voltage across a $1\text{-}\Omega$ resistor) is

$$N = \frac{n_0}{2} \int_{-\infty}^{\infty} |H(\omega)|^2 \, df \tag{5-64}$$

$\sqrt{N}$ is the rms output noise in the absence of a signal.

We would now like to choose $H(\omega)$ such that the ratio $A/\sqrt{N}$ is maximized. This is the same as maximizing the square of the ratio, or A^2/N. This squared ratio is just the ratio of instantaneous peak signal power at $t = t_0$ to mean noise power, and will henceforth be referred to as the peak power signal-to-noise ratio, or peak SNR.

Since the input signal $f(t)$ is assumed given, its energy content $\int_{-\infty}^{\infty} f^2(t) \, dt$ is a constant. Calling the energy E, it is left to the reader as an exercise to show the following Fourier-transform identity is a valid one[23]:

$$E = \int_{-\infty}^{\infty} f^2(t) \, dt = \int_{-\infty}^{\infty} |F(\omega)|^2 \, df \tag{5-65}$$

Dividing the peak signal power to mean noise power by the constant E will obviously not affect the determination of the maximum ratio. So we can take as our problem that of maximizing the ratio

$$\frac{A^2}{EN} = \frac{\left| \int_{-\infty}^{\infty} F(\omega)H(\omega)e^{j\omega t_0} \, df \right|^2}{(n_0/2) \int_{-\infty}^{\infty} |F(\omega)|^2 \, df \int_{-\infty}^{\infty} |H(\omega)|^2 \, df} \tag{5-66}$$

This is readily done by means of *Schwarz's inequality*, relating the integral of products of complex functions:

[23] This relation is a basic theorem in Fourier-integral theory and is frequently called Parseval's theorem. It may be easily derived from Eq. (2-19b) in Chap. 2 by taking $(-T/2, T/2)$ as limits in the integral and letting $T \to \infty$. Alternatively, it may be proved by applying the convolution theorem. [Assume a function $f(t)$ passed through a linear filter with transfer function $H(\omega) = F^*(\omega)$.]

$$\left| \int_{-\infty}^{\infty} X(\omega)Y(\omega)\, d\omega \right|^2 \le \int_{-\infty}^{\infty} |X(\omega)|^2\, d\omega \int_{-\infty}^{\infty} |Y(\omega)|^2\, d\omega \tag{5-67}$$

Schwarz's inequality for integrals of complex functions is just an extension of an inequality for real integrals, given by

$$\left[\int_{-\infty}^{\infty} f(t)g(t)\, dt \right]^2 \le \int_{-\infty}^{\infty} g^2(t)\, dt \int_{-\infty}^{\infty} f^2(t)\, dt \tag{5-68}$$

It might be termed a generalization of the familiar distance relation among vectors that the magnitude of the sum of two vectors is less than or equal to the sum of the magnitudes of the two vectors:

$$|\bar{a} + \bar{b}| \le |\bar{a}| + |\bar{b}| \tag{5-69}$$

In the vector case the equality is satisfied if $\bar{a} = K\bar{b}$, or $\bar{a}$ and $\bar{b}$ are collinear. Similarly in Eq. (5-68) the equality is satisfied if $f(t) = Kg(t)$. In the case of complex functions the equality is satisfied if

$$Y(\omega) = KX^*(\omega) \tag{5-70}$$

K is a real number.

How do we apply Schwarz's inequality to our problem of maximizing peak signal-to-noise ratio?

Note that the ratio of Eq. (5-66) contains exactly the integrals of Eq. (5-67) if we let

$$X(\omega) = F(\omega)e^{j\omega t_0} \qquad Y(\omega) = H(\omega)$$

The ratio $n_0 A^2/2EN$ must then be less than or equal to 1, and

$$\left| \int_{-\infty}^{\infty} F(\omega)H(\omega)e^{j\omega t_0}\, df \right|^2 \le \int_{-\infty}^{\infty} |F(\omega)|^2\, df \int_{-\infty}^{\infty} |H(\omega)|^2\, df \tag{5-71}$$

In particular, the ratio is a *maximum* when the equality holds, or

$$H(\omega) = K[F(\omega)e^{j\omega t_0}]^* = KF^*(\omega)e^{-j\omega t_0} \tag{5-72}$$

As an example, if $f(t)$ is the rectangular pulse of width τ and height V, $F(\omega) = V\tau\{[\sin(\omega\tau/2)]/(\omega\tau/2)\}$ and

$$H(\omega) = K \frac{\sin(\omega\tau/2)}{\omega\tau/2} e^{-j\omega t_0}$$

for maximum ratio of peak signal to rms noise.

Filters possessing the characteristic of Eq. (5-72) are said to be *matched filters*. The response at the output of such a filter, to $f(t)$ applied at the input, is

$$s(t) = \int_{-\infty}^{\infty} F(\omega)H(\omega)e^{j\omega t}\, df$$

$$= K \int_{-\infty}^{\infty} |F(\omega)|^2 e^{j\omega(t-t_0)}\, df \tag{5-73}$$

In particular $s(t)$ will have amplitude A when

$t = t_0$:
$$\left| s(t_0) \right| = K \int_{-\infty}^{\infty} \left| F(\omega) \right|^2 df = A$$

Note that the amplitude A is thus proportional to the signal energy E.

An interesting relation for the matched-filter output signal-to-noise ratio $A/\sqrt{N}$ may be derived by applying the matched-filter condition of Eq. (5-72) to Eq. (5-66). Specifically, it is apparent that at the output of the matched filter the peak power SNR is

$$\frac{A^2}{N} = \frac{2E}{n_0}$$

or
$$\frac{A}{\sqrt{N}} = \sqrt{\frac{2E}{n_0}} \qquad (5\text{-}74)$$

The signal-to-noise ratio is thus a function solely of the energy in the signal and the white-noise spectral density. *The dependence on the signal input waveshape $f(t)$ has been obliterated by use of the matched filter.* Two different signal waveshapes will provide the same probability of error in the presence of additive white noise, providing they contain the same energy and are filtered by the appropriate matched filter in each case. It is the *energy* in the signal that provides its ultimate detectability in noise. This point will be pursued further in Chap. 6.

If the input time function $f(t)$ is symmetrical in time $[f(t) = f(-t)]$, $F(\omega)$ will be a real function of frequency. From Eq. (5-72) $H(\omega) = KF(\omega)e^{-j\omega t_0}$ for this case. This means that the impulse response $h(t)$ is $h(t) = Kf(t - t_0)$. The impulse response of a filter matched to a symmetrical input is a delayed replica of such an input.

If $f(t)$ is not symmetrical, $F(\omega)$ is complex. By using the Fourier-integral relations it may be shown that

$$h(t) = Kf[-(t - t_0)] \qquad (5\text{-}75)$$

in general. The proof is left to the reader as an exercise since it is identical to the procedure used to prove Parseval's theorem above. Since $f(t)$ is normally defined for positive t, $h(t)$ in the general case will be defined for negative t. As pointed out in Chap. 2, such a filter is physically not realizable. In the general case, then, the matched filter is *not realizable.*

Of what value then is this entire analysis leading to the matched-filter result? Just as in the case of the ideal filter of Chap. 2, we can approximate the matched-filter characteristics by those of an actual filter. We can also compare practical filters with the matched filter so far as the ratio of peak output signal to noise is concerned and can optimize their shape and bandwidth as far as practicable. This is the procedure we shall follow in the remainder of this section.

The input time function $f(t)$ is assumed to be a rectangular pulse of width τ seconds and height V. We assume that $V\tau = 1$. With $f(t)$ symmetrically located about $t = 0$, $F(\omega) = [\sin (\omega\tau/2)]/(\omega\tau/2)$, a real function as noted above. The impulse response of the matched filter is then also a rectangular pulse of τ

seconds duration. [$H(\omega) = F(\omega)$ here.] The response of this matched filter to the rectangular pulse will then be the convolution of two rectangular pulses. This gives a triangular-pulse output. We can check this by noting that for the matched filter $F(\omega)H(\omega) = \{[\sin(\omega\tau/2)]/(\omega\tau/2)\}^2 e^{-j\omega t_0}$ ($V\tau = 1$). This is just the Fourier transform of a triangular pulse of width 2τ seconds as shown in Chap. 2. This output pulse is τ seconds wide at the half-amplitude points.

Note that such a triangular pulse is not too different in shape from the output of the ideal low-pass filter of Sec. 2-4, with $B = 1/\tau$ (see Fig. 2-35). In fact the ideal-filter output for $B = 1/\tau$ could very well have been approximated by such a triangle. If the bandwidth of the matched filter is assumed to be the frequency of the first zero in its amplitude characteristic, the bandwidth is also just $1/\tau$ ($\sin \omega\tau/2 = 0$; $\omega\tau/2 = \pi$; $f = 1/\tau$).

This is an interesting point, for it agrees with our previous results that for producing a recognizable pulse a filter bandwidth of the order of $B = 1/\tau$ should be used. We shall see below that for an ideal low-pass filter $B\tau = 0.7$ actually gives maximum signal-to-noise ratio.

Just how significant the shapes of the filter characteristic and bandwidth are in determining the output peak signal-to-noise ratio for a rectangular-pulse input can be found by applying Eq. (5-66) to various filters. The resulting signal-to-noise ratio can then be compared with that for the optimum matched filter. We shall actually calculate $n_0 A^2/2EN$ so that the optimum value is normalized to 1. In all cases the rectangular-pulse input is assumed to have unit area ($V\tau = 1$), so that

$$F(\omega) = \frac{\sin(\omega\tau/2)}{\omega\tau/2}$$

Ideal low-pass filter, variable bandwidth. Here

$$\begin{aligned} H(\omega) &= e^{-j\omega t_0} & |\omega| &\leq 2\pi B \\ H(\omega) &= 0 & |\omega| &> 2\pi B \end{aligned} \tag{5-76}$$

The peak power SNR for this case, as obtained from Eq. (5-66), becomes

$$\frac{\left| \int_{-2\pi B}^{2\pi B} \frac{\sin(\omega\tau/2)}{\omega\tau/2}\,d\omega \right|^2}{\int_{-\infty}^{\infty} \left[\frac{\sin(\omega\tau/2)}{\omega\tau/2}\right]^2 d\omega \int_{-2\pi B}^{2\pi B} d\omega} = \frac{\left[\left(\frac{2}{\tau}\right) \int_{-a}^{a} \frac{\sin x}{x}\,dx \right]^2}{\frac{2}{\tau} \int_{-\infty}^{\infty} \left(\frac{\sin x}{x}\right)^2 dx\, 4\pi B}$$

with $x \equiv \omega\tau/2$ and $a = \pi B\tau$. Using the relation $\int_{-\infty}^{\infty} [(\sin x)/x]^2\,dx = \pi$, and recalling the sine integral definition, $\mathrm{Si}\,a = \int_0^a [(\sin x)/x]\,dx$, the ratio squared becomes

$$\frac{2}{\pi a}\,(\mathrm{Si}\,a)^2$$

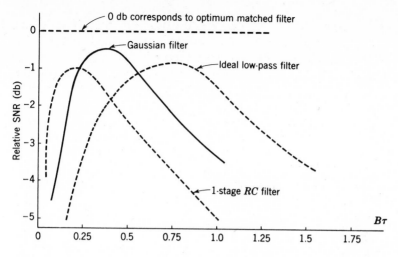

Figure 5-29 Peak SNR for various filters, compared with matched filter (rectangular-pulse input).

This expression may be plotted by using tables of the sine integral[24] and is found to have a maximum at $a = \pi B\tau = 2.2$. This corresponds to $B\tau = 2.2/\pi = 0.7$. At this bandwidth the peak power SNR is found to be 0.83, as compared with 1 for the optimum filter. This corresponds to a relative deterioration of 0.8 dB.

The ideal-low-pass-filter case is plotted in Fig. 5-29 as a function of $B\tau$. The decibel scale used is relative to the 0-dB case for the optimum matched filter. Although the maximum ratio is found for $B\tau = 0.7$, the maximum is very broad and varies less than 1 dB from $B\tau = 0.4$ to $B\tau = 1$. For a pulsed carrier or OOK signal the bandwidth would be twice the bandwidth shown here, so that $B\tau = 1.5$ would be optimum for a rectangular filter.

One-stage RC filter, variable bandwidth

$$H(\omega) = \frac{1}{1 + j\omega RC} \tag{5-77}$$

with $2\pi B = 1/RC$ the 3-dB radian bandwidth. The peak response of this filter to a rectangular-pulse input occurs at $t = \tau$ and is given by

$$s(\tau) = 1 - e^{-\tau/RC}$$

[This procedure is much simpler than the frequency-response method of Eq. (5-61) in this case.] The normalized signal-to-noise peak power ratio becomes for this case, after some manipulations,

---

[24] S. Goldman, *Frequency Analysis, Modulation, and Noise*, McGraw-Hill, New York, 1948; E. Jahnke and F. Emde, *Tables of Functions*, Dover, New York, 1945.

$$\frac{(1 - e^{-2a})^2}{a} \qquad a = \frac{\tau}{2RC} = \pi B\tau$$

The details of this calculation are left to the reader as an exercise.

The signal-to-noise ratio for this case has also been plotted in Fig. 5-29 and shows a maximum value at $B\tau = 0.2$ ($B = 1/2\pi RC$). At this bandwidth the filter output is only 1 dB worse than that for the matched-filter case. For $B\tau = 0.5$ the S/N ratio is 2.3 dB worse than the matched-filter case so that the variation with bandwidth is again small.

Multistage RC filters. A signal would normally be amplified by several stages of amplifiers, with filtering included in each amplifier. It is thus of interest to compare the matched-filter signal-to-noise output with that of a multistage amplifier. We assume a simple RC filter in each stage. (For a pulsed carrier we would use a single-tuned circuit for each stage.)

The peak SNR could of course be found by determining the overall frequency response of the multistage amplifiers. This becomes unwieldy mathematically. It can be shown,[25] however, that the transfer function of a large number of isolated RC sections approaches the form

$$H(\omega) = e^{-0.35(f/B)^2} e^{-jt_0\omega} \tag{5-78}$$

where B is the 3-dB bandwidth of the overall filter.

Note that this is the same mathematical form as the gaussian density function discussed earlier in this chapter, and first included as an example among the filter shapes shown in Chap. 2. This gaussian filter is often used as a convenient mathematical model for systems analyses.

The gaussian filter response curve has a characteristic "bell" shape and is symmetrical about $f = 0$. It is shown sketched in Fig. 5-30.

We can again use Eq. (5-66) to calculate the peak power SNR for a rectangular pulse applied to a network with this amplitude characteristic. The result will then approximate the output of a multistage RC amplifier. The analysis is identical to that carried out for the ideal filter. The details will not be presented here, but the results are plotted in Fig. 5-29, again compared with the optimum matched-filter case.

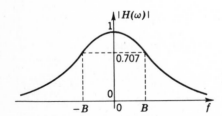

Figure 5-30 Frequency response, multistage amplifier (gaussian error curve).

[25] C. Cherry, *Pulses and Transients in Communication Circuits,* Chapman & Hall, London, 1949; Dover, New York, 1950, p. 311.

Actual calculations for two and three stages of amplification with RC filtering included produce results which do not differ substantially from those for the gaussian filter, so that the gaussian-error-curve analysis will be a good approximation to a multistage RC amplifier for two or more stages.

Note that the peak SNR occurs in this case for $B\tau = 0.4$. (For a two-stage amplifier the peak occurs at $B\tau = 0.3$.) The maximum is quite broad, however, varying by no more than 1 dB from $B\tau = 0.2$ to $B\tau = 0.7$.

At the maximum the ratio is only 0.5 dB less than that for the optimum matched filter. This would be quite negligible in most practical applications. The overall filter characteristic of a multistage amplifier will thus be close to optimum for transmission of rectangular pulses, with the overall bandwidth chosen as $B = 0.5/\tau$.

The filters considered here are all low-pass, and the signal and noise are both considered to be present at baseband. In practice, of course, sine-wave carrier transmission is used with all pulse systems. As shown in Chap. 4 the high-frequency circuits equivalent to the low-pass circuits here generally require twice the low-pass bandwidth ($\pm B$ about the carrier frequency). If the matched filters discussed here are included in the i-f section of a receiver, for example, all bandwidths shown in Fig. 5-29 must be multiplied by two. A multistage i-f amplifier would thus be designed to have a bandwidth of $2 \times 0.5/\tau$, or $1/\tau$ hertz, to optimize pulsed signal detection in noise.

Such a bandwidth choice is common practice in the design of pulse radars. It is also the optimum choice in those digital or PCM systems in which noise is a more crucial factor than intersymbol interference. If intersymbol interference is a problem as well, the system bandwidth may be widened somewhat to narrow the pulses transmitted and hence decrease their overlap. The broad maxima in Fig. 5-29 indicate that the optimum bandwidth choice is not critical in combating noise, and hence widening the bandwidths somewhat will not deteriorate the noise performance too much.[26]

The design of filters to minimize both intersymbol interference and noise will not be considered here. The subject has received increasing attention in recent years and the interested reader is referred to the literature for theoretical investigations of this problem.[27]

The question of i-f matched filtering as contrasted with baseband filtering will be further pursued in the sections following after a discussion of noise representations at high frequencies.

5-7 NARROWBAND NOISE REPRESENTATION

In the discussion of noise through linear systems in Sec. 5-5 we stressed noise passed through low-pass devices. We assumed these low-pass structures to

[26] M. Schwartz, W. R. Bennett, and S. Stein, *Communication Systems and Techniques,* McGraw-Hill, New York, 1966, pp. 310–313.

[27] See, e.g., R. W. Lucky, F. Salz, and E. J. Weldon, Jr., *Principles of Data Communication,* McGraw-Hill, New York, 1968, chap. 5, for a discussion and list of references.

have an effective bandwidth B hertz, centered at 0 Hz (dc), and showed, for example, that the output noise power with white noise applied at the input was proportional to B. In the last section, in discussing the maximization of SNR by matched filtering, we also stressed, for simplicity's sake, low-pass filtering.

We did point out that the results obtained were applicable at carrier frequencies as well, with the equivalent bandpass filters having twice the low-pass bandwidth. As a matter of fact, the discussion in Sec. 5-5 of noise (and random signals) through linear systems is general enough to enable us to handle high-frequency carrier transmission and the various narrowband bandpass circuits encountered in practice. Thus the transfer function $H(\omega)$ that appears in the spectral-density relation for input-output noise,

$$G_{n_0}(f) = |H(\omega)|^2 G_{n_i}(f) \tag{5-50}$$

is, in the case of high-frequency transmission, simply that of a filter centered at the desired center frequency.

We shall find it useful to develop a representation of noise particularly appropriate to narrowband transmission, however. This will enable us to realistically discuss the problem of detecting, at a receiver, high-frequency signals in the presence of noise. In particular, we shall use this representation for narrowband noise to discuss, in a comparative way, the detection process in various types of digital and analog systems.

Thus we shall answer questions related to binary carrier transmission and reception that were first raised earlier in connection with our discussion of digital carrier systems: What *are* the reasons for selecting between OOK, PSK, and FSK transmission in a particular situation? What *are* the quantitative differences between synchronous (coherent) and envelope (noncoherent) detection of binary signals in noise?

In particular, we shall find that PSK systems with synchronous detection offer a distinct improvement in either probability of error or signal-to-noise ratio over the other schemes and are therefore favored *if* phase coherence may be maintained. If envelope detection *must* be used (phase coherence is either not available or the cost does not justify the additional circuitry required to maintain it), FSK is found to be superior to OOK, but again with the requirement of somewhat more complex circuitry.

Using the narrowband representation of noise we shall also discuss AM and FM detection in the presence of noise, obtaining the well-known SNR improvement of wideband FM over AM (this above the so-called FM threshold).

Consider then noise $n(t)$ at the output of a narrowband filter. Its spectral density $G_n(f)$ is centered about f_0 as in Fig. 5-31. For simplicity's sake we shall

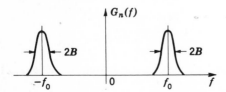

Figure 5-31 Spectral density, narrowband noise.

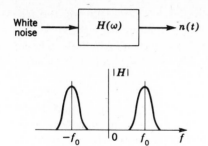

Figure 5-32 Generation of narrowband noise.

assume $G_n(f)$ symmetrical about the frequency f_0, with bandwidth $2B \ll f_0$. (The noise may be assumed to have been generated by white noise passed through a narrowband filter, as shown in Fig. 5-32. These assumptions are not necessary in a more general approach to narrowband noise representation, but are used here to simplify the discussion.[28] They are, of course, frequently encountered in practice.) It is then apparent that the noise $n(t)$, although random, will be oscillating, on the average, at frequency f_0. (As the bandwidth $2B$ is made smaller and smaller, the output should approach more and more that of a pure sine wave at frequency f_0.) We indicate this by writing $n(t)$ in the narrowband form

$$n(t) = r(t) \cos [\omega_0 t + \theta(t)] \tag{5-79}$$

We would expect $r(t)$ and $\theta(t)$ to be varying, in a random fashion, roughly at the rate of B hertz, representing, respectively, the "envelope" and "phase" of the noise. This is indicated in Fig. 5-33.

To actually develop $n(t)$ in the form of Eq. (5-79) we use a simple artifice. We visualize the noise to be represented as the sum of many closely spaced sine waves, the spacing $\Delta f \ll B$. Thus, let

$$n(t) = \sum_{l=-\infty}^{\infty} a_l \cos [(\omega_0 + l \, \Delta\omega)t + \theta_l] \tag{5-80}$$

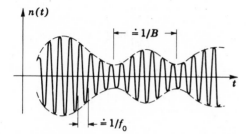

Figure 5-33 Narrowband noise.

[28] See Schwartz, Bennett, and Stein, *op. cit.*, pp. 35–45, for a more systematic and general approach.

The variation about the center frequency f_0 is indicated by measuring the frequency of the different sine waves with respect to f_0. One would thus expect the coefficients a_l to be large in the vicinity of f_0 (i.e., small l), and small elsewhere.

Since we are not interested here in a unique representation of noise, but rather a model that will be useful in analysis, we now assume the θ_l's to be independent, uniformly distributed random variables. By the central limit theorem, $n(t)$ *then has gaussian statistics,* just the property we assumed in the error calculations earlier in this chapter. It is also apparent that $E(n) = 0$, averaging statistically over the random θ_l's. To find the coefficients a_l we now note that the sine-wave expansion of Eq. (5-80) is equivalent to assuming the continuous power spectral distribution of Fig. 5-34a to be replaced by a discrete spectrum of the same shape and power. This is shown in Fig. 5-34b. [We have concentrated on positive frequencies only, as shown, because of the form of Eq. (5-80). We then simply double the power spectral density at each frequency, as shown in the figure. The resultant spectral density, defined for positive frequencies only, is often called the one-sided spectral density, as contrasted to the *two-sided* $G_n(f)$, symmetrical in positive and negative frequencies.]

It is apparent from both figures that the total noise power N must be given by

$$N = 2\int_0^\infty G_n(f)\,df = \sum_{l=-\infty}^\infty 2G_n(f_0 + l\,\Delta f)\,\Delta f \qquad (5\text{-}81)$$

Again $l\,\Delta f$ represents the variation away from the center frequency f_0. Although $l\,\Delta f$ is indicated as ranging between $-\infty$ and ∞, it is of course only significant within the range $\pm B$ about f_0.

Now note that the average power in the representation $n(t)$ of Eq. (5-80) must be given by

$$N = E(n^2) = \sum_{l=-\infty}^\infty \frac{a_l^2}{2} \qquad (5\text{-}82)$$

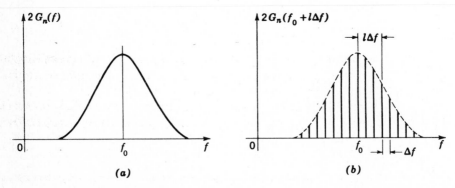

Figure 5-34 Discrete representation of noise spectral density. (a) One-sided (positive-frequency) spectrum. (b) Discrete equivalent.

a familiar result. {One may check this by multiplying together two series representations for $n(t)$ to get $n^2(t)$. Then ensemble averaging over the random variables θ_l, and noting that $E[\cos (l + m) \Delta t + \theta_l + \theta_m] = 0$, one gets Eq. (5-82).} Comparing with Eq. (5-81), we must have

$$a_l^2 = 4G_n(f_0 + l \Delta f) \Delta f$$
$$a_l = \sqrt{4G_n(f_0 + l \Delta f) \Delta f}$$

$$(5\text{-}83)$$

The a_l coefficients are thus uniquely known, once the θ_l's are assumed independent and uniformly distributed.

To get Eq. (5-80) more specifically in the narrowband form of Eq. (5-79), we now expand a typical sine-wave term about f_0 in the following manner:

$$\cos [\omega_0 t + (l \Delta\omega t + \theta_l)] = \cos (l \Delta\omega t + \theta_l) \cos \omega_0 t$$
$$- \sin (l \Delta\omega t + \theta_l) \sin \omega_0 t \quad (5\text{-}84)$$

Grouping the *low-frequency* terms $\cos (l \Delta\omega t + \theta_l)$ and $\sin (l \Delta\omega t + \theta_l)$ together, we then get

$$n(t) = \left[\sum_l a_l \cos (l \Delta\omega t + \theta_l) \right] \cos \omega_0 t$$
$$- \left[\sum_l a_l \sin (l \Delta\omega t + \theta_l) \right] \sin \omega_0 t \quad (5\text{-}85)$$

The resulting *low-frequency sums* shown in brackets we denote as $x(t)$ and $y(t)$, respectively:

$$x(t) = \sum_l a_l \cos (l \Delta\omega t + \theta_l)$$
$$y(t) = \sum_l a_l \sin (l \Delta\omega t + \theta_l)$$

$$(5\text{-}86)$$

We then have, finally,

$$n(t) = x(t) \cos \omega_0 t - y(t) \sin \omega_0 t$$
$$= r(t) \cos [\omega_0 t + \theta(t)]$$

$$(5\text{-}87)$$

as in Eq. (5-79), with

$$r^2 = x^2 + y^2 \qquad \theta = \tan^{-1} \frac{y}{x} \qquad (5\text{-}88)$$

Invoking the central limit theorem it is apparent from Eq. (5-86) that both $x(t)$ and $y(t)$ are gaussian random processes. As a matter of fact, it is readily shown that by writing $x^2(t)$ and $y^2(t)$ as double sums and statistically averaging over the θ_l's,

$$E(x^2) = E(y^2) = E(n^2) = N \qquad (5\text{-}89)$$

{The details are left to the reader, but note, as a hint, that

$$E[\cos (l\Delta\omega t + \theta_l) \cos (m \Delta\omega t + \theta_m)] = 0$$

with θ_l and θ_m *independent*, unless $l = m$, in which case $E(\ ) = \frac{1}{2}$.}

Both the inphase noise term $x(t)$ and the quadratic term $y(t)$ individually have the same variance or power as the original noise $n(t)$. [It is apparent by appropriate averaging that $E(x) = E(y) = 0$.]

In addition, it is readily shown by calculating $E(xy)$, using the same series representations of Eq. (5-86), that x and y are uncorrelated and, being gaussian, *independent*.

It is of interest to discuss the power spectral densities of $x(t)$ and $y(t)$. Comparing Eqs. (5-80) and (5-86), it is apparent that the noise terms $x(t)$ and $y(t)$ may be visualized as the original noise $n(t)$ shifted down to zero frequency. With $n(t)$ a random noise wave, oscillating, with a bandwidth B hertz, about the center frequency f_0, the inphase and quadrature terms $x(t)$ and $y(t)$ are both noise processes centered at dc; hence they are slowly varying at the bandwidth B. The calculation of the spectral densities in fact verifies this. It is readily shown that $G_x(f)$ and $G_y(f)$ are equal, and correspond to the original noise spectral density $G_n(f)$ shifted down to dc (zero frequency). In particular, they are found to be given by

$$G_x(f) = G_y(f) = G_n(f + f_0) + G_n(f - f_0) \quad -f_0 < f < f_0$$
$$= 0 \qquad\qquad\qquad\qquad\qquad \text{elsewhere} \qquad (5\text{-}90)^{29}$$

A sketch of a typical high-frequency noise spectral density and the low-frequency $G_x(f)$ are shown in Fig. 5-35.

If, as a special and very common case, $G_n(f)$ is symmetrical about the carrier frequency f_0, the positive- and negative-frequency contributions of $G_n(f)$ may be simply shifted down to zero frequency and added to give

$$G_x(f) = G_y(f) = 2G_n(f + f_0) \qquad (5\text{-}91)$$

(This is the case where the i-f filtering is symmetrical about f_0.) As an example, if $G_n(f)$ is gaussian-shaped and given by

$$G_n(f) = \frac{N/2}{\sqrt{2\pi\sigma^2}}\, e^{-(f-f_0)^2/2\sigma^2+} + \frac{N/2}{\sqrt{2\pi\sigma^2}}\, e^{-(f+f_0)^2/2\sigma^2} \qquad (5\text{-}92)$$

with $\sigma \ll f_0$, we have

$$G_x(f) = G_y(f) = \frac{N}{\sqrt{2\pi\sigma^2}}\, e^{-f^2/2\sigma^2} \qquad (5\text{-}93)$$

The result is sketched in Fig. 5-36. [Note that σ may be defined as an rms bandwidth. For from the properties of gaussian functions[30], it is apparent that

$$\sigma^2 = \frac{\displaystyle\int_{-\infty}^{\infty} f^2 G_x(f)\, df}{\displaystyle\int_{-\infty}^{\infty} G_x(f)\, df} \qquad (5\text{-}94)$$

[29] Schwartz, Bennett, and Stein, *op. cit.*, pp. 39, 40.

[30] Recall in a probabilistic context that this is just the definition of variance if the average value is zero.

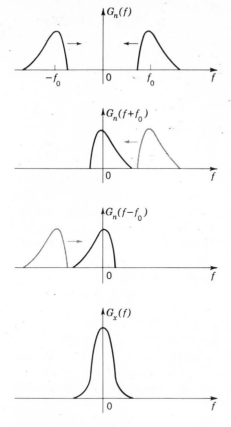

Figure 5-35 Inphase (low-pass) noise spectral density (asymmetric noise spectrum).

This is often used as the definition of bandwidth in spectral analysis, and extends the various possible definitions considered in Chap. 2.]

5-8 BINARY r-f TRANSMISSION COMPARED: SYNCHRONOUS DETECTION OF BINARY SIGNALS IN NOISE

Using the narrowband representation of noise, we are now in a position to study the comparative SNR properties of PSK, OOK, and FSK techniques (see Chap. 4), as well as others if so desired. To do this we shall assume gaussian

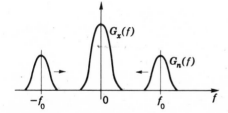

Figure 5-36 Low-pass noise spectral density, symmetrical passband.

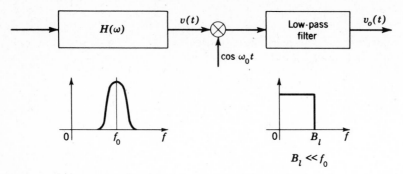

Figure 5-37 Synchronous detector for PSK and OOK signals.

noise added during transmission and assume decisions to be made after synchronous detection at the receiver. We shall then show, using the narrowband representation of noise, that probabilities of error may be written down directly using the error calculations of Sec. 5-1. In the next section we extend this analysis to those cases where envelope detection is used instead.

Recall from Chap. 4 that synchronous detection requires carrier phase coherence to be maintained. The process of synchronous detection consists of multiplication of a received carrier signal by a locally generated sine wave of the same frequency and phase, the resultant product term then passed through a low-pass filter to eliminate second harmonic terms.

For a PSK binary sequence of the form $\pm A \cos \omega_0 t$, or for an OOK sequence consisting of either $A \cos \omega_0 t$ or 0,[31] we simply multiply by $\cos \omega_0 t$ and filter. A synchronous detector for these signals is shown in Fig. 5-37.

For the FSK sequence consisting of $A \cos \omega_1 t$ or $A \cos \omega_2 t$, *two* sets of synchronous detectors are needed, one operating at frequency f_1, the other at frequency f_2. The resultant FSK detector is shown in Fig. 5-38. Note that in

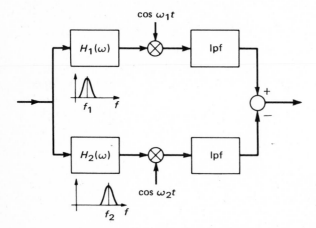

Figure 5-38 Synchronous detection of FSK signals.

[31] Rectangular pulse shaping is assumed here for simplicity. Any shape $f(t)$ could be written in place of A.

both figures predetection (i-f) filters are shown. These are narrowband filters with bandwidths chosen wide enough to pass the respective carrier signals (generally $2B_t$ hertz, if the low-pass filter bandwidths are B_l). Actually, we shall show quite simply that they should be matched filters or reasonable approximations to these if additive noise is the primary source of error in detection.

Now assume that signal plus noise appear at the output of a particular narrowband filter. [This is labeled $v(t)$ in Fig. 5-37.] It is apparent, as pointed out in the discussion of Chap. 4, that all three binary cases are covered by assuming a signal of the form $f(t) \cos \omega_0 t$. [In the PSK case $f(t)$ is $\pm A$, in the OOK case, $+A$ or 0. In the FSK case ω_0 is either ω_1 or ω_2, and $f(t)$ is A if a signal is present in one of the two parallel channels, 0 if it is absent.] The composite signal plus noise at the input to the detector may thus be written

$$v(t) = f(t) \cos \omega_0 t + n(t)$$
$$= [f(t) + x(t)] \cos \omega_0 t - y(t) \sin \omega_0 t \tag{5-95}$$

using the narrowband noise representation of the last section.

The noise $n(t)$ is narrowband in form, its spectral density dependent on the i-f filters $H(\omega)$ shown in Figs. 5-37 and 5-38. The low-pass noise terms $x(t)$ and $y(t)$ thus have half the i-f bandwidth, assumed small compared to the center frequency f_0. Multiplication of $v(t)$ by the locally generated $\cos \omega_0 t$ therefore results in a low-frequency term $[f(t) + x]/2$ passed by the low-pass filter, as well as a double carrier, $2f_0$, term rejected by the filter.

The multiplicative factor of $\frac{1}{2}$ may be assumed absorbed in any gain factors. (For simplicity we have been ignoring these, since signal and noise are then both multiplied by the same constants.) The detector output $v_o(t)$ is then simply

$$v_o(t) = f(t) + x(t) \tag{5-96}$$

Decision circuits then sample the low-pass $v_o(t)$ and decide on the binary symbol transmitted. We showed in the last section, however, that $x(t)$ has the same gaussian statistics, as well as the same variance (power) N as the noise $n(t)$. From the form of Eq. (5-96), it is then apparent that the binary decision problem here is *identical* with that first considered in Sec. 5-1. The $x(t)$ term here is the same as the low-pass $n(t)$ included there!

The process of synchronous detection performs the same operation on noise as on signal. It merely serves to translate the center frequency down, from f_0 to 0 frequency.

In particular, in the OOK case, the detector output is just

$$v_{o,\text{OOK}}(t) = \begin{matrix} A \\ \text{or} \\ 0 \end{matrix} + x(t) \tag{5-97}$$

Since this output is identical with that discussed in Sec. 5-1, it is apparent that the probability of error is just

$$P_{e,\text{OOK}} = \frac{1}{2}\left(1 - \text{erf}\, \frac{A}{2\sqrt{2N}}\right) = \frac{1}{2}\, \text{erfc}\, \frac{A}{2\sqrt{2N}} \tag{5-98}$$

with erf x the error function, and erfc x, the *complementary error function,* given by

$$\text{erfc } x = 1 - \text{erf } x$$

$$\text{erf } x \equiv \frac{2}{\sqrt{\pi}} \int_0^x e^{-y^2} \, dy \tag{5-99}$$

[See Eq. (5-7). We assume here, as in deriving that equation, that 1's and 0's in the binary sequence are equally likely, and that the decision level has been set at $A/2$.]

Figure 5-5, showing the probability of error as a function of A/σ, or $A/\sqrt{N}$ in this case, may thus be used directly in determining the performance of the OOK system in additive gaussian noise. The amplitude A here is just the amplitude of the sine wave representing the 1 (or mark) symbol. Since minimum probability of error corresponds to maximizing $A/\sqrt{N}$, it is apparent that the $H(\omega)$ filter preceding the synchronous detector should be a matched filter (or a good approximation to this), matched to the pulsed sine wave $A \cos \omega_0 t$. Some thought indicates that this is just the low-pass matched filter translated up to the center frequency f_0. This assumes white noise at the input to the i-f filter $H(\omega)$.

Alternatively, a little thought will indicate that the $H(\omega)$ filter may be widened considerably if desired, and the matched filtering performed in the low-pass filter following the multiplier. The synchronous detector merely serves to translate frequencies down, so that the *overall* filtering is effectively due to the cascaded effect of $H(\omega)$ and the low-pass filter. It is this overall filter that should be matched to the signal. The low-pass filter is often called a *post-detection* filter.

In the PSK case the synchronous detector output consists of a polar signal $\pm A$ plus noise. This thus corresponds exactly to the polar signal analysis in Sec. 5-1. Here, however, we have $\pm A$ as the signal, rather than $\pm A/2$, as assumed there. Again choosing 0 as the decision level ($v_o > 0$ is called a 1, $v_o < 0$ a 0 signal), and assuming equally likely binary symbols, the probability of error is just

$$P_{e,\text{PSK}} = \frac{1}{2} \text{ erfc } \frac{A}{\sqrt{2N}} \tag{5-100}$$

As noted in Sec. 5-1, and as is apparent by comparing Eqs. (5-98) and (5-100), the PSK system requires only half the signal amplitude that the OOK system does, for the same probability of error. There is thus a 6-dB peak SNR improvement. On an average power basis, however, the improvement is only 3dB because the OOK system is off half the time, on the average, requiring only half as much power.

In the case of the FSK system the outputs of two detectors are compared. At any one time one detector has signal plus noise, the other noise only.

Calling the noise output of one channel x_1, that of the other x_2, we have, on subtracting the two channel outputs, the FSK output given by

$$v_{o,\text{FSK}} = \begin{matrix} +A \\ \text{or} \\ -A \end{matrix} \; + (x_1 - x_2) \qquad (5\text{-}101)$$

The output signal is again polar: $+A$ appears if a 1 has been transmitted, $-A$ for a 0 transmitted. The total noise output is, however, $x_1 - x_2$. If the noises in the two channels are *independent* [true, if the system input noise is white, and the two bandpass filters $H_1(\omega)$ and $H_2(\omega)$ do not overlap,[32] the usual case], the variances *add*.[33] We have effectively *doubled the noise* by subtracting the two outputs! However, since the output signal is polar, the effective signal excursion, as in the PSK case, is twice that of the OOK case. The FSK system thus provides results intermediate between the OOK and PSK cases:

$$P_{e,\text{FSK}} = \frac{1}{2} \, \text{erfc} \, \frac{A}{2 \, \sqrt{N}} \qquad (5\text{-}102)$$

For a specified probability of error the FSK system requires 3 dB more signal power than the equivalent PSK system with the same noise power, but is 3 dB better than the OOK system on a peak power basis. (Recall from Chap. 4 that FSK requires wider transmission or channel bandwidths than either of the other two systems. The channel bandwidth is measured prior to the H_1 and H_2 filters of Fig. 5-38.)

Again a minimum probability of error requires maximization of the ratio $A/\sqrt{N}$ at the input to the synchronous detector. For this purpose the two filters H_1 and H_2 in Fig. 5-38 should be matched as closely as possible to their respective signal inputs.

It is apparent from this simple analysis of binary signals in noise that PSK transmission is to be preferred if *phase coherence* is available. Some of the deep-space probes have used PSK modulation successfully in their telemetry systems. Rather sophisticated techniques are of course required to establish and maintain the necessary phase synchronism. More commonly, FSK transmission with envelope detection at the receiver has been used in commercial data (teletype) transmission. The analysis of high-frequency binary transmission with envelope detection will consequently be considered in the next section. We shall find there, as expected, that envelope detection results in a somewhat higher probability of error, or a corresponding loss in SNR.

---

[32] Schwartz, Bennett, and Stein, *op. cit.,* pp. 44, 45.

[33] If $y = x_1 - x_2$, and $E(x_1) = E(x_2) = 0$, as here,

$$\text{var} \, (y) = E(y^2) = E(x_1 - x_2)^2 = E(x_1{}^2) - 2E(x_1 x_2) + E(x_1{}^2) = E(x_1{}^2) + E(x_2{}^2)$$

since x_1 and x_2 are independent and hence uncorrelated. More generally, readers are asked to show for themselves that if $y = a_1 x_1 + a_2 x_2$, x_1 and x_2 independent, var $(y) = a_1{}^2$ var $(x_1{}^2) + a_2{}^2$ var $(x_2{}^2)$, *independent* of the first moments or mean values.

The probability of error results appearing in Eqs. (5-98), (5-100), and (5-102) may be written in a unified way by recalling the matched filter result of Sec. 5-6: It was shown there, in the case of the detection of a pulse in noise, that the matched filter SNR output is given by

$$\frac{A^2}{N} = \frac{2E}{n_0} \tag{5-103}$$

[see (5-74)]. Here E represents the energy in the signal at the point where the white gaussian noise of spectral density $n_0/2$ is added.

A little thought will indicate that the pulse to be detected in Sec. 5-6 didn't necessarily have to be baseband in nature. It could just as well have been a high-frequency pulse, corresponding to the OOK case here. The optimum matched filter would then be a bandpass filter, centered at the carrier frequency f_0 and shaped to correspond to the high-frequency pulse shape. This is precisely one of the two possibilities mentioned above in the design of the two filters of Fig. 5-37. The result of Sec. 5-6, given by (5-103), thus goes over directly to the OOK case here: the minimum probability of error result, Eq. (5-98), for OOK transmission may be written by replacing $A/\sqrt{N}$ by the equivalent ratio $\sqrt{2E/n_0}$ from (5-103).

One important comment must be made, however. The energy E in the present case is the *high-frequency* signal energy, measured again at the point where the high-frequency signal and noise are added.

Using a similar argument, $A/\sqrt{N}$ in each of the other cases considered, (5-100) for PSK transmission and (5-102) for FSK transmission, may be replaced by $\sqrt{2E/n_0}$ if matched filtering is used. All three results can then be encompassed by the following single probability-of-error expression:

$$P_e = \frac{1}{2} \operatorname{erfc} \frac{a}{2\sqrt{2}} \tag{5-104}$$

The parameter a written here has the following definitions:

For OOK transmission,

$$\frac{a^2}{8} = \frac{E}{4n_0}$$

For FSK transmission,

$$\frac{a^2}{8} = \frac{E}{2n_0}$$

For PSK transmission,

$$\frac{a^2}{8} = \frac{E}{n_0}$$

E is again the *high-frequency* signal energy at the input to the receiver; $n_0/2$ is the gaussian white noise spectral density measured at the same point.

In the case of PSK or OOK transmission,

$$E = \int_0^T [f(t) \cos \omega_0 t]^2 \, dt \tag{5-105}$$

with T the binary interval over which the pulse is defined, and $f(t)$ the pulse shape. In the case of FSK transmission the same definition of the energy holds, except that ω_0 is replaced by ω_1 and ω_2, the FSK carrier frequencies in rad/s. If the shape factor $f(t)$ is varying slowly compared to the carrier frequencies, $(B \ll f_0, f_1, f_2)$, and many carrier cycles are encompassed in the binary interval T $(Tf_0 \gg 1)$, the energy E is given approximately by

$$E \doteq \frac{1}{2} \int_0^T f^2(t) \, dt \tag{5-106}$$

Independent arguments leading to (5-104) as the error probability expression for OOK, PSK, and FSK transmission, using synchronous detection with matched filtering, appear in several problems at the end of this chapter. A more general approach introduced in Chap. 6 demonstrates that (5-104) is applicable to the case of any two binary wave shapes to be detected in the presence of additive white gaussian noise. Specifically, let $s_1(t)$ be the explicit expression for one signal and $s_2(t)$ the expression for the other, each lasting T seconds in time. Then the minimum probability of error of reception of these two waveshapes, assumed transmitted with equal probability, is found to be given by (5-104), with a^2 defined by the following expression:

$$a^2 \equiv \int_0^T \frac{(s_1 - s_2)^2 \, dt}{n_0/2} \tag{5-107}$$

It is left to the reader to show that the results for OOK, PSK, and FSK transmission are encompassed by this definition.

The unified expression for the minimum binary (bit) error probability, (5-104), is shown plotted in Fig. 5-39.

5-9 NONCOHERENT BINARY TRANSMISSION

If phase coherence cannot be maintained, or if it is just uneconomical to incorporate phase control circuits in the receiver, one generally resorts to envelope detection of high-frequency carriers. The resultant system is generally referred to as noncoherent. The envelope-detection process has been discussed in previous chapters. We limit ourselves here to recalling that if we have a high-frequency sine wave of the form $r(t)(\cos \omega_0 t + \theta)$, with $r(t)$ a positive quantity, the envelope detector provides $r(t)$ at its output. In practice, a nonlinear device plus low-pass filtering is needed to recover $r(t)$.

It is apparent that PSK signals require phase coherence to be demodulated. We therefore consider here only OOK or FSK signals, detected with envelope detectors. As in Fig. 5-38 the FSK receiver consists of *two* channels, one tuned

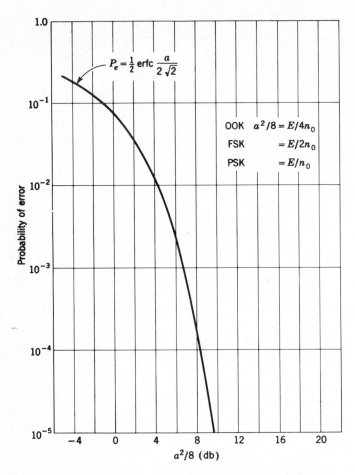

Figure 5-39 Probability of error, optimum binary transmission.

to frequency f_1, the other to frequency f_2. Each synchronous detector in Fig. 5-38 is replaced by an envelope detector. The outputs of the two detectors are then compared to determine whether one binary symbol or the other was transmitted. The FSK envelope-detection scheme is shown in Fig. 5-40b. The OOK receiver consists of one channel, tuned to the carrier frequency of f_0, with one envelope detector providing the desired output. The OOK receiver is shown in Fig. 5-40a. A decision level on the output then decides whether a 1 or a 0 was transmitted.

For simplicity's sake we shall discuss first the simple OOK system. Results for the FSK case are found in a similar manner, by an extension of the analysis. If we assume an OOK sequence of symbols, the output $v(t)$ of a narrowband filter centered at f_0 is again, adding signal and noise,

$$v(t) = [f(t) + x(t)] \cos \omega_0 t - y(t) \sin \omega_0 t \tag{5-108}$$

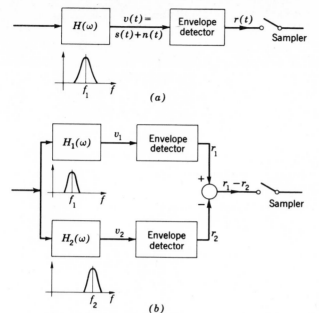

Figure 5-40 Noncoherent envelope detection. (*a*) OOK receiver. (*b*) Noncoherent FSK detection.

Here $f(t) = A$ or 0. Rewriting this in the equivalent form,

$$v(t) = r(t) \cos [\omega_0 t + \theta(t)] \tag{5-109}$$

with $\qquad r = \sqrt{(f + x)^2 + y^2} \qquad$ and $\qquad \theta = \tan^{-1} \dfrac{y}{f + x}$

it is apparent that an envelope detector will produce $r(t)$ at its output. Sampling $r(t)$ once each binary period, we then make a decision about whether a 1 or a 0 is present by noting whether the sampled $r(t) > b$, or $r(t) < b$, respectively, b a specified decision level.

The probability of error depends on the statistics of r in the two cases, $f = A$ or 0. We develop the probability density function for these two cases in the paragraph following.

Rayleigh and Rician Statistics

Consider first the case where the signal is absent. This is the noise-only case with $A = 0$. With x and y independent and gaussian the problem is to determine the statistics of the random envelope r. We do this by first finding the statistics, jointly, of r and θ, and then integrating out over θ to find the density function of r.

A little thought will indicate that from the definitions of r and θ, in (5-109), in terms of x and y, one can write $x = r \cos \theta, y = r \sin \theta$. The variables x and y correspond to two-dimensional rectangular coordinates, the variables r and θ, to the equivalent polar coordinates (Fig. 5-41). There is thus a unique transformation from one set of variables to the other. We use the known transformations to find the probability-density functions for r and θ. This is the extension to two dimensions (two random variables) of the one-dimensional transformation shown in the Appendix.

In the one-variable case we equate areas under the probability-density curves. Here too we equate probabilities:

$$\text{Prob } (x_1 < x < x_2, \; y_1 < y < y_2) = \text{Prob } (r_1 < r < r_2, \; \theta_1 < \theta < \theta_2) \quad (5\text{-}110)$$

This corresponds, however, to equating volumes under the joint probability-density curves. As noted above we are effectively converting from rectangular (x,y) coordinates to polar (r,θ) coordinates in this case. With $f_{r\theta}(r,\theta)$ the probability-density function for the polar coordinates, we must have

$$f_{xy}(x,y) \; dx \; dy = f_{r\theta}(r,\theta) \; dr \; d\theta \quad (5\text{-}111)$$

From the independence and gaussian statistics of x and y,

$$f_{xy}(x,y) = f_x(x)f_y(y) = \frac{e^{-(x^2+y^2)/2\sigma^2}}{2\pi\sigma^2} = \frac{e^{-r^2/2\sigma^2}}{2\pi\sigma^2} \quad (5\text{-}112)$$

using σ^2 for the variances and $r^2 = x^2 + y^2$. Transforming differential areas, we have

$$dx \; dy = r \; dr \; d\theta \quad (5\text{-}113)$$

(see Fig. 5-41).

From Eq. (5-111), then, with Eqs. (5-112) and (5-113),

$$f_{r\theta}(r,\theta) \; dr \; d\theta = \frac{re^{-r^2/2\sigma^2}}{2\pi\sigma^2} \; dr \; d\theta \quad (5\text{-}114)$$

and

$$f_{r\theta}(r,\theta) = \frac{re^{-r^2/2\sigma^2}}{2\pi\sigma^2} \quad (5\text{-}115)$$

To find the density function $f_r(r)$ for the envelope alone, we simply average Eq. (5-115) over all possible phases. Since the phase angle θ varies between 0 and 2π, we get

$$f_r(r) = \int_0^{2\pi} f_{r\theta}(r,\theta) \; d\theta = \frac{re^{-r^2/2\sigma^2}}{\sigma^2} \quad (5\text{-}116)$$

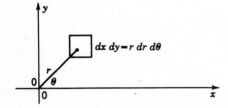

Figure 5-41 Rectangular and polar coordinates.

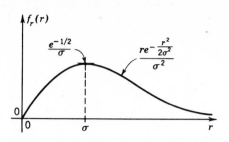

Figure 5-42 Rayleigh distribution.

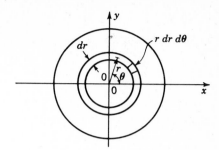

Figure 5-43 A target.

This is called the *Rayleigh distribution* and is shown sketched in Fig. 5-42. The peak of this distribution occurs at $r = \sigma$ and is equal to $e^{-1/2}/\sigma$. As σ (the standard deviation of the gaussian variables x and y) increases, the distribution flattens out, the peak decreasing and moving to the right. It is easily seen that $f_r(r)$ is properly normalized, so that $\int_0^\infty f_r(r)\, dr = 1$. Note that the normalization is from 0 to ∞ here, instead of from $-\infty$ to ∞ for the gaussian distribution. The envelope can have only *positive* values.

The Rayleigh distribution appears in many other applications of statistics. Another simple example involves the firing of bullets at a target. Assume that the distribution of the bullets hitting the target is gaussian along the horizontal, or x, axis of the target and also gaussian along the vertical, or y, axis (i.e., a two-dimensional gaussian distribution) (Fig. 5-43). The average location of the hits is at the origin, and the standard deviation in each direction is σ. The probability that the hits will lie within an annular ring dr units wide and r units from the origin is just $f_r(r)\, dr$, with $f_r(r)$ the Rayleigh distribution. [Note from Fig. 5-43 that, if $f_{r\theta}(r,\theta)\, dr\, d\theta$ represents the probability that the bullets will lodge within the differential area $r\, dr\, d\theta$, the probability that the bullets will lodge within the concentric annular ring is found by integrating over all values of θ. This is exactly what was done in Eq. (5-116).]

The envelope of the noise-only case ($A = 0$) thus obeys Rayleigh statistics. Now consider the case where the signal is present during OOK transmission. What is the probability distribution of the envelope in this case? Equation (5-108) is now given by

$$v(t) = [A + x(t)]\cos \omega_0 t - y(t)\sin \omega_0 t \qquad (5\text{-}117)$$

where x and y are the previous gaussian-distributed terms, each with variance σ^2. Considering the term $x + A$ alone, we note that the sum represents a gaussian variable with A the average value and σ^2 still the variance. Calling the sum a new parameter x',

$$x' \equiv x + A \qquad (5\text{-}118)$$

we have[34]

$$f(x') = \frac{e^{-(x'-A)^2/2\sigma^2}}{\sqrt{2\pi\sigma^2}} \tag{5-119}$$

The envelope of $v(t)$ is now given by

$$r^2 = x'^2 + y^2 = (x + A)^2 + y^2 \tag{5-120}$$

and the phase is

$$\theta = \tan^{-1}\frac{y}{x'} = \tan^{-1}\frac{y}{x + A} \tag{5-121}$$

We can again find the probability distributions for both the envelope and phase by a transformation to polar coordinates. This will give us the probability distribution at the output of an envelope detector, as well as the distribution at the output of a phase detector.

With x' and y independent random variables related to r and θ by the transformations $x' = r\cos\theta$, $y = r\sin\theta$, we have

$$f(r,\theta)\,dr\,d\theta = f(x',y)\,dx'\,dy = \frac{e^{-[(x'-A)^2+y^2]/2\sigma^2}}{2\pi\sigma^2}\,dx'\,dy$$

$$= \frac{e^{-A^2/2\sigma^2}re^{-(r^2-2rA\cos\theta)/2\sigma^2}}{2\pi\sigma^2}\,dr\,d\theta \tag{5-122}$$

Note that we cannot write $f(r,\theta)$ as a product $f(r)f(\theta)$, since a term in the equation appears with both variables multiplied together as $r\cos\theta$. This indicates that r and θ are *dependent* variables. They are connected together in this case by the term $rA\cos\theta$. This is apparent from Eqs. (5-120) and (5-121) as well as Eq. (5-122). If $A \to 0$, the two variables again become independent and $f(r,\theta)$ reduces to the product $f(r)f(\theta)$ found for the zero-signal case.

We can find $f(r)$ again by integrating over all values of θ. This gives us

$$f(r) = \frac{e^{-A^2/2\sigma^2}\,re^{-r^2/2\sigma^2}}{2\pi\sigma^2}\int_0^{2\pi} e^{(rA\cos\theta)/\sigma^2}\,d\theta \tag{5-123}$$

The integral in Eq. (5-123) cannot be evaluated in terms of elementary functions. Note, however, its similarity to the defining integral for the Bessel function of the first kind and zero order given by Eq. (4-73). It is in fact related to the Bessel function of the first kind. In particular,

$$I_0(z) \equiv \frac{1}{2\pi}\int_0^{2\pi} e^{z\cos\theta}\,d\theta \tag{5-124}$$

is called the *modified* Bessel function of the first kind and zero order. In terms of $I_0(z)$, Eq. (5-123) becomes

$$f(r) = \frac{r}{\sigma^2}\,e^{-(r^2+A^2)/2\sigma^2}\,I_0\left(\frac{rA}{\sigma^2}\right) \tag{5-125}$$

---

[34] We henceforth drop the subscript on the density functions for ease in printing.

The modified Bessel function can be written as an infinite series, just as in the case of the Bessel function of the first kind. This series can be shown to be

$$I_0(z) = \sum_{n=0}^{\infty} \frac{z^{2n}}{2^{2n}(n!)^2} \tag{5-126}$$

For $z \ll 1$,

$$I_0(z) \doteq 1 + \frac{z^2}{4} + \cdots \doteq e^{z^2/4} \tag{5-127}$$

Letting $A \to 0$ in Eq. (5-125) we get the Rayleigh distribution again, checking our previous result for the zero-signal case.

The envelope distribution of Eq. (5-125) is often called the *Rician distribution* in honor of S. O. Rice of Bell Telephone Laboratories, who developed and discussed the properties of this distribution in a pioneering series of papers on random noise.[35]

We have indicated that for $A^2/2\sigma^2 \to 0$, the envelope of the received signal follows the Rayleigh distribution. For large $A^2/2\sigma^2$, however, it again approaches the original gaussian distribution of the inphase (x) signal term. This is apparent from Eq. (5-117) or Eq. (5-120). For as A increases relative to σ, the inphase term dominates, the quadrature term contribution to the envelope becomes negligible, and the envelope becomes just

$$r \doteq A + x \qquad A \gg \sigma \tag{5-128}$$

This is just a gaussian random variable with average value A_c.

This can also be demonstrated directly from the Rician density function [Eq. (5-125)] itself. To show this we make use of a known property of the modified Bessel function—that is, that it approaches asymptotically, for large values of the argument, an exponential function. Thus for $z \gg 1$,

$$I_0(z) \doteq \frac{e^z}{\sqrt{2\pi z}} \tag{5-129}$$

Letting $rA \gg \sigma^2$ in Eq. (5-125), we make use of Eq. (5-129) to put $f(r)$ in the form

$$f(r) \doteq \frac{r}{\sqrt{2\pi rA\sigma^2}} \, e^{-(r-A)^2/2\sigma^2} \tag{5-130}$$

This function peaks sharply about the point $r = A$, dropping off rapidly as we move away from this point. Most of the contribution to the area under the $f(r)$ curve (or the largest values of the probability of a range of r occurring) comes from points in the vicinity of $r = A$. In this range of r, then, we can let $r = A$ in the nonexponential (and slowly varying) portions of $f(r)$ and get

$$f(r) = \frac{1}{\sqrt{2\pi\sigma^2}} \, e^{-(r-A)^2/2\sigma^2} \qquad rA \gg \sigma^2 \tag{5-131}$$

[35] S. O. Rice, "Mathematical Analysis of Random Noise," *Bell System Tech. J.*, vol. 23, pp. 282–333, July 1944; vol. 24, pp. 96–157, January 1945 (reprinted in N. Wax, *Selected Papers on Noise and Stochastic Processes,* Dover, New York, 1954).

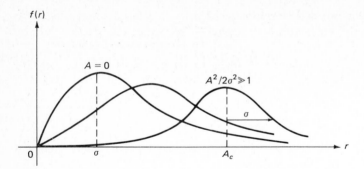

Figure 5-44 Rician distribution.

In the vicinity of the point $r = A$, then, the distribution approximates the normal (gaussian) distribution as noted above.

The Rician distribution is shown plotted in Fig. 5-44 for different values of $A^2/2\sigma^2$.

Probability of Error Calculations

We now use the Rayleigh and Rician distributions just derived to determine the probability of error in the case of envelope detection. The two statistical distributions are rewritten below for convenience. The noise power N has now been written in place of the variance σ^2.

1. Noise-only case, $A = 0$:

$$f_n(r) = \frac{re^{-r^2/2N}}{N} \qquad r \geq 0 \tag{5-132}$$

2. Signal-plus-noise case, $A > 0$:

$$f_s(r) = \frac{re^{-r^2/2N}}{N}e^{-A^2/2N}I_0\left(\frac{rA}{N}\right) \qquad r \geq 0 \tag{5-133}$$

(The subscript n stands for noise, s for signal plus noise.) Recall again that $I_0(x)$ is the modified Bessel function. N is of course the noise variance or the mean noise power measured at the output of the narrowband filter $H(\omega)$ in Fig. 5-40a. (Here $A^2/2$ is the average signal power with a 1 transmitted. $A^2/2N$ is thus a power signal-to-noise ratio.)

A typical sketch of noise at the output of the narrowband filter is shown in Fig. 5-45. Note that it resembles a sine wave at frequency f_0, slowly varying in amplitude and phase at a rate determined by the filter bandwidth. The envelope $r(t)$ shown dashed in Fig. 5-45 has the Rayleigh statistics of Eq. (5-132). The functional form of the noise shown is of course given by Eq. (5-108) with $A = 0$. The random phase angle $\theta(t)$ is uniformly distributed between 0 and 2π.

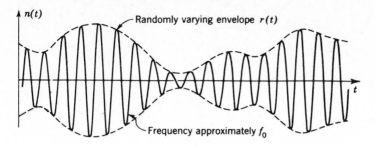

Figure 5-45 Noise at output of narrowband filter.

Using the two density functions, that of Eq. (5-132) for noise only (a 0 or *space* transmitted), and that of Eq. (5-133) for signal plus noise (a 1 or *mark* transmitted), we are now in a position to calculate the probability of error for OOK signaling, with envelope detection. The analysis is similar to that of the error analysis of Sec. 5-1, the only difference being that we must now consider envelope statistics rather than the gaussian statistics assumed there.

Here we decide on a 0 transmitted if $r < b$, a 1 transmitted if $r > b$, as noted earlier. The decision level b thus corresponds to the decision level d of Sec. 5-1. Although d could take on any value, positive or negative, b is of course restricted to positive values only because of the envelope characteristics. Figure 5-46 shows the two decision regions introduced by defining the decision level b.

Assume, as in Sec. 5-1, that the a priori probabilities of transmitting a 0 and a 1 are, respectively, P_0 and $P_1 = 1 - P_0$. The overall probability of error is then given by

$$P_e = P_0 \int_b^\infty f_n(r) \, dr + P_1 \int_0^b f_s(r) \, dr \tag{5-134}$$

with the Rayleigh-density function of Eq. (5-132) used in the first integral and the Rician-density function of Eq. (5-133) used in the second. The two integrals are indicated by the crosshatching in Fig. 5-46.

Although the first integral is directly evaluable, giving $e^{-b^2/2N}$, the second integral cannot be evaluated in closed form. It has, however, been numerically

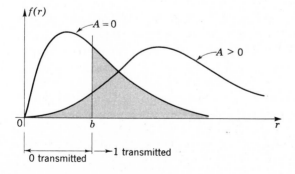

Figure 5-46 Decision regions with envelope-detected OOK signals.

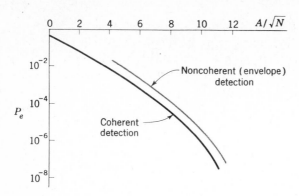

Figure 5-47 Binary error probabilities, OOK transmission.

evaluated and tabulated by numerous investigators.[36] Carrying out the indicated integrations of Eq. (5-134) in the special case where $P_0 = P_1 = \frac{1}{2}$, and the two integrals are the same (i.e., the two crosshatched areas of Fig. 5-46 are the same), one obtains the curve of Fig. 5-47. Also shown is the comparable curve as obtained earlier for coherent (synchronous) detection. Note that the envelope-detection process is somewhat inferior to coherent detection. For the same probability of error P_e it requires somewhat more signal power (higher $A/\sqrt{N}$), or for the same SNR $A/\sqrt{N}$, P_e is somewhat larger. This is as might be expected since we are essentially throwing away useful information by ignoring the phase in the envelope-detection case.

Actually one can again optimize the decision level b, choosing it to minimize the probability of error P_e from Eq. (5-134), as was done in Sec. 5-1. The result is not too different from that of Fig. 5-47. (As will be shown in Chap. 6 the optimum corresponds to the point at which the two density functions in Fig. 5-46 intersect. For $A^2/N \gg 1$, which is necessary to ensure low probability of error, that optimum point does not differ substantially from the values of b used in calculating Fig. 5-47. A discussion of the optimum b, and the effect on the probability of error in deviations from it, appears in the reference.[37])

Note one very important point, however. Again *the probability of error depends uniquely on the ratio $A/\sqrt{N}$*, the ratio of *peak signal* at the *narrowband filter output* to the *rms noise* measured *at the same point*. The only way to decrease the probability of error is to increase $A/\sqrt{N}$. How does one do this, aside from the obvious ways of increasing transmitter power and trying to decrease the additive noise? It is apparent that a matched filter is again called for. The *narrowband filter* $[H(\omega)]$ should thus be *matched to the signal* representing the "1" symbol or mark, in this case a sinusoidal burst at frequency f_0, lasting the binary interval.

A word of caution should again be injected here, however. As noted earlier, in first discussing matched filters, we are neglecting intersymbol interference

[36] See Schwartz, Bennett, and Stein, *op. cit.*, pp. 27, 28; 289, 290; appendix A.

[37] Schwartz, Bennett, and Stein, *op. cit.*, pp. 291, 292.

and are assuming that additive noise is the prime culprit in causing errors in detection. The point made in first introducing matched filters can now be repeated. In all digital transmission systems with additive gaussian noise the prime cause of detection errors, matched filters or reasonable approximations to them will normally be used to minimize the probability of error. This includes baseband digital systems, coherent or synchronous carrier systems (as studied in the previous section), the FSK system to be discussed briefly in the paragraph following, search radar systems, etc. If individual pulse decisions may be considered, independent of streams of data preceding or following the pulse in question, a matched filter is usually called for. If intersymbol interference is the major problem, however, the filter design of Chap. 3, or that presented in the references, should be considered instead.[38]

In practice, of course, the matched-filter condition is usually met by designing the filter to have the appropriate bandwidth. This is usually the reciprocal of the signal pulse width (or the reciprocal of the binary interval), as noted earlier in discussing matched filters. Again, as noted earlier, some widening of the bandwidth may be tolerated and in fact is often employed to decrease pulse overlap (intersymbol interference) without affecting the SNR critically.

How does one now determine the probability of error of the noncoherent, or envelope-detected, FSK system of Fig. 5-40b? Assume here that frequency f_1 corresponds to a 1 (mark) transmitted, f_2 to a 0 (space). If a mark is actually transmitted, signal plus noise will appear on channel 1, noise alone on channel 2. The sampled value of $(r_1 - r_2)$ should be positive for a correct decision to be made. Similarly, $(r_1 - r_2)$ should be negative if a space is transmitted. (Note that although phase synchronism is no longer assumed between transmitter and receiver, binary-interval synchronism must be maintained. This is true as well with the OOK receiver of Fig. 5-40a.)

An error will obviously be made if $(r_1 - r_2)$ is negative, with a mark transmitted, or positive with a space transmitted. If 1's and 0's are assumed equally likely to be transmitted, the probability of either type of error is the same by the symmetry of the system of Fig. 5-40b. Assuming as an example a mark transmitted, the probability of error is just the probability that the noise causing r_2 will exceed the signal-plus-noise envelope signal r_1. This is given by

$$P_e = \int_{r_1=0}^{\infty} f_s(r_1) \left[\int_{r_2=r_1}^{\infty} f_n(r_2)\, dr_2 \right] dr_1 \qquad (5\text{-}135)$$

with the density functions of Eqs. (5-132) and (5-133) used where indicated. [The inner integral provides the probability of error for a *fixed* value of r_1. Averaging over all possible values of r_1, one then obtains Eq. (5-135).] Since $f_n(r)$ is just the Rayleigh density, the inner integral integrates readily to $e^{-r_1^2/2N}$. The expression for the probability of error is then just

$$P_e = \int_0^{\infty} \frac{r_1}{N}\, e^{-r_1^2/N} e^{-A^2/2N} I_0\left(\frac{r_1 A}{N}\right) dr_1 \qquad (5\text{-}135a)$$

[38] Lucky, Salz, and Weldon, *op. cit.*

To integrate this expression we use a simple trick. We define a new (dummy) variable $x = \sqrt{2}\,r_1$. Equation (5-135a) then becomes, with a little manipulation,

$$P_e = \tfrac{1}{2}e^{-A^2/4N}\int_0^\infty \frac{x}{N}\, e^{-x^2/2N}e^{-A^2/4N}I_0\left(\frac{xA}{\sqrt{2}\,N}\right) dx \qquad (5\text{-}135b)$$

But the integrand is exactly in the form of the Rician-density function of Eq. (5-133) if A there is replaced by $A/\sqrt{2}$ here. The integral must then just equal 1, and we have, quite simply,

$$P_e = \tfrac{1}{2}e^{-A^2/4N} \qquad (5\text{-}136)$$

The probability of error of the noncoherent FSK system thus decreases exponentially with the power SNR $A^2/2N$. The error curve corresponding to Eq. (5-136) is plotted in Fig. 5-48, together with the corresponding curves for

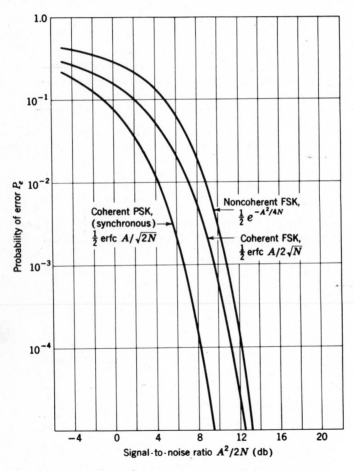

Figure 5-48 Performance curves, binary transmission.

synchronous (coherent) PSK and FSK transmission, using the results of the previous section and Sec. 5-1. Note that just as in the case of noncoherent versus coherent OOK (Fig. 5-47), there is a penalty paid for using envelope rather than synchronous detection. The noncoherent systems require slightly more signal power for the same probability of error.

This loss in SNR due to envelope detection becomes negligible at high SNR, however, as is apparent from Fig. 5-48, and as may be shown from the synchronous-detection results of the last section. For it is readily shown[39] that the asymptotic ($x \gg 1$) form for the complementary error function is given by

$$\operatorname{erfc} x \doteq \frac{e^{-x^2}}{x \sqrt{\pi}} \qquad x \gg 1 \tag{5-137}$$

The coherent FSK error probability derived in the last section and indicated in Fig. 5-48 is then just

$$P_e = \frac{1}{2} \operatorname{erfc} \frac{A}{2 \sqrt{N}} \doteq \frac{1}{\sqrt{\pi} \, A/2 \sqrt{N}} \frac{e^{-A^2/4N}}{2} \tag{5-138}$$

Note that the dominating exponential behavior is just that of the noncoherent FSK expression of Eq. (5-136). At high $A^2/2N$ the other terms in the denominator rapidly become negligible and noncoherent and coherent FSK approach one another.

It is similarly possible to show that the asymptotic (high SNR) error probability for both coherent and noncoherent OOK is given by[40]

$$P_e \doteq \tfrac{1}{2} e^{-A^2/8N} \tag{5-139}$$

Comparing this with Eqs. (5-136) and (5-138), it is apparent that the OOK systems require twice the SNR ($A^2/2N$) that the FSK systems require. Since the OOK signals use power only during mark transmission, the two systems achieve the same error rate at the same *average signal power*.

Again, minimum probability of error is ensured by maximizing $A^2/2N$, and hence using matched filters. In the case of noncoherent FSK this means that the two narrowband filters (Fig. 5-40b) are *each* matched to their respective channel.

The stress here in the binary FSK case has been on two-filter receivers; the two frequencies transmitted are separately detected, as in Figs. 5-38 and 5-40b. In practice, discriminators and zero-crossing detectors (Chap. 4) are commonly used for binary FM detection, as well as for analog FM. The error calculations for these detectors become rather involved and are left to the references.[41]

[39] See, for example, B. O. Peirce, *A Short Table of Integrals*, Ginn, Boston, 1929, p. 29.

[40] Schwartz, Bennett, and Stein, *op. cit.*, p. 292.

[41] W. R. Bennett and J. R. Davey, *Data Transmission*, McGraw-Hill, New York, 1965, chap. 9; Lucky, Salz, and Weldon, *op. cit.*, chap. 8.

5-10 SIGNAL-TO-NOISE RATIOS IN FM AND AM

We alluded in Chap. 4 to the fact that wideband FM gives a significant improvement in noise or interference rejection over AM. We would now like to demonstrate this statement and see just how well, and under what conditions, FM provides an improvement over AM.

In previous sections we were able to determine system performance quite uniquely by calculating probabilities of errors. An analogous approach is more difficult to adopt here because of the *continuous* nature of the signals with which we deal. (The calculation of probabilities of errors requires discrete levels of transmission.) Instead, we shall use here a less satisfactory, although still useful, approach: continuous-wave (c-w) systems such as AM and FM will be compared on the basis of SNR at the receiver input and output.[42] This will essentially hinge on a comparative discussion of the detection process: the envelope detector in the AM case, and the discriminator in the FM case. By comparing SNR at the input and output of the detector in the two cases, we shall find that the wide-band FM system produces the well-known SNR improvement with bandwidth, while in the AM system the input and output SNR can at best be the same.[43]

This exchange of bandwidth for SNR in the case of FM, with SNR improvement obtained at the expense of increased transmission bandwidth, is not as efficient, however, as the optimum exponential exchange predicted by the Shannon capacity expression. Since the form of the information-capacity expression for PCM is similar to that of the Shannon optimum, PCM systems provide a SNR bandwidth exchange much more efficient than that for FM. For the purpose of comparison of FM and AM we shall assume the same carrier power and noise spectral density at the input to each system. We shall calculate the SNR at the system outputs and compare.

Amplitude Modulation

A typical AM receiver to be analyzed is shown in Fig. 5-49. The amplitude-modulated carrier at the input to the envelope detector has the form

$$v(t) = A_c[1 + mf(t)] \cos \omega_0 t \qquad |mf(t)| \le 1 \qquad (5\text{-}140)$$

---

[42] Other measures of c-w system performance, including, for example, mean-squared error between a random signal input and the noisy output of a receiver, are considered in J. M. Wozencraft and I. M. Jacobs, *Principles of Communication Engineering*, Wiley, New York, 1965, chap. 8; see also D. J. Sakrison, *Communication Theory: Transmission of Waveforms and Digital Information*, Wiley, New York, 1968, chaps. 7 and 9, and A. J. Viterbi, *Principles of Coherent Communication*, McGraw-Hill, New York, 1966, part II.

[43] D. Middleton, *Introduction to Statistical Communication Theory*, McGraw-Hill, New York, 1960, and J. L. Lawson and G. E. Uhlenbeck, *Threshold Signals*, McGraw-Hill, New York, 1950, contain additional extensive discussions of both AM and FM. See also Schwartz, Bennett, and Stein, *op. cit.*, chap. 3.

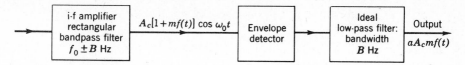

Figure 5-49 Idealized AM receiver.

if noise is assumed absent. A_c is the unmodulated carrier amplitude measured at the same point.

If the modulating signal $f(t)$ is band-limited to B hertz, the intermediate-frequency (i-f) amplifier preceding the detector should have a bandwidth of at least $2B$ Hz, centered about f_0 hertz. The amplifier is assumed to have the characteristics of an ideal rectangular filter.

This AM signal is now envelope-detected and passed through an ideal filter B hertz wide. As shown in Chap. 4, the output of the envelope detector $f_d(t)$ will be proportional to $f(t)$, or

$$f_d(t) = aA_c m f(t) \tag{5-141}$$

with a a constant of proportionality of the detector. (This constant will henceforth be set equal to 1, since amplifiers can always be introduced to change the gain arbitrarily. In addition, we shall be interested in *ratios* of signal to noise, and such constants cancel out anyway.)

The actual envelope-detection process was shown in Chap. 4 to be that of a nonlinear operation on the input AM signal followed by a low-pass filter as in Fig. 5-49. Two types of nonlinearity were examined in Chap. 4: a piecewise-nonlinear characteristic and one possessing smooth curvature. Both were shown to provide the desired envelope-detected output term. It turns out that the signal-to-noise analysis is most readily carried out assuming a detector with quadratic curvature. We shall for this reason concentrate on such a *quadratic envelope detector* here. Analyses for other types of detectors lead to similar results and so will not be considered in detail here.[44]

Our study of the signal-to-noise properties of AM detection is simplified still further by considering the rather artificial case of an *unmodulated* carrier in the presence of additive noise. This zero-modulating-signal case is a common artifice (we shall use the same approach in discussing FM noise), and is useful because the major results found apply in the modulated carrier case as well.[45] Interestingly, this implies modifying the low-pass filter somewhat. The envelope output of Eq. (5-141) assumes zero dc transmission at the filter, blocking the unmodulated carrier portion of the AM wave of Eq. (5-140). For the unmodulated carrier model we shall be considering, however, it is just this blocked term that will represent the output signal. But this apparent inconsis-

[44] Middleton, *op. cit.;* W. B. Davenport, Jr., and W. L. Root, *Introduction to Random Signals and Noise,* McGraw-Hill, New York, 1958.

[45] Middleton, *op. cit.;* Davenport and Root, *op. cit.;* Schwartz, Bennett, and Stein, *op. cit.*

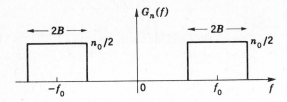

Figure 5-50 Noise spectral density at i-f amplifier output.

tency notwithstanding, the approach does provide an understanding of, and useful quantitative information about, the signal-to-noise properties of AM detection.

Assume accordingly for a signal-to-noise analysis that the instantaneous voltage $v(t)$ at the detector input (i-f output) consists of the unmodulated carrier portions of Eq. (5-140) plus gaussian noise $n(t)$. Then, as in previous sections, we may write $v(t)$ in the form

$$
\begin{aligned}
v(t) &= A_c \cos \omega_0 t + n(t) \\
&= (x + A_c) \cos \omega_0 t - y(t) \sin \omega_0 t \\
&= r(t) \cos [\omega_0 t + \theta(t)]
\end{aligned}
\tag{5-142}
$$

Here

$$r^2 = (x + A_c)^2 + y^2$$

and

$$\theta = \tan^{-1} \frac{y}{x + A_c}$$

With the noise assumed white at the input to the narrowband rectangular-shaped i-f filter, the noise spectral density $G_n(f)$ at the filter output must have the rectangular shape shown in Fig. 5-50. The mean power N is thus given by

$$E(n^2) = N = \int_{-\infty}^{\infty} G_n(f)\, df = 2n_0 B$$

We now assume $v(t)$ passed through a quadratic envelope detector, and ask for the SNR at the low-pass filter output (Fig. 5-51). Recalling from Chap. 4 that the quadratic envelope detector squares the input $v(t)$ and then passes low-frequency components only ($f \ll f_0$), we have, at the low-pass filter output,

$$z = v_{\text{lowpass}}^2 = \frac{r^2}{2} = \frac{(x + A_c)^2 + y^2}{2}
\tag{5-143}$$

from Eq. (5-142). (Recall that x, y, and r are *slowly varying* random functions, with bandwidth $B \ll f_0$.) We have again ignored a detector constant with dimensions of volts/volts^2, since we shall shortly take the ratio of signal to noise powers, the constant then dropping out anyway. We shall in fact henceforth

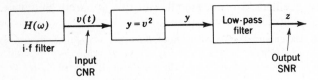

Figure 5-51 Quadratic envelope detector.

ignore the $\frac{1}{2}$ factor appearing in front of r^2, since it is, similarly, an immaterial constant for our purposes.

It is apparent that the SNR analysis of the AM detector must be tied up with the statistics of the envelope $r(t)$. The properties of the Rician distribution discussed earlier can in fact be utilized in determining the output SNR. The use of the quadratic detector, however, enables us to find the output SNR much more simply in terms of the statistics of the gaussian-distributed x and y.

Specifically, we note the randomly varying output voltage z has a signal-dependent part (providing the output signal term), as well as a noise-dependent part. We now *define* the output SNR S_o/N_o to be the ratio of the output signal power in the absence of noise to the mean noise power at the output.[46] Since z is a *voltage,* z^2 must be used to find the powers.

Setting $n(t)$ in Eq. (5-142), or, equivalently, x and y in Eq. (5-143), equal to zero, it is apparent the output signal power in the absence of noise is

$$S_o = A_c{}^4 \tag{5-144}$$

(Recall that we are neglecting the immaterial constant $\frac{1}{2}$.)

The mean noise power at the output must then be the average power or second moment of the random variable z less the signal term:

$$N_o = E(z^2) - A_c{}^4 \tag{5-145}$$

From Eq. (5-143), with the $\frac{1}{2}$ factor again dropped,

$$E(z^2) = E(r^4) = E[(x + A_c)^2 + y^2]^2 \tag{5-146}$$

It is here that the simplicity of the quadratic detector becomes apparent. For the desired second moment of the output z is for this detector type just the fourth moment of the envelope r, found either by appropriate integration of the Rician density function, or, more simply, by expanding the right-hand side of Eq. (5-146) and taking the indicated average of the *gaussian* variables term by term. In this latter case we make use of the following identities:

1. The mean noise power at the i-f output, as shown previously, is

$$E(x^2) = E(y^2) = N$$

2. A known property of gaussian functions (this is easily checked either by direct integration, or by the moment-generating property of characteristic functions[47]) is

$$E(x^4) = E(y^4) = 3N^2$$

---

[46] This is unfortunately not a unique definition. This is one of the difficulties with using a SNR formulation at the output of nonlinear devices. However, other possible definitions provide similar results so that the specific definition of SNR to be used is not critical. One simply selects a definition simple enough to be evaluable and yet meaningful at the same time. See Schwartz, Bennett, and Stein, *op. cit.,* pp. 102–120, for various approaches.

[47] A. Papoulis, *Probability, Random Variables, and Stochastic Processes*, McGraw-Hill, New York, 1965.

3. By assumption

$$E(x) = E(y) = 0$$

4. Since x and y are *independent* gaussian variables,

$$E(x^2 y^2) = E(x^2)E(y^2) = N^2$$

Carrying out the indicated averaging of Eq. (5-146) then (the details are left as an exercise for the reader), we get the very simple expression

$$E(z^2) = E(r^4) = 8N^2 + 8NA_c^2 + A_c^4 \tag{5-147}$$

The mean noise power at the envelope-detector output is thus

$$N_o = \underbrace{8N^2}_{n \times n} + \underbrace{8NA_c^2}_{s \times n} \tag{5-148}$$

Note the two terms appearing. The first is often called the $n \times n$ term, which is due, as we shall see shortly, to the detector input noise nonlinearly beating with itself. The second is the so-called $s \times n$ term, which is due to the noise non-linearly mixing with the carrier (or with a modulating signal if present). This second term obviously disappears when the unmodulated carrier goes to zero. Interestingly, this also predicts an *increase* in output noise level when an un-modulated AM carrier is tuned in. This is in fact easily noticed on commercial AM receivers. An opposite effect, a noise *suppression* or quieting effect, will be found to occur in FM.

Combining Eqs. (5-144) and (5-148), the output SNR is found to be given by

$$\frac{S_o}{N_o} = \frac{A_c^4}{8N^2 + 8NA_c^2} \tag{5-149}$$

This may be written in a more illuminating form by defining the carrier-to-noise ratio as

$$\text{CNR} \equiv \frac{A_c^2}{2N} \tag{5-150}$$

($A_c^2/2$ is of course the average power in the unmodulated sine-wave carrier.) Then the output SNR becomes

$$\frac{S_o}{N_o} = \frac{1}{2} \frac{(\text{CNR})^2}{1 + 2\text{CNR}} \tag{5-151}$$

Note that for the CNR $\ll$ 1 (0 dB), the output SNR drops as the *square* of the carrier-to-noise ratio. This is the *suppression characteristic* of envelope detection. The output SNR deteriorates rapidly as the carrier-to-noise ratio drops below 0 dB. This quadratic SNR behavior below 0 dB is characteristic of all envelope detectors, and is due specifically to the $n \times n$ noise term dominating at low CNR.

For high CNR (CNR $\gg$ 1), on the other hand,

$$\frac{S_0}{N_0} \doteq \tfrac{1}{4}\text{CNR} \qquad \text{CNR} \gg 1 \qquad (5\text{-}152)$$

The output SNR is then linearly dependent on the carrier-to-noise ratio, again a common characteristic of envelope detectors. (Here the $s \times n$ noise term dominates so far as the output noise is concerned.) Thus *no SNR improvement is possible with AM systems*.

With FM, on the other hand, we shall find it possible to get significant improvement in output S_0/N_0 by increasing the modulation index at the expense of course of increased transmission bandwidth. Here in the AM case an increase in the transmission bandwidth beyond the bandwidth $2B$ needed to pass the AM signals serves only to increase the noise N and hence decrease the output SNR.

The complete signal-to-noise characteristic for the envelope detector, showing both the asymptotic high and low CNR cases, is shown sketched in Fig. 5-52. Although derived here only for the quadratic detector case, similar characteristics may be derived for other types of nonlinearity or detector law. The exact intersection of the two asymptotic lines depends on the detector law assumed, but the slopes, or shape of the detector signal-to-noise characteristic, will be the same for all detector laws.

It is apparent from the discussion of previous sections that a synchronous detector does not produce the quadratic suppression characteristics, due to envelope detection, at small CNR. The inphase or coherent carrier injection provides the same signal and noise powers at the output as at the input. Output SNR is thus everywhere linearly proportional to the input CNR. The reader is asked to demonstrate this himself to his own satisfaction.

The change from the nonlinear (quadratic) to the linear portion of the detector characteristic of Fig. 5-52 may be accounted for in an alternative and rather instructive fashion by considering the envelope Rayleigh- and Rician-density functions discussed previously. Referring for example to Fig. 5-44, we note that for low $A/\sqrt{N}$ (low CNR) the envelope will not deviate too far on the average from the $r = 0$ axis. Since r can never become negative by definition of an envelope, we would expect a lopsided probability distribution with most of

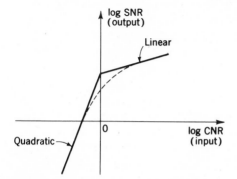

Figure 5-52 Asymptotic signal-to-noise characteristics, envelope detector.

the contribution coming in the vicinity of $r = 0$. This is just the Rayleigh distribution. As the carrier A_c increases, however, we would eventually expect the envelope to be symmetrically distributed about A_c. Although variations of the envelope above A_c have theoretically no limit, while below A_c the envelope is constrained never to become negative, the $r = 0$ axis is so remote for $A_c \gg \sqrt{N}$ that this nonzero constraint becomes insignificant and the probability-density curve approaches just the symmetrical bell-shaped characteristic of the gaussian function.

As far as the signal plus noise at the output is concerned, then, the nonlinear operation of the envelope detector has no effect on the distribution for large CNR ratio. The same result is found to hold true for other nonlinear demodulators with high CNR ratio: the output distribution of signal plus noise remains the same as that of the input.

This is exactly the reason why we have been assuming the noise distribution to remain gaussian as the noise progresses from the r-f stages of a receiver down to the narrowband i-f output. Even with no transmitted signal present the local oscillator injects a large enough signal voltage at the nonlinear mixer to ensure a gaussian noise distribution at the mixer output. The mixer then serves only to translate r-f energy down to the i-f spectrum. The signal and noise properties remain relatively unchanged.

As noted earlier, the local carrier injection required for suppressed-carrier demodulation produces the same result; the noise distribution remains gaussian at the demodulator output.

5-11 AM DETECTOR SPECTRAL ANALYSIS

The envelope-detector signal-to-noise characteristic may be obtained quantitatively in a completely different fashion by focusing attention on the spectral aspects of the detection process. Specifically, we determine the noise density at the output of the nonlinear device, and then investigate the filtering effect of the low-pass filter.

Not only is this approach valid in its own right, enabling us to specifically consider the low-pass filter design and its effect on the output noise, but it also enables us to introduce an extremely powerful tool in random signal and noise analysis—the use of correlation functions in determining the spectral properties of random signals passed through nonlinear devices or in calculating the spectra of various random signals.

We defined the power spectral density and the autocorrelation function in Sec. 5-4 to be Fourier-transform pairs. We did not pursue this point further (until the present section of course) because most of the emphasis in preceding sections has either been on noise passed through linear devices or on probability-of-error calculations. The spectral density at the output of linear filters is of course obtained by multiplying the input spectrum by the square of the magnitude of the transfer function. But how does one handle spectra at the output of nonlinear devices such as the envelope detector under consideration here? How does one calculate the spectrum of particular classes of random signals? Many techniques have been developed for handling these problems.[48] One of the most common and most useful is that of calculating the autocorrelation function at the nonlinear device output, and then taking its Fourier transform to find the spectrum at the same point. We shall demonstrate this technique here. Other applications, to FM spectral analysis for example, appear in the references just cited, as well as in the current periodical literature.[49]

The quadratic detector is again stressed here for simplicity's sake. The output voltage $z(t)$ will thus again contain the low-pass components of $v^2(t)$ (Fig. 5-51). We shall focus attention on the quadratic term $y = v^2$, however, from which $z(t)$ can be found by filtering.

Recall from Sec. 5-4 that the autocorrelation function of a random process $y(t)$ is simply given by taking the expected value of the product of $y(t)$ and the delayed term $y(t + \tau)$:

$$R_y(z) \equiv E[y(t)y(t + \tau)] \tag{5-153}$$

Expressing $y(t)$ here in terms of the input $v(t)$, we have

$$R_y(\tau) = E[v^2(t)v^2(t + \tau)] \tag{5-153a}$$

Evaluating this expression and formally taking its Fourier transform, we find the spectral density $G_y(f)$. Passing this through the low-pass filter $H_L(\omega)$, we then find of course that $G_z(f) = |H_L|^2 G_y(f)$, the desired output spectral density. Here we shall simply say that the filter is an ideal low-pass filter of bandwidth $<f_0$, but high enough to pass all low-frequency components of $y(t)$. Instead of formally writing out $G_y(f)$, we shall then go directly to $G_z(f)$ by ignoring all high-frequency terms in $G_y(f)$.

Before formally evaluating Eq. (5-153a), however, we introduce one modification into the expression previously used for $v(t)$. Recall that the reason for using envelope detection in the first place was that we assumed lack of phase coherence. This implies that the received carrier signal must be of the form $A_c \cos(\omega_0 t + \phi)$, with ϕ a random, unknown phase. We have tacitly ignored this phase term so far, writing the unmodulated carrier signal as $A_c \cos \omega_0 t$, because it played no real role in the analysis, and could arbitrarily be set equal to zero without destroying the validity of the treatment. Here, however, where we are taking statistical averages to find autocorrelation functions, the fact that ϕ is random does affect the result. The indicated statistical averaging will thus be taken over the random carrier phase angle, as well as the noise. We shall assume, as previously, that ϕ is uniformly distributed over 2π radians, and statistically independent of the noise.

---

[48] Middleton, *op. cit.;* Davenport and Root, *op. cit.;* Lawson and Uhlenbeck, *op. cit.;* Schwartz, Bennett, and Stein, *op. cit.,* sec. 3-2.

[49] See Schwartz, Bennett, and Stein, *op. cit.,* chap. 3, for a summary of applications to FM noise and signal bandwidth determination.

With this modification we now write the composite signal at the i-f filter output as

$$v(t) = A_c \cos (\omega_0 t + \phi) + n(t) \tag{5-154}$$

The expression for $v(t + \tau)$ is then of course[50]

$$v(t + \tau) = A_c \cos [\omega_0(t + \tau) + \phi] + n(t + \tau) \tag{5-155}$$

Squaring both $v(t)$ and $v(t + \tau)$ and multiplying them together, the autocorrelation function for $y(t)$ becomes

$$R_y(\tau) = E\{[A_c \cos (\omega_0 t + \phi) + n(t)]^2[A_c \cos [\omega_0(t + \tau) + \phi] + n(t + \tau)]^2\} \tag{5-156}$$

Expanding the terms in brackets, performing the indicated ensemble averages term by term over ϕ and n, and ignoring terms that will lead to high-frequency components of the spectral density, we get, as the autocorrelation function of the low-pass z,

$$R_z(\tau) = \underbrace{\frac{A_c{}^4}{4}}_{s \times s} + \underbrace{NA_c{}^2 + 2R_n(\tau)A_c{}^2 \cos \omega_0 \tau}_{s \times n} + \underbrace{R_{n^2}(\tau)}_{n \times n} \tag{5-157}$$

Note again the various terms appearing: the signal component $A_c{}^4/4$ that we have labeled $s \times s$, the $s \times n$ term corresponding to signal beating (or mixing) with noise, and the $n \times n$ term $R_{n^2}(\tau)$ corresponding to noise mixing nonlinearly with itself. The $n \times n$ term is defined as

$$R_{n^2}(\tau) \equiv E[n^2(t)n^2(t + \tau)] \tag{5-158}$$

This particular autocorrelation term may be further simplified by again using a known property of two dependent gaussian variables [$n(t)$ and $n(t + \tau)$ here]: Calling them, for ease in writing, n_1 and n_2, it may be shown[51] that

$$E(n_1{}^2 n_2{}^2) = N^2 + 2R_n{}^2(\tau) \tag{5-159}$$

---

[50] As a check the reader may find $R_v(\tau) = E[v(t)v(t + \tau)]$. Inserting Eqs. (5-154) and (5-155) into this expression, multiplying, performing the indicated ensemble average over ϕ and n and discarding terms centered at $2f_0$, it is easy to show that

$$R_v(\tau) = \frac{A_c{}^2}{2} \cos \omega_0 \tau + R_n(\tau)$$

Taking Fourier transforms, we then get

$$G_v(f) = \frac{A_c{}^2}{4} [\delta(f - f_0) + \delta(f + f_0)] + G_n(f)$$

which is just the spectral density corresponding to the sine wave at frequency f_0 plus the additive noise.

[51] Schwartz, Bennett, and Stein, *op. cit.*, p. 118; Papoulis, *op. cit.* More generally, for two dependent gaussian variables x_1 and x_2, $E(x_1{}^2 x_2{}^2) = \mu_{11}\mu_{22} + 2\mu_{12}\mu_{21}$, with the μ's the central moments of these variables.

Then $R_z(\tau)$ simplifies to

$$R_z(\tau) = \left(\frac{A_c^2}{2} + N\right)^2 + \underbrace{2R_n(\tau)A_c^2 \cos \omega_0\tau}_{s \times n} + \underbrace{2R_n^2(\tau)}_{n \times n} \qquad (5\text{-}160)$$

Taking the Fourier transform of this expression term by term to find the desired spectral density $G_z(f)$, we obtain the following interesting results.

1. The first term gives rise to an impulse at dc, of area $(A_c^2/2 + N)^2$:

$$G_1(f) = \left(\frac{A_c^2}{2} + N\right)^2 \delta(f) \qquad (5\text{-}161)$$

This dc term is shown sketched in Fig. 5-53 for the rectangular-shaped $G_n(f)$ assumed here.
2. The second term results in a shift down to dc of the noise spectrum $G_n(f)$, originally centered at f_0:

$$G_2(f) = 2A_c^2 \int_{-\infty}^{\infty} R_n(\tau) \cos \omega_0\tau e^{-j\omega\tau} \, d\tau$$

$$= A_c^2 \int_{-\infty}^{\infty} R_n(\tau)[e^{-j(\omega-\omega_0)\tau} + e^{-j(\omega+\omega_0)\tau}] \, d\tau$$

$$= A_c^2[G_n(f - f_0) + G_n(f + f_0)] \qquad (s \times n) \quad (5\text{-}162)$$

Here use is made of the Fourier-transform relation between $G_n(f)$ and $R_n(\tau)$. (The shift indicated includes one up to $2f_0$ hertz as well, but we ignore this, concentrating on the low-pass expressions.)

This shift in frequency is as expected, since this $s \times n$ spectral term corresponds precisely to the spectrum of the term $n(t) \cos (\omega_0 t + \phi)$, obtained when squaring $v(t)$. Just as in previous chapters, multiplication by $\cos \omega_0 t$ corresponds to a shift up and down by f_0. Since $n(t)$ is itself centered at f_0, the resultant multiplication shifts $n(t)$ down to 0 Hz and up to $2f_0$ hertz, as shown here. This $s \times n$ contribution to the output spectral density is shown sketched in Fig. 5-53 as well.

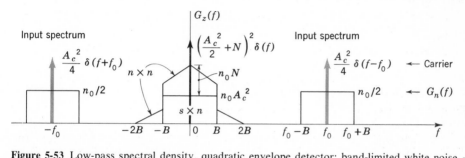

Figure 5-53 Low-pass spectral density, quadratic envelope detector: band-limited white noise + carrier at input.

3. The third term, $2R_n{}^2(\tau)$, results in a spectral contribution obtained by convolving the input noise spectral density with itself:

$$G_3(f) = 2\int_{-\infty}^{\infty} G_n(f')G_n(f - f')\,df' \qquad (n \times n) \qquad (5\text{-}163)$$

This is easily demonstrated by recalling from Chap. 2 that the Fourier transform of a product is just the convolution of the two individual Fourier transforms. Thus, if $G(\omega) = F(\omega)H(\omega)$, we have

$$g(t) = \int_{-\infty}^{\infty} f(\tau)h(t - \tau)A\tau$$

Here the two functions corresponding to F and H are both $R_n(\tau)$, and we are going from the τ domain to the f domain, rather than from f to t, but by the symmetry of Fourier transforms it is apparent Eq. (5-163) is valid.

Specifically, for the rectangular $G_n(f)$ centered at f_0 (and $-f_0$ as well), convolution results in the triangular-shaped $n \times n$ term shown centered at 0 Hz in Fig. 5-53. (Another contribution at $2f_0$ is again not shown.)

Physically, this $n \times n$ term, the power spectrum of $n^2(t)$, appears because of the multiplication of the input $n(t)$ by itself in forming v^2. The triangular spectrum may be verified qualitatively by visualizing $n(t)$ represented by a large number of equal-amplitude sine waves, all in the vicinity of f_0 hertz. Each one beating with itself, as well as with all the others, gives rise to difference frequencies extending from 0 to a maximum of $2B$ hertz. (This maximum contribution is due to the two extreme frequency terms located at $f_0 - B$ and $f_0 + B$, respectively.) Some thought will indicate that most contributions to the $n \times n$ spectrum come from sine waves closely spaced to one another. (There are proportionately more of these.) As the spacing between sine waves to be beat together increases, there are correspondingly fewer terms involved, and the overall contribution drops. This method of approximating the spectrum by discrete lines, and determining the beat frequencies and the number of contributions to each, is an alternative (albeit much more tedious) way of determining the noise spectra at the output of this quadratic envelope detector.[52]

The dc spectral contribution $G_1(f)$ may be checked quite easily by considering the output random process $z(t)$. We have

$$E(z) = E(v^2)_{\text{lowpass}} = \frac{A_c{}^2}{2} + N \qquad (5\text{-}164)$$

from Eq. (5-154) directly. {Again $E(n) = 0$; $E[\cos^2(\omega_0 t + \phi)] = \frac{1}{2}$.} Then the *dc power* is just $(A_c{}^2/2 + N)^2$, as found here.

The total spectrum $G_z(f) = G_1(f) + G_2(f) + G_3(f)$. As first noted in Sec. 5-4 the total output power is then obtained by summing $G_z(f)$ over the

---

[52] Schwartz, Bennett, and Stein, *op. cit.*, pp. 108–117.

entire frequency range. In this case the indicated integration can be done by inspection, using Fig. 5-53, and adding up the three contributions term by term. Thus

$$E(z^2) = \int_{-\infty}^{\infty} G_z(f) \, df$$

$$= \underbrace{\left(\frac{A_c^2}{2} + N\right)^2}_{\text{dc}} + \underbrace{2Bn_0 A_c^2}_{s \times n} + \underbrace{2Bn_0 N}_{n \times n}$$

$$= \left(\frac{A_c^2}{2} + N\right)^2 + NA_c^2 + N^2 \tag{5-165}$$

(In the last line we have replaced $2Bn_0$ by its equivalent N.)

Expanding out and collecting like terms, we have, finally,

$$4E(z^2) = A_c^4 + 8NA_c^2 + 8N^2 \tag{5-166}$$

just as in Eq. (5-147) previously! [The apparent factor of 4 difference is just due to the $\frac{1}{2}$ factor neglected in deriving Eq. (5-147).] The spectral approach thus provides us with the same result obtained previously. [As another check, using Eq. (5-160), $R_z(0)$ gives the same answer, noting that $R_n(0) = N$.]

But not only do we have the total noise power at the output, we also have its spectral distribution as well. As pointed out earlier, this enables us to determine the effects of different filters on the result. For example, it is apparent that the output noise we have calculated—previously, and again just now using the spectral approach—is actually somewhat more than one would get in practice using the quadratic detector and low-pass filter. For it includes noise spectral contributions out to $2B$ hertz (Fig. 5-53), whereas all that is required is a low-pass filter cutting off at B Hz. The power due to the $n \times n$ noise term is then

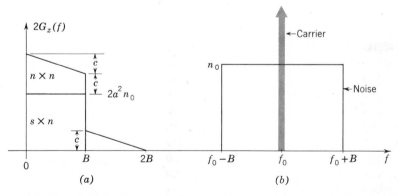

Figure 5-54 Low-frequency power spectrum, noise plus carrier, at output of piecewise-linear detector. (*a*) Output spectrum. (*b*) Input spectrum. (S. O. Rice, "Mathematical Analysis of Random Noise," *Bell System Tech. J.*, vol. 24, pp. 46–156, January 1945, fig. 9, part 4. Copyright, 1945, The American Telephone and Telegraph Co., reprinted by permission.)

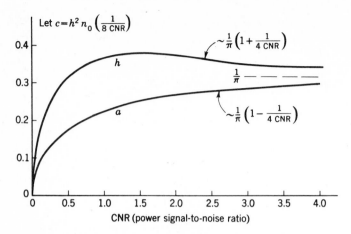

Figure 5-55 Coefficients for linear detector output shown in Fig. 5-54. (S. O. Rice, "Mathematical Analysis of Random Noise," *Bell System Tech. J.*, vol. 24, pp. 45–156, January 1945, fig. 10, part 4. Copyright, 1945, The American Telephone and Telegraph Co., reprinted by permission.)

reduced from N^2 to $0.75N^2$ (Fig. 5-53), increasing the output SNR somewhat. With a modulating signal introduced, the dc terms can also be eliminated and S_o/N_o (as redefined to include the modulated output) improved further.

A similar analysis for a piecewise-linear detector (often called a *linear envelope detector*) results in the low-frequency spectrum shown in Fig. 5-54.[53] (The negative-frequency terms have been folded over and combined with the positive ones.) Again the two types of noise, $s \times n$ and $n \times n$, appear. As shown in the figure the parameter c determines the $n \times n$ contribution, the parameter a the $s \times n$ contribution. The two parameters depend on CNR in a rather complicated way, as shown in Fig. 5-55. For CNR > 4, however,

$$c \doteq \frac{(1/\pi)^2}{8\text{CNR}} n_0$$

and

$$a \doteq \frac{1}{\pi}$$

So even for CNR = 4, the $s \times n$ contribution predominates, as in the quadratic detector case. For smaller values of CNR, the curves of Fig. 5-55 may be used to obtain c and a.

The dc contributions to the spectral density are not shown here. Note that the dimensions of the noise spectral components are not the same as those of Fig. 5-53 for the quadratic detector. But this is as it should be, since the output voltage in the linear detector case is proportional to r, the envelope itself, rather than r^2, as in the quadratic case. In particular, the output signal here is just $S_o =$

[53] S. O. Rice, "Mathematical Analysis of Random Noise," *Bell System Tech. J.*, vol. 24, pp. 46–156, part 4, January 1945.

$A_c{}^2/2$, rather than $A_c{}^4$, as previously. The ratio of S_o and N_o, however, has the same form in both cases.

5-12 FREQUENCY-MODULATION NOISE IMPROVEMENT

We now consider the case of frequency modulation. Again assuming the received signal to have gaussian noise added to it, we shall show that, contrary to the AM case, widening the transmission bandwidth (as is required for wideband FM) *does* improve the output SNR. As previously the SNR will be defined as the ratio of mean signal power with noise absent to the mean noise power in the presence of an unmodulated carrier. For simplicity's sake the analysis will be confined to the case of large carrier-to-noise ratio (CNR).[54]

The idealized FM receiver to be discussed here is shown in Fig. 5-56. A frequency-modulated signal of transmission bandwidth B_T hertz is first passed through an ideal limiter which removes all amplitude variations. The limiter output, after filtering, goes to the discriminator, assumed to give an output directly proportional to the instantaneous frequency of the signal, and then to an ideal low-pass filter of bandwidth B hertz $(B < B_T/2)$. B is the maximum bandwidth of the actual information signal $f(t)$ being transmitted.

Assuming, as in Chap. 4, a sine-wave signal of the form

$$f(t) = \Delta\omega \cos \omega_m t \qquad (5\text{-}167)$$

the frequency-modulated carrier measured at the *output* of the *i-f amplifier* is given by

$$f_c(t) = A_c \cos (\omega_0 t + \beta \sin \omega_m t) \qquad (5\text{-}168)$$

Noise is assumed absent here. $\beta = \Delta f/B$ is the modulation index and $\Delta f = \Delta\omega/2\pi$ the maximum-frequency deviation.

The average power of the FM wave is simply

$$S_c = \tfrac{1}{2}A_c{}^2 \qquad (5\text{-}169)$$

independent of the modulating signal.

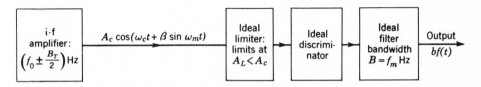

Figure 5-56 Block diagram, idealized FM receiver.

[54] See Schwartz, Bennett, and Stein, *op. cit.*, chap. 3, for a discussion of the complete FM noise problem, including references to the literature.

The instantaneous frequency is given by

$$\omega = \frac{d\theta}{dt} = \omega_0 + \beta\omega_m \cos \omega_m t = \omega_0 + \Delta\omega \cos \omega_m t$$
$$= \omega_0 + f(t) \tag{5-170}$$

The discriminator output is proportional to the instantaneous frequency deviation away from ω_0, or $\omega - \omega_0$. This frequency deviation is just $f(t)$. The output signal is then

$$f_d(t) = b \, \Delta\omega \cos \omega_m t = bf(t) \tag{5-171}$$

where b is a constant of the discriminator. As in the AM case this constant will be set equal to 1.

The discriminator output must be filtered to eliminate higher-frequency distortion terms. Filtering will also reduce the output noise when present. The filter bandwidth is chosen as $B = f_m$ hertz in order to pass all frequency components of $f(t)$.

From Eq. (5-171) the average output signal power is simply

$$S_o = \frac{(\Delta\omega)^2}{2} \quad \text{watts} \tag{5-172}$$

if a 1-Ω normalized load resistor is assumed.

The case of noise plus unmodulated carrier (signal absent) can now be treated in a manner directly analogous to that for AM. We again assume fluctuation noise of spectral density $n_0/2$ watts/Hz uniformly distributed (i.e., band-limited white noise) at the output of the i-f amplifier. Here, as shown in Fig. 5-57, the noise is uniformly distributed over the range $\pm B_T/2$ hertz about the carrier frequency f_0. (Compare with Fig. 5-50 for the AM case, where the transmission bandwidth $2B$ is generally less than the FM transmission bandwidth B_T.) The noise power at the i-f output (detector input) is thus

$$N = B_T n_0 \tag{5-173}$$

As previously, we may use the narrowband representation for the noise to write the unmodulated carrier plus noise in the form

$$v(t) = A_c \cos \omega_0 t + n(t)$$
$$= (A_c + x) \cos \omega_0 t - y(t) \sin \omega_0 t$$
$$= r(t) \cos [\omega_0 t + \Phi(t)] \tag{5-174}$$

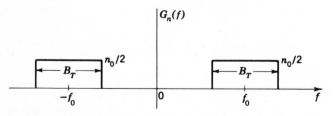

Figure 5-57 Noise spectral density, FM receiver.

In the AM case we focused attention on the properties of the envelope $r(t)$. It is apparent that in the FM case we focus attention on the phase term $\Phi(t)$. For in this unmodulated carrier case it is apparent that $\Phi(t)$ represents the noise at the discriminator output. In particular, since

$$\Phi = \tan^{-1} \frac{y}{x + A_c} \tag{5-175}$$

the discriminator output is given by

$$\dot{\Phi} = \frac{(x + A_c)\dot{y} - y\dot{x}}{y^2 + (x + A_c)^2} \tag{5-176}$$

This is a rather formidable-looking expression and is the reason why FM noise analysis, even in the unmodulated carrier case, is so difficult to carry out. [The straightforward, albeit mathematically difficult, way of determining the spectral density of the noise at the discriminator output is to find the autocorrelation function of the random noise process $\dot{\Phi}(t)$, relating it to the known input correlation function $R_n(\tau)$. One then takes the Fourier transform to find the output spectral density $G_{\dot{\phi}}(f)$.[55]]

For large CNR this expression for the discriminator output, $\dot{\Phi}(t)$, simplifies considerably. Thus, recalling that $E(x^2) = E(y^2) = N$, and assuming that $A_c^2 \gg N$ (the CNR is actually $S_c/N = A_c^2/2N$), the discriminator output is given simply by

$$\dot{\Phi} \doteq \frac{1}{A_c} \dot{y} \qquad A_c^2 \gg N \tag{5-177}$$

This is apparent from either Eq. (5-175) or Eq. (5-176). The discriminator output noise is thus proportional to the time derivative of the quadrature noise term $y(t)$. (If the standard deviation $\sqrt{N}$ of x is small compared to A_c, it is apparent that the probability is small that the gaussian function x will be comparable to A_c in magnitude. It may then be neglected in comparison with A_c. Similarly, y/A_c is then small with a high probability, so that $\dot{\Phi} \doteq \dot{y}/A_c$.)

Since the discriminator is followed by a low-pass filter (Fig. 5-56), we must now find the spectral density of the discriminator output noise $\dot{\Phi}$ in order to take into account the effect of the filter. To do this we simply note that differentiation is a linear operation. Hence Eq. (5-177) indicates that $\dot{\Phi}$ may be considered the response at the output of a (linear) differentiator $H(\omega)$ with y

---

[55] See Schwartz, Bennett, and Stein, *op. cit.*, chap. 3, for a discussion of this approach. Detailed analyses also appear in S. O. Rice, "Statistical Properties of a Sine Wave Plus Random Noise," *Bell System Tech. J.*, vol. 27, secs. 7 and 8, pp. 138–151, January 1948; Lawson and Uhlenbeck, *op. cit.*, chap. 13; Middleton, *op. cit.*, chap. 15. Rice has also developed a simpler approach to FM noise analysis, using the so-called "clicks" phenomenon, that is readily extended to other types of FM receivers. See S. O. Rice, "Noise in FM Receivers," in M. Rosenblatt (ed.), *Proceedings, Symposium of Time Series Analysis*, Wiley, New York, 1963, chap. 25, pp. 375–424. This approach is also summarized in Schwartz, Bennett, and Stein, *op. cit.*, pp. 144–154.

applied at the input. From our discussion of random signals and noise we then write directly

$$G_{\dot{\phi}}(f) = |H(\omega)|^2 G_y(f) \tag{5-178}$$

But differentiation of a time function corresponds to multiplication of its Fourier transform by $j\omega$. Then $H(\omega) = j\omega/A_c$, $|H(\omega)|^2 = \omega^2/A_c^2$, and we have, very simply,

$$G_{\dot{\phi}}(f) = \frac{\omega^2}{A_c^2} G_y(f) \qquad A_c^2 \gg N \tag{5-179}$$

This is shown schematically in Fig. 5-58.

Recall from our initial discussion of the narrowband noise representation, however, that for symmetrical bandpass filters,

$$G_x(f) = G_y(f) = 2G_n(f + f_0) \tag{5-180}$$

Thus we simply find the spectral density of the quadrature (or inphase) noise component by shifting the noise spectral density $G_n(f)$ down to 0 frequency. Then we also have

$$G_{\dot{\phi}}(f) = \frac{2\omega^2}{A_c^2} G_n(f + f_0) \qquad A_c^2 \gg N \tag{5-181}$$

In particular, for the band-limited white-noise case assumed here (Fig. 5-57),

$$G_{\dot{\phi}}(f) = \frac{\omega^2 n_0}{A_c^2} \qquad -\frac{B_T}{2} < f < \frac{B_T}{2}$$

$$= 0 \qquad \text{elsewhere} \tag{5-182}$$

This is shown schematically in Fig. 5-59.

We are now in a position to readily determine the total noise at the FM receiver output. For assuming a low-pass filter of known spectral shape $H_L(\omega)$ following the discriminator, the noise spectral density at the low-pass filter output is simply $|H_L(\omega)|^2 G_{\dot{\phi}}(f)$. We then integrate this spectral density over all frequencies to find the output noise N_o. Again for simplicity's sake assume this final filter to be an ideal low-pass one of bandwidth $B < B_T/2$. The bandwidth B should just be sufficient to pass all signal components, yet no larger to avoid increasing the noise passed. B is thus just the bandwidth of the original modulating signal (see Chap. 4). In the case of the sine-wave signal of frequency f_m, B is just f_m.

The output noise N_o is thus found by integrating $G_{\dot{\phi}}(f)$ from $-B$ to B (Fig. 5-59).

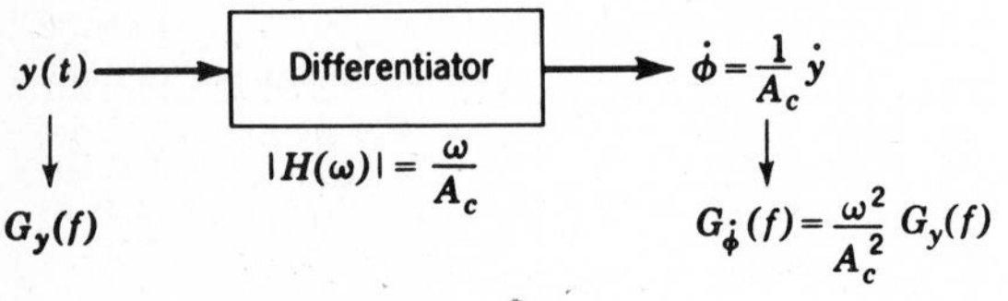

Figure **5-58** Output-noise spectral density, FM receiver, high CNR.

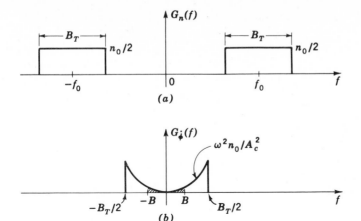

Figure 5-59 FM noise spectrum (high-CNR case). (*a*) Spectrum of i-f noise. (*b*) Spectral density at FM detector output.

$$N_o = \int_{-B}^{B} G_{\dot\phi}(f) \, df$$

$$= \frac{2(2\pi)^2 n_0}{A_c^2} \int_0^B f^2 \, df$$

$$= \frac{(2\pi)^2 n_0}{3 S_c} B^3 \tag{5-183}$$

Note that the output noise, although proportional to B^3, is *inversely* proportional to the carrier power $S_c = A_c^2/2$. As the carrier increases, therefore, the noise power drops. This *noise quieting* effect is of course well known in FM, and is just the opposite of the effect encountered in AM. (Recall that there the introduction of the carrier *increased* the noise.)

By using Eq. (5-172), the expression for the output signal power (the signal is assumed to be a cosine wave with noise absent), the mean SNR at the output becomes

$$\frac{S_o}{N_o} = 3 \left(\frac{\Delta f}{B}\right)^2 \frac{S_c}{2 n_0 B} \tag{5-184}$$

But $2 n_0 B = N_c$, the average power in the AM sidebands, and $\beta = \Delta f / B$ is the modulation index. Then

$$\frac{S_o}{N_o} = 3 \beta^2 \frac{S_c}{N_c} \tag{5-185}$$

Note that S_c / N_c corresponds to the CNR of an *AM system* with the same carrier power and noise spectral density. The total noise power over the transmission bandwidth B_T of the FM system is greater than N_c since $B_T > 2B$.

For a specified carrier amplitude and noise spectral density at the i-f amplifier output Eq. (5-185) shows that the output SNR increases with the modulation index or the transmission bandwidth. This is in contrast to the AM case, where increasing the bandwidth beyond $2B$ was found only to deteriorate the output SNR.

We may specifically compare the FM and AM systems by assuming the same unmodulated carrier power and noise spectral density n_0 for both. (These quantities are measured here at the output of the i-f amplifier.) For a 100 percent modulated AM signal we have

$$\left(\frac{S_o}{N_o}\right)_{AM} = \frac{S_c}{N_c}$$

Equation (5-185) may thus be modified to read

$$\left(\frac{S_o}{N_o}\right)_{FM} = 3\beta^2 \left(\frac{S_o}{N_o}\right)_{AM} \tag{5-186}$$

For large modulation index (this corresponds to a wide transmission bandwidth, for, with $\beta \gg 1$, the bandwidth approaches $2\Delta f$), we can presumably increase S_o/N_o significantly over the AM case. As an example, if $\beta = 5$, the FM output SNR is 75 times that of an equivalent AM system. Alternatively, for the same SNR at the output in both receivers the power of the FM carrier may be reduced 75 times. But this requires increasing the transmission bandwidth from $2B$ (AM case) to $16B$ (FM case); see Fig. 4-51. Frequency modulation thus provides a substantial improvement in SNR, but at the expense of increased bandwidth. This is of course characteristic of all noise-improvement systems.

Can we keep increasing the output SNR indefinitely by increasing the frequency deviation and hence the bandwidth? If we keep the transmitter power fixed, S_c is fixed. With the noise power per unit bandwidth $(n_0/2)$ fixed and the audio signal bandwidth B fixed N_c presumably remains constant, *but*, as the frequency deviation increases and the bandwidth with it, more noise must be accepted by the limiter. Eventually, the noise power at the limiter becomes comparable with the signal power. The above simplified analysis, which assumes large carrier-to-noise power ratios, does not hold any more, and noise is found to "take over" the system.

This effect is found to depend very sharply upon FM carrier-to-noise ratio S_c/N and is called a *threshold* effect. For this ratio greater than a specified threshold value FM functions properly and shows the significant improvement in SNR predicted by Eq. (5-186). For the ratio below this threshold level the noise improvement is found to deteriorate rapidly, and Eq. (5-186) no longer holds. The actual threshold level depends upon the FM carrier-to-noise ratio and upon β. For large β the level is usually taken as 10 dB.

This threshold phenomenon is a characteristic of all wideband noise-improvement systems and was encountered previously in discussing PCM systems.

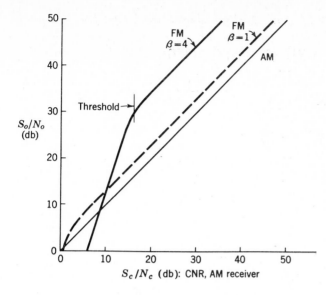

Figure 5-60 Measured characteristics, FM and AM receivers. (Adapted from M. G. Crosby, "Frequency Modulation Noise Characteristics," *Proc. IRE*, vol. 25, pp. 472–514, April, 1937, fig. 10, by permission.)

Two conditions must thus ordinarily be satisfied for an FM system to show significant noise-"quieting" properties:

1. Frequency-modulation carrier-to-noise ratio > 10 db to avoid the threshold effect.
2. With FM carrier-to-noise ratio > 10 db, $\beta > 1/\sqrt{3}$ if $S_o/N_o > S_c/N_c$ [see Eq. (5-186)].

But $\beta > 1\sqrt{3} \doteq 0.6$ corresponds to the transition between narrowband and wideband FM. *Narrowband FM thus provides no* SNR *improvement over AM.* This is of course as expected, for the improvement is specifically the result of restricting the noise phase deviations of the carrier to small values, while the signal variations are assumed to be large.

Experimental and theoretical studies of FM noise characteristics show the threshold phenomenon very strikingly and also bear out the validity of Eq. (5-186) above the threshold value. Figure 5-60 is taken from some experimental work of M. G. Crosby[56] and shows a comparison of AM and FM receivers for $\beta = 4$ and $\beta = 1$. Note that for $\beta = 4$ the FM signal-to-noise ratio deteriorates rapidly for $S_c/N_c < 13$ dB. In fact, for $S_c/N_c < 8$ dB, the AM system becomes superior. For $S_c/N_c > 15$ dB, however, the FM system shows an improvement of 14 dB. For $\beta = 4$ we would expect the theoretical improvement to be $3\beta^2 = 48$, or 17 dB. For $\beta = 1$ the threshold level is experimentally found to occur at 2 dB. Above this value of AM carrier-to-noise ratio the FM improvement over

[56] M. G. Crosby, "Frequency Modulation Noise Characteristics," *Proc. IRE*, vol. 25, pp. 472–514, April 1937, fig. 10.

AM is 3 dB. The theoretical improvement would be expected to be $3\beta^2 = 3$, or 4.8 dB.

These measured characteristics of M. G. Crosby are of historical significance because they were among the first obtained in quantitative studies of FM noise. They have of course since been reproduced countless times by many investigators. The detailed theoretical analyses noted earlier bear these results out as well.

Much engineering time has been devoted in recent years to the development of threshold-improvement receivers designed to reduce threshold in FM receivers. These include the FM receiver with feedback (FMFB), the phase-locked loop, and the frequency-locked loop.[57]

Signal-to-Noise Improvement through Deemphasis

We showed in Chap. 4 that the transmission bandwidth of an FM system is determined by the maximum-frequency deviation Δf produced by the highest modulating frequency f_m to be transmitted. In particular, for $f_m = 15$ kHz and a maximum-frequency deviation Δf of 75 kHz we found that the required bandwidth was 240 kHz.

In practice the higher-frequency components of the modulating signal rarely attain the amplitudes needed to produce a 75-kHz frequency deviation. Audio signals of speech and music are found to have most of their energy concentrated in the lower-frequency ranges. The instantaneous signal amplitude, limited to that required to give the 75-kHz deviation, is due most of the time to the lower-frequency components of the signal. The smaller-amplitude high-frequency (h-f) components will, on the average, provide a much smaller frequency deviation. The FM signal thus does not fully occupy the large bandwidth assigned to it.

The spectrum of the noise introduced at the receiver does, however, occupy the entire FM bandwidth. In fact, as we have just noted, the noise-power spectrum at the output of the discriminator is emphasized at the higher frequencies. (The spectrum is proportional to f^2 for a large carrier-to-noise ratio.)

This gives us a clue as to a possible procedure for improving the SNR at the discriminator output: we can artificially *emphasize* the h-f components of our input audio signal at the transmitter, *before the noise is introduced,* to the point where they produce a 75-kHz deviation most of the time. This *equalizes* in a sense the l-f and h-f portions of the audio spectrum and enables the signal fully to occupy the bandwidth assigned.

Then, at the output of the receiver discriminator, we can perform the

[57] Schwartz, Bennett, and Stein, *op. cit.,* pp. 157–163; A. J. Viterbi, "Principles of Coherent Communications," McGraw-Hill, New York, 1966; K. K. Clarke and D. F. Hess, "Frequency-locked Loop Demodulator," *IEEE Trans. Commun. Technol.,* pp. 518–524, August 1967. The equivalence of all three devices under limiting conditions is discussed by D. T. Hess, "Equivalence of FM Threshold Extension Receivers," *IEEE Trans. Commun. Technol.,* October 1968.

inverse operation, or *deemphasize* the higher-frequency components, to restore the original signal-power distribution. But in this deemphasis process we reduce the h-f components of the noise also and so effectively increase the SNR.

Such a preemphasis and deemphasis process is commonly used in FM transmission and reception and provides, as we shall see, 13 to 16 dB of noise improvement. Note that this procedure is a simple example of a signal-processing scheme which utilizes differences in the characteristics of the signal and the noise to process the signal more efficiently. The entire FM process is itself an example of a much more complex processing scheme in which use is made of the fact that random noise alters the instantaneous frequency of a carrier much less than it does the amplitude of the carrier (for large carrier-to-noise ratio). The noise-improvement properties of PCM and other wideband systems are again due to differences in the characteristics of random noise and signal.

A simple frequency transfer function that emphasizes the high frequencies and has been found very effective in practice is given by

$$H(\omega) = 1 + j\,\frac{\omega}{\omega_1}. \tag{5-187}$$

An example of an RC network that approximates this response very closely is shown in Fig. 5-61a. The asymptotic logarithmic amplitude-frequency plot for this network is shown in Fig. 5-61b.

With $r \gg R$ the amplitude response has two break frequencies given by $\omega_1 = 1/rC$ and $\omega_2 \doteq 1/RC$. Signals in the range between ω_1 and ω_2 are thus emphasized. (Actually, the higher-frequency components are passed unaltered, and the lower-frequency components are attenuated. The attenuation can of course be made up by amplification.) The choice of $f_1 = \omega_1/2\pi$ is not critical, but 2.1 kHz is ordinarily used in practice ($rC = 75\ \mu s$). $f_2 = \omega_2/2\pi$ should lie above the highest audio frequency to be transmitted. $f_2 \geq 30$ kHz is a typical number. In the range between these two frequencies $|H(\omega)|^2 \doteq 1 + (f/f_1)^2$, and all audio frequencies above 2.1 kHz are increasingly emphasized.

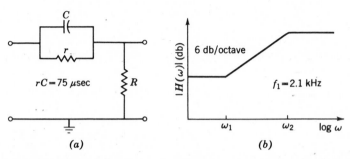

Figure 5-61 Example of a preemphasis network. (a) Preemphasis network, $r \gg R$, $rC = 75\ \mu$sec. (b) Asymptotic response, $\omega_1 = 1/rC$, $\omega_2 \doteq 1/RC$.

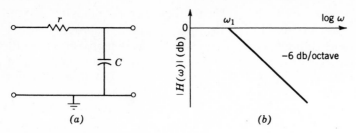

Figure 5-62 Example of a deemphasis network. (*a*) Deemphasis network, $rC = 75 \mu$sec. (*b*) Asymptotic response, $f_1 = 2.1$ kHz.

The receiver deemphasis network, following the discriminator, must have the inverse characteristic given by

$$H(\omega) = \frac{1}{1 + jf/f_1} \tag{5-188}$$

with $f_1 = 2.1$ kHz as before. This then serves to restore all signals to their original relative values. The simple RC network of Fig. 5-62 ($rC = 75 \mu$s) provides this deemphasis characteristic. (Note incidentally that the two networks are identical with the lead and lag networks, respectively, of the servomechanisms and feedback amplifier fields.)

How much does the deemphasis network improve the SNR ratio at the discriminator output?

From Eq. (5-182) of the FM noise analysis, the noise spectral density at the discriminator output (for large carrier-to-noise ratio) is

$$G_{\dot\phi}(f) = \frac{n_0\omega^2}{A_c^2} = \frac{n_0\omega^2}{2S_c} \tag{5-189}$$

where n_0 is the input noise spectral density, A_c the carrier amplitude, and $S_c = A_c^2/2$ the mean carrier power at the discriminator input.

If this noise is now passed through the RC deemphasis network of Fig. 5-62, the modified spectral density at the network output is

$$G_H(f) = G_{\dot\phi}(f)|H(\omega)|^2 = \frac{n_0\omega^2}{2S_c}\frac{1}{1 + (f/f_1)^2} \tag{5-190}$$

The original noise-power spectrum $G_{\dot\phi}(f)$ at the output of the discriminator and the modified spectrum $G_H(f)$ are shown sketched in the logarithmic plots of Fig. 5-63. (Only the one-sided spectra, defined for positive f only, are shown because of the logarithmic plots.)

Note that for $f > f_1$ the noise spectrum with deemphasis included becomes a uniform spectrum; the deemphasis network has canceled out the ω^2, increasing noise frequency, factor of Eq. (5-189).

The total mean noise power at the output of an ideal low-pass filter of bandwidth B hertz is given by

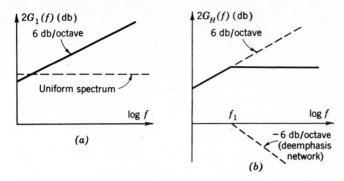

Figure 5-63 Logarithmic FM noise-power spectrum: output of discriminator ($S_c \gg N_c$). (a) Spectrum without deemphasis. (b) Spectrum with deemphasis.

$$N_{oD} = \int_{-B}^{B} G_H(f) \, df$$

$$= \int_{-B}^{B} \frac{n_0 \omega^2}{2S_c} \frac{df}{1 + (f/f_1)^2} \tag{5-191}$$

where N_{oD} represents the mean noise power with the deemphasis network included, as compared with the symbol N_o used previously for the noise power with no deemphasis network.

Equation (5-191) is readily integrated to give

$$N_{oD} = \frac{n_0}{2\pi S_c} \omega_1{}^3 \left(\frac{B}{f_1} - \tan^{-1} \frac{B}{f_1} \right) \tag{5-191a}$$

Multiplying numerator and denominator by $3/B^3$, and recalling from Eq. (5-183) that $N_o = (2\pi)^2 n_0 B^3/3S_c$, the mean noise power can be written in the form

$$N_{oD} = N_o D \tag{5-192}$$

where

$$D = 3 \left(\frac{f_1}{B} \right)^3 \left(\frac{B}{f_1} - \tan^{-1} \frac{B}{f_1} \right) \tag{5-193}$$

The parameter D represents the effect of the deemphasis network and is readily seen to be less than or equal to 1 in value. The deemphasis network thus reduces the output noise. For $B \ll f_1$ (the deemphasis effect is then introduced beyond the range of the low-pass filter) $D \to 1$, as is to be expected. For $B \gg f_1$, $\tan^{-1}(B/f_1) \to \pi/2$, and

$$D \to 3 \left(\frac{f_1}{B} \right)^2 \qquad f_1 \ll B \tag{5-194}$$

The output noise power thus decreases rapidly with increasing B.

Increasing B indefinitely provides no *absolute* improvement in the output noise, however, for N_o increases as B^3. The output filter bandwidth B should thus be restricted to just the bandwidth required to pass the highest audio frequency and no more.

With $f_1 = 2.1$ kHz, as noted previously, and $B = 15$kHz, $D = \frac{1}{20}$. The noise is thus reduced by a factor of 20, or 13 dB. If $B = 10f_1 = 21$ kHz, the improvement due to the deemphasis network is 16 dB.

The signal power at the output of the discriminator is $S_o = (\Delta\omega)^2/2$ [Eq. (5-172)], for a single sine wave, independent of the preemphasis and deemphasis procedure. The SNR ratio with deemphasis is thus

$$\frac{S_o}{N_{oD}} = \frac{1}{D}\frac{S_o}{N_o} = \frac{3\beta^2}{D}\frac{S_c}{N_c} \tag{5-195}$$

The signal-to-noise improvement in decibels is the same as the noise reduction: 13 dB for $B = 15$ kHz, 16 dB for $B = 21$ kHz.

Preemphasis and deemphasis techniques are obviously not restricted to FM only. They are possible because the audio signals to be transmitted in practice are concentrated at the low end of the spectrum. All modulation systems can thus use deemphasis techniques to improve the SNR at the receiver output.

As an example, consider AM. Here we showed that for large carrier-to-noise ratio (CNR $\gg$ 1) the noise-power spectrum at the detector output was uniform. (See Figs. 5-53 and 5-54. Recall that for large CNR the $s \times n$ terms dominate.) In particular,

$$G_z(f) = K \tag{5-196}$$

at the detector output, where K is, for example, $n_0 A_c{}^2$ in the quadratic detector case and $a^2 n_0$ in the linear detector case.

Using the preemphasis and deemphasis networks of the FM analysis,

$$G_H(f) = G_z(f)|H(\omega)|^2 = \frac{K}{1 + (f/f_1)^2} \tag{5-197}$$

is the spectral density at the output of the deemphasis network. Both $G_H(f)$ and $G_z(f)$ are sketched in Fig. 5-64. The total mean noise power at the output of an ideal low-pass filter of bandwidth B is

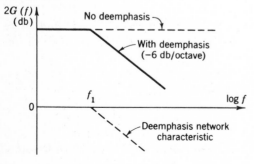

Figure 5-64 Amplitude-modulation spectral density, detector output (CNR $\gg$ 1).

$$N_{oD} = \int_{-B}^{B} G_H(f)\, df$$

$$= 2Kf_1 \tan^{-1} \frac{B}{f_1}$$

$$= N_o D' \tag{5-198}$$

where $N_o = 2BK$ is again the mean noise power with no deemphasis network and

$$D' = \frac{f_1}{B} \tan^{-1} \frac{B}{f_1} \tag{5-199}$$

is the improvement factor due to deemphasis in the AM case.

For $B \ll f_1$, $D' \to 1$, so that there is no improvement in this case (the filter bandwidth is well within the deemphasis range). For $B/f_1 \gg 1$, $\tan^{-1} (B/f_1) \doteq \pi/2$, and

$$D' \doteq \frac{\pi}{2} \frac{f_1}{B} \qquad \frac{B}{f_1} \gg 1 \tag{5-200}$$

The noise power thus decreases inversely with the filter bandwidth B. In the FM case the noise power decreased inversely as the square of the bandwidth [Eq. (5-194)], so that the RC deemphasis network provides relatively greater improvement for FM than for AM.

As an example, if $f_1 = 2.1\,\text{kHz}$ and $B = 15\,\text{kHz}$, D' is 6 dB for the AM case and D is 13 dB for the FM case. With $B = 21\,\text{kHz}$, D' is 8 dB for AM, while D is 16 dB for FM. For $B = 5\,\text{kHz}$, $D' = 3\,\text{dB}$.

Although this deemphasis technique is not as efficient for AM as for FM, it still would provide some improvement. Why then has it not been adopted for AM systems? Several reasons can be listed[58]: Its use would require modification of all existing receivers. Emphasizing the higher-frequency sidebands would increase interference between adjacent AM channels. An increase in m, the modulation factor, at higher audio frequencies would put severe requirements on detector design.

Deemphasis networks are commonly used in FM receivers, however, with the corresponding preemphasis networks built into the audio section of FM transmitters. The same "equalization" principle is also used quite commonly in sound recording to reduce "scratch."

The discussion of SNR improvement by deemphasis has proceeded on an *ad hoc* basis. A specific deemphasis network was postulated and noise improvement demonstrated by calculation. The discussion can, however, be handled on a more general level by postulating the use of preemphasis and deemphasis networks and then finding the networks that minimize the output noise. Specifically, we shall use an approach similar to that adopted previously in

[58] L. B. Arguimbau and R. D. Stuart, *Frequency Modulation,* Methuen, London, 1956, p. 5.

discussing matched filters. Here too the Schwarz inequality will be found to play a role.[59] The one distinction is that the signal as well as the noise is now assumed random.

Specifically, let the signal at the transmitter have a known spectral density $G_s(f)$. The signal is then passed through a preemphasis network $H(\omega)$, and the resultant signal actually transmitted. This output signal now has a spectral density $G_s(f)|H|^2$, and the average transmitter power is

$$P = \int_{-\infty}^{\infty} G_s(f)|H(\omega)|^2 \, df \tag{5-201}$$

We assume, as is usually the case in practice, that P is *fixed* at some maximum level.

Noise of spectral density $G_n(f)$ is added during transmission, and signal plus noise passed through a deemphasis network with transfer function $H^{-1}(\omega)$, just the inverse of the preemphasis network. The output signal power is then just

$$S_o = \int_{-\infty}^{\infty} G_s(f) \, df \tag{5-202}$$

a fixed quantity, and the output noise power is

$$N_o = \int_{-\infty}^{\infty} G_n(f)|H(\omega)|^{-2} \, df \tag{5-203}$$

How do we now choose $|H(\omega)|$ to maximize S_o/N_o (or minimize N_o, since S_o is fixed), with P a known constant? (Note that the phase of the two correction networks is arbitrary, provided, however, that the phase of the deemphasis network corrects for that of the preemphasis network to avoid signal distortion.)

Since P is constant, we can just as well minimize

$$PN_o = \int_{-\infty}^{\infty} G_s(f)|H|^2 \, df \int_{-\infty}^{\infty} G_n(f)|H|^{-2} \, df \tag{5-204}$$

But recall that the Schwarz inequality for real integrals may be written

$$\int A^2(f) \, df \cdot \int B^2 \, df \geq [\int A(f)B(f) \, df]^2 \tag{5-205}$$

If we let

$$G_s(f)|H|^2 \equiv A^2(f)$$

and
$$G_n(f)|H|^{-2} \equiv B^2(f)$$

we have immediately

$$PN_o \geq [\int \sqrt{G_s(f)G_n(f)} \, df]^2 \tag{5-206}$$

---

[59] This approach to deemphasis network analysis was suggested to the author in a private communication by Dr. Robert Price.

with the minimum output noise attained when $A(f) = B(f)$, or

$$|H(\omega)|^2_{\text{opt}} = \sqrt{\frac{G_n(f)}{G_s(f)}} \qquad (5\text{-}207)$$

Providing that the signal and noise have *different* spectral shapes (an almost obvious consideration), preemphasis does pay off (i.e., optimum networks in the sense indicated here do exist).

As an example assume that $G_s(f)$ varies as $1/f^2$ (this is a particularly simple model for indicating relatively lower energy at higher frequencies) and $G_n(f)$ increases over a limited frequency range as f^2. This is then the high CNR FM case discussed earlier, where attention is focused just on the baseband portions of the system. Then the optimum $|H| \propto f$, just the preemphasis network assumed earlier (Fig. 5-61). (The nonlinear modulation portions of the FM system may be ignored here, since the discussion is based solely on the optimization of the baseband preemphasis and deemphasis networks, not on the entire system.)

For an equivalent AM system, with the output-noise spectral density flat, and $G_s(f)$ again assumed dropping off as $1/f^2$, at least over a limited range, the optimum $H(\omega) \propto \sqrt{f}$, somewhat different than the networks considered.

This optimization may be carried one step further without assuming any particular structure for the system (i.e., the preemphasis-deemphasis network pair assumed here), and complete receivers can be designed that demodulate analog signals in some optimum sense in the presence of noise. Such criteria as least mean-squared error between transmitted and received signals, maximum SNR, maximum a posteriori probability, etc., have been adopted, and optimum, albeit not necessarily realizable, structures obtained. But approaches such as these are beyond the scope of this book.[60]

5-13 THERMAL NOISE CONSIDERATIONS[61]

Up to this point in this chapter we have been assuming the white-noise spectral density $n_0/2$ either given to us, or measured using some of the techniques discussed earlier. In most communication systems this noise may be grouped into either of two types—*thermal noise,* which has a fundamental thermodynamic origin, and *shot noise* due to an average current flow in the circuits making up the systems. (Note again that we are focusing only on *spontaneous fluctuation noise,* which is ever present in the physical environment, and which results in basic limitations on the transmission of information.) We shall discuss

---

[60] See H. Van Trees, *Detection, Estimation, and Modulation Theory: I,* Wiley, New York, 1968; Viterbi, *op. cit.,* chap. 5.

[61] Comprehensive references to the material of this section include the following books: W. R. Bennett, *Electrical Noise,* McGraw-Hill, New York, 1960; D. A. Bell, *Electrical Noise, Fundamentals and Physical Mechanism,* Van Nostrand, London, 1960; A. Van der Ziel, *Noise,* Prentice-Hall, Englewood Cliffs, N.J., 1954. See also M. Schwartz, *Information Transmission, Modulation, and Noise,* 2nd ed., McGraw-Hill, New York, 1970, chap. 7.

the origin of the thermal noise at some length, because of its fundamental nature, coming up with a very simple and specific formula for the noise spectral density. Thus we shall show that the thermal noise power spectral density may be taken as white at radio frequencies as high as 10^{13} Hz, the formula for the spectral density being given by

$$G_n(f) = \frac{n_0}{2} = \frac{kT}{2} \qquad \text{watts/Hz} \qquad (5\text{-}208)$$

or $n_0 = kT$.

Here $k = 1.38 \times 10^{-23}$ J/K is the Boltzmann constant, and T is the absolute temperature of the thermal noise source, in degrees Kelvin. Shot noise will be mentioned only briefly. We shall show that its effect, as well as the effect of other noise sources, may be included by assuming an effective noise temperature higher than the actual T in the thermal noise equation.

In the next section we tie this material together with that of the previous chapters by working out some actual calculations of SNR and E/n_0 for space and satellite communications. This then enables us to actually evaluate the performance of some typical communication systems.

We can summarize the distinction between thermal and shot noise in the following way. *Thermal noise* is associated with random motion of particles in a force-free environment. For example, the air pressure in a given room is due to the summed effect of countless air molecules moving chaotically in all directions. The molecules are in continuous turbulent motion, striking and rebounding from one to another. When we talk of the "pressure" at a point we refer in effect to the resultant force per unit area of all molecules striking a surface located at the point in question. This force will fluctuate or vary in time as fewer or more molecules strike the wall from time to time. Since the numbers of molecules involved are ordinarily tremendously large for normal-sized surfaces, the *average* force in time will remain constant as long as the average molecular energy remains constant. The instantaneous force as a function of time, however, will vary randomly about this average value. Increasing the temperature increases the molecular energy, and the average pressure goes up, as do the fluctuations about the average. Thermodynamics and the kinetic theory of heat applied to this problem indicate that for an ideal gas (one for which intermolecular forces may be neglected), the average kinetic energy of motion per molecule in any one direction is $kT/2$, with $k = 1.38 \times 10^{-23}$ J/K the Boltzmann constant, and T the absolute temperature in degrees Kelvin. This is the reason why we find the mean-squared thermal noise proportional to kT.

In electrical work we encounter similar fluctuations which are thermally induced. Conductors contain a large number of "free" electrons and ions strongly bound by molecular forces. The ions vibrate randomly about their normal (average) positions, however, this vibration being a function of temperature. Collisions between the free electrons and the vibrating ions continuously take place. There is a continuous transfer of energy between electrons and ions. This is the source of the resistance in the conductor. The freely moving electrons constitute a current, which over a long period of time averages to zero, since as many electrons on the average move in one direction as another. (This is the analog of the pressure case noted above, in which the *average* molecular velocity is zero, although chaotic motion exists.) There are random fluctuations about this average, however, and, in fact, we shall see that the mean-squared fluctuations in current are proportional to kT also.

Both cases noted—pressure fluctuations and current fluctuations—deal with the chaotic motion of particles (molecules or electrons) possessing thermal energy. There are no forces present "organizing" this motion in preferred directions. Both cases may therefore be treated by equilibrium thermodynamics, with the mean-squared fluctuations found proportional to kT.

The second type of noise noted above, *shot noise,* is also due to the discrete nature of matter, but here we assume an average flow in some direction taking place: electrons flowing between cathode and anode in a cathode-ray oscilloscope, electrons and holes flowing in semiconductors, photons emitted in some laser systems, photoelectrons emitted in photodiodes, fluid moving continuously under the action of a pressure gradient, etc. Although averaging over many particles we find the average flow, or average number moving per unit time, to be a constant, there will be fluctuations about this average. The mechanism of the fluctuations depends on the particular process; in a vacuum-tube case it is the random emission of the electrons from the cathode, in a semiconductor it is the randomness in the number of electrons that continually recombine with holes, or in the number that diffuse, etc. Thus the processes that give rise to an average flow have statistical variations built in, producing fluctuations about the average. It is found that in all these cases the mean-squared fluctuations about the average value are proportional to the average value itself, so that *shot noise* is characterized by a dependence of the *noise* on the *average value*. (Because these processes involve forces producing the flow of particles, they are in nonequilibrium, thermodynamically speaking, so that they are difficult to treat using classical thermodynamics.)

Thermal noise was first thoroughly studied experimentally by J. B. Johnson of Bell Laboratories in 1928. His experiments, together with the accompanying theoretical studies by H. Nyquist, demonstrated that a metallic resistor could be considered the source of spontaneous fluctuation voltages with mean-squared value

$$\overline{v^2} = 4kTRB \tag{5-209}$$

where T is the temperature in degrees Kelvin of the resistor, R its resistance in ohms, k the Boltzmann constant already referred to (1.38×10^{-23} J/K), and B any arbitrary bandwidth. Johnson was able to measure the value of k fairly accurately using this equation and thus demonstrated its validity. He also showed that $\overline{v^2}$ was proportional to temperature.

This expression for the mean-squared thermal noise due to a resistor R implies that the noise is white. We shall see shortly that it is valid up to extremely high frequencies of the order of 10^{13} Hz. At these high frequencies

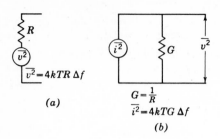

$\overline{v^2} = 4kTR\,\Delta f$

(a)

$G = \dfrac{1}{R}$

$\overline{i^2} = 4kTG\,\Delta f$

(b)

Figure 5-65 Thermal-noise circuit models. (a) Voltage model. (b) Current model.

quantum-mechanical effects set in. (This assumes of course that R is independent of frequency over this tremendous range.) We shall develop a more general relation later which includes the quantum-mechanical effects at high frequencies. (The more general expression is necessary when considering communication at optical frequencies.)

For white noise $\overline{v^2}$ may be written as $n_0 B$, with $n_0/2$ the noise spectral density in volts² (see Sec. 5-5). From Eq. (5-209) the voltage thermal-noise spectral density is thus given by the simple expression

$$G_r(f) = \frac{n_0}{2} = \frac{\overline{v^2}}{2B} = 2kTR \tag{5-210}$$

We shall develop this white-noise spectral density as a special case of the more general relation to which reference has just been made.

Nyquist's original derivation of Eq. (5-209)[62] was based on thermodynamic reasoning, assuming temperature equilibrium. The actual mechanism of thermal-noise generation—assumed to be due to the random interaction between the conduction electrons and ions in a metallic conductor—is not necessary for the derivation. Although this at first appears disconcerting, it is actually a blessing in disguise. For using the same thermodynamic reasoning it may be shown that *any linear passive* device, mechanical, electromechanical, microphones, antennas, etc., has associated with it thermal noise of one form or another. In some cases this may be due to random agitation of the air molecules, in others to random electrical effects in the ionosphere and atmosphere, etc. This is so because the $\frac{1}{2}kT$ term occurs generally in thermodynamics as the energy associated with any mode of oscillation (as temperature increases, gas molecules move more rapidly, ions vibrate more violently in a lattice structure, etc.). This was the basis of Nyquist's derivation. We shall use a somewhat similar approach here.

Because of this reasoning the *resistors in the electrical analog of a passive linear physical device* may be considered *sources of noise voltage* as given by Eq. (5-209). This concept is commonly used in antenna work, acoustics, etc.

A voltage-model representation of Eq. (5-209) is shown in Fig. 5-65a. R is assumed noise-free, with the noise effect lumped into the noise-voltage source

[62] H. Nyquist, *Phys. Rev.,* no. 32, p. 110, 1928.

shown. An application of Norton's theorem gives the current-source equivalent of Fig. 5-65b. (Since $i = v/R$, $\overline{i^2} = \overline{v^2}/R^2 = 4kTGB$; $G = 1/R$.)

Either model may be used, although the current-source model is often more convenient, especially when calculating noise voltages across parallel elements. Some typical numbers are of interest. Let B be 5 kHz, T equal 293 K (this is normal room temperature, or 20°C), and R equal 10 kΩ. Then $\overline{v^2} = 0.8 \times 10^{-12}$ V², or

$$\sqrt{\overline{v^2}} = 0.90 \ \mu\text{V rms}$$

$\sqrt{\overline{i^2}} = 0.90 \times 10^{-10}$ A rms. If the bandwidth is quadrupled to 20 kHz, the rms noise voltage and current are doubled to 1.8 μV and 1.8×10^{-10} A, respectively. The rms noise voltage is proportional to the square root of the resistance and to the square root of the bandwidth.

If the temperature is increased, the resistance value used refers to the new temperature, as does T. Since the derivation of Eq. (5-209) depends on thermal equilibrium, the equation holds only after a steady-state temperature has been reached, not during the heating or cooling period.

Recapitulating, both theory and experiment indicate that a resistance R at temperature T is found to be the source of a fluctuation (noise) voltage with mean-squared value

$$\overline{v^2} = 4kTRB \tag{5-209}$$

and voltage spectral density

$$G_r(f) = 2kTR \equiv \frac{n_0}{2} \tag{5-210}$$

The corresponding mean-squared value and spectral density for the current-source model may be written, respectively,

$$\overline{i^2} = 4kTGB \tag{5-211}$$

and

$$G_i(f) = 2kTG \equiv \frac{n_0}{2} \tag{5-212}$$

As previously, B represents the noise-equivalent bandwidth of the measuring instrument or circuit used to measure these quantities.

Depending on whether we deal with voltage or current in a network we use the spectral density forms of (5-210) or (5-212). The same symbol $n_0/2$ is used here for both cases, just as was done in earlier sections of this chapter. The dimensions of $n_0/2$ depend on which physical quantity is being measured. (In earlier sections we assumed a unit resistance. Since we commonly take ratios of mean-squared signal to mean-squared noise to form the signal-to-noise ratio, the actual dimensions of $n_0/2$ do not really matter, as long as we are consistent.)

The distinction between voltage or current spectral density disappears completely when we discuss actual *power* generated. As is true with any power source the power generated depends on the load impedance. In particular, maximum power is transferred when the load impedance is matched to the generator impedance, in this case just R in Fig. 5-65. The maximum power available, under matched conditions, is called the *available power*. For the noise source of Fig. 5-65a, the available power in a bandwidth of B hertz is just

$$N = \frac{\overline{v^2}}{4R} = kTB \equiv n_0 B \qquad (5\text{-}213)$$

and the corresponding power spectral density in watts/Hz, is just

$$G_n(f) = \frac{n_0}{2} = \frac{kT}{2} \qquad (5\text{-}208)$$

and $n_0 = kT$. This is of course the expression written earlier. In carrying out the SNR calculations of the next section we shall use this expression to compute the noise power.

The discussion to this point of the noise spectral density, whether in units of volt^2 (5-210), current^2 (5-212), or power (5-208), has been based on (5-209), an expression found experimentally by J. B. Johnson. It is now of interest to digress somewhat to discuss the actual derivation of thermal noise spectral density. This will enable us to determine the validity of the white-noise approximation and the frequency (in the infrared range) at which the spectral density begins to vary with frequency. It will also enable us to readily consider extensions to masers and lasers—i.e., thermal-noise calculations at infrared and optical frequencies—as well as to thermal noise generated in space and received at an antenna.

Recall again that we assume we have a resistor R sitting at an equilibrium temperature of T degrees Kelvin. This resistor is the source of random current (and hence voltage) fluctuations produced by the heat energy assumed provided by its surroundings. As the temperature T increases, the electrons moving freely inside the conducting material are raised to higher energy levels and the mean-squared current flow increases. (The average current of course remains zero with no electric field applied, since currents in opposite directions cancel, on the average.) The actual physical mechanism of current flow in the conductor depends on the interaction of the electrons with the thermally vibrating ions in the metallic crystal. Since the noise derivation using an assumed model for electrical conduction through a metal is rather complicated,[63] we resort instead to the stratagem first used by H. Nyquist[64] in deriving the thermal-noise equation for a resistor. This technique is based on simple equilibrium thermodynamics, avoids the problem of the specific physical mechanism involved in the noise generation, and has the advantage of easily being extended to

[63] J. L. Lawson and G. E. Uhlenbeck, *Threshold Signals*, McGraw-Hill, New York, 1950.
[64] Nyquist, *op. cit.*

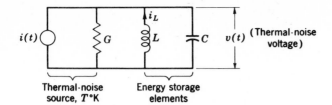

Thermal-noise source, $T°K$

Energy storage elements

Figure 5-66 Tuned circuit for thermal-noise calculations.

thermal-noise calculations where physical resistors may not be present, e.g., blackbody radiation, thermal noise in microwave circuits, masers, lasers, etc.

The stratagem consists of assuming the resistor connected electrically to one or more energy-storage elements. The elements store up the thermal energy provided by the resistors. By knowing the energy stored and equating it to the thermal energy supplied by the resistor, one then finds the mean-squared thermal-noise current (or voltage) available at the resistor terminals.

Although various combinations of energy-storage elements may be used for this calculation, we shall select the oscillatory circuit consisting of a parallel inductor and capacitor, as shown in Fig. 5-66. We do this for two reasons:

1. The stored thermal energy of the tuned circuit is readily written down.
2. The approach used enables us to extend the results to many other situations: nonelectrical systems where G, L, C are the electrical analogs (e.g., a mass-spring combination, with G representing the dissipative elements assumed at temperature T); microwave and optical circuits, with LC representing a resonant cavity, G its dissipation, etc.

This oscillatory circuit, which oscillates at the resonant frequency $f_0 = 1/2\pi\sqrt{LC}$, is the prototype of the *harmonic oscillator* of modern physics. In most books on modern physics and quantum mechanics[65-67] it is shown that the harmonic oscillator, resonant at frequency f_0, can possess discrete stored energies only, its discrete energy levels given by

$$E_n = (nh + \tfrac{1}{2})f_0 \qquad n = 0, 1, 2, \ldots \qquad (5\text{-}214)$$

(Fig. 5-67).

The constant h is Planck's constant, $h = 6.6257 \times 10^{-34}$ J-s. If one now energizes this harmonic oscillator thermally at T degrees Kelvin, one finds that all energy levels may be occupied, but with differing probabilities. In particular, the probability of exciting the nth energy level is found to exponentially decrease with n, and is proportional to $e^{-E_n/kT}$, k again the Boltzmann constant,

[65] R. B. Leighton, *Principles of Modern Physics,* McGraw-Hill, New York, 1959.
[66] R. L. Sproull, *Modern Physics,* 2nd ed., Wiley, New York, 1963.
[67] R. T. Weidner and R. L. Sells, *Elementary Modern Physics,* Allyn and Bacon, Boston, 1960, pp. 433–435, 491–494.

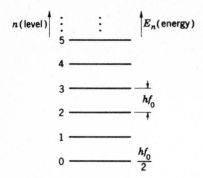

Figure 5-67 Harmonic-oscillator energy levels.

$k = 1.38 \times 10^{-23}$ J/K. A simple statistical average to provide the average energy of the harmonic oscillator then gives

$$\bar{E} = \frac{\sum\limits_{n=0}^{\infty} E_n e^{-E_n/kT}}{\sum\limits_{n=0}^{\infty} e^{-E_n/kT}} = \frac{hf_0}{2} + \frac{hf_0}{e^{hf_0/kT} - 1} \qquad (5\text{-}215)$$

after substituting in Eq. (5-214) and performing the indicated summations.

As an example, for very low temperatures, $kT \ll hf_0$, $\bar{E} = hf_0/2$, the zero-point energy or lowest energy level. If the thermal energy provided is very small, the harmonic oscillator will on the average remain at its lowest energy level. If the temperature is now *high*, $kT \gg hf_0$, $e^{-hf_0/kT} \doteq 1 - hf_0/kT$, and

$$\bar{E} = kT \qquad kT \gg f_0 \qquad (5\text{-}216)$$

This is the "classical" result, which says that the average energy of the oscillator is proportional to the absolute temperature. It is instructive to note that by average we again mean either of two possibilities—an ensemble or time average:

1. Many identical oscillators, oscillating independently of each other, are available. Each one will randomly occupy a particular energy level. The average energy of the *ensemble* is then given by Eq. (5-215).
2. Alternatively, *one* oscillator will occupy only one level at any one time. But over many observation times it will occupy different levels with different probabilities, the *average* being given by Eq. (5-215).

The *LC* circuit of Fig. 5-66 is just one of many possible "harmonic oscillators" to which Eq. (5-215) applies. (In most physics texts the symbol ν, rather than f_0, is used to denote the frequency of oscillation.) The "oscillator" may be made up of two bound atoms forming a molecule such as H_2. Heat applied to a gaseous system of such molecules tends to pull the atoms apart, restoring forces tend to keep them together, the two atoms then vibrating at a characteristic frequency determined by their masses and restoring force constants. $\bar{E}$ is then the average vibrational energy of the molecules. The thermally vibrating ions in a crystal, bound to their specific locations by molecular forces but vibrating due to heat energy applied, may be modeled by harmonic oscillators. The application of Eq. (5-215) then enables one to quite accurately calculate the specific heat of a solid.[68,69] Thermal radiation produced by individual atoms radiating in a given substance is also calculable using Eq. (5-215). Here one assumes the radiation "captured" by a resonant structure of arbitrary size. Standing electromagnetic waves are set up in this structure, at frequencies appropriate to its boundary conditions. Each frequency is then assumed to correspond to a fictitious harmonic oscillator at that frequency, and has average energy given by Eq. (5-215). Summing over all possible frequencies, one finds the thermal energy distribution is found to be exactly that of *blackbody radiation*[70,71]: the radiant energy emanated by many heated sources at T degrees Kelvin (a furnace, the sun, stars, the atmosphere, etc.). It is this blackbody radiation, produced by the sun, stars, sources of radiation in the earth's atmosphere, etc., that appears as thermal noise at an antenna input in any high-frequency radio receiver.

Consider now specifically the parallel G, L, C combination of Fig. 5-66. The resistor at temperature T degrees Kelvin may be an actual resistor connected across the LC circuit, or it could be the equivalent dissipation resistance of the LC circuit, assumed connected in parallel, or it could be the equivalent resistance of an antenna tuned to frequency $f_0 = 1/2\pi \sqrt{LC}$, etc. We visualize the resistor kept at temperature T degrees Kelvin by literally being "immersed in a heat bath" at that temperature; e.g., if the resistor is part of a circuit in a room, T is the normal room temperature. On being connected to the LC circuit the resistor provides the energy $\bar{E}$, given by Eq. (5-215), eventually stored in the L and C after the connection has been made for a while. The connection also enables a current to flow through the resistor, producing a power loss in that element. We relate all of these by saying that there is an average stored energy in the capacitor,

$$\bar{E}_1 = \tfrac{1}{2}C\overline{v^2}$$

equal to the stored energy $\bar{E}_2 = \tfrac{1}{2}L\overline{i_L^2}$ in the inductor. The total stored energy is then

$$\bar{E} = C\overline{v^2} = \frac{hf_o}{2} + \frac{hf_o}{e^{hf_o/kT} - 1} \tag{5-217}$$

from Eq. (5-215). The quantity $\overline{v^2}$ is just the mean-squared fluctuation of the voltage $v(t)$ appearing across the tuned circuit, and thus represents the thermal

---

[68] Leighton, *op. cit.*

[69] F. Reif, *Fundamentals of Statistical and Thermal Physics*, McGraw-Hill, New York, 1965.

[70] Leighton, *op. cit.*

[71] Reif, *op. cit.*

noise measured at that point. If we visualize this noise term having a spectral density $G_r(f)$, we must have

$$\overline{v^2} = \int_{-\infty}^{\infty} G_r(f)\, df \qquad (5\text{-}218)$$

Assuming now that the source of the thermal noise in the resistor is represented by the current source $i(t)$ in Fig. 5-66, we must have the following relation connecting the spectral densities of $v(t)$ and $i(t)$:

$$G_r(f) = |H(\omega)|^2 G_i(f) \qquad (5\text{-}219)$$

From Eqs. (5-217) to (5-219), we then get the following equation, which enables us to find the spectral density $G_i(f)$ of the noise source in terms of the known stored thermal energy of the tuned circuit:

$$\overline{v^2} = \frac{\bar{E}}{C} = \int_{-\infty}^{\infty} |H(\omega)|^2 G_i(f)\, df \qquad (5\text{-}220)$$

The transfer function appearing in (5-220) is the impedance of the parallel combination of the G, L, and C in Fig. 5-66. Letting its bandwidth be very narrow compared to the center frequency, $G_i(f)$ may be assumed to be slowly varying over that range of frequencies and may be taken out of the integral of (5-220). Carrying out the indicated integration, and then solving for $G_i(f)$, using (5-215) for the stored energy of the circuit, we get, finally,

$$G_i(f) = 2\left(\frac{hf}{2} + \frac{hf}{e^{hf/kT} - 1}\right) G \qquad (5\text{-}221)$$

as the spectral density of the resistive noise-current source. Here we have dropped the "0" subscript in the frequency term, since f_0 was an arbitrary frequency to which we assumed the oscillatory circuit tuned.

Equation (5-221) represents the general spectral-density expression for the thermal-noise-current source associated with the resistor. It can of course be converted to a voltage equivalent by simply replacing G by R.

This general form for spectral density now holds for *any* dissipative element at $T°K$, at any frequency, with an equivalent resistance R, or equivalent conductance $G = 1/R$.

The first term $hf/2$ in the expression for spectral density, due to the zero-point energy of the harmonic oscillator, is a strictly quantum-mechanical one. This quantum-mechanical noise is negligible at frequencies $f \ll kT/h \sim 10^{13}$ Hz. At infrared and optical frequencies, however, it begins to play a dominant role.

Note that in general the noise spectral density is *not* white. It varies with frequency, and, in particular, begins to decrease at frequencies the order of kT/h, or 10^{13} Hz at room temperature. For frequencies considerably below 10^{13} Hz, however, the quantity within the parentheses of (5-221) is approximated quite closely by kT. At these frequencies the noise is effectively white, and the noise spectral density is given by the simple expressions (5-208), (5-210), and (5-212).

As an example of the applicability of (5-221), or its approximate, white-

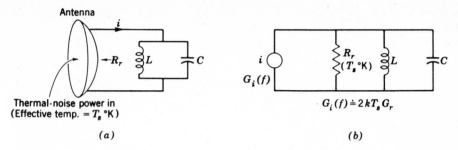

Figure 5-68 Thermal radiation power at antenna. (*a*) Actual system. (*b*) Noise model.

noise form, Eq. (5-212), assume that the antenna of a radio receiver is connected directly to a tuned circuit at the desired r-f (carrier) frequency. In addition to the desired signals coming in at the antenna, thermal noise also appears. Assume that this noise is primarily due to blackbody radiation from space at an average temperature of T_s degrees Kelvin. It is well known[72] that the antenna may be assumed to have an effective radiation resistance R_r such that $I^2 R_r$ = power at antenna input, with I the rms current actually measured at the antenna terminals. This is shown in Fig. 5-68*a*.

Using the same argument as previously, the tuned circuit must have an average stored energy $\bar{E} = kT_s$ (for frequency $f_0 \ll kT_s/h$), since it is the thermal or blackbody radiation from space, at effective temperature T_s degrees Kelvin, that provides the stored energy, by hypothesis. This is important to stress. It is *not* the temperature of the tuned circuit that determines the thermal-noise energy stored, but the temperature of whatever mechanism is responsible for providing the noise power.

Since the tuned circuit sees, effectively, the radiation resistance R_r in parallel with it, it is apparent that the noise source acts as if it were a resistance R_r at a temperature T_s degrees Kelvin. This results in the noise model shown in Fig. 5-68*b*.

How does one determine the effective temperature T_s? This depends of course on what the antenna "sees": it depends on the actual thermal radiators in the solid angle subtended by the antenna, their actual temperatures (these could be the earth's surface reradiating thermal energy, the atmosphere, radiation from the sun if the antenna is aimed in that direction, galactic radiation, etc.), the frequency of the antenna system and the thermal noise received at this frequency, etc. All these must be measured. In practice one measures an effective rms noise current at the antenna terminals (or, perhaps, an rms noise voltage across the tuned circuit). If this is essentially due to incoming thermal radiation, and *not* to actual dissipation in the antenna itself, one must have

$$\overline{i^2} = 4kT_s G_r B \qquad (5\text{-}222)$$

[72] E. C. Jordan, *Electromagnetic Waves and Radiating Systems*, Prentice-Hall, Englewood Cliffs, N.J., 1950.

with B the bandwidth of the measuring apparatus. With G_r, B, and $\overline{i^2}$ known (measured), this provides a measure of T_s.

How *does* one take into account actual antenna dissipation in the tuned circuit (including a resistor possibly put in parallel)? Let R represent the equivalent resistance corresponding to these quantities. Letting the actual temperature of the antenna system be T degrees Kelvin, and the effective temperature of space T_s degrees Kelvin, *each* source must *independently* provide energy to the tuned circuit. Assuming the frequency f low enough so that $f \ll kT/h$ and $f \ll kT_s/h$, to simplify the analysis [otherwise we simply use the more complicated form, Eq. (5-221) for the spectral densities], the spectral density at the tuned circuit output (or *any* circuit output for that matter) is given by

$$G_v(f) = |H(\omega)|^2(2kT_sG_r + 2kTG) \qquad G = \frac{1}{R},\ G_r = \frac{1}{R_r} \qquad \text{(5-223)}$$

One simply *adds* the independent noise contributions. (Recall that if independent or uncorrelated random variables are added, their variances add.) This is shown in Fig. 5-69.

More commonly, one deals with the *power* spectral density form of (5-208) in carrying out calculations involving antenna receiving systems. The noise power output of a typical receiving system then consists essentially of two quantities—the radiation from space at some equivalent source temperature T_s degrees Kelvin, plus all other noise generated within the receiving system itself. This additional noise generally contains both thermal and shot noise components. It has become common to lump all these additional noise sources together by defining an effective *noise temperature* T_e degrees Kelvin. This is the temperature of a (fictitious) thermal-noise source at the system input that would be required to produce the same added noise power at the output. If this effective temperature T_e is much less than the equivalent temperature T_s of

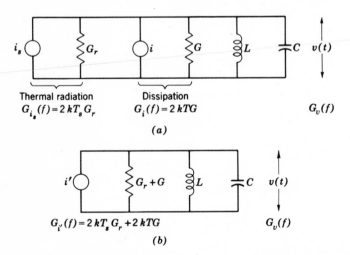

(a)

(b)

Figure 5-69 Multiple thermal-noise sources. (*a*) Complete model. (*b*) Reduced model.

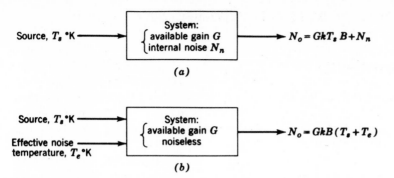

Figure 5-70 Definition of effective noise temperature. (*a*) Actual system noise. (*b*) Effective noise temperature.

noise sources actually producing noise power at the system input, this implies the system introduces no noise of its own. The total noise power spectral density, as measured at the antenna input, is then given by

$$G_n(f) = \frac{k(T_s + T_e)}{2} = \frac{kT}{2} \qquad (5\text{-}224)$$

with T the overall system noise temperature. Figure 5-70 shows schematically how the noise temperature T_e is defined. A system with power gain G and bandwidth B is shown producing at its output the sum of two noise contributions: one due to the amplification (or attenuation, if $G < 1$) of the input source noise, the other, N_n, representing the noise power due to all the system noise sources. Reflecting these back to the input, the system noise sources are represented as arising from an equivalent thermal noise source at temperature T_e at the input.

In practice, low-noise receiving systems (e.g., maser amplifiers) are often used to reduce the effective noise temperature. In some space communication systems the resultant T_e may be reduced to as low as 2 K. In others it ranges from 10 to 30 K.

The term *noise figure* is sometimes also used to measure the noisiness of a receiving system. The noise figure F is defined as the ratio of noise power appearing at the system output to that that would appear if the system generated no noise of its own, the output noise power then being that due to the thermal noise power at the input only. A noise figure of 2 (3 dB) thus means the system noise contribution equals that introduced at the input to the system. The power spectral density, using the noise figure concept, is then simply

$$G_n(f) = \frac{FkT_s}{2} \qquad F \geq 1 \qquad (5\text{-}225)$$

Alternately, noise figure and noise temperature are related by the expression

$$F = 1 + \frac{T_e}{T_s} \qquad (5\text{-}226)$$

5-14 APPLICATIONS TO SPACE AND SATELLITE COMMUNICATIONS

In previous sections of this chapter we evaluated the performance of both digital and analog communication systems in the presence of additive noise, most often taken as white. We showed that the performance of digital systems depended on E/n_0, the ratio of received signal energy to white noise spectral density, while the performance of the analog systems depended on the SNR, involving the ratio of signal power to noise power $n_0 B$.

In this section we carry out some specific performance calculations, using the thermal-noise results of the previous section, for a few typical space and satellite communication examples. This will enable us to see how noise limits the rate of transmission in the digital case, and requires the use of higher-gain antennas and/or lower-noise receivers in the analog FM case to ensure above-threshold operation. Trade-offs among signal power, antenna sizes, noise temperature, bandwidth and/or digital data rate arise naturally out of the discussion. This then leads directly into the material of Chap. 6, where, focusing on digital communications, we ask whether the limitations on the rate of transmission due to noise can be reduced by more sophisticated coding. We shall in fact show there that coding can be used to improve the performance of digital communications systems limited by noise, but that the E/n_0 ratio still remains the limiting factor.

The examples to be worked out in this section deal with transmission through space from a transmitter on a space vehicle (either involved in a deep-space probe, or a synchronous satellite in orbit about the earth). The results obtained are more general, however, since they refer to radio communication between any two bodies located a known distance apart. Figure 5-71 describes the general picture. The transmitter on the space vehicle communicates with the earth station d meters away. The system operates at a known carrier fre-

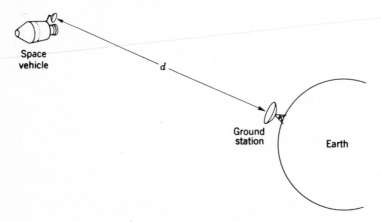

Figure 5-71 Space communication.

quency of f hertz (or equivalent wavelength λ meters), and has an average transmitter-output power S_T watts at that frequency. The problem is to determine the signal power and/or energy received at the earth and use this to calculate both, E/n_0 and the SNR.

If the power on the vehicle were radiated isotropically, the power density (watts/meter²) at a distance d would be $S_T/4\pi d^2$. An antenna on the vehicle serves to concentrate or focus the electromagnetic energy transmitted into a beam. The larger the antenna size (in wavelengths) the narrower the beam and the greater the energy concentration. Specifically, this focusing effect is denoted by assigning to the antenna a gain G_T over isotropic radiation. The effective power density at a distance d from the transmitting antenna is then $(S_T/4\pi d^2)G_T$. At high frequencies aperture-type antennas (parabolas or lens antennas are examples) are commonly used, in which case it may be shown that the maximum gain attainable is proportional to the aperture area A_T (the solid angle subtended by the beam is inversely proportional to the area) and is given by $G_T = 4\pi\eta_T A_T/\lambda^2$.[73] Here λ is the wavelength and $\eta_T < 1$ is an efficiency parameter. At the receiver the receiving antenna of aperture area A_R ideally presents an effective area equal to A_R in picking up the received power. This area is less than A_R because of losses. An efficiency parameter η_R is introduced to account for these losses. The received power in watts is then

$$S_R = \frac{S_T}{4\pi d^2}\, G_T A_R \eta_R \qquad (5\text{-}227)$$

This simple expression relates received and transmitted powers. (This expression ignores transmission losses due to attenuation along the propagation path. It assumes free space transmission.) Quite commonly actual power calculations are carried out in decibels. This of course simplifies the calculation, and enables any additional loss or gain factors to be simply added or subtracted. (These would include antenna feed losses, propagation losses, etc.)

Consider some numbers from the Mariner 10 deep-space mission to Mercury in 1974.[74] One of the prime objectives was to return relatively high resolution images of Mercury via a high-bit-rate telemetry system. $S_T = 16.8$ W of transmitter power was available for this purpose. (A 20 W-transmitter was used, but 15 percent of the power was required for another, low-data-rate digital channel, as well as for power to transmit a carrier signal used for synchronous detection at the receiver.)

The telemetry signal was transmitted at a frequency of 2,300 MHz (S-band), or a wavelength of 0.13 m. The transmitter antenna had a diameter of 53 in, or 1.35 m. Using an antenna efficiency parameter $\eta_T = 0.54$, the gain of the

[73] D. J. Angelakos and T. E. Everhart, *Microwave Communications*, McGraw-Hill, New York, 1968, sec. 5-7.

[74] M. G. Easterling, "From 8⅓ Bits/s to 100,000 Bits/s in Ten Years," *Proceedings, National Telecommunications Conference*, Dallas, Texas, 1976; reprinted in *IEEE Commun. Soc. Mag.*, vol. 15, no. 6, pp. 12–15, November 1977.

transmitting antenna is readily found to be $G_T = 575$, or, in decibels, 27.6 dB. The distance to Mercury at the time of the encounter was 99×10^6 mi, or 1.6×10^{11} m. The distance loss factor $4\pi d^2$, in (5-227), is then calculated to be 235 dB, again using decibel measure. The earth station antennas were 64-m parabolic dishes, with an efficiency $\eta_R = 0.575$. $\eta_R A_R$ in decibels is then calculated to be $\eta_R A_R = 32.5$ dB.

Putting all of these numbers together, the received power, in decibels relative to a watt (dBW), is

$$S_R\big|_{\text{dBW}} = S_T\big|_{\text{dBW}} + G_T\big|_{\text{dB}} + \eta_R A_R\big|_{\text{dB}} - 4\pi d^2\big|_{\text{dB}} \qquad (5\text{-}228)$$

For the numbers given here this is calculated to be -162.6 dBW. This is an astoundingly small power (5.44×10^{-17} W), due of course to the extraordinarily long distances (160×10^6 km) over which communication must be carried out. But, as has been noted many times in this book, it is the ratio of the signal to the noise that determines the effectiveness of communications. In particular, the very small power by itself poses no problem—one can always introduce amplification to raise the signal to any desired level. (This is of course precisely what is done.)

The noise spectral density is thus critical. In this example of the Mariner 10 mission very low maser receivers, with effective noise temperatures of 2.1 K, had been specially designed. As a result, the overall system temperature (space temperature plus noise temperature) was calculated to be 13.5 K. Again using decibel measure, it is readily shown that $n_0 = kT = -217.3$ dB at this temperature. (It is left to the reader to show that k, Boltzmann's constant, is in decibel measure -228.6 dB.) Finally, then, $S_R/n_0\big|_{\text{dB}} = 54.7$ dB.

We now note that the average received power S_R is simply the average energy E received in a bit interval divided by the length of the bit interval. Alternatively, letting R be the bit rate in bits per second,

$$\frac{E}{n_0} = \frac{S_R}{n_0 R} \qquad (5\text{-}229)$$

Since S_R/n_0 is known in this example, E/n_0 may be calculated for any bit rate. More appropriately, choosing an acceptable probability of bit error P_e, we can calculate the E/n_0 required for a given digital modulation scheme, and from this determine the allowable binary transmission rate R. In particular, for the Mariner 10 mission it was decided that $P_e = 0.05$ was acceptable for the image telemetry channel under consideration. This relatively high error rate was specifically chosen to allow higher bit rates to be transmitted, with corresponding increases in image resolution. The fact that 5 bits in 100, on the average, would be in error was deemed acceptable for the application (picture transmission) involved in this example.

PSK transmission was used for the telemetry data. Using Fig. 5-39 to determine E/n_0 for this case, one finds that for $P_e = 0.05$, $E/n_0 = 1.4$ dB. Hence from (5-229), $R = 214{,}000$ bits/s is the bit transmission rate allowed. In the

Mariner 10 example, a 117,600-bits/s data rate was actually used. (Additional power losses in the system not accounted for here result in the difference in transmission rates.)

The fact that it is possible to transmit digital data at rates exceeding 100 kbits/s over distances of 160×10^6 km or more must be considered a remarkable achievement. It is the result of the coordinated efforts of many engineers working jointly over a period of many years to develop higher power transmitters at the frequency in question, larger antennas on both the space vehicle and on the ground, lower noise receivers, better modulation and coding techniques, etc. The E/n_0 ratio as seen on the ground reflects all these inputs.

The power dependence on distance, going inversely as d^2 [see Eq. (5-227)], is a particularly sensitive one. Note that a factor-of-2 change in distance results in a factor-of-4 (6-dB) change in power. This in turn reflects itself in a change of four in the allowable data rate.

To show the interplay of the various factors determining the performance of these space communication systems, consider as a contrast to the Mariner 10 system the telemetry system used in the 1964 Mariner 4 mission to Mars.[75] The transmitting antenna diameter was 31 in, or 0.79 m., with a gain of 23.4 dB, compared to the 27.6 dB gain of the Mariner 10 antenna. The transmitter power was 10 W, of which only 29 percent could be used directly for data transmission, the remainder being used for transmitting the carrier and a synchronizing signal. The ground antennas were 26-m (85-ft) dishes, rather than the 64-m (210-ft) antennas used with Mariner 10. This alone reduces the allowable bit rate by a factor of 6. The noise temperature at the receiver was 55 K. The distance to Mars was 216×10^6 km. All of these differences, together with other loss factors, result in a combined reduction in E/n_0 of close to 9,000, as compared to the Mariner 10 system! The Mariner 4 system was required to operate at a lower error probability as well. As a result, the data rate in 1964, in transmissions from Mars, was $8\frac{1}{3}$ bits/s. Contrast this with the 117,600-bits/s data transmission rate from Mercury just 10 years later.

Both of these space missions were to the nearby planets. The Voyager missions, to the outer planets, required additional improvements and changes to allow data to be transmitted at 100-kbits/s rates over the long distances involved.[76] The transmitting antenna in particular was increased in size to 3.66 m (12 ft). Its efficiency parameter was improved to 0.65. The transmitting frequency was increased from 2295 MHz (S band) to 8415 MHz (X band) to take advantage of the $1/\lambda^2$ gain improvement. As a result, the gain of the Voyager antenna is 48 dB, as compared to 27.6 dB for the Mariner 10 antenna. This enabled the data rate at a distance of 750×10^6 km to be established at a design value of 115 kbits/s, with $P_e = 0.005$.

[75] *Ibid.*, pp. 14, 15.

[76] J. R. Kolden and V. L. Evanchuk, "Planetary Telecommunications Development During the Next Ten Years," *Proceedings, National Telecommunications Conference*, Dallas, Texas, 1976; reprinted in *IEEE Commun. Soc. Mag.*, vol. 15, no. 6, pp. 16–19, 24, November 1977.

Satellite Performance Calculations

Equation (5-227) for the received power at a distance d meters from a transmitting antenna is directly applicable to the calculation of the signal-to-noise ratio at the receiver in satellite communication systems as well. It is customary in satellite communication applications to rewrite the equation somewhat in terms of the gain of the receiving antenna, however, to focus attention on design parameters that are under the control of individual users on the earth. Recall that the Intelsat worldwide communications system has multiple users at various earth stations accessing a given synchronous satellite (see Sec. 4-12). Since the satellite parameters and frequency of transmission are fixed, the performance of the communications system as seen by individual users can only be modified by varying the receiver noise temperature and the receiver antenna size, or, equivalently, its gain G_R. The SNR at the receiver and hence the performance will be seen to depend on G_R/T. This ratio has thus come to be considered a figure of merit of a given earth station.

Equation (5-227) is easily rewritten by noting again that the antenna gain is related to the area by the expression $G_R = 4\pi\eta_R A_R/\lambda^2$. Hence replacing $\eta_R A_R$ in (5-227) by $\dfrac{\lambda^2}{4\pi} G_R$, one gets

$$S_R = \frac{S_T}{4\pi d^2} G_R G_T \frac{\lambda^2}{4\pi} \tag{5-230}$$

as an equivalent expression. The power-to-noise spectral density is then given by dividing S_R in (5-230) by $n_0 = kT$, just as done in the space communications examples. Equation (5-230) does not include various losses occurring at the transmitter, during transmission, and at the receiver.

We now focus very specifically on the Intelsat IV system. The down-link frequency for this system is 4 GHz, as noted earlier in Sec. 4-12. The transponder power output is 3.2 W, and the satellite antenna gain is $G_T = 20$ dB for the 17°-wide global beam that provides maximum earth coverage.[77] (These numbers refer to points on the beam axis, directly below the satellite. A 3-dB loss is incurred by users located at the beam edge.) The satellite-earth distance, again considering users located directly below the satellite, is 35,788 km. It is left to the reader to show, using these numbers, that the received signal-to-noise spectral density, in decibels, is given by

$$\left.\frac{S_R}{n_0}\right|_{dB} = 58 + \left.\frac{G_R}{T}\right|_{dB} - \text{losses (dB)} \tag{5-231}$$

This points up the significance of G_R/T as a figure of merit, as noted previously.

Recall from Sec. 4-12 that frequency modulation is used as the FDMA modulation technique in the Intelsat IV system. Assuming the received

---

[77] "The Intelsat IV Communications System," P. L. Bargellini (ed.), *Comsat Tech. Rev.,* vol. 2, no. 2, pp. 437–572, Fall 1972.

power S_R is distributed uniformly across the 36-MHz transponder bandwidth, the carrier-to-noise ratio CNR is just $S_R/(36 \times 10^6)n_0$. The carrier-to-noise ratio in decibels is then given by

$$\left. \text{CNR} \right|_{\text{dB}} = \left. \frac{G_R}{T} \right|_{\text{dB}} - 17.7 - \text{losses (dB)} \tag{5-232}$$

FM reception requires the carrier-to-noise ratio to be above a threshold of about 10 dB for FM performance to be satisfactory (Sec. 5-12). This puts a lower limit on G_R/T, from (5-232). A suggested value for G_R/T that should provide good performance is 40.7 dB. Note from (5-232) that for this number $\text{CNR}_{\text{dB}} = 23 - \text{losses (dB)}$. This thus provides a loss margin of about 10 dB to ensure above-threshold operation.

Typical large earth station antennas that provide this 40.7-dB design value for G/T include a 26-m. (85.3-ft) antenna with 50 percent efficiency, operated in conjunction with a low-noise receiver at 50 K and a 29.5-m (96.7-ft) reflector with 70 percent efficiency and a noise temperature of 78 K.[78] Note that very large antennas are called for. The same trade-offs are at work here as in the space communications case: were higher-power satellite transmitters to become available the earth station costs could be reduced correspondingly. Going to higher frequencies would also increase the satellite gain, thereby requiring correspondingly smaller earth antennas.

5-15 SUMMARY

In this chapter we have explored in some depth the effect of additive noise on communication signal transmission. In the case of digital signals the noise occasionally results in a mistaken digit, and the performance is thus evaluated in terms of error probability. In the case of analog signal transmission, with AM and FM as the prime examples, the signal-to-noise ratio (SNR) serves as a comparable measure of system performance.

In order to study the effect of noise on system performance we showed how one analyzes both noise and random signals passing through systems. The mechanism of doing this was to define the autocorrelation function and its Fourier transform, the power spectral density. These two functions are central to the study of random processes and recur over and over again in more advanced treatments than that given here.

The power spectral density represents, as implied by the name, the signal (or noise) power distribution over all frequencies. If the power is concentrated in a definable range of frequencies, we can talk of the signal (or noise) bandwidth B, just as in earlier chapters. By the usual Fourier relations, the width of the corresponding autocorrelation function is proportional to $1/B$. This

[78] *Ibid.*, p. 482.

in turn is, by definition of the autocorrelation function, a measure of the correlation between signal samples: for a time separation $\tau < 1/B$, the samples are essentially correlated; for $\tau > 1/B$, the correlation usually goes to zero. This in turn describes, in an intuitive way, the rate of change of the signal (noise) in time.

The effect of passing a random signal through a linear system $H(\omega)$ is to multiply the signal spectral density by $|H(\omega)|^2$. This simple relation then enabled us to determine both the band-limiting effect of systems on signals, and the actual signal (noise) power at the system output.

With these basic concepts of random processes as an introduction we were able to discuss in a systematic way the comparative signal-to-noise properties of a representative group of high-frequency binary transmission systems as well as more traditional AM and FM systems. Thus, after first introducing the *matched filter* both as an application of the noise spectral analysis and because of its importance in its own right, we went on to discuss the reception of high-frequency binary signals in noise. Both synchronous and noncoherent (envelope) detection were considered, the discussion including the PSK, FSK, and OOK signals first introduced in Chap. 4. The narrowband representation of noise enabled us to actually follow through the detection process. We found that matched filtering was appropriate to *all* binary receivers to minimize probability of error, that synchronous detection provides lower error rates (for the same SNR) than envelope detection, and that, ideally, coherent PSK was the system to be preferred.

The matched filter will be encountered again in Chap. 6, where it will be shown to arise quite naturally in discussing, from a fundamental statistical viewpoint, ways of processing digital signals in the presence of additive gaussian noise to minimize probability of error. We showed here that the effect of the matched filter is to make the probability of error dependent on the ratio E/n_0 of signal energy to white-noise spectral density. This ratio will again be shown in Chap. 6 to provide the basic limitation on digital transmission.

In the case of analog modulation systems we were able to show that FM, as an example of a "wideband" modulation scheme, provides a SNR improvement over AM systems. This in turn provides an improvement in the ability to transmit information. In particular, in FM systems, an increase in bandwidth provides a proportional increase in output SNR, effectively suppressing the noise more and more relative to the signal. This is only possible above the threshold region of carrier-to-noise ratio, however. Thus, below a CNR of approximately 10 dB the output SNR plunges rapidly.

This bandwidth-SNR exchange does not occur in AM-type systems. There, above a 0-dB CNR, the output SNR is strictly proportional to the input SNR. (Below 0 dB a suppression effect is also encountered in AM.)

The SNR-bandwidth exchange of FM systems is of course not as efficient as the exponential exchange encountered with coded PCM systems in Sec. 5-2.

To actually evaluate the performance of communication systems in the presence of noise, one must know precisely what is meant by the noise appearing at the input to a receiving system. We showed at the end of the chapter that ever-present thermal noise provides the ultimate limitation on information transmission. We were able to demonstrate that thermal noise spectral density is proportional to the absolute temperature at frequencies well below the optical range. Other sources of noise are then commonly introduced by defining an equivalent noise temperature that incorporates their effect.

Some sample calculations drawn from space and satellite communications enabled us to apply some of the results of this chapter to real systems. We were able to show how trade-offs in signal power, noise temperature, and antenna gains affected the performance of such systems.

PROBLEMS

5-1 The output rms noise voltage of a given linear system is found to be 2 mV. The noise is gaussian fluctuation noise. What is the probability that the instantaneous noise voltage at the output of the system lies between -4 and $+4$ mV?

5-2 Repeat Prob. 5-1 if a dc voltage of 2 mV is added to the output noise.

5-3 Show that the probability of error in mistaking a binary pulse in noise for noise alone or of mistaking noise for a binary pulse is given by Eq. (5-7). Calculate this error for various values of pulse height to rms noise, and check the curve of Fig. 5-5.

5-4 A binary transmission system transmits 50,000 digits per second. Fluctuation noise is added to the signal in the process of transmission, so that at the decoder, where the digits are converted back to a desired output form, the signal pulses are 1 V in amplitude, with the rms noise voltage 0.2 V. What is the average time between mistakes of this system? How is this average time changed if the signal pulses are doubled in amplitude? *Note:* Assume that 1's and 0's are equally likely to be transmitted.

5-5 Show that the optimum decision level for polar signals in additive gaussian noise is given by Eq. (5-10).

5-6 The laplacian distribution is given by $f(x) = ke^{-|x|/c}$ (Fig. P5-6).
   (*a*) Determine k such that $f(x)$ is properly normalized.
   (*b*) Show the standard deviation $\sigma = \sqrt{2}\,c$.
   (*c*) Sketch the cumulative distribution $F(x)$.
Note: This distribution is sometimes used to model the amplitude distribution of burst-type or impulse noise, occurring in high-frequency digital communications.

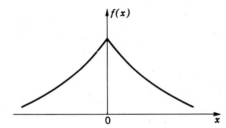

Figure P5-6 Laplacian density function.

5-7 Noise having the laplacian distribution of Prob. 5-6 is added to a polar signal sequence of amplitude $\pm A$. Find the probability of error in terms of A/σ (σ the noise standard deviation or rms value), if the decision level for one signal-plus-noise sample is set at zero. Plot the error curve and compare with Fig. 5-5 for gaussian noise.

5-8 The input to a binary communications channel consists of 0's with a priori probability $P_0 = 0.8$, and 1's with a priori probability $P_1 = 0.2$. The transition probabilities on the channel, i.e., the probabilities during transmission that 1's will be received as 1's, 1's received as 0's, etc., are indicated in Fig. P5-8.

(a) Find the probability of error P_e.

(b) Find P_e if the decisions are *reversed*, i.e., received 1's are called 0's and vice versa.

(c) The decision is made to *always* call a received signal a 0. Calculate the probability of error. Which is the best *decision rule* of the three described in (a), (b), (c)?

(d) Repeat (a), (b), (c) if $P_0 = P_1 = 0.5$.

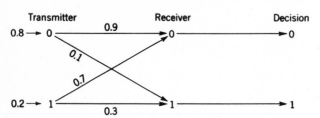

Transmitter   Receiver   Decision

Figure P5-8 Transition probabilities.

5-9 Consider a binary PCM system transmitting 1's with a probability $P_1 = 0.6$ and 0's with a probability $P_0 = 0.4$. The receiver recognizes 0's, 1's, and a third symbol E, called an erasure symbol. There is a probability $P(0|1) = 0.1$ that the 1's will be received (mistakenly) as 0's, $P(E|1) = 0.1$ that they will be received as E's, and $P(1|1) = 0.8$ that they will be received (correctly) as 1's. Similarly, with 0 assumed transmitted, the appropriate probabilities of events at the receiver are given by $P(1|0) = 0.1$, $P(E|0) = 0.1$, $P(0|0) = 0.8$.

(a) Sketch a diagram indicating two transmit and three receive levels, show the appropriate transitions between them, and indicate the appropriate probabilities. *Note:* In the symbolism used above, all conditioning refers to the *transmitter*.

(b) Calculate the probability of receiving a 0, a 1, and an E, respectively. Show these sum to 1, as required.

(c) Show that the probability of an error is 0.1, the probability of a correct decision at the receiver is 0.8, and the probability of an erasure is 0.1.

(d) Repeat (b) and (c) if the transition probabilities, with a 0 transmitted, are changed to $P(1|0) = 0.05$, $P(E|0) = 0.05$, $P(0|0) = 0.9$.

5-10 Refer to Prob. 5-9. The symbol 1 is received. What is the probability it came from a 0? From a 1? Repeat for the symbols 0, and E, as received. (It may pay to adopt new symbols such as T_0, T_1, and R_0, R_1, R_E, or A_1, A_2, and B_1, B_2, B_3 to keep the appropriate conditional probabilities straight.) Check your results by summing appropriate probabilities.

5-11 A binary source outputs bits at a 2,400-bits/s rate. Gaussian noise is added during transmission. The signal-to-noise ratio at the receiver (amplitude of received signal pulse to noise standard deviation) is 10 dB. Find the probability of error of the system in the two cases of (1) NRZ on-off transmission, and (2) polar transmission. 1's and 0's are equally likely to appear. *Hint:*

$$\frac{1}{2}(1 - \operatorname{erf} x) \doteq \frac{e^{-x^2}}{2\sqrt{\pi}\,x} \qquad x > 3$$

5-12 It is common practice in digital transmission systems to strive for an error probability due to noise of 10^{-5}. Find the necessary signal-to-noise at the receiver (signal amplitude to rms noise) in the two cases of NRZ on-off, and polar, transmission. *Hint:* See Prob. 5-11 for a useful approximation to $\frac{1}{2}(1 - \operatorname{erf} x)$.

5-13 A digital transmission channel has a bandwidth of 2,400 Hz and provides a signal-to-noise ratio (SNR or S/N) of 20 dB. Find the information transmission rate (in bits/s) possible over this channel if PCM transmission is used and an error probability of 10^{-5} is specified. Compare with the Shannon capacity for the same channel. Repeat for SNR = 10 dB; 30 dB.

5-14 NRZ polar signals are transmitted over a long line with 99 repeaters used. Find the overall probability of error at the output of the line if the probability of error at any repeater is $p = 10^{-7}$. Compare with the probability of error if amplifiers are used instead. (Additive gaussian noise is assumed.)

5-15 A random signal $s(t)$ of zero average value has the triangular spectral density of Fig. P5-15.
    (a) What is the average power (mean-squared value) $S \equiv E(s^2)$ of the signal?
    (b) Show that its autocorrelation function is

$$R_s(\tau) = S \left(\frac{\sin \pi B\tau}{\pi B\tau}\right)^2$$

Hint: Refer back to Chap. 2 for the Fourier transform of a triangular pulse.
    (c) $B = 1$ MHz, $K = 1 \mu V^2/Hz$. Show that the rms value of the signal is $\sqrt{S} = 1$ mv and that samples of $s(t)$ spaced 1 μs apart are uncorrelated.

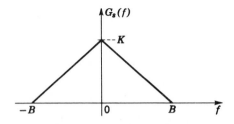

    Figure P5-15

5-16 Band-limited white noise $n(t)$ has spectral density $G_n(f) = 10^{-6}$ V^2/Hz, over the frequency range -100 to $+100$ kHz.
    (a) Show the rms value of the noise is approximately 0.45 V.
    (b) Find $R_n(\tau)$. At what spacings are $n(t)$ and $n(t + \tau)$ uncorrelated?
    (c) $n(t)$ is assumed gaussian. What is the probability at any time t that $n(t)$ will exceed 0.45 V? 0.9 V?
    (d) The noise, again assumed gaussian, is added to polar signals of amplitude $\pm A$. What is the probability of error if the binary signals are equally likely, the decision level is taken as 0, and $A = 0.9$ V? Repeat for $A = 1.8$ V, and 4.5 V.

5-17 $R_n(\tau) = N \cos \omega_0\tau$. Show $G_n(f) = (N/2)\delta(f - f_0) + (N/2)\delta(f + f_0)$. Sketch both $R_n(\tau)$ and $G_n(f)$.

5-18 Consider a random signal $s(t) = A \cos (\omega_0 t + \theta)$, with θ a uniformly distributed random variable. Show that

$$E[s(t)s(t + \tau)] = \frac{A^2}{2} \cos \omega_0\tau$$

by averaging over the random variable θ, as indicated. Compare this result with that of Prob. 5-17. Can you explain this result?

5-19 Consider a random signal

$$s(t) = \sum_{i=1}^{n} a_i \cos (\omega_i t + \theta_i)$$

with the frequencies all distinct, and the θ_i's random and independent. Calculate the autocorrelation function of $s(t)$ by averaging $s(t)s(t + \tau)$ over the random θ_i's, and show that

$$R_s(\tau) = \sum_{i=1}^{n} \frac{a_i^2}{2} \cos \omega_i \tau$$

Calculate the spectral density $G_s(f)$ and sketch. Compare this result with that of Probs. 5-17 and 5-18.

5-20 Consider the periodic rectangular pulse train of Fig. P5-20. Timing jitter causes the entire train to be shifted relative to a fixed time origin by a random variable Δ as shown. Δ may be assumed uniformly distributed over the range 0 to T. Show by averaging $s(t)s(t + \tau)$ over Δ that the autocorrelation function is a periodic sequence of triangular pulses of peak amplitude $A^2 t_0/T$, base width $2t_0$, and period T.

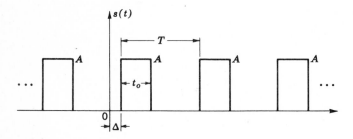

Figure P5-20

5-21 An autocorrelation function that frequently arises in practical problems is given by

$$R(\tau) = R(0)e^{-\alpha|\tau|} \cos \beta\tau$$

(a) Calculate the power spectrum $G(f)$.
(b) Typical values for α and β are $\alpha = 1$, $\beta = 0.6$. Plot $R(\tau)/R(0)$ and $G(f)$.
(c) Check the results of (a) by considering the two limiting cases (1) $\alpha = 0$, (2) $\beta = 0$.

5-22 White noise of spectral density $G_n(f) = n_0/2$ watts/Hz is applied to an ideal low-pass filter of bandwidth B hertz and transfer amplitude A. Find the correlation function of noise at the output. Calculate the total average power at the filter output from the spectral density directly, and compare with $R(0)$.

5-23 White noise of spectral density $G(f) = 10 \ \mu V^2/Hz$ is passed through a noiseless narrowband amplifier centered at 400 Hz. The amplifier may be represented by an ideal bandpass filter of 50-Hz bandwidth about the 400-Hz center frequency and amplification factor of 1,000.

(a) Write an expression for the autocorrelation function $R(\tau)$ at the amplifier output.
(b) The output noise voltage is to be sampled at intervals far enough apart so as to ensure uncorrelated samples. How far apart should the samples be taken?

5-24 White noise of spectral density $10^{-6}V^2/Hz$ is applied at the input of the circuit in Fig. P5-24, as shown.

(a) Find the noise spectral densities $G_1(f)$ and $G_2(f)$, at points 1 and 2, respectively, in terms of the bandwidth $B \equiv 1/2\pi RC$.
(b) Find the output rms noise voltage $\sqrt{N_2}$ in terms of the bandwidth B. $R = 10 \ k\Omega$. What is $\sqrt{N_2}$ if $C = 0.0001 \ \mu F$? $0.01 \ \mu F$?

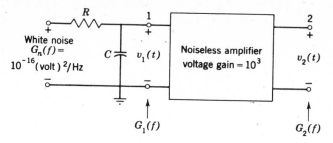

Figure P5-24

5-25 Consider the circuit of Fig. P5-25.

(a) Show that the power spectral density of $v(t)$ at the amplifier input is $G_r(f) = 10^{-14}/[1 + (f/B)^2]$ volts²/Hz; $B = G/2\pi C = 1,600$ Hz.

(b) Show that the rms noise voltage at the output of the amplifier is 7 mV.

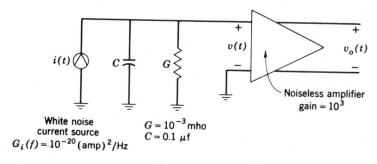

White noise
current source
$G_i(f) = 10^{-20}$ (amp)²/Hz

$G = 10^{-3}$ mho
$C = 0.1\ \mu f$

Figure P5-25

5-26 Consider the random signal shown in Fig. P5-26. $s(t)$ switches randomly between $+1$ and -1. The probability a switch will take place in any small interval Δt is $\lambda \Delta t \ll 1$, independent of switches

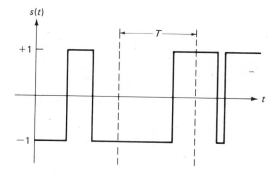

Figure P5-26

in any other time interval. The probability that no switch will take place is $1 - \lambda \Delta t$. The probability of k switches in T seconds is then given by the Poisson probability

$$P(k) = \frac{1}{k!} (\lambda T)^k e^{-\lambda T} \qquad k = 0, 1, 2, \ldots$$

The average number of switches in T seconds is λT, and the parameter λ represents the average number of switches per unit time. λ thus measures the rapidity of change of $s(t)$ and one would expect the bandwidth of $s(t)$ to depend on λ. In this problem we show how one determines the spectral density of a random signal such as $s(t)$ from a calculation of the autocorrelation function.

(a) Show that

$$\begin{aligned} R_s(\tau) &\equiv E[s(t)s(t + \tau)] \\ &= P[s(t)s(t + \tau) > 0] - P[s(t)s(t + \tau) < 0] \end{aligned}$$

Hint: What are the four possible values of $s(t)s(t + \tau)$ that one must average over?

(b) Show that the result of (a) is equivalent to $P[$even number of switches in τ seconds$]$ − $P[$odd number of switches in τ seconds$]$. Use the Poisson distribution to calculate this difference and show that $R_s(\tau) = e^{-2\lambda\tau}$, $\tau > 0$. Repeating for negative τ, show that $R_s(\tau) = e^{-2\lambda|\tau|}$.

(c) Show that

$$G_s(f) = \frac{1/\lambda}{1 + (\pi f/\lambda)^2} \cdot$$

What is the bandwidth of this signal? How does it relate to λ? Compare this result to white noise passed through an RC filter. (See Prob. 5-24, for example.)

5-27 A source $n(t)$ has an autocorrelation given by $R_n(\tau) = 3e^{-a|\tau|}$.

(a) Find $G_n(f)$. Sketch $R_n(\tau)$ and $G_n(f)$ for $a/2\pi = 10^4$ and 10^6. Compare the two sets of curves. What is the rms noise $\sqrt{N}$ in each of the two cases?

(b) Refer to Fig. P5-27. Find the output rms noise power $\sqrt{N_0}$ if $B = a/2\pi = 10^4$.

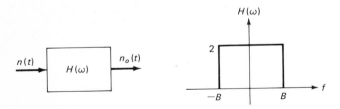

Figure P5-27

5-28 Refer to Fig. 5-19 describing the input-output notation for a linear system. Set up the expression for the output autocorrelation function $R_{n_0}(\tau)$ in terms of the expectation of the product of $n_0(t)$ and $n_0(t + \tau)$. Replace $n_0(t)$ and $n_0(t + \tau)$ by the respective convolution integrals relating the output time function to the input time function and the impulse response $h(t)$. Find $R_{n_0}(\tau)$ in terms of $R_{n_i}(\tau)$ and $h(t)$. Take Fourier transforms and show that (5-50), relating input and output spectral densities, results.

5-29 An on-off binary sequence $s(t)$ has white gaussian noise $n(t)$ added to it, as shown in Fig. P5-29. 1,000 bits/s are transmitted. A typical binary 1 has the shape shown in the figure (50 percent roll-off sinusoidal shaping is used). $H(\omega)$ may be taken to have zero phase. The decision circuit shown outputs a 1 if $v > A/2$ and a 0 if $v < A/2$, with v the value of the voltage at the output of $H(\omega)$, taken once every millisecond at the expected maximum A of the signal component of $v(t)$. Find the probability of error of this system, if 1's and 0's are equally likely to be transmitted.

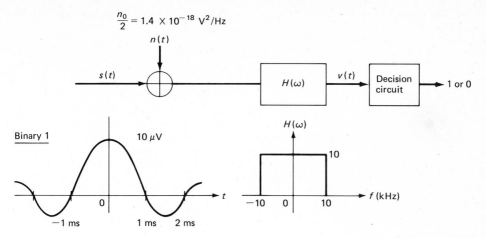

Figure P5-29

5-30 A rectangular pulse of amplitude V volts and width τ seconds is applied to a matched filter. Show that the output is a triangular-shaped pulse. Find the peak value of this pulse. Calculate the total noise power at the output of the filter (assumed noiseless) if white noise of spectral density $n_0/2$ volts^2/Hz is applied at the input. Calculate the output signal-to-noise ratio at the peak of the signal pulse if signal and noise appear together at the filter input.

5-31 (*a*) The signal pulse of Prob. 5-30 is applied to a single RC section, with $RC = 2\tau/3$. Sketch the output pulse, and compare with that of Prob. 5-30. Calculate the total output noise power if white noise of spectral density $n_0/2$ volts^2/Hz is again added to the input signals. Calculate the peak SNR at the output, and compare with the result of Prob. 5-30, checking a point on the appropriate curve of Fig. 5-29.

(*b*) Let the pulse width τ vary, and repeat (*a*). Calculate the peak SNR at the output, and plot as a function of τ. Compare with the appropriate curve of Fig. 5-29.

5-32 A signal pulse of unit peak amplitude and half-power width τ is given mathematically by the "gaussian" error curve $f(t) = e^{-0.35(2t/\tau)^2}$. Added to the signal pulse is white noise of spectral density $n_0/2$ volts^2/Hz. The signal plus noise is applied to a filter.

Show that the optimum filter characteristic for maximizing the peak SNR at the filter output is given by a gaussian curve in frequency,

$$\left| H(\omega) \right| = \sqrt{\frac{\pi}{1.4}} \, \tau e^{-(\omega\tau)^2/5.6}$$

(Recall that $\displaystyle\int_{-\infty}^{\infty} e^{-x^2} \, dx = \sqrt{\pi}$.) Sketch $f(t)$ and $\left| H(\omega) \right|$.

5-33 The gaussian pulse and white noise of Prob. 5-32 are passed through an ideal low-pass filter of cutoff frequency $2\pi B$ radians/s. Show that the maximum peak SNR at the output occurs at $2\pi B\tau = 2.4$. Show that this maximum value of the peak SNR is only 0.3 dB less than that found by using the optimum filter characteristic of Prob. 5-32.

5-34 (*a*) Prove that the impulse response of a matched filter is given in general by Eq. (5-75).

(*b*) A signal pulse given by $f(t) = e^{-\alpha t}$, $t \geq 0$, is mixed with white noise of spectral density $n_0/2$ volts^2/Hz. Find the impulse response of the matched filter, and compare with $f(t)$.

5-35 The i-f section of a communications receiver consists of four identical tuned amplifiers with an overall 3-dB bandwidth of 1 MHz. The center frequency is 15 MHz. The amplifiers represent the bandpass equivalent of RC-coupled amplifiers. The input to the i-f strip consists of a 1-μs rectangular signal pulse plus fluctuation noise generated primarily in the r-f stage and mixer of the receiver. Calculate the relative difference in decibels of the peak SNR at the receiver output for the amplifier given, as compared with an optimum matched filter. Repeat for overall amplifier bandwidths of 500 kHz and 2 MHz.

5-36 (*a*) Find and sketch the matched-filter output to the two waveforms shown in Fig. P5-36.

(*b*) Compare the output of matched filter 2 at times $t = T, t = T - T/4n, t = T - T/2n$. How critical is the *phase* synchronization between transmitter and receiver in this case? Explain.

(*c*) The received signal $s(t)$ is a sinusoidal pulse at frequency $f_c = 1{,}000$ MHz, and pulse width $T = 1$ μs. The output of a filter matched to $s(t)$ is to be sampled precisely at the end of the pulse, $t = T = 1$ μs. Using (*b*), discuss the synchronization required.

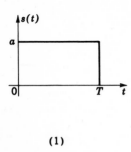

(1)

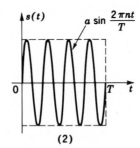

(2)

Figure P5-36

5-37 (*a*) Prove the Fourier-transform relationship

$$ E = \int_{-\infty}^{\infty} |f(t)|^2 \, dt = \int_{-\infty}^{\infty} |F(\omega)|^2 \, df $$

Hint: Consider the special case $f(t)$ real. Let $f(t)$ be the input to a matched filter. Find and equate the outputs using the convolution integral, and then Fourier transforms.

(*b*) Check this equality for $f(t)$ a rectangular pulse, a gaussian pulse (see Prob. 5-32), and $f(t) = e^{-a|t|}$.

5-38 The rectangular-shaped polar binary sequence of Fig. P5-38 is received in the presence of additive gaussian white noise. A' is in units of volts. The bit rate is 1 Mbits/s, equally likely to be 1's and 0's. The noise spectral density is $n_0/2 = \frac{1}{2} \times 10^{-20}$ V²/Hz.

(*a*) Matched filter detection is used. $A' = 0.22$ μV. Find the probability of error P_e. What is the average time between errors?

(*b*) Repeat (*a*) if $A' = 0.3$ μV.

(*c*) An RC filter, adjusted to be 1 dB from the matched filter result, is used at the receiver. Find P_e if $A' = 0.3$ μV.

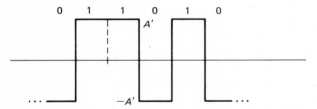

Figure P5-38

5-39 (*a*) Using Eq. (5-86), show by averaging over θ_l that

$$E(x^2) = E(y^2) = N$$

(*b*) Show, similarly, that $E(x) = E(y) = 0$, and that $E(xy) = 0$, as well.

5-40 A noise process $n(t)$ has the spectral density $G_n(f)$ shown in Fig. P5-40. $[E(n) = 0.]$

(*a*) Find the autocorrelation function and sketch for $f_0 \gg B$.

(*b*) Find the mean-squared value (power) of the process.

(*c*) $n(t)$ is written in the narrowband form $n(t) = x(t) \cos \omega_0 t - y(t) \sin \omega_0 t$. Sketch the spectral densities $G_x(f)$ and $G_y(f)$. What are $E(x^2)$ and $E(y^2)$?

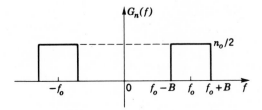

Figure P5-40

5-41 A binary communications system uses on-off keyed (OOK) transmission: the transmitted signal $s(t) = f(t) \cos \omega_0 t$ or 0, repeating every $T = 1$ ms. $f_0 = 1$ MHz. $f(t)$ is the triangle in Fig. P5-41. White gaussian noise, $n_0/2 = 10^{-20}$ V^2/Hz, is added during transmission. The receiver block diagram is shown in the figure.

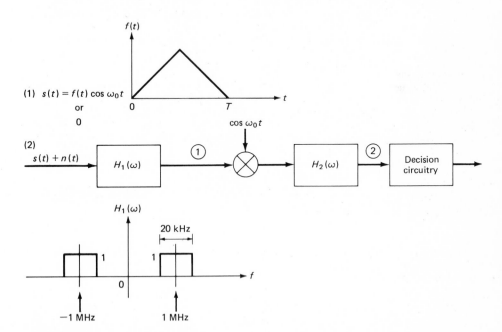

Figure P5-41

(a) Find the average noise power (mean-squared noise) at point 1.

(b) $H_2(\omega)$ is to be designed to maximize the signal-to-noise ratio at point 2. Write an expression for $H_2(\omega)$.

(c) An RC filter is used for $H_2(\omega)$:

$$|H_2(\omega)| = \frac{1}{\sqrt{1 + (f/f_c)^2}}$$

Find the average noise power at 2 for the two cases, $f_c = 1$ kHz and $f_c = 10$ kHz.

5-42 Find and sketch the spectral density of the derivative of the noise processes in Probs. 5-16 and 5-40. *Hint:* Differentiation is a linear operation and hence may be represented by a linear filter. What is $H(\omega)$ for this filter?

5-43 Compare the correlation functions for $n(t)$ and $x(t)$ in Prob. 5-40.

5-44 Consider the random process $z(t) = x \cos \omega_0 t - y \sin \omega_0 t$, where x and y are independent random gaussian variables with zero expected value and variance σ^2.

(a) Show that $z(t)$ is also normal (gaussian) with zero expected value and variance σ^2. Show the autocorrelation function $R_z(t, t + \tau) = \sigma^2 \cos \omega_0 \tau$, independent of t. (The process is therefore *wide-sense stationary*.) *Hint:* Set up formally the expression $E[z(t)z(t + \tau)]$, and then average over both x and y.

(b) Show that $z(t)$ can also be written as $r \cos (\omega_0 t + \theta)$. Find the joint density function of r and θ.

(c) Repeat (a) for $z(t) = x \cos (\omega_0 t + \varphi) - y \sin (\omega_0 t + \varphi)$ with φ a uniformly distributed $(0,2\pi)$ random variable independent of x and y.

5-45 Consider $z(t) = x(t) \cos \omega_0 t - y(t) \sin \omega_0 t$, where $x(t)$ and $y(t)$ are normal (gaussian), zero-mean, independent random *processes* with $R_x(\tau) = R_y(\tau)$. (They are therefore wide-sense stationary.)

(a) Show $R_z(\tau) = R_x(\tau) \cos \omega_0 \tau$. (It is therefore wide-sense stationary as well.) Distinguish between this problem and Prob. 5-44. *Hint:* Again average over both $x(t)$ and $y(t)$, as in Prob. 5-44.

(b) Find the power spectrum $G_z(f)$ in terms of the spectrum $G_x(f) = G_y(f)$.

(c) With $R_x(\tau) = \sigma^2 e^{-\alpha|\tau|}$ evaluate the spectra of $x(t)$ and $z(t)$. Sketch. Compare with Prob. 5-21.

(d) Show the density function of $z(t)$, at any time t, is zero-mean gaussian, with variance $R_x(0)$.

5-46 A high-frequency PSK signal $f(t) \cos \omega_0 t$ plus gaussian white noise appear at the input to the receiver shown in Fig. P5-46. The receiver consists of narrowband bandpass circuitry, a synchronous detector, and a low-pass filter before the decision circuitry at the output.

(a) The bandpass filter has a bandwidth large compared to the signal bandwidth (although small compared to the center frequency f_0). Characterize and sketch the amplitude characteristic $H_2(\omega)$ of the low-pass filter if the probability of error at the receiver output is to be minimized. [$f(t) = \pm a$: rectangular pulses T seconds long.]

(b) The low-pass filter $H_2(\omega)$ has a bandwidth $\gg 1/T$. Characterize and sketch $H_1(\omega)$ if the probability of error at the receiver output is to be minimized.

(c) In both (a) and (b) show that $A^2/N = 2E/n_0$. A is the peak value of the output signal $s(t)$ and N is the mean-square value of the output noise $n(t)$. E is the mean energy in the high-frequency signal:

$$E = \int_0^T [f(t) \cos \omega_0 t]^2 \, dt \doteq \tfrac{1}{2} \int_0^T f^2(t) \, dt$$

Show that with equally likely binary signals the probability of error is given by

$$P_{e,\text{PSK}} = \tfrac{1}{2} \text{ erfc } \sqrt{\frac{E}{n_0}}$$

This is then the matched-filter error probability for PSK signals agreeing with (5-104) and the discussion relating to that equation.

(d) Filter $H_1(\omega)$ has a rectangular amplitude characteristic of bandwidth $\pm B$ about f_0. Filter $H_2(\omega)$ has a rectangular (ideal low-pass) amplitude characteristic of bandwidth B hertz. Find the peak SNR at the receiver output for $B = 0.5/T$, $0.75/T$, and $1/T$. *Hint:* Show that the ideal low-pass curve in Fig. 5-29 is applicable.

(e) $E/n_0 = 8$ dB. Find the probability of error in (c) and for the three bandwidths of (d).

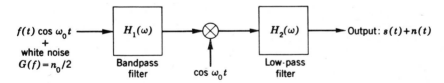

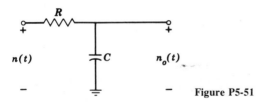

Figure P5-46

5-47 Repeat Prob. 5-46 for an OOK signal plus gaussian white noise at the input to the receiver of Fig. P5-46. Show that the matched-filter error probability is in this case

$$P_{e,\text{OOK}} = \tfrac{1}{2}\,\text{erfc}\,\left(\frac{1}{2}\,\sqrt{\frac{E}{n_0}}\right)$$

5-48 Draw a block diagram for an FSK matched-filter receiver with synchronous detection. Show that with additive white gaussian noise at the receiver input and equally likely binary signals the probability of error for the system is

$$P_{e,\text{FSK}} = \tfrac{1}{2}\,\text{erfc}\,\sqrt{\frac{E}{2n_0}}$$

E and n_0 are the same as in Probs. 5-46 and 5-47. Calculate the probability of error for $E/n_0 = 8$ dB and compare with the PSK and OOK results in Probs. 5-46 and 5-47 respectively. Replace the matched filter with the rectangular filters of Prob. 5-46d and evaluate the probability of error for the three bandwidths given there. Compare with the PSK and OOK results in Probs. 5-46 and 5-47.

5-49 An FSK system transmits 2×10^6 bits/s. White gaussian noise is added during transmission. The amplitude of either signal at the receiver input is 0.45 μV, while the white-noise spectral density at the same point is $n_0/2 = \tfrac{1}{2} \times 10^{-20}$ V^2/Hz. Compare the probability of error for a receiver using synchronous detection with one using envelope detection. Assume matched-filter detection in both cases. Repeat for received signal amplitudes of 0.9 μV.

5-50 (a) Gaussian bandpass noise $n(t)$ with power spectral density $G_n(f) = n_0/2$, $f_0 - B < |f| < f_0 + B$, zero elsewhere, as in Fig. P5-40, is applied to an envelope detector. Find the probability-density function of the detector output $r(t)$, as well as its dc and rms values.

(b) Repeat for another noise process $n(t)$ with spectral density n_0 over the range of frequencies $f_0 - 2B < |f| < f_0 + 2B$, zero elsewhere. Sketch the two probability-density functions on the same scale, and compare dc and rms values.

5-51 Zero-mean bandpass gaussian noise $n(t)$, as in Prob. 5-50a, is applied to the RC filter shown in Fig. P5-51. The output noise $n_0(t)$ is then gaussian as well. Find and sketch the power spectral

Figure P5-51

density of $n_0(t)$ for various values of RC. Find the mean and mean-squared value of $n_0(t)$ as a function of RC. Write the probability-density function of $n_0(t)$ for some value of RC.

5-52 (a) Show the Rayleigh distribution of (5-116) is properly normalized. Show its expected value and variance are $\sqrt{\pi/2}\,\sigma$ and $[2 - (\pi/2)]\sigma^2$, respectively.

(b) What is the probability that the envelope of narrowband gaussian noise will exceed three times its rms value? *Hint:* Use the result of part (a).

5-53 Gaussian bandpass noise with mean-squared value N is detected by a quadratic (or square-law) envelope detector, whose output is proportional to the square of the instantaneous envelope voltage. Thus the detector output voltage $z(t)$ is cr^2, as shown in Fig. P5-53, with c a constant of proportionality and r the envelope voltage.

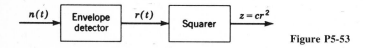

Figure P5-53

(a) Show that the probability-density function of z at any time t is

$$f(z) = \frac{e^{-z/2cN}}{2cN} \qquad z \geq 0$$

(b) Calculate the voltages at the output of the detector that would be read by a long-time-constant dc meter, a long-time-constant rms meter, and a long-time-constant rms meter preceded by a blocking capacitor.

5-54 A non-phase-coherent communication receiver consists of a narrowband tuned amplifier and a quadratic envelope detector as in Prob. 5-53. The receiver accepts signal pulses of fixed amplitude. To distinguish between incoming signals and noise present in the receiver, a specific voltage level at the output of the detector is chosen such that all voltages exceeding this level are called signal.

(a) What is the probability that noise in the absence of signal will be mistaken for signal if the level is set at $2N$, with N the mean-squared noise voltage at the detector input? (Assume that the detector constant of proportionality is 1 V/V².)

(b) Plot the probability of error due to noise if the voltage level is varied from 0 to $20N$.

5-55 *Computer generation of gaussian random variables from uniformly distributed random numbers.*

(a) Assume that random numbers x_i, uniformly distributed from 0 to 1, are available. Let $y = (b/n) \sum_{i=1}^{n} (x_i - \frac{1}{2})$, n a fixed number. Show y approximates a gaussian random variable of zero average value, and variance $\sigma^2 = b^2/12n$.

(b) The gaussian approximation of (a) is poor on the tails of the distribution. Why? A better approximation, using two independent uniform random numbers x and y, is obtained as follows:

(1) Let $r = \sqrt{-2\sigma^2 \log_e x}$. Show

$$f(r) = \frac{re^{-r^2/2\sigma^2}}{\sigma^2}$$

i.e., the Rayleigh distribution of Eq. (5-116).

(2) Show $z = r \cos 2\pi y$ is zero-mean gaussian, with variance σ^2. *Hint:* Refer to the derivation of the Rayleigh density function in Sec. 5-9.

5-56 Carry out the indicated statistical averaging in Eq. (5-146) to obtain Eq. (5-147).

5-57 Assume that an unmodulated carrier plus narrowband noise are synchronously detected. Show that the output SNR is proportional to the input CNR for *all* values of CNR.

5-58 (*a*) Gaussian bandpass noise $n(t)$ with power spectral density $G_n(f) = n_0/2, f_0 - B < |f| < f_0 + B$, zero elsewhere, is applied to the quadratic envelope detector of Fig. P5-53. (Assume the detector constant $c = 1$.) Find the spectral density of the output noise $z(t)$ by first finding the correlation function $R_z(\tau)$ in terms of the correlation function of the input $x(t)$ and $y(t)$. *Hint:* $r^2 = x^2 + y^2$. Assume x and y independent. Recall that $R_x(\tau) = R_y(\tau)$. Use Eq. (5-159) for gaussian processes to relate $R_{x^2}(\tau)$ to $R_x(\tau)$. The spectral density of $G_x(f)$ is given in terms of the input noise spectral density by Eq. (5-91).

(*b*) Repeat part (*a*), using the procedure of Sec. 5-11. Thus assume that quadratic envelope detection is equivalent to squaring the input $n(t)$ and passing $n^2(t)$ through a low-pass filter. Show that the resultant output spectrum is the same as that found in (*a*), except for the $\frac{1}{4}$ constant factor noted in Secs. 5-10 and 5-11.

5-59 An FM receiver consists of an ideal bandpass filter of 225-kHz bandwidth centered about the unmodulated carrier frequency, an ideal limiter and frequency discriminator, and an ideal low-pass filter of 10-kHz bandwidth in the output. The ratio of average carrier power to total average noise power at the input to the limiter is 40 dB. The modulating signal is a 10-kHz sine wave that produces a frequency deviation Δf of 50 kHz.

(*a*) What is the signal-to-noise ratio S_o/N_o at the output of the low-pass filter?

(*b*) A deemphasis network with a time constant $rC = 75$ μs is inserted just before the output filter. Calculate S_o/N_o again.

(*c*) Repeat (*a*) and (*b*) if the modulating signal is a 1-kHz sine wave of the same amplitude as the 10-kHz wave. Repeat with the amplitude reduced by a factor of 2. (The carrier amplitude and filter bandwidths are unchanged.)

5-60 An audio signal is to be transmitted by either AM or FM. The signal consists of either of two equal-amplitude sine waves: one at 50 Hz, the other at 10 kHz. The amplitude is such as to provide 100 percent carrier modulation in the AM case and a frequency deviation $\Delta f = 75$ kHz in the FM case.

The average carrier power and the noise-power density at the detector input are the same for the AM and FM systems.

(*a*) Calculate the transmission bandwidth and output filter bandwidth required in the AM case and the FM case. The carrier-to-noise ratio is 30 dB for the AM system. Calculate the output S_o/N_o for each sine wave for each of the systems. (Preemphasis and deemphasis networks are not used.)

(*b*) Repeat (*a*) if the amplitude of the 10-kHz wave is reduced by a factor of 2.

5-61 The ratio of average carrier power to noise spectral density, S_c/n_0, at the detector input is 4×10^6 for both an FM and an AM receiver. The AM carrier is 100 percent modulated by a sine-wave signal, while a sine-wave signal produces a maximum deviation of 75 kHz of the FM carrier.

Calculate and compare the output signal-to-noise ratio for both the FM and the AM receivers if the bandwidth B of the low-pass filter in each receiver is successively 1, 10, and 100 kHz. Deemphasis networks are not used. (Assume that the r-f bandwidths are always at least twice the low-pass bandwidth.)

5-62 The low-pass filter bandwidth of an AM and an FM receiver is 1 kHz. The ratio of average carrier power to noise spectral density, S_c/n_0, is the same in both receivers and is kept constant. The AM carrier is 100 percent modulated.

Calculate the relative SNR improvement of the FM over the AM system if the frequency deviation Δf of the FM carrier is 100 Hz; 1 kHz; 10 kHz.

5-63 The time constant of an RC deemphasis network is chosen as 75 μs. Plot the improvement in output SNR due to the deemphasis network for both FM and AM detectors as a function of the bandwidth of the output low-pass filter. (Assume that S_c/n_0 is the same and constant in both cases.) What is the minimum filter bandwidth to be used if audio signals from 0 to 10 kHz are to be passed?

5-64 The received signal $s(t) = m(t) \cos(\omega_c t + \theta) - \hat{m}(t) \sin(\omega_c t + \theta)$ is embedded in white gaussian noise with spectral density $n_0/2$. [$m(t)$ is the message; $\hat{m}(t)$ its Hilbert transform.]

(*a*) Show that $s(t)$ is a SSB signal.

(*b*) Assuming that θ is known, draw a block diagram of a synchronous demodulator, including appropriate low-pass filtering.

(*c*) Find the output SNR if $m(t)$ is random, $E(m) = 0$, $E(m^2) = S$.

5-65 The output of a "jittery" oscillator is described by $g(t) = \cos(\omega t + \theta)$, where ω and θ are independent random variables with the following probability-density functions:

1. θ uniform $(0, 2\pi)$.
2. ω has a probability-density function $f_\omega(\omega)$.

Show that the power spectral density of $g(t)$ is given by

$$G_g\left(\frac{\omega}{2\pi}\right) = \frac{\pi}{2} f_\omega(\omega)$$

Sketch for $f_\omega(\omega)$ gaussian, with expected value ω_0. *Hint:* Write the autocorrelation function for $g(t)$, averaging over *both* ω and θ. Show $R_g(\tau) = \frac{1}{2} E(\cos \omega \tau)$, referring to expectation over ω. Write out the form for this expectation in terms of $f_\omega(\omega)$, then compare with the expression for the Fourier transform of $G_g(f)$.

5-66 The output SNR and input SNR of an FM discriminator *above threshold* are related by the equation

$$\frac{S_o}{N_o} = 3\beta^2(\beta + 1)\frac{S_i}{N_i}$$

with β the modulation index. The complete SNR characteristic is shown in Fig. P5-66. An FM signal at the discriminator input has a power of 55 mW. The input noise is white with spectral density $n_0/2 = 0.25 \times 10^{-10}$ W/Hz. The maximum modulation frequency is 5 MHz. The frequency deviation is 25 MHz. If the bandwidth following the discriminator is B hertz, while that preceding the discriminator is $2(\beta + 1)B$, find

(*a*) S_i/N_i in decibels.

(*b*) S_o/N_o in decibels.

(*c*) Repeat if the deviation is increased to 50 MHz.

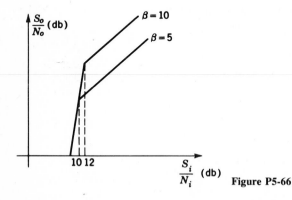

Figure P5-66

5-67 A communication system produces at its output an impulse term plus random noise given by

$$e_0(t) = 2\pi\delta(t - t_i) + n(t)$$

The random noise has power spectral density

$$G_n(f) = kf^2$$

up to very high positive and negative frequencies. (This is a model of impulse and random noise occurring at the output of FM receivers.)

This output voltage is to be passed through a linear filter $H(\omega)$ designed to optimize the recognition of the impulse term. Determine the expression for $H(\omega)$ which maximizes, at time $t_1 + t_0$, the ratio of the output due to the impulse to the rms output due to $n(t)$. *Hint:* Use the Schwarz inequality here.

5-68 Band-limited white gaussian noise with spectral density $n_0/2$, $-B < f < B$, is passed through a square-law detector whose output is $n^2(t)$, and then through an ideal low-pass filter of bandwidth B.

Determine the power spectrum of $n^2(t)$, the spectrum at the low-pass filter output, and the rms value of the output voltage.

5-69 Calculate the rms noise voltage in a 10-kHz bandwidth across
  (*a*) a 10-kΩ resistor at room temperature.
  (*b*) a 100-kΩ resistor at room temperature.
Repeat for a 5-kHz bandwidth. Repeat for $T = 30$ K.

5-70 Consider a "noisy" resistor R at temperature T degrees Kelvin and an inductor L connected in series.
  (*a*) Find the spectral density of the current through the inductor.
  (*b*) Show that the average energy stored in the inductor is $\frac{1}{2}kT$.

5-71 Using Eq. (5-214), show that the average energy of the harmonic oscillator is given by Eq. (5-215).

5-72 Starting with Eq. (5-220), derive (5-221) for the spectral density of the resistive noise-current source in Fig. 5-66. To do this, first show that if the bandwidth of $H(\omega)$ is very narrow compared to the center frequency $f_0 = 1/2\pi\sqrt{LC}$, $|H(\omega)|^2$ may be written

$$|H(\omega)|^2 \doteq \frac{1/G^2}{1 + (\delta/\alpha)^2}$$

with $\delta = \omega - \omega_0 \ll \omega_0$ and $\alpha = G/2C$.

This represents a translation of $H(\omega)$ down to $\delta = 0$. Since $G_i(f)$ may be assumed constant over all values of δ, take it out of the integral, integrate over all possible values of δ, and show that (5-221) results.

5-73 A simplified model of an optical communications system is shown in Fig. P5-73. The transmitter produces pulses of light which travel down the optical path to the photodetector. The photodetector current i may be written

$$i = \bar{I} + i_s + i_B$$

$\bar{I}$, the *signal*, is the average current due to the transmitted light. i_s is a shot-noise component with spectral density $G_s(f) = e\bar{I}$, due to the arrival of the photons at discrete but random instants. i_B is a white-noise component with spectral density $G_B(f) = \eta$, due to background radiation.

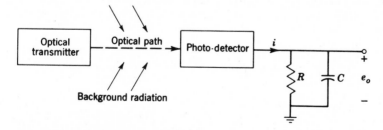

Figure P5-73 Optical communications system.

(a) Derive an expression for the signal-to-noise power ratio at the output of the RC filter. Assume that the resistor R is noiseless.

(b) Repeat if R is a source of thermal noise.

5-74 *Noise figure of lossy transmission line (cable)*. Consider signals transmitted along a cable of length l with voltage attenuation constant α (Fig. P5-74). The ratio of power out to power in is then $L = e^{-2\alpha l}$. It can then be shown[79] that the line introduces thermal noise due to the line losses given by $N_l = (1 - L)kTB$, $hf \ll kT$, with T the temperature of the line and B a specified bandwidth.

Show that the noise figure of the lossy cable at T degrees Kelvin is

$$F = 1 + \frac{1 - L}{L} = \frac{1}{L}$$

For a line with $L = \frac{1}{2}$, then, $F = 2$.

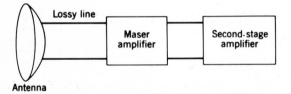

$$P_o = P_i e^{-2\alpha l} = P_i L$$

Figure P5-74 Lossy transmission line.

5-75 *Maser receiving systems*. Consider the low-noise maser receiver shown in Fig. P5-75. Although the low-noise (low-temperature) maser amplifier is used to decrease the system noise as much as possible, it is found in practice that the line feed from antenna to the maser amplifier provides the limitation on the reduction of the system noise. To demonstrate this, consider the following examples:

(a) The line loss is $L \equiv -0.1$ dB. The maser has an effective noise temperature of 3 K. Show that the effective noise temperature at the antenna input = 10 K. (The second-stage noise contribution may be neglected.)

(b) The maser amplifier noise is now 30 K. Show that the effective noise temperature at the antenna input is $\doteq 37$ K.

```
Lossy line
Maser amplifier
Second-stage amplifier
Antenna
```

Figure P5-75 Maser receiver.

5-76 *Measurement of effective noise temperature*

(a) Consider a system denoted by H in Fig. P5-76 whose effective noise temperature T_e is to be measured. Two thermal-noise sources, one at temperature T_1 degrees Kelvin, the other at temperature T_2 degrees Kelvin, as shown, are separately connected to the input of H. The output power, denoted respectively by P_1 and P_2, is measured in each case. Show that the effective noise temperature is then given by

$$T_e = \frac{T_1 - yT_2}{y - 1}$$

with $y \equiv P_1/P_2$ the ratio of the two powers.

[79] A. E. Siegman, *Microwave Solid-State Masers*, McGraw-Hill, New York, 1964, pp. 373–375.

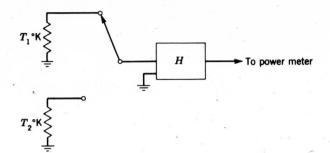

Figure P5-76 Measurement of effective noise temperature.

(b) Show how a calibrated attenuator plus some sort of power-indicating device at the output of H may be used to measure P_1/P_2. (Absolute power readings are then not necessary.)

5-77 Refer to the discussion of the Mariner 10 deep-space mission to Mercury discussed at the beginning of Sec. 5-14. Show, for the numbers given, that the received power at the earth is 5.4×10^{-17} W. Show that for an overall system temperature of 13.5 K, $S_R/n_0 = 54.7$ dB. Finally, verify that for bit error probability $P_e = 0.05$, $R = 213,000$ bits/s is the bit transmission rate allowed using PSK transmission. What bit rate would be allowed if FSK transmission were used instead? What bit rate would be allowed if $P_e = 10^{-5}$ were required?

5-78 The Voyager space mission has the following design features: $f = 8415$ MHz, $S_T = 23$ W, transmitting antenna diameter is 3.66 m and $\eta_T = 0.65$. $d = 7.58 \times 10^8$ km, receiver antenna diameter is 64 m, and $\eta_R = 0.575$, $T = 21$ K. Other losses total 5.5 dB. Show that data may be transmitted at a bit rate of 115.2 kbits/s using PSK transmission with an acceptable bit error probability of $P_e = 0.005$.

5-79 Consider a deep-space telemetry system with the following parameters: $S_T = 20$ W, with 50 percent of this power used for telemetry. The transmitting antenna gain is $G_T = 24$ dB. The distance to the earth is 1.6×10^8 km (10^8 mi). The receiving antenna is a 64-m dish, with 57.5 percent efficiency. The system noise temperature is 25 K. Additional power losses equal 3 dB. Find the allowable bit rates using PSK transmission for the three cases $P_e = 10^{-3}, 0.05$, and 10^{-5}. Repeat if the distance is increased to 3.2×10^8 km.

5-80 Repeat Prob. 5-79 if the transmitting antenna size is doubled, all the power is devoted to telemetry, and the noise temperature is reduced to 12.5 K.

5-81 Derive Eqs. (5-231) and (5-232), the expressions for signal-to-noise spectral density and CNR, respectively, for the Intelsat IV communications system.

5-82 Refer to the Intelsat IV system. Show that a 30-m ground antenna, with an efficiency of 70 percent, requires a low-noise receiver with $T = 85$ K, to achieve $G_R/T = 40.7$ dB at a 4-GHz frequency.

5-83 Refer to the satellite system calculations in Sec. 5-14. The down-link frequency is increased to 12 GHz, all other satellite parameters remaining the same. Comment on the G_R/T requirement now to maintain good performance. Find the earth station antenna size required if $\eta_R = 0.5$ and $T = 50$ K. Compare with existing designs.

5-84 A 5-MHz baseband signal is to be transmitted from a synchronous satellite to the earth 3.6×10^7 m away, using FM transmission at a 4-GHz carrier frequency. The signal-to-noise ratio SNR at the receiver discriminator output is to be at least 30 dB. The FM carrier-to-noise ratio CNR, at the input to the discriminator, is to be at least 10 dB. The system noise temperature is taken as 100 K. Determine appropriate values for the transmitter power S_T, the FM modulation index β, and the transmission bandwidth B_T in hertz. The transmitter and receiver antenna gains are, respectively, 10 dB and 50 dB.

5-85 Real-time transmission of TV pictures from a lunar spacecraft to earth (400,000 km) is to be investigated. A receiver CNR of 20 dB at a bandwidth of 4 MHz is required. A 500-MHz carrier is to be used, with a 2 meter × 2 meter antenna on the spacecraft and an antenna with 45-dB gain on the earth. What is the power requirement of the transmitter on the space vehicle?

5-86 Investigate the possibility of maintaining voice contact with a space mission 8×10^9 km from the earth. (This represents the distance to the farthest planet of the solar system.) A receiver CNR of 20 dB is desired. A low-noise receiving system with a 5 K effective noise temperature at 2 GHz is available. The spacecraft may be allowed up to 250 W power output and can carry a large unfurlable parabolic antenna.

STATISTICAL COMMUNICATION THEORY AND DIGITAL COMMUNICATIONS

We have now come full circle in our discussion of information transmission. In Chap. 1 we pointed out qualitatively that the rate of transmission of information, in bits per second, was limited by two basic quantities, the time response or bandwidth of systems through which the information was to pass, and noise innately present in all systems. We then went on to discuss in detail system frequency response, noise, and their connection in various types of communication systems.

We have thus far concentrated on the analysis and comparison of the most commonly used communication schemes, whether for the transmission of digital or analog signals. We now return to the question of information transmission from a more fundamental viewpoint. We ask whether it is possible to *optimize* the design of systems in the sense of maximizing the information-transmission rate with prescribed constraints on error probability, signal power, noise, and bandwidth. This leads us into the realm of statistical communication theory.

We have already considered some aspects of system optimization at various points throughout this book. Thus, in discussing binary communications, we discussed the optimum setting of a decision threshold. The concept of matched filtering arose out of the discussion of maximizing SNR in binary transmission. We applied the Schwarz inequality in considering optimum emphasis networks for FM and AM analog transmission. We discussed the SNR-bandwidth exchange in pulse-code-modulation (PCM) systems, and indicated that it was similar to that found by Shannon for a hypothetical optimum digital transmission system.

These were essentially isolated cases, however, useful in developing familiarity with existing systems and their performance. We now attack the problem

453

of optimum information transmission in a more systematic way. We use here as a tool the elements of statistical decision theory. The emphasis throughout will be on *digital* communications, first because of its rapidly growing importance in modern technology and second because the optimization procedures, based primarily on the minimization of probability of error, are much simpler to carry out and interpret than for analog signal transmission.[1]

After a necessary introduction to statistical decision theory with specific reference to binary communication, we consider in detail the optimum design of binary communication systems, designed to perform with minimum probability of error in the presence of additive gaussian noise.[2] The specific questions to be answered are: (1) What is the optimum decision procedure at the receiver? and (2) Is it possible to optimally design signals at the transmitter?

Surprisingly, we shall find both questions answered simultaneously by our statistical decision approach. Polar signal transmission with matched-filter decision threshold detection will be found to be the optimum binary transmission scheme! This is, of course, exactly the system analyzed earlier—the system we found theoretically superior to both frequency-shift-keyed (FSK) and on-off-keyed (OOK) transmission.

In Chap. 4 we discussed briefly the possibility of using quadrature amplitude modulation (QAM) to increase digital transmission rates over fixed bandwidth channels. Following the discussion of binary transmission we extend the analysis to show how one optimally processes QAM signals received over an additive white noise channel. It turns out that matched filter detection is again called for, and, as predicted in Chap. 4, the error performance of these systems deteriorates from the binary case.

QAM transmission is a special case of M-ary transmission, with n binary digits encoded into one of $M = 2^n$ possible signals. It turns out that by encoding into *orthogonal* signals (M different frequencies provide one example), one can reverse the error performance deterioration and in fact drive the error probability down as far as one likes, thus improving on the binary system performance. The price paid is an increase in the transmission bandwidth. We show that this improvement in error performance is a special case of the Shannon capacity expression first introduced in Chap. 1, and then referred to again in Chap. 5 in studying PCM system performance. The Shannon capacity expression shows that there must exist coding schemes that reduce the probability of error.

We conclude this chapter by considering coding techniques that provide error detection and error correction as a means of improving digital transmis-

---

[1] Optimum analog transmission relies on the concepts of statistical *estimation* theory. See A. J. Viterbi, *Principles of Coherent Communication,* McGraw-Hill, New York, 1966, part 2, for an introduction and comprehensive bibliography. A detailed discussion appears in H. L. Van Trees, *Detection, Estimation, and Modulation Theory,* vol. I, Wiley, New York, 1968.

[2] As noted earlier in the book, this model is particularly appropriate for space communications. We ignore intersymbol interference here, the major problem in telephone data transmission. See R. W. Lucky, J. Salz, and E. J. Weldon, Jr., *Principles of Data Communication,* McGraw-Hill, New York, 1968, for a detailed treatment of both noise and intersymbol interference.

sion performance. Here the addition of extra (redundant) data symbols to the signal binary stream enables error detection and correction to be carried out at the cost, of course, of increased system complexity and reduction of data rate (or increased bandwidth).

6-1 STATISTICAL DECISION THEORY

It is apparent from all our discussions in this book that the problems of deciding between either of two signals transmitted in a binary communication case, or of appropriately processing analog signals in the AM or FM case, are essentially statistical in nature. The signals transmitted are, of course, random to begin with, the noise added enroute or at the receiver can generally only be described statistically, the fading and multiplicative noise possibly encountered enroute can likewise only be described statistically, etc. We must then look to the realm of statistics for techniques that may be directly carried over to the communication field to help develop schemes for optimumly transmitting and processing signals.

The fields of statistical decision and estimation theory have proven particularly fruitful in handling the problem of optimum information transmission. Statistical decision theory, as is apparent from the name, deals specifically with the problem of developing statistical tests for optimumly deciding between several possible hypotheses. One would expect techniques developed here to be particularly useful in digital communications, detection radar, and other systems where discrete *decisions* have to be made (which one of two signals transmitted, which one of M signals transmitted, is it a target or noise, etc.). Statistical estimation theory, on the other hand, deals with methods of estimating as "best" as possible continuous random parameters or time-varying random functions. It is thus applicable to problems of analog transmission. We concentrate here only on statistical decision theory and its application to digital communications.[3] By "optimum" or "best" system we shall mean here for the most part one that minimizes the probability of error.[4]

---

[3] Many fine books exist on statistical theory. Included are A. M. Mood and F. A. Graybill, *Introduction to the Theory of Statistics*, 2nd ed., McGraw-Hill, New York, 1963; H. Cramer, *Mathematical Methods of Statistics*, Princeton, Princeton, N.J., 1946; M. Fisz, *Probability Theory and Mathematical Statistics*, 3rd ed., Wiley, New York, 1963. Books in communications that incorporate and expand on the material in this chapter include D. J. Sakrison, *Communication Theory: Transmission of Waveforms and Digital Information*, Wiley, New York, 1968, chap. 8; J. M. Wozencraft and I. M. Jacobs, *Principles of Communication Engineering*, Wiley, New York, 1965; Viterbi, *op. cit.*, part 3; C. W. Helstrom, *Statistical Theory of Signal Detection*, Pergamon, New York, 1960. Van Trees, *op. cit.*, as well as M. Schwartz and L. Shaw, *Signal Processing*, McGraw-Hill, New York, 1975, discuss communications and other application areas.

[4] It is important to keep in mind that the word *optimum* is usually meant in a restricted sense; it refers to the "best" system according to the particular criterion adopted for evaluating the system performance.

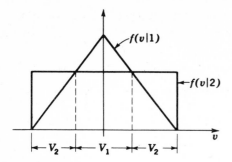

Figure 6-1 Choice of binary decision regions.

To bring out the elements of statistical decision theory and its applicability to digital communication, consider the problem of distinguishing between either one of two possible signals received. Assume that s_1 or s_2 has been transmitted and a voltage v received. We sample v, and, based on the value of the sample measured, determine as "best" as possible which of the two signals was transmitted. (Later, we shall extend this to the case of many sequential samples.)

In statistical terminology, we are given the value of a statistical sample (or group of samples) and wish to select between either of two alternative hypotheses. Call these H_1 and H_2. Hypothesis H_1 corresponds to the decision s_1 transmitted; H_2 of course to the decision s_2 transmitted.

How does one establish a rule for deciding between the two hypotheses? Note that we can make two types of error. Assuming H_1 true, s_2 may very well have been transmitted, or alternatively, assuming H_2 true, s_1 may have been transmitted. It is apparent that the total probability of error to be minimized is based on both these errors. It is also apparent that the rule to be chosen will consist of splitting up the one-dimensional space corresponding to all possible values of v into two nonoverlapping (mutually exclusive) regions V_1 and V_2 such that the overall error probability is minimized.[5] We assume that we know the a priori probabilities P_1 and P_2 of transmitting s_1 and s_2, respectively ($P_1 + P_2 = 1$), as well as the *conditional* probability densities $f(v|1)$ and $f(v|2)$, corresponding respectively to the probability of receiving v given s_1 transmitted, and v given s_2 transmitted. A typical example is shown in Fig. 6-1.

We set up the expression for overall probability of error and then minimize it by adjusting V_1 and V_2. This then provides the desired rule.

Just as in the error-probability calculations of Chap. 5, we may find the probability of error by considering first the probability that v will fall in region V_2 even though a 1 has been transmitted. This is simply

$$\int_{V_2} f(v|1)\ dv$$

---

[5] In most of the references cited, more general cost functions than simple error probability are considered. Generally, the resultant rule is a simple extension of the one developed here, however.

Similarly, the probability that v will fall in V_1 even though signal 2 has been transmitted is given by

$$\int_{V_1} f(v|2)\, dv$$

Again, as in Chap. 5 the overall probability P_e that is to be minimized by adjusting V_1 (or V_2) is found by weighting each of the integrals above by its respective a priori probability and then summing the two:

$$P_e = P_1 \int_{V_2} f(v|1)\, dv + P_2 \int_{V_1} f(v|2)\, dv \qquad (6\text{-}1)$$

We use a little trick now to perform the minimization quite directly. Since $V_1 + V_2$ covers all possible values of v, we have

$$\int_{V_1+V_2} f(v|1)\, dv = 1 = \int_{V_1} f(v|1)\, dv + \int_{V_2} f(v|1)\, dv \qquad (6\text{-}2)$$

We can then eliminate the integral over V_2 in Eq. (6-1), writing instead

$$P_e = P_1 + \int_{V_1} [P_2 f(v|2) - P_1 f(v|1)]\, dv \qquad (6\text{-}3)$$

Since P_1 is a specified number and assumed known, P_e is minimized by choosing the region V_1 appropriately. A little thought indicates that this is done by adjusting V_1 to have the integral term in Eq. (6-3) negative and as large numerically as possible. But we recall that probabilities and density functions are always positive. The solution then corresponds quite simply to picking V_1 as the regions of v corresponding to

$$P_1 f(v|1) = P_2 f(v|2) \qquad (6\text{-}4)$$

This is then the desired decision rule. As an example, let $P_1 = P_2 = \frac{1}{2}$. The region V_1 then corresponds to all values of v where $f(v|1) > f(v|2)$. Three examples are shown in Fig. 6-2.

Note that the first example resembles that of choosing between two gaussian distributions, and shows the optimum decision level occurring precisely at the intersection of the two functions. This agrees of course with the optimum threshold location found in Chap. 5. Some simple mathematics will verify this result. Assume that the two signals transmitted are A and 0, with a priori probabilities P_1 and P_2, respectively. Gaussian noise of zero average value and variance N (mean noise power) is added on reception. As in Chap. 5 the two conditional density functions are, respectively,

$$f(v|1) = \frac{1}{\sqrt{2\pi N}} e^{-(v-A)^2/2N} \qquad (6\text{-}5)$$

and

$$f(v|2) = \frac{1}{\sqrt{2\pi N}} e^{-v^2/2N} \qquad (6\text{-}6)$$

The optimum threshold level is now found by substituting into Eq. (6-4).

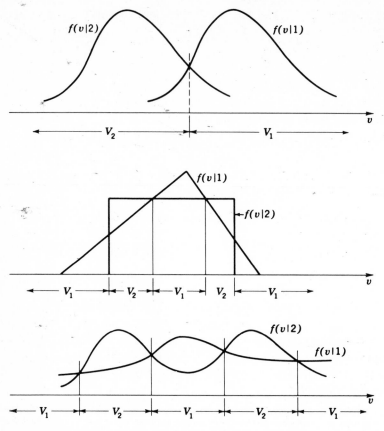

Figure 6-2 Examples of binary decision regions, $P_1 = P_2 = \frac{1}{2}$.

The optimum decision rule often is written in the equivalent form

$$l \equiv \frac{f(v|1)}{f(v|2)} > \frac{P_2}{P_1} \tag{6-7}$$

The parameter l, a function of the sample value v, is called the *likelihood ratio*. The region V_1 corresponding to the decision that S_1 was transmitted then corresponds to those values of v for which $l > P_2/P_1$. More general approaches to optimum decision theory, in which specified costs rather than probability of error are to be minimized, lead to the likelihood ratio solution as well. The number P_2/P_1 is then replaced by suitable combinations of the costs and P_1 and P_2. This is discussed in some of the references cited earlier.

Since the logarithm of a function is monotonic with that function, it is apparent that the boundaries V_1 or V_2 will not change if we take the logarithm of both sides of Eq. (6-7). This simplifies the algebra considerably, particularly in those cases where the density functions are exponential. Using the

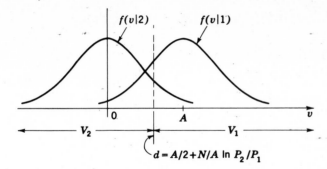

Figure 6-3 On-off signals plus gaussian noise.

gaussian functions of Eqs. (6-5) and (6-6) as specific examples, it is apparent, after taking logarithms, that the region V_1 is given by

$$v^2 - (v - A)^2 > 2N \ln \frac{P_2}{P_1} \tag{6-8}$$

Expanding the term in parentheses and rewriting, we find as the desired region V_1 all those values of v corresponding to

$$v > \frac{A}{2} + \frac{N}{A} \ln \frac{P_2}{P_1} \tag{6-9}$$

Note that for equally likely signals, the optimum threshold is $A/2$, just the value found in Chap. 5. If signal s_2 is transmitted more often, the threshold moves to the right, indicating a greater willingness to choose s_2. This is shown in Fig. 6-3.

As another example, assume OOK transmission, additive gaussian noise, and envelope detection at the receiver. From our discussion in Chap. 5 it is apparent that the detected envelope r is Rayleigh-distributed in the case when a zero is transmitted, Rician-distributed when a 1 is transmitted. The two envelope-density functions are shown sketched in Fig. 6-4, with the optimum decision level shown at their intersection in the case of equally probable transmission.

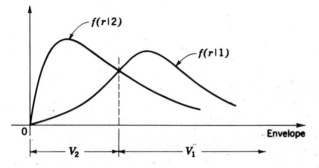

Figure 6-4 Envelope-detected signals, optimum decision regions.

6-2 SIGNAL VECTORS—MULTIPLE-SAMPLE DETECTION

We have shown how one optimally decides between either of two signals received on the basis of one received sample. It is apparent that one should be able to improve the signal detectability by making available more received samples. The obvious questions then are:

1. How much improvement is to be expected?
2. What is the optimum decision rule?

In this latter case one can alternatively ask: What is the optimum way of processing the samples?

We shall assume for simplicity that we have n *independent* samples on which to base the decision. (For the case of binary pulses in additive gaussian noise this corresponds generally to sampling at intervals $>1/$bandwidth of the noise. As shown in Chap. 5, the correlation rapidly decreases to zero beyond this point.) These are generated by sequentially sampling the received waveform as in Fig. 6-5. Alternatively, a particular pulse could be repeated n times, each one then being sampled once.

The n samples define n-fold density functions. In particular, if signal s_1 is transmitted, we get the n-dimensional conditional-density function

$$f(v_1, v_2, \ldots, v_j, \ldots, v_n | 1) = \prod_{j=1}^{n} f(v_j | 1)$$

with successive samples assumed independent. Similarly, if s_2 is transmitted, we may write the alternative n-fold conditional-density function

$$f(v_1, v_2, \ldots, v_j, \ldots, v_n | 2) = \prod_{j=1}^{n} f(v_j | 2)$$

As in the one-sample case, we have to assume these functions known. Note that the n samples may be visualized as describing an n-dimensional space, the possible range of values of each serving to define the range of that particular dimension. A little thought indicates that our statistical decision problem now

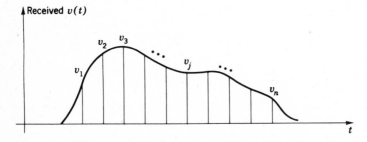

Figure 6-5 Generation of n samples for processing.

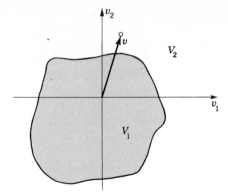

Figure 6-6 Two-dimensional signal space.

boils down to dividing the n-dimensional space into two mutually exclusive regions V_1 and V_2 ($V_1 + V_2$ corresponding to the entire space of the n samples). We would like to choose the boundary between these two regimes such that the probability of error is minimized. If the n samples fall into region V_1, we declare signal s_1 present, if into region V_2, s_2 is declared present.

The composite samples v_1, v_2, . . . , v_j, . . . , v_n now constitute an n-dimensional vector **v**. This geometric and vector approach often simplifies quite considerably the analysis and optimum design of communication systems, and has been widely adopted by communication theorists. An example of the two-dimensional region corresponding to $n = 2$ samples is shown in Fig. 6-6. Received vector **v** is shown falling in region V_2, so that signal s_2 would be declared present.

How do we choose V_1 and V_2 optimumly now, in the sense of minimizing the error probability P_e? We again write the probability of error explicitly in terms of the two conditional-density functions. Since V_1 and V_2 are now n-dimensional, the integration must be carried out over the n-dimensional volumes:

$$P_e = P_1 \int_{V_2} f(\mathbf{v}\,|\,1)\; dv_1\, dv_2 \cdots dv_n + P_2 \int_{V_1} f(\mathbf{v}\,|\,2)\; dv_1\, dv_2 \cdots dv_n \quad (6\text{-}10)$$

Here we have written the two conditional-density functions using the short-hand vector notation for **v**.

But note now that aside from the extension to n dimensions the minimization process here is identical to that carried out earlier for one dimension. The optimum choice for the region V_1 is thus given by the likelihood ratio

$$l \equiv \frac{f(\mathbf{v}\,|\,1)}{f(\mathbf{v}\,|\,2)} > \frac{P_2}{P_1} \quad (6\text{-}11)$$

As an example assume we again transmit an on-off signal of amplitude A or 0, and gaussian noise is added at the receiver. The received signal plus noise voltage $v(t)$ is sampled n times (alternatively, the signal may be assumed repeated n times) and the n samples $v_1 \cdots v_n$ used to make the decision. The conditional-density functions for the jth sample (assuming the samples independent) are given respectively by

$$f(v_j|1) = \frac{e^{-(v_j-A)^2/2N}}{\sqrt{2\pi N}} \qquad (6\text{-}12)$$

and

$$f(v_j|2) = \frac{e^{-v_j^2/2N}}{\sqrt{2\pi N}} \qquad (6\text{-}13)$$

The n-dimensional density functions are products of these individual sample functions. Again, taking logs to simplify, we find

$$\ln l = \frac{1}{2N}\left[\sum_{j=1}^{n} v_j^2 - \sum_{j=1}^{n} (v_j - A)^2\right] > \ln \frac{P_2}{P_1} \qquad (6\text{-}14)$$

In the special case where $P_2 = P_1 = \frac{1}{2}$, we have the region V_1 defined by

$$\sum_{j=1}^{n} v_j^2 > \sum_{j=1}^{n} (v_j - A)^2 \qquad (6\text{-}15)$$

Note, however, that not only can we define a vector $\mathbf{v} \equiv (v_1, v_2, \ldots, v_n)$, but we can talk of two n-dimensional vectors $\mathbf{s}_1 \equiv (s_1^{(1)}, s_2^{(1)}, \ldots, s_n^{(1)})$ and $\mathbf{s}_2 \equiv (s_1^{(2)}, s_2^{(2)}, \ldots, s_n^{(2)})$, where the subscripts represent the sample number and the superscripts the particular signal, 1 or 2, under consideration. In this special case $\mathbf{s}_1 \equiv (A, A, A, \ldots, A)$ and $\mathbf{s}_2 \equiv (0, 0, 0, 0, \ldots, 0)$, since we have assumed a rectangular pulse of height A throughout the n-sample interval (Fig. 6-7).

It is now apparent that the right-hand side of Eq. (6-15) represents the squared length of the vector $\mathbf{v} - \mathbf{s}_1$, and the left-hand side the squared length of the vector $\mathbf{v} - \mathbf{s}_2$.

In terms of the usual vector notation we have the squared length of a vector $\mathbf{a}$ given by the dot product $\mathbf{a} \cdot \mathbf{a}$, so that Eq. (6-15) may be equally well given by

$$(\mathbf{v} - \mathbf{s}_2) \cdot (\mathbf{v} - \mathbf{s}_2) > (\mathbf{v} - \mathbf{s}_1) \cdot (\mathbf{v} - \mathbf{s}_1) \qquad (6\text{-}16)$$

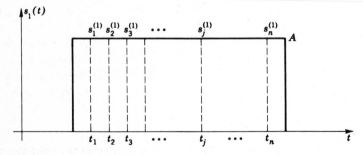

Figure 6-7 n-dimensional signal s_1.

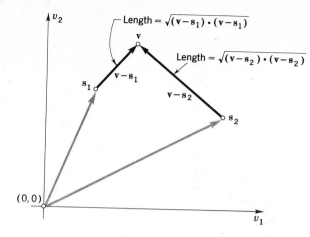

Figure 6-8 Vector signals and decision rule in two dimensions (additive gaussian noise): received signal v is "closer to" signal s_1.

The n-dimensional vector notation and dot product is just the extension to n dimensions of the common three-dimensional vector notation. [The reader may find it instructive to use arbitrary but known pulse shapes $s_1(t)$ and $s_2(t)$ for the two signals. Vectors s_1 and s_2 can then again be defined as the composite of the n samples for each, and Eq. (6-16) obtained in this more general case.] The interpretation of the optimum decision rule in this case of two signals plus gaussian noise is now apparent from Eq. (6-16). Pick signal s_1 if the distance between the received vector v and the known vector s_1 is less than the distance between v and s_2. This is shown graphically in Fig. 6-8 for the special case of two dimensions.

Note that the receiver must have stored replicas of both vectors s_1 and s_2 [or, equivalently, the n sample values of $s_1(t)$ and of $s_2(t)$]. It measures the received vector v, and then forms the necessary dot products and decides on s_1 or s_2, following the rule of Eq. (6-16).

The rule may be simplified in this case, however. For note that in both Eqs. (6-15) and (6-16) a common factor

$$\mathbf{v} \cdot \mathbf{v} = \sum_{j=1}^{n} v_j^2$$

may be canceled out on left- and right-hand sides. We then have remaining the rather simple expression

$$\sum_{j=1}^{n} v_j > \frac{nA}{2} \tag{6-17}$$

for the special case of on-off signals. In words: Sum the n received samples and see if the sum exceeds the specified threshold $nA/2$. Note that this is just the extension to n dimensions of the one-dimensional result obtained earlier. The interesting point here is that the rule calls for the *sum* of the received samples. This is specifically due to the assumption of known binary signals in additive gaussian noise. In examples to follow we shall find different rules for different assumed statistics.

More generally, if instead of on-off binary signals we assume two *arbitrary* equally likely binary signals $s_1(t)$ and $s_2(t)$, the n samples of each define, respectively, the two vectors $\mathbf{s}_1$ and $\mathbf{s}_2$. In this case, the optimum decision rule from Eq. (6-16) becomes, after expanding the dot products and dropping the common term $\mathbf{v} \cdot \mathbf{v}$,

$$\mathbf{v} \cdot (\mathbf{s}_1 - \mathbf{s}_2) > \frac{\mathbf{s}_1 \cdot \mathbf{s}_1 - \mathbf{s}_2 \cdot \mathbf{s}_2}{2} \tag{6-18}$$

In terms of the samples themselves, we have

$$\sum_{j=1}^{n} v_j [s_j^{(1)} - s_j^{(2)}] > \sum_{j=1}^{n} \frac{s_j^{(1)2} - s_j^{(2)2}}{2} \tag{6-19}$$

The receiver now performs a weighted sum, weighting each received sample v_j with the stored samples $s_j^{(1)} - s_j^{(2)}$ before adding. The final sum is then compared to a known threshold.

The boundary between regions V_1 and V_2 is found by replacing the inequalities in the equations above by equal signs. In n dimensions the boundary between regions V_1 and V_2, given for additive gaussian noise by either Eq. (6-17), Eq. (6-18), or Eq. (6-19), is called a *hyperplane*. For the special case of on-off signals in two dimensions this degenerates into the line $v_1 + v_2 = A$, shown sketched in Fig. 6-9.

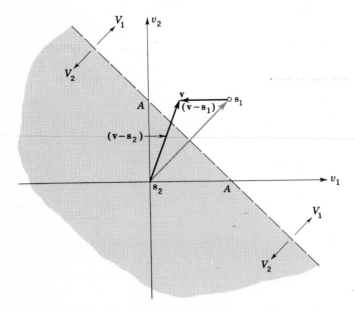

Figure 6-9 Decision rule for on-off signals ($s_2 = 0$) in two dimensions (additive gaussian noise).

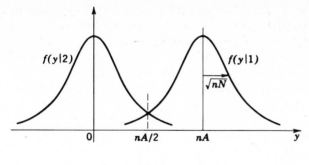

Figure 6-10 Alternate error calculation, additive gaussian noise.

How much improvement does the use of the n samples provide over one sample? The answer of course depends on the particular signal shapes and noise statistics assumed. Consider, however, the on-off case as a specific example. If both signals are assumed equally likely for simplicity, either signal is equally likely as well to be mistaken for the other. The probability of an error is then the probability that vector $\mathbf{v}$, with $\mathbf{s}_1$ transmitted, appears in region V_2. From Eq. (6-17) this corresponds to the sum of the samples falling *below* the threshold level $nA/2$. Rather than calculate the probability of the n-dimensional gaussian vector $\mathbf{v}$ falling into the region V_2 we can deal with the much simpler sum of the v_j's. We note that with the v_j's each gaussian, the sum must be gaussian as well. In particular, the variance of the sum is the sum of the variances, or just nN in this case, while the expected values add as well.

Letting

$$y \equiv \sum_{j=1}^{n} v_j$$

we then have the two conditional-density functions, one for signal s_1, the other for s_2,

$$f(y\,|\,1) = \frac{e^{-(y-nA)^2/2nN}}{\sqrt{2\pi nN}} \tag{6-20}$$

and

$$f(y\,|\,2) = \frac{e^{-y^2/2nN}}{\sqrt{2\pi nN}} \tag{6-21}$$

These are shown sketched in Fig. 6-10. The decision level $nA/2$ is of course just at the intersection of the two curves. The probability of error is then just that of the single sample case, but with the signal amplitude A replaced by nA, the noise power N replaced by nN! The improvement obtained by using the n samples optimumly (adding the received samples and requiring the sum to exceed a threshold level) corresponds to a net increase in the single sample signal-to-noise ratio $A/\sqrt{N}$ by a factor of $\sqrt{n}$. The n-sample system thus corresponds to a single-sample system with effective SNR given by $\sqrt{n} A/\sqrt{N}$. The SNR in power improves linearly with n.

Alternatively, for the same probability of error the peak signal power A^2 may be reduced by a factor of n if n samples are added. There is thus an exchange of power and time (or bandwidth). For to obtain large numbers of independent samples, the signal duration must be increased accordingly. The bit rate must thus be decreased. (The same is true if instead of lengthening the signal duration the signal pulses are always repeated n times.) Note, however, that the signal *energy* $E = \int s^2(t)\ dt$ remains *fixed*. For if the basic bit interval is τ and the pulse height $\sqrt{n}\ A$, $E = nA^2\tau$. If pulses of height A and duration $n\tau$ are now transmitted, E remains the same (Fig. 6-11). Ultimately, then, it is the signal energy that determines the probability of error. This was first noted in discussing matched filters in Chap. 5. We shall return to these points and tie them together once and for all in the next section.

We have spent substantial time on the gaussian case because of its great utility and importance in communication problems. We shall also extend the ideas further in the next section as well as those following. We now consider some other examples of optimum signal processing for minimum probability of error.

Assume as the first example that the two signals s_1 and s_2 are zero-mean gaussian-distributed variables, but with differing variances σ_1^2 and σ_2^2, respectively. Again n independent samples of the received wave $v(t) = s_1(t)$ or $s_2(t)$ are taken, and we require the optimum way of combining the n samples in the sense of minimum error probability. Here we have as the conditional-density functions for the jth sample,

$$f(v_j|1) = \frac{e^{-v_j^2/2\sigma_1^2}}{\sqrt{2\pi\sigma_1^2}} \qquad (6\text{-}22)$$

and

$$f(v_j|2) = \frac{e^{-v_j^2/2\sigma_2^2}}{\sqrt{2\pi\sigma_2^2}} \qquad (6\text{-}23)$$

These are sketched in Fig. 6-12.

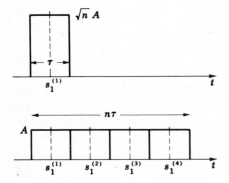

Figure 6-11 Exchange of power for time.

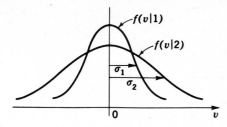

Figure 6-12 Binary signals—another example.

The n-dimensional density functions needed for calculating the likelihood ratio of Eq. (6-11) are then, with independent samples assumed,

$$f(\mathbf{v}\,|\,1) = \frac{\exp\left[-(1/2\sigma_1{}^2)\sum_{j=1}^{n} v_j{}^2\right]}{(2\pi\sigma_1{}^2)^{n/2}} = \frac{\epsilon^{-(1/2\sigma_1{}^2)\mathbf{v}\cdot\mathbf{v}}}{(2\pi\sigma_1{}^2)^{n/2}} \tag{6-24}$$

and

$$f(\mathbf{v}\,|\,2) = \frac{\exp\left[-(1/2\sigma_2{}^2)\sum_{j=1}^{n} v_j{}^2\right]}{(2\pi\sigma_2{}^2)^{n/2}} = \frac{e^{-(1/2\sigma_2{}^2)\mathbf{v}\cdot\mathbf{v}}}{(2\pi\sigma_2{}^2)^{n/2}} \tag{6-25}$$

Here vector notation has again been introduced. Again taking the log of the likelihood ratio to simplify results, we now find as the rule for deciding signal s_1 present,

$$n\log\left(\frac{\sigma_2{}^2}{\sigma_1{}^2}\right) + \mathbf{v}\cdot\mathbf{v}\left(\frac{1}{\sigma_2{}^2} - \frac{1}{\sigma_1{}^2}\right) > 2\log\frac{P_2}{P_1} \tag{6-26}$$

If $\sigma_1{}^2 < \sigma_2{}^2$, as in Fig. 6-12, we get as the rule for deciding s_1 is present,

$$\mathbf{v}\cdot\mathbf{v} < \underbrace{\frac{n\log(\sigma_2{}^2/\sigma_1{}^2) + 2\log(P_1/P_2)}{1/\sigma_1{}^2 - 1/\sigma_2{}^2}}_{d^2} \tag{6-27}$$

The square of the length of vector $\mathbf{v}$ is to be compared to a threshold. If

$$\mathbf{v}\cdot\mathbf{v} \equiv \sum_{j=1}^{n} v_j{}^2 > d^2$$

declare s_2 present; if $< d^2$, declare s_1 present. The geometry here is again shown in the two-dimensional case of Fig. 6-13. In the n-dimensional case the boundary between V_1 and V_2 corresponds to a hypersphere concentric with the origin. Points inside the sphere correspond to s_1, those outside to s_2. This is consistent with our intuition, for with s_1 and s_2 the random gaussian variables of Fig. 6-12, it is apparent polarity should play no role in detection here. One would expect signal s_2 to have larger amplitudes, on the average, than signal s_1. Squaring the received samples eliminates polarity from consideration. We then compare the sum of the squared values with the threshold d^2.

As a second additional example we assume polar signals of amplitude

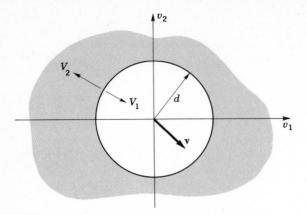

Figure 6-13 Decision regions for case of Fig. 6-12.

$\pm A$ transmitted. Noise is again added, but this time the noise probability-density function is given by the laplacian function

$$f(n) = \frac{1}{2c} e^{-|n|/c} \tag{6-28}$$

This is sketched in Fig. 6-14. It is left to the reader to show that the function is properly normalized and that the standard deviation (rms noise) is given by

$$\sigma = \sqrt{2}\, c \tag{6-29}$$

This density function is sometimes used to model additive impulse or burst-type noise. The simple exponential behavior, rather than the quadratic exponential of the gaussian-density function, means higher amplitudes have correspondingly higher probabilities of appearing, a characteristic of this type of impulse noise.

The received signal sample v is then

$$v = \begin{matrix} s_1 \\ \text{or} \\ s_2 \end{matrix} + n \tag{6-30}$$

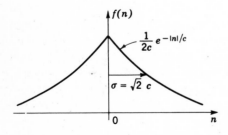

Figure 6-14 Laplacian noise.

The two conditional-density functions are given by

$$f(v|1) = \frac{e^{-|v-A|/c}}{2c} \tag{6-31}$$

and

$$f(v|2) = \frac{e^{-|v+A|/c}}{2c} \tag{6-32}$$

as sketched in Fig. 6-15.

We now sample the received signal v n times and again ask for the optimum processing procedure. With independent samples assumed we again set up the likelihood ratio as in Eq. (6-11) and proceed to crank out the result. Here we have to be somewhat careful, however, because of the shape of $f(n)$. To take into account the abrupt change in the form of $f(v|1)$ and $f(v|2)$ at $v = \pm A$, we break the range of v into three regions and consider each separately. Consider a typical term in the likelihood ratio, $f(v_j|1)/f(v_j|2)$:

1. $v_j < -A$. Then

$$\frac{f(v_j|1)}{f(v_j|2)} = \frac{e^{(v_j-A)/c}}{e^{(v_j+A)/c}} = e^{-2A/c} \tag{6-33}$$

Note that the dependence on v_j cancels out! The only knowledge retained is the fact that $v_j < -A$.

Assume now that n_1 of the n samples available fall in this range of v_j. There are then n_1 terms like Eq. (6-33) in the likelihood ratio, all to be multiplied together. Again taking the natural log of the likelihood ratio to simplify results, it is apparent that these n_1 terms in the likelihood ratio contribute

$$- \frac{2n_1 A}{c}$$

to the total log l.

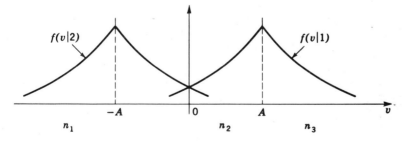

Figure 6-15 Polar signals in Laplacian noise.

2. $v_j > A$. Here

$$\frac{f(v_j|1)}{f(v_j|2)} = \frac{e^{-(v_j-A)/c}}{e^{-(v_j+A)/c}} = e^{2A/c} \tag{6-34}$$

Again the dependence on v_j cancels out, and each term in log l that corresponds to $v_j > A$ contributes $+2A/c$ to the total. Assuming that n_3 of n samples have $v_j > A$, we get as the contribution to log l,

$$\frac{+2n_3A}{c}$$

3. $-A < v_j < A$. Here

$$\frac{f(v_j|1)}{f(v_j|2)} = \frac{e^{(v_j-A)/c}}{e^{-(v_j+A)/c}} = e^{2v_j/c} \tag{6-35}$$

Assume that n_2 of the n samples fall in this range. Taking logs, we find these n_2 terms in the likelihood ratio contribute the term

$$\frac{2}{c} \sum_{j=1}^{n_2} v_j$$

to the total.

The three numbers n_1, n_2, and n_3 are indicated in Fig. 6-15. It is apparent that we have the constraint

$$n_1 + n_2 + n_3 = n \tag{6-36}$$

Combining all three terms, the rule for deciding on signal s_1 now becomes

$$(n_3 - n_1)A + \sum_{j=1}^{n_2} v_j > \frac{c}{2} \ln \frac{P_2}{P_1} \tag{6-37}$$

The interpretation here is quite interesting. The optimum processor consists of three-level threshold circuitry followed by an appropriate counter and an adding device. If the threshold circuitry detects a sample in the $v_j > A$ region, a positive count of A is added to the counter. If a sample in the range $v_j < -A$ appears, A is *subtracted* from the counter. If $-A < v_j < A$, the sample value is stored in the adding circuit. At the end of the n samples, the counter and adder outputs are summed. If they exceed the decision level $(c/2) \ln (P_2/P_1)$, s_1 is declared present; otherwise s_2 is assumed present. Note that such a processor lends itself nicely to digital circuitry. Note also that if A/σ is small (small SNR), the number of samples n_2 falling in the central region would tend to be small, the adder output would become negligible, and the counter alone would suffice.

6-3 OPTIMUM BINARY TRANSMISSION

We now focus attention on one particular case of binary transmission: the transmission of binary symbols $s_1(t)$ or $s_2(t)$ in additive gaussian noise. We assume a fixed binary interval T sec long and ask for both the optimum receiver structure and optimum signal shapes $s_1(t)$ and $s_2(t)$. Part of the answer is already available to us. We showed in the previous section that with n samples of the received signal plus noise the optimum processing was that described by the vector Eq. (6-18), or its equivalent Eq. (6-19). Thus choose $s_1(t)$ if

$$\mathbf{v} \cdot (\mathbf{s}_1 - \mathbf{s}_2) > \frac{\|\mathbf{s}_1\|^2 - \|\mathbf{s}_2\|^2}{2} \tag{6-38}$$

(Here we use the symbol $\|\quad\|$ for magnitude of a vector. Also recall the assumption that the a priori probabilities P_1 and P_2 are equal.)

We now ask the obvious question: In a given interval T, how many samples n should we use? To answer this we assume the noise is band-limited white noise, with spectral density $G_n(f) = n_0/2$, over the range of frequencies $\pm B$ hertz. The total noise power is then $N = n_0 B$. We shall also assume the two signals $s_1(t)$ and $s_2(t)$ band-limited over the same band B.

For this model of noise we recall the autocorrelation function, the Fourier transform of the spectral density, is just

$$R_n(\tau) = N \frac{\sin 2\pi\tau B}{2\pi\tau B} \tag{6-39}$$

Both $G_n(f)$ and $R_n(\tau)$ are shown sketched in Fig. 6-16. Note that there is zero correlation between samples spaced multiples of $1/2B$ seconds apart. For gaussian noise this further indicates the samples are *independent*. There are thus $n = 2BT$ independent samples available to us. Recall also from the sampling theorem in Chap. 3 that samples spaced $1/2B$ seconds apart suffice to uniquely characterize a band-limited signal. In fact, we indicated that with the sample

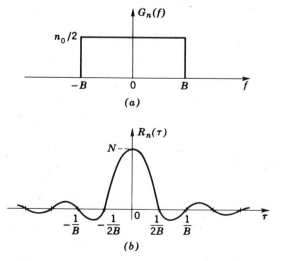

Figure 6-16 Band-limited white noise. (*a*) Spectral density. (*b*) Autocorrelation function.

values $n(j/2B), j = 0, \pm 1, \pm 2, \ldots$, known, we could reproduce the original wave $n(t)$ as

$$n(t) = \sum_j n\left(\frac{j}{2B}\right) \frac{\sin 2\pi B(t - j/2B)}{2\pi B(t - j/2B)} \qquad (6\text{-}40)$$

The one catch here is that all samples $j = (-\infty, +\infty)$ have to be known, whereas we only have $n = 2BT$ of them. For $BT \gg 1$, however, Eq. (6-40) provides a good approximation to $n(t)$ in the binary interval T, and we shall make this assumption.

The $(\sin x)/x$ functions in Eq. (6-40) have an interesting property that proves extremely useful to us. They are examples of orthogonal functions, with the property that

$$\int_{-\infty}^{\infty} \frac{\sin \pi(2Bt - k)}{\pi(2Bt - k)} \cdot \frac{\sin \pi(2Bt - m)}{\pi(2Bt - m)} dt = \frac{1}{2B} \delta_{km} \qquad (6\text{-}41)$$

Here δ_{km} is the Kronecker delta

$$\begin{aligned} \delta_{km} &= 1 & k = m \\ &= 0 & k \neq m \end{aligned}$$

The integral of the product of two $(\sin x)/x$ functions displaced in time is thus zero. The Fourier series of Chap. 2 is another example of a set of orthogonal functions, in which the integral of the product over a specified interval (finite or infinite) is zero.

The proof of the orthogonality of the $(\sin x)/x$ functions is left as an exercise for the reader. [As a hint, consider the integral of Eq. (6-41) to be a convolution integral. It is then equal to the integral containing the product of the Fourier transforms of the two $(\sin x)/x$ functions. These are just constants over the range $\pm B$, with appropriate exponential phase factors. It is then easy to show the integral of the resultant exponential terms goes to zero.]

If we now assume the two binary signals $s_1(t)$ and $s_2(t)$ band-limited to B hertz as well, with $BT \gg 1$, it is apparent that they can equally well be expressed as the series of Eq. (6-40) in $(\sin x)/x$, with the sample values of s_1 and s_2 appropriately inserted. The received signal $v(t) = [s_1(t) \text{ or } s_2(t)] + n(t)$ is thus also band-limited and equally expressible in the same series. The series expansions in terms of the orthogonal $(\sin x)/x$ functions have the same properties as those of the Fourier series of Chap. 2. Specifically, a generalized form of the Parseval theorem mentioned earlier is easily derived, relating sums of sample values to an integral of the analog time function. Thus, consider the integral $\int n^2(t) \, dt$. [Although we should restrict the integration to the range $(-T/2, T/2)$, over which $n(t)$ is defined, with $BT \gg 1$ the integration may be taken over the infinite range $(-\infty, \infty)$.] Replacing each $n(t)$ by the sum of Eq. (6-40), interchanging the order of integration and summation, and noting the orthogonality relation of Eq. (6-41), we find

$$\int_0^T n^2(t) \, dt \doteq \frac{1}{2B} \sum_{j=1}^{2BT} n^2\left(\frac{j}{2B}\right) \qquad BT \gg 1 \qquad (6\text{-}42)$$

This is the same as the Parseval relation found earlier, and a similar relation may readily be derived the same way for *all* orthogonal series expansions.

If we now consider two time functions $v(t)$ and $s(t)$, band-limited over $\pm B$, we readily demonstrate in the same way that

$$\int_0^T v(t)s(t) \, dt \doteq \frac{1}{2B} \sum_{j=1}^{2BT} v\left(\frac{j}{2B}\right) s\left(\frac{j}{2B}\right) \tag{6-43}$$

But this is quite interesting, for if we look at the optimum processor for binary signals in gaussian noise, Eq. (6-38), we note that the vector products shown are just in the form of the right-hand side of Eq. (6-43). Specifically, then, applying Parseval's theorem to both sides of Eq. (6-38), we find the optimum processor to be equally well given by

$$\int_0^T v(t)[s_1(t) - s_2(t)] \, dt > \frac{1}{2} \int_0^T [s_1{}^2(t) - s_2{}^2(t)] \, dt \tag{6-44}$$

Alternatively, note that the integrals on the right-hand side above are just the respective energies in the signals. The inequality can thus be written

$$\int_0^T v(t)[s_1(t) - s_2(t)] \, dt > \frac{E_1 - E_2}{2} \tag{6-44a}$$

with
$$E_1 \equiv \int_0^T s_1{}^2(t) \, dt \qquad E_2 \equiv \int_0^T s_2{}^2(t) \, dt$$

The interpretation here is quite interesting. It says that instead of the digital operations on the samples indicated by Eq. (6-38), one may equally well take the incoming signal $v(t)$, multiply it by two stored replicas of $s_1(t)$ and $s_2(t)$, integrate, and sample the resultant output every T seconds to see if it exceeds a specified threshold. Note that the $s_1(t)$ and $s_2(t)$ inserted at the receiver must be precisely in phase with the $s_1(t)$ or $s_2(t)$ portion of the received $v(t)$. The resultant operation is nothing more than our old friend coherent or synchronous detection! In the general form of Eq. (6-44a) it is also often called correlation detection. A further alternative form may be obtained by separating the s_1 and s_2 terms:

$$\int_0^T v(t)s_1(t) - \frac{E_1}{2} > \int_0^T v(t)s_2(t) \, dt - \frac{E_2}{2} \tag{6-44b}$$

The E_1 and E_2 thus serve as fixed-bias terms to equalize the detector outputs. The correlation detector and sampler of Eq. (6-44b) is shown sketched in Fig. 6-17.

As an example, let $s_1(t) = \cos \omega_0 t$, $s_2(t) = -\cos \omega_0 t$, just the PSK signals of Chap. 4. Then $E_1 = E_2$, and the correlation detector of Fig. 6-17 becomes the synchronous detector discussed previously. This is shown in Fig. 6-18a. (The integrator of course provides the necessary low-pass filtering.) If $s_1(t) = A \cos \omega_0 t$, $s_2(t) = 0$, we have the OOK signal of Chap. 4. Then

$$E_1 = \int_0^T A^2 \cos^2 \omega_0 t \doteq \frac{A^2 T}{2}$$

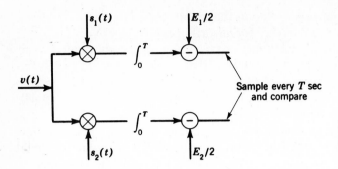

Figure 6-17 Optimum binary processor—correlation detection.

if $\omega_0 T \gg 2\pi$, and the processing called for is just

$$\frac{2}{T} \int_0^T v(t) \cos \omega_0 t \, dt > \frac{A}{2}$$

This corresponds of course to synchronous detection, followed by the $A/2$ decision level (Fig. 6-18b). Finally, if $s_1(t) = \cos \omega_1 t$, $s_2(t) = \cos \omega_2 t$, $E_1 = E_2$, we get the FSK synchronous detector of Fig. 6-18c.

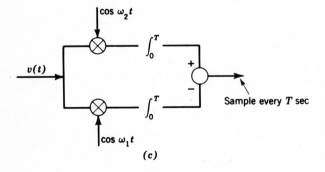

Figure 6-18 Examples of Fig. 6-17. (a) PSK synchronous detector. (b) OOK synchronous detector. (c) FSK synchronous detector.

There is still a further interpretation of the optimum processor that leads directly to the matched filter of Chap. 5. The integral of Eq. (6-44a) may be rewritten as the following convolution integral,

$$\int_0^T v(t)[h_1(T - t) - h_2(T - t)]\, dt > \frac{E_1 - E_2}{2} \tag{6-44c}$$

where
$$h_1(T - t) \equiv s_1(t)$$
$$h_1(t) = s_1(T - t)$$
and
$$h_2(T - t) \equiv s_2(t) \tag{6-45}$$
$$h_2(t) = s_2(T - t)$$

By taking Fourier transforms of both sides of Eq. (6-45), it is apparent that we must have

$$H_1(\omega) = e^{-j\omega T} S_1^*(\omega)$$
and
$$H_2(\omega) = e^{-j\omega T} S_2^*(\omega) \tag{6-46}$$

just our earlier (Chap. 5) conditions for a matched filter. The optimum processor may thus be drawn in the matched-filter form of Fig. 6-19. This, in the special cases referred to above, is identical with the matched-filter formulation we obtained in Chap. 5. There we wanted to maximize a SNR in order to decrease the probability of error. Here we have shown that one can do no better—provided that the assumptions made in Chap. 5 are valid. Thus we assume negligible intersymbol interference, with band-limited white gaussian noise the only possible source of binary error.

Although we have stressed the analog processors of Figs. 6-17 and 6-19 it is apparent that with modern integrated circuits digital processing is often to be preferred. One then seeks ways of implementing Eq. (6-38), or its digital equivalents, directly. One may consider carrying out the process

$$\frac{1}{2B} \sum_{j=1}^{n=2BT} v_j \left[s_1 \left(\frac{j}{2B} \right) - s_2 \left(\frac{j}{2B} \right) \right] > \frac{E_1 - E_2}{2} \tag{6-47}$$

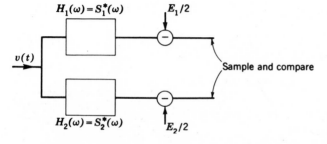

Figure 6-19 Matched-filter form of optimum binary processor.

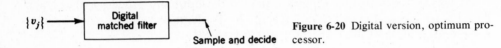

Figure 6-20 Digital version, optimum processor.

directly, using the incoming samples v_j and stored signal samples, or, alternatively, pass the successive incoming samples v_j through a *digital* matched filter for processing. This is indicated in Fig. 6-20.

Optimum Waveshapes

It is apparent that the calculation of probability of error for the optimum binary processor must lead to the same erfc x curves obtained in Chap. 5, since we have just demonstrated the equivalence of the *ad hoc* detectors discussed there to the optimum processors studied here. It is still useful to repeat the calculation, however, for in the process we shall find a simple answer to a second question we raised earlier. What are the optimum waveshapes to be used for $s_1(t)$ and $s_2(t)$? We shall show the answer to be simply given by $s_1(t) = -s_2(t)$, just the condition satisfied by a polar or PSK sequence! So unless we find other waveshapes to be simpler to generate, it is apparent we need look no further for "better" waveshapes.

The probability of error is readily calculated using either Eq. (6-44a) or the digital equivalent Eq. (6-47). Assume, for example, that we know $s_1(t)$ has been transmitted. The probability that Eqs. (6-44a) or (6-47) are not satisfied gives us the probability of error in this case. By symmetry it is also the same as the probability of mistaking $s_2(t)$ for $s_1(t)$, and therefore gives the total probability of error of the system P_e. (Recall again that we have assumed $P_1 = P_2 = \frac{1}{2}$ here for simplicity. The analysis is readily extended to the more general case $P_1 \neq P_2$ as well.)

With $s_1(t)$ transmitted, the received signal $v(t)$ is

$$v(t) = s_1(t) + n(t) \tag{6-48}$$

The probability of error is then simply the probability that

$$\int_0^T [n(t) + s_1(t)][s_1(t) - s_2(t)] \, dt < \frac{E_1 - E_2}{2}$$

or, alternatively, using the Parseval identity,

$$\frac{1}{2B} \sum_{j=1}^{2BT} (n_j + s_{1_j})(s_{1_j} - s_{2_j}) < \frac{E_1 - E_2}{2}$$

[We have used the simpler notation s_{1_j} and s_{2_j} to represent the jth samples of $s_1(t)$ and $s_2(t)$, respectively.] Simplifying by leaving the $n(t)$ term only on the left-hand side, and recalling the definitions of E_1 and E_2,

$$E_1 \equiv \int_0^T s_1{}^2(t)\,dt \doteq \frac{1}{2B} \sum_{j=1}^{2BT} s_{1_j}{}^2$$

$$E_2 \equiv \int_0^T s_2{}^2(t)\,dt \doteq \frac{1}{2B} \sum_{j=1}^{2BT} s_{2_j}{}^2$$

(6-49)

we get as the condition for an error to occur

$$y \equiv \int_0^T n(t)[s_1(t) - s_2(t)] < -b \equiv -\tfrac{1}{2} \int_0^T [s_1(t) - s_2(t)]^2\,dt \qquad (6\text{-}50)$$

Equivalently, using the sample values, an error occurs if

$$y \equiv \frac{1}{2B} \sum_{j=1}^{2BT} n_j(s_{1_j} - s_{2_j}) < -b \qquad (6\text{-}51)$$

But the noise $n(t)$ is assumed gaussian. The random variable y is then gaussian as well. [This is apparent either from Eq. (6-51), where y is defined as the weighted sum of $2BT$ gaussian variables n_j, or equally well from Eq. (6-50), describing y as the output of a *linear* matched filter with $n(t)$ applied at the input.] The expected value and variance of y are given very simply by

$$E(y) = \frac{1}{2B} \sum_{j=1}^{2BT} E(n_j)(s_{1_j} - s_{2_j}) = 0 \qquad (6\text{-}52)$$

(we have assumed zero-mean noise), and

$$\sigma_y{}^2 = E[y - E(y)]^2$$

$$= \left(\frac{1}{2B}\right)^2 E\left[\sum_i \sum_j n_i n_j (s_{1_i} - s_{2_i})(s_{1_j} - s_{2_j})\right]$$

$$= \left(\frac{1}{2B}\right) \sum_i \sum_j E(n_i n_j)(s_{1_i} - s_{2_i})(s_{1_j} - s_{2_j}) \qquad (6\text{-}53)$$

interchanging summation and expectation.

Recall, however, that for band-limited white noise, the samples spaced $1/2B$ seconds apart are *uncorrelated*. Also, $E(n_i{}^2) = N$ [see Eq. (6-39)]. Therefore, $E(n_i n_j) = N\delta_{ij}$. Equation (6-53) then simplifies to

$$\sigma_y{}^2 = \frac{N}{(2B)^2} \sum_{j=1}^{2BT} (s_{1_j} - s_{2_j})^2 = \frac{N}{2B} \int_0^T [s_1(t) - s_2(t)]^2\,dt$$

$$= \frac{n_0}{2} \int_0^T [s_1(t) - s_2(t)]^2\,dt \qquad (6\text{-}53a)$$

again using the Parseval relation. Here we have also put $N = n_0 B$ for the band-limited white noise. (Recall that $n_0/2$ is the band-limited spectral density. See Fig. 6-16a.)

The probability density of the variable y defined by Eqs. (6-50) and (6-51) is then

$$f(y) = \frac{1}{\sqrt{2\pi\sigma_y^2}} e^{-y^2/2\sigma_y^2} \tag{6-54}$$

and the probability of error is just

$$
\begin{aligned}
P_e &= \int_{-\infty}^{-b} \frac{e^{-y^2/2\sigma_y^2}\, dy}{\sqrt{2\pi\sigma_y^2}} \\
&= \int_{-\infty}^{-b/\sqrt{2}\sigma_y} \frac{e^{-x^2}\, dx}{\sqrt{\pi}} \\
&= \frac{1}{2}\operatorname{erfc} \frac{b}{\sqrt{2}\,\sigma_y} \tag{6-55}
\end{aligned}
$$

The appropriate integration to obtain P_e is indicated by the shaded area in Fig. 6-21.

Using the definition of b in Eq. (6-50) and that of σ_y^2 in Eq. (6-53a), we have as the final result for P_e,

$$P_e = \frac{1}{2}\operatorname{erfc} \frac{a}{2\sqrt{2}} \tag{6-55a}$$

where the new parameter a is defined by

$$a^2 \equiv \int_0^T \frac{(s_1 - s_2)^2\, dt}{n_0/2} \tag{6-56}$$

This is precisely the result obtained earlier in Chap. 5, in discussing high-frequency binary transmission. There it was obtained in a rather *ad hoc* fashion, by showing that the three cases considered, OOK, PSK, and FSK transmission, could all be subsumed in a probability of error expression identical to (6-55a) [see Eq. (5-104)], with the parameter a defined appropriately in each case. Here we have shown that (6-55a) is in fact more general: *any* form of binary communication in the presence of additive gaussian noise will have a probability of error expression given by (6-55a), with the parameter a defined in terms of the two signals by (6-56).

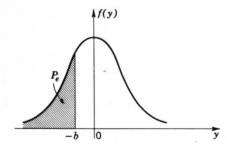

Figure 6-21 Probability of error in binary wave-shape transmission.

As a check, consider the three types of high-frequency binary transmission considered in Chap. 5. For each of these we first write the probability of error expression obtained, assuming matched filtering, in Chap. 5, and then show it is in fact given by Eqs. (6-55a) and (6-56). This thus repeats the calculations of Chap. 5. We then point out more generally that these two equations enable us to determine the *optimum* waveshapes to be used in the presence of additive gaussian noise.

In all three examples below we assume rectangular shaping for simplicity. Sinusoidal roll-off shaping, or Nyquist shaping more generally, as described in Chaps. 3 and 4, could be handled just as well by incorporating it in the expressions for $s_1(t)$ and $s_2(t)$.

1. OOK:

$$P_e = \frac{1}{2} \operatorname{erfc} \left(\frac{1}{2} \sqrt{\frac{E}{n_0}} \right) \tag{6-57}$$

To show this is included in (6-55a), let $s_2(t)$ in Eq. (6-56) $= 0$, and $s_1(t) = A \cos \omega_0 t$, just the OOK case. Then the parameter

$$a^2 = (2/n_0) \int_0^T A^2 \cos^2 \omega_0 t \, dt = 2E/n_0$$

and

$$\frac{a}{2\sqrt{2}} = \frac{1}{2\sqrt{2}} \sqrt{\frac{2E}{n_0}} = \frac{1}{2} \sqrt{\frac{E}{n_0}}$$

just as in Eq. (6-57).

2. FSK:

$$P_e = \frac{1}{2} \operatorname{erfc} \sqrt{\frac{E}{2n_0}} \tag{6-58}$$

Here we have $s_1(t) = A \cos \omega_1 t$ and

$$s_2(t) = A \cos \omega_2 t$$

If $\omega_1 T \gg 2\pi$, $\omega_2 T \gg 2\pi$, and $(\omega_1 - \omega_2)T \geq 2\pi$, it is readily shown that $\int_0^T s_1(t)s_2(t) \, dt \doteq 0$. [The signals $s_1(t)$ and $s_2(t)$ are examples of orthogonal signals.] Equation (6-56) then simplifies to

$$a^2 = \frac{2}{n_0} \int_0^T [s_1{}^2(t) + s_2{}^2(t)] \, dt = \frac{4E}{n_0}$$

since here $E_1 = E_2 = E$. Then we also have $a/2\sqrt{2} = \sqrt{E/2n_0}$, just as in Eq. (6-58).

3. PSK:

$$P_e = \frac{1}{2} \operatorname{erfc} \sqrt{\frac{E}{n_0}} \tag{6-59}$$

Here we have $s_1(t) = -s_2(t)$, and $a^2 = 8E/n_0$, from Eq. (6-56). Then $a/2 \sqrt{2} = \sqrt{E/n_0}$, agreeing of course with Eq. (6-59).

The optimum waveshapes to be used in binary signal transmission in additive gaussian white noise are also readily obtained from Eqs. (6-55a) and (6-56), as already noted. Since erfc $(a/2 \sqrt{2})$ decreases with a, the larger a is, the smaller the probability of error P_e. A little thought will indicate that a is maximized and P_e minimized by setting $s_1(t) = -s_2(t)$. The PSK signals of course satisfy these conditions. Other pairs of polar signals would serve just as well. Since $a^2 = 8E/n_0$ for this case, with

$$E = \int_0^T s_1{}^2 \, dt = \int_0^T s_2{}^2 \, dt$$

it is again the ratio of average energy E to noise spectral density n_0 that determines the probability of error. This is the point we made in discussing matched filters in Chap. 5. To minimize error probability, one should use signals with as high an average energy content as allowable. In passing the signals through the matched filter, the specific dependence on the details of the waveshape disappears. The matched-filter signal-power output is proportional to the average energy E.

A universal probability-of-error curve for optimum binary transmission in additive white gaussian noise is shown in Fig. 6-22. This is precisely the curve plotted earlier as Fig. 5-39 and is repeated here for ease of access. It is just a plot of Eq. (6-55a). As indicated above and again in the figure, the parameter $a^2/8$ on which the probability of error depends is a function only of E/n_0, the ratio of signal energy to noise spectral density.

In this discussion, synchronous or phase-coherent detection is again assumed available, as in Chap. 5. If it is not possible to maintain phase synchronism, one must again resort to envelope detection with its attendant SNR deterioration. The need for synchronous detection is implicit in the correlation or matched-filter detection circuits of Figs. 6-17 and 6-19. As an example assume the input to the matched filter is $A \cos \omega_0 t$, defined over the binary interval T as in the examples just cited. The matched filter must then be a bandpass filter centered at f_0, and of bandwidth $\sim 1/T$. Alternatively, its impulse response is the same high-frequency rectangular pulse $\cos \omega_0 t$, of pulse width T although turned around in time. The output of the filter, the convolution of signal and impulse response, is then just a high-frequency sinusoidal pulse $2T$ seconds long, at the same frequency f_0, but with a *triangular* envelope, reaching its peak at the T-second interval. (Recall that the convolution of two rectangles is a triangle.) This is shown in Fig. 6-23.

Note that the output peaks up, as expected, at the sampling time T. But it is crucial that sampling take place *exactly* at T. If sampling takes place a quarter of a cycle, or $1/4f_0$ seconds early or late, the output drops to zero! This synchronism must be maintained to within a fraction of $1/4f_0$ seconds, which is not

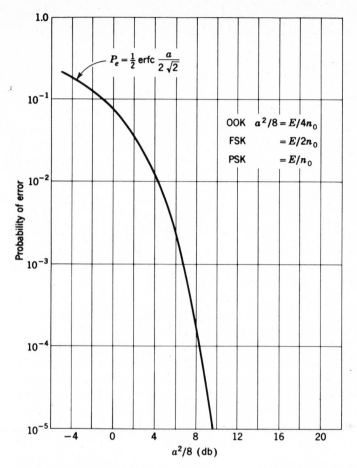

Figure 6-22 Probability of error, optimum binary transmission.

a mean task if the carrier frequency f_0 is in the VHF, UHF, or microwave range.

Actually, in practice one would not allow the output pulse to extend $2T$ seconds in length. For then successive binary pulses spaced $1/T$ seconds apart would interfere with one another. One way of avoiding this problem is to use as the matched filter an *integrate and dump circuit*.[6] This consists of a resonant circuit tuned to frequency f_0, with provision for shorting out the tuned circuit every T seconds just before receiving a signal input, to *dump* energy stored from previous intervals.

[6] J. M. Wozencraft and I. M. Jacobs, *Principles of Communication Engineering*, Wiley, New York, 1965, pp. 235, 236.

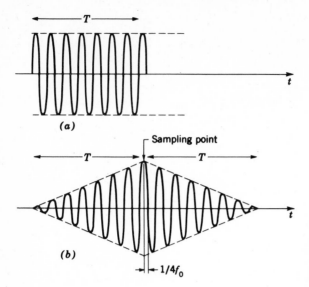

Figure 6-23 Matched-filter output to input sinusoidal pulse. (*a*) Input pulse. (*b*) Output pulse (ignoring a constant time delay).

6-4 M-ary TRANSMISSION, ADDITIVE WHITE GAUSSIAN NOISE CHANNEL

The material of the previous section can be readily extended to the more general case of M-ary signal transmission using a somewhat different approach. Recall that in Chap. 4 (Sec. 4-3) we introduced quadrature amplitude modulation (QAM) as a means of transmitting at higher bit rates over fixed-bandwidth channels. We showed, as an example, that the use of a 16-point QAM signal constellation allows 9,600 bits/s to be transmitted over a 2,700-Hz-bandwidth telephone channel. In this case four successive binary digits must be stored and recoded as one of the 16 signal shapes to be transmitted. The QAM technique is just one example of M-ary signal transmission, in which successive binary digits are stored and recoded to allow one of $M = 2^k$ waveshapes to be transmitted. This process is repeated every T seconds, T being the M-ary signal interval. If the original binary rate is R bits/s, the binary interval is $1/R$ seconds. In T seconds, then, the time of transmission of an M-ary signal, $k = RT$ binary digits are stored for coding into one of $M = 2^{RT}$ possible signals. As an example, two binary symbols in succession result in $M = 4$ possible waveshapes. If four binary signals are stored, they give rise to $M = 16$ possible waveshapes. Details appear in Sec. 4-3 in the discussion of QAM signaling. Figure 6-24 reviews the general concept of M-ary transmission. Figure 6-24*b* provides an example of five successive binary digits being stored, for which $M = 2^5 = 32$ possible output signals.

One problem noted in Chap. 4 with the use of the QAM technique was that as more signal points in the two-dimensional constellation are used, for higher rate transmission over the same bandwidth channel, they crowd more closely

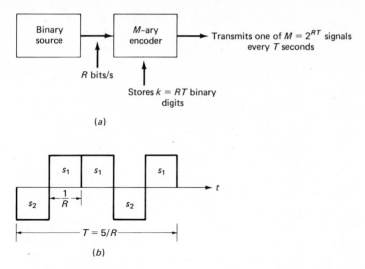

Figure 6-24 Binary to M-ary conversion. (a) M-ary encoding. (b) Example with $M = 2^5 = 32$.

together and become more vulnerable to noise and other disturbances on reception. We shall show in this section how probability-of-error calculations can be carried out for various QAM signal sets verifying quantitatively the conjectures made in Chap. 4.

The general idea of M-ary transmission can be extended in another direction as well. We shall show that by choosing M *orthogonal* signals (one example is M-ary FSK, a set of M carrier frequencies spaced wide enough apart to ensure orthogonality), one can actually *reduce* the probability of error below that of optimum binary (PSK) transmission. The price paid is that the transmission bandwidth and complexity of the system increases as M increases. The use of M orthogonal signals is thus suggested for relatively wideband channels, where the increased bandwidth required may not pose too severe a problem, but where power limitations necessitate some such coding of binary signals to reduce the error probability below that obtained for the binary signals in the previous section. Space communication channels serve as one possible model of such channels. The M-ary QAM signals, on the other hand, which show an *increase* in error rate as M increases are appropriate for such channels as the band-limited telephone channel where there may not be a severe E/n_0 limitation, but the bandwidth limitation restricts high-speed binary symbol transmission.

To discuss the performance of M-ary signal sets in detail, we choose the simplest model of a transmission channel: one that introduces additive white gaussian noise only. This turns out to be a fairly realistic model for many space communication systems. Although telephone channels are limited more by intersymbol interference and impulsive noise than by additive gaussian noise, the results obtained for the QAM signals are useful in the telephony case, since they do provide an upper limit on the performance and do suggest useful receiver structures.

The additive white gaussian noise model is often referred to as the AWGN channel. To study the performance of M-ary signals transmitted over this channel we again apply the basic concepts of statistical decision theory, but approach the problem somewhat differently than in the previous section. The approach adopted here does not rely on sampling the signal plus noise to generate independent statistical samples, as assumed in the previous section. In the case of $M = 2$ or binary transmission, the results turn out to be identical to those of the previous section. This is to be expected since the optimum binary processor and its error performance, as expressed by (6-44), (6-55a), and (6-56), depend neither on the bandwidth B of the band-limited noise model chosen, nor on the samples of the various waveshapes. These were in a sense artificially introduced to enable the calculations to be made. The approach of this section thus provides an alternative way of generating the matched filter-correlation detector results of the previous section, allowing extension to more general cases as well. The key result of this section is that we demonstrate that M-ary signal detection may be studied by working with the signal waveshapes directly.

To motivate the approach to be used, consider first, as an example, the case of four QAM signals. (Recall from Chap. 4 that these can also be labeled QPSK signals.) As shown in Chap. 4, the set of four signals may be expressed as

$$s_1(t) = \sqrt{\frac{2}{T}} \, a \cos \omega_0 t + \sqrt{\frac{2}{T}} \, a \sin \omega_0 t$$

$$s_2(t) = \sqrt{\frac{2}{T}} \, a \cos \omega_0 t - \sqrt{\frac{2}{T}} \, a \sin \omega_0 t$$

$$s_3(t) = -\sqrt{\frac{2}{T}} \, a \cos \omega_0 t - \sqrt{\frac{2}{T}} \, a \sin \omega_0 t \qquad (6\text{-}60)$$

$$s_4(t) = -\sqrt{\frac{2}{T}} \, a \cos \omega_0 t + \sqrt{\frac{2}{T}} \, a \sin \omega_0 t$$

(Rectangular waveshapes are again assumed for simplicity.) The $\cos \omega_0 t$ carrier term is called the *inphase term,* and the $\sin \omega_0 t$ term the *quadrature term.* The information carried in the signal is then the sign of each of these terms.

The carriers in (6-60) have each been written with a $\sqrt{2/T}$ term modifying them for normalization purposes: $\sqrt{2/T} \cos \omega_0 t$ and $\sqrt{2/T} \sin \omega_0 t$ are then orthonormal with respect to one another. To make this more explicit, we can rewrite (6-60) in the form

$$s_i(t) = \sum_{j=1}^{2} s_{ij} \phi_j(t) \qquad i = 1, 2, 3, 4 \qquad (6\text{-}61)$$

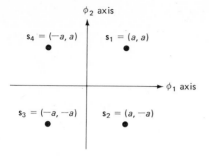

Figure 6-25 QAM signal constellation, $M = 4$.

Here $\phi_1(t) = \sqrt{2/T} \cos \omega_0 t$ and $\phi_2(t) = \sqrt{2/T} \sin \omega_0 t$ are orthonormal functions, defined by the usual integral expression,

$$\int_0^T \phi_i(t)\phi_j(t) \; dt = \delta_{ij} \tag{6-62}$$

The coefficients s_{ij} are just the appropriate values of a or $-a$, from (6-60). As an example, $s_{11} = a$, $s_{12} = a$, $s_{21} = a$, $s_{22} = -a$, etc. Just as in Chap. 4, we may define a signal plane whose two axes represent the coefficients, respectively, of $\phi_1(t)$ and $\phi_2(t)$, in (6-61) or (6-60). The four signals $s_1(t)$, $s_2(t)$, $s_3(t)$, $s_4(t)$, may then be represented by four points in the plane. The set of points, as noted in Chap. 4, comprises a signal constellation. We denote them, in this example, by $\mathbf{s}_1 = (s_{11}, s_{12}) = (a, a)$, $\mathbf{s}_2 = (s_{21}, s_{22}) = (a, -a)$, $\mathbf{s}_3 = (s_{31}, s_{32}) = (-a, -a)$, and $\mathbf{s}_4 = (s_{41}, s_{42}) = (-a, a)$. These are shown plotted in Fig. 6-25.

The critical thing to note now is that the information carried in the particular signal, $s_i(t)$, transmitted, is given by the value of the corresponding vector $\mathbf{s}_i$, or, equivalently, the two numbers, (s_{i1}, s_{i2}). Since the orthonormal carriers $\phi_1(t)$ and $\phi_2(t)$ are common to all four signals and hence carry no information, they can be stripped away at the receiver without affecting in any way the optimum processing of the received signal. This is exactly the extension to this somewhat more general case of the processes of frequency conversion and synchronous detection discussed in both Chaps. 4 and 5. Shifting baseband signals up in frequency for transmission and then back again to baseband at reception does not affect the system performance in noise.

How does one carry out the process of stripping away the orthonormal carriers? This is obviously done, making use of the orthogonality property of (6-62), and generalizing the concept of synchronous detection, by multiplying the particular $s_i(t)$ transmitted by both $\phi_1(t)$ and $\phi_2(t)$, and then integrating over the T-second interval. The resultant correlator is shown in Fig. 6-26. The output of each of the two branches is $+a$ or $-a$, depending on the signal $s_i(t)$ transmitted.

Now assume that noise has been added during transmission. The received signal over a T-second interval is then

$$v(t) = s_i(t) + n(t) \qquad 0 \le t \le T \tag{6-63}$$

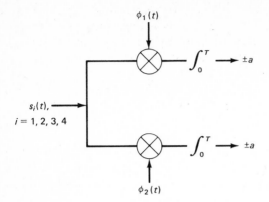

Figure 6-26 Correlator, $M = 4$ signals.

if $s_i(t)$ is the particular one of the four signals transmitted. Applying $v(t)$ to the correlator of Fig. 6-26, it is apparent that there are two noisy outputs. These are shown in Fig. 6-27, labeled v_1 and v_2 respectively. From the form of $v(t)$ in (6-63), and from the correlator operations of Fig. 6-26, it is apparent that these two outputs may be written

$$v_1 = s_{i1} + n_1$$

and
$$v_2 = s_{i2} + n_2 \tag{6-64}$$

with s_{i1} and s_{i2} precisely the components of $s_i(t)$ appearing in (6-61). The two noise terms, n_1 and n_2, are given by

$$n_j = \int_0^T n(t)\phi_j(t) \, dt \tag{6-65}$$

They are thus, mathematically, the orthonormal projections of the noise wave $n(t)$.

Since no information has been lost in carrying out the correlation operation of Fig. 6-26, the optimum way of processing the received waveshape $v(t)$ must

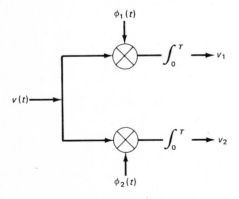

Figure 6-27 Correlator output, noisy signal at input.

correspond to optimum processing of the two numbers v_1 and v_2. By "optimum" we of course mean the processing procedure that minimizes the probability of error in transmitting these QAM signals. By the simple expedient of forcing a correlator to be used at the receiver (with no attendant loss of information), we have reduced the problem of distinguishing between one of four possible waveshapes on the basis of complete analog information, $v(t)$, $0 \leq t \leq T$, to one involving the best way of handling two numbers v_1 and v_2. We thereby avoid the need (at least theoretically) to sample $v(t)$ and process the samples, as was done in the preceding section for the binary transmission case.

It is apparent that the two outputs, v_1 and v_2, represent a vector $\mathbf{v} = (v_1, v_2)$ that can be plotted on the same two-dimensional plane as the four transmitted signal vectors $\mathbf{s}_1, \mathbf{s}_2, \mathbf{s}_3, \mathbf{s}_4$. This is shown in Fig. 6-28. It is tempting to now say that the optimum processor is one that declares that signal transmitted that is closest to $\mathbf{v}$. In the example of Fig. 6-28, this would be signal $s_4(t)$. This will in fact be shown to be the case for equally likely signals transmitted over the AWGN channel, extending the equivalent result of the binary case (see Fig. 6-8).

Before proceeding to carry out the calculations involving additive white gaussian noise, we generalize the QAM example to the M-ary transmission case. Consider now that one of M signals, $s_i(t)$, $i = 1, 2, \ldots, M$, is transmitted every T seconds, as shown in Fig. 6-24. [These can correspond to the general QAM case, to M different frequencies, to baseband transmission, or to any choice of M known waveshapes. Signal shaping is also included since it reflects itself in the specific form of the $s_i(t)$.] Let these M signals be transmitted over a channel using $N \leq M$ carriers. Without loss of generality we take these carriers to be orthonormal signals, $\phi_1(t), \phi_2(t), \ldots, \phi_N(t)$. We then have, as a representation of $s_i(t)$,

$$s_i(t) = \sum_{j=1}^{N} s_{ij}\phi_j(t) \qquad i = 1, 2, \ldots, M \qquad (6\text{-}66)$$

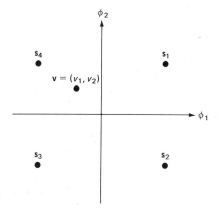

Figure 6-28 Vector representation of correlator output.

Here, as previously,

$$\int_0^T \phi_i(t)\phi_j(t) \, dt = \delta_{ij} \tag{6-67}$$

This then extends the QAM example, with $N = 2$ carriers (cos $\omega_0 t$ and sin $\omega_0 t$) needed, to more general cases. For example, if the $s_i(t)$'s are themselves orthogonal signals (as noted, M-ary FSK is exactly this case), $N = M$, $s_i(t) = a\phi_i(t)$, and $s_{ij} = 0$, $i \neq j$. Time-displaced (nonoverlapping) pulses as well as other types of waveshapes could be used as these carriers.

That only $N \leq M$ orthogonal carriers are needed in general may be proven by invoking the standard Gram-Schmidt orthogonalization procedure.[7] The precise form of the carriers is not important at this point. (The specific choice *is* important, as is apparent from previous chapters of the book, in determining the transmission bandwidth, peak power requirements, cost and ease of implementation, etc.) In fact, the Gram-Schmidt procedure may be invoked to generate a set of orthonormal waveshapes from a given set of M signal shapes.[8]

Each signal is distinguished by its N coefficients s_{ij}, $j = 1, 2, \ldots, N$. These are then the generalizations of the coefficients s_{i1} and $s_{i2} = +a$ or $-a$ defined in the QAM example. From the orthonormal property of the ϕ_j's, it is apparent from (6-66) that

$$s_{ij} = \int_0^T s_i(t)\phi_j(t) \, dt \tag{6-68}$$

[Recall from Chap. 2 that this was the method used to generate the Fourier coefficients. The Fourier series is of course one example of an infinite orthogonal series. The (sin $x)/x$ series of the previous section is another example of an infinite orthogonal series. Here only a finite number, $N \leq M$, of orthonormal functions is needed.]

As in the QAM case, the information as to which of the M signals is being transmitted is carried by the N coefficients s_{ij}. For a particular signal $s_i(t)$ the N coefficients may be visualized as representing an N-dimensional vector $\mathbf{s}_i = (s_{i_1}, s_{i_2}, \ldots, s_{iN})$. They can also be plotted as points in an N-dimensional space, each of whose axes corresponds to one of the orthonormal functions. s_{ij} is then the projection of $s_i(t)$ along the $\phi_j(t)$ axis.

Generalizing the QAM approach, then, the carriers $\phi_j(t)$, $i = 1, 2, \ldots, N$, may be stripped off at the receiver by carrying out a correlation process. This leaves just the N coefficients s_{ij} at the correlator output. With noise $n(t)$ added during transmission, the received signal, just as in the QAM case, is given by

$$v(t) = s_i(t) + n(t) \qquad 0 \leq t \leq T \tag{6-69}$$

[7] Wozencraft and Jacobs, *op. cit.*, p. 266 et seq.
[8] *Ibid*.

and the output of the correlation receiver is given by N numbers $v_1, v_2, \ldots,$ v_N. They make up the vector

$$\mathbf{v} = (v_1, v_2, \ldots, v_N) = \mathbf{s}_i + \mathbf{n} \tag{6-70}$$

The noise vector $\mathbf{n} = (n_1, n_2, \ldots, n_N)$, with

$$n_j = \int_0^T n(t)\phi_j(t)\,dt \tag{6-71}$$

exactly as in (6-65). A block diagram of this correlator receiver, consisting of a bank of N correlation operations, with $\mathbf{v}$ appearing at the output, is shown in Fig. 6-29. This is shown followed by an optimum processor for determining, with minimum probability of error, which of the M signals is transmitted every T seconds. The N numbers making up the vector $\mathbf{v}$ play the role of the n samples described in Sec. 6-2.

How does one now extend the optimum binary detection case of the previous sections to the case of M-ary detection? This is done quite simply by first returning to the optimum likelihood ratio test for binary hypothesis testing, given by (6-11). It is apparent that the optimum choice for region V_1 in that case is equally well given by

$$P_1 f(\mathbf{v}|1) > P_2 f(\mathbf{v}|2) \tag{6-72}$$

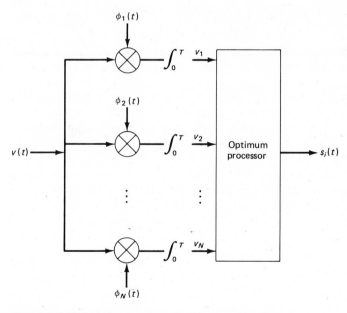

Figure 6-29 Correlation receiver, N orthonormal carriers.

From probability theory each side of the inequality in (6-72) is the joint probability of two events. Letting $i = 1$ or 2, we may thus write

$$P_i f(\mathbf{v}|i) = P(\mathbf{v}, i) = P(i|\mathbf{v}) f(\mathbf{v}) \tag{6-73}$$

Introducing the right-hand side of (6-73) into (6-72), and canceling the term $f(\mathbf{v})$ common to both sides, we have, as an alternative optimum decision procedure, the rule: choose hypothesis H_1 if

$$P(1|\mathbf{v}) > P(2|\mathbf{v}) \tag{6-74}$$

The two probabilities, $P(1|\mathbf{v})$ and $P(2|\mathbf{v})$, are called *a posteriori* probabilities. They represent, respectively, the probability that signal s_1 was transmitted, and that s_2 was transmitted, *given* the vector $\mathbf{v}$ received. The rule of (6-74) says simply that one is to select the signal more likely to have been transmitted. This guarantees minimum probability of error. Now extend this to one of M signals transmitted. Calculate the a posteriori probabilities $P(i|\mathbf{v})$, $i = 1, 2, \ldots, M$, for each of the signals. The obvious extension of the optimum rule to this case is to select the most probable event, or the signal with *maximum a posteriori* probability. Signal $s_i(t)$ is thus declared present, at the receiver, if

$$P(i|\mathbf{v}) > P(j|\mathbf{v}) \qquad \text{all } j \neq i \tag{6-75}$$

Alternatively, by reversing the process by which (6-74) was derived, we calculate $P_i P(\mathbf{v}|i)$ for all i and select the signal for which this is maximum. This provides the optimum processor of Fig. 6-29. This optimum maximum a posteriori (MAP) processor thus calculates $P(i|\mathbf{v})$, or, equivalently, $P_i P(\mathbf{v}|i)$ for all the signals, using the value of $\mathbf{v}$ appearing at its input, and outputs that signal for which this is largest. This is shown diagramatically in Fig. 6-30.

In the special case of the AWGN channel, the optimum processor is particularly easy to implement and is just the extension of the binary results of Secs. 6-2 and 6-3 to the M-ary case. We now show how to obtain this processor, by first finding the joint statistics of $\mathbf{n}$ and from this, using (6-70), the statistics of $\mathbf{v}$.

Recall from Chap. 5 that the spectral density of zero mean white noise is given by

$$G_n(f) = \frac{n_0}{2} \text{ watts/Hz} \tag{6-76}$$

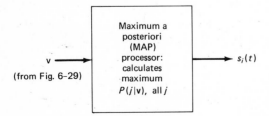

Figure 6-30 Optimum processor of Fig. 6-29.

The corresponding autocorrelation function is

$$R_n(t, s) = E[n(t)n(s)] = \frac{n_0}{2}\delta(t - s) \tag{6-77}$$

We have already noted in the previous section that with $n(t)$ gaussian the noise coefficient n_j given by a linear operation on $n(t)$ of the form of (6-71) is gaussian as well. [It may be readily shown, from the theory of random processes, that any linear operation on a gaussian process produces another gaussian process.[9] This was utilized in Chap. 5 in describing the statistics of narrowband noise, as an example. The linear operation of (6-71) thus produces a gaussian random variable.] All one needs to write its probability density function explicitly is its mean value and variance. In like manner the joint statistics of the N random variables $n_j, j = 1, 2, \ldots, N$, making up $\mathbf{n}$ are completely defined by their mean values, variances, and cross-correlations. To find these quantities we use (6-71) and the zero-mean, white-noise attribute of $n(t)$. Specifically, taking the expectation of n_j in (6-71), interchanging the expectation and integration operations,[10] and noting that $n(t)$ is by definition zero mean, we have first that n_j is a zero-mean random variable:

$$E(n_j) = 0$$

Also,
$$\sigma_j{}^2 = E(n_j{}^2) = E\left[\int_0^T \int_0^T n(t)n(s)\phi_j(t)\phi_j(s)\ dt\ ds\right]$$

$$= \int_0^T \int_0^T E[n(t)n(s)]\phi_j(t)\phi_j(s)\ dt\ ds$$

$$= \frac{n_0}{2} \tag{6-78}$$

again interchanging expectation and integration operations, and then using both the white-noise definition of (6-77) and the orthonormal property [Eq. (6-67)] of the $\phi_j(t)$'s. Details are left to the reader. It is also left to the reader to show in a similar manner that

$$E[n_i n_j] = 0 \qquad i \neq j \tag{6-79}$$

The white gaussian noise assumption thus leads to very simple results for the statistics of the noise samples at the outputs of the N correlators of Fig. 6-29. The noise samples are all gaussian, zero mean, have equal variance, and are pairwise uncorrelated. From the properties of gaussian random variables as well, they are then *independent* random variables. The joint statistics of the N noise samples are then simply written. They represent

---

[9] A. Papoulis, *Probability, Random Variables, and Stochastic Processes*, McGraw-Hill, New York, 1965.

[10] *Ibid.*, chap. 9.

a multidimensional set of independent gaussian random variables. Thus,

$$f(n_1, n_2, \ldots , n_N) = f(\mathbf{n}) = \prod_{j=1}^{N} f(n_j)$$

$$= \frac{e^{-\mathbf{n}\cdot\mathbf{n}/n_0}}{(\pi n_0)^{N/2}} \tag{6-80}$$

The vector dot product $\mathbf{n} \cdot \mathbf{n}$ is simply the shorthand notation for $\sum_{j=1}^{N} n_j^2$. The simplicity of the form for the joint statistics of $\mathbf{n}$ is the reason for the desire to model channels as having additive white gaussian noise.

Since $\mathbf{v} = \mathbf{s}_i + \mathbf{n}$ is the vector representation of the correlator output in Fig. 6-29, it is apparent that the probability density of $\mathbf{v}$, *conditioned* on $\mathbf{s}_i$ being present, is given simply by replacing $\mathbf{n}$ in (6-80) by $\mathbf{v} - \mathbf{s}_i$. Hence we have, for the AWGN channel,

$$f(\mathbf{v}\,|\,\mathbf{s}_i) = \frac{e^{-(\mathbf{v}-\mathbf{s}_i)\cdot(\mathbf{v}-\mathbf{s}_i)/n_0}}{(\pi n_0)^{N/2}} \tag{6-81}$$

This is exactly the expression needed to determine the a posteriori probabilities of (6-75).

Now consider the special but common case in which the M signals are equally likely to appear. We then have $P_1 = P_2 = \cdots P_M = 1/M$. From (6-75) the MAP detector in this case thus simply selects the signal with the largest conditional probability density function $f(\mathbf{v}\,|\,\mathbf{s}_i)$. Note from (6-81) that this is just the signal with the smallest exponent, or the one that *minimizes*

$$(\mathbf{v} - \mathbf{s}_i) \cdot (\mathbf{v} - \mathbf{s}_i) \equiv \|\mathbf{v} - \mathbf{s}_i\|^2 = \sum_{j=1}^{N} (v_j - s_{ij})^2 \tag{6-82}$$

This is precisely the expression for the distance squared between $\mathbf{v}$ and $\mathbf{s}_i$ in the N-dimensional Euclidean sense. We have thus shown that for the AWGN channel with one of M equally likely signals transmitted the optimum detector is the one that selects the signal "closest to" the received signal vector $\mathbf{v}$. The MAP processor of Fig. 6-30 thus calculates the distance between $\mathbf{v}$ and all stored signal vectors $\mathbf{s}_j, j = 1, 2, \ldots , M$ and selects the $\mathbf{s}_i$ that is closest to $\mathbf{v}$. This is shown schematically in Fig. 6-31. In the two-dimensional example of Fig. 6-28 with one of four signals transmitted, $\mathbf{s}_4$ would be selected as the one most likely to have been transmitted. The results of this section extend those of Sec. 6-3.

Matched-Filter Detection

The optimum receiver structure of Fig. 6-31 may be redrawn in two alternative but equivalent ways. One structure replaces the distance calculation of the MAP processor by a dot-product calculation. The second structure is completely different: the correlator receiver is replaced by a completely equivalent *matched-filter* structure, with the received signal $v(t)$ correlated against stored

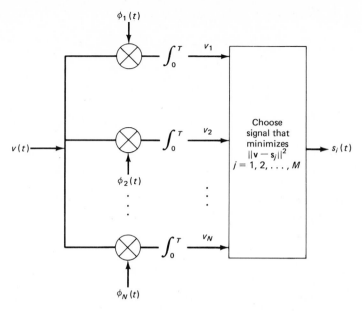

Figure 6-31 Optimum correlation receiver, equally likely signals, AWGN channel.

versions of each of the M possible signals transmitted, rather than the N orthogonal carriers of Fig. 6-31. Both of these structures extend the results of the previous sections. The matched-filter structure has of course already been discussed in connection with binary signaling in Sec. 6-3. The dot-product processing is similar to that given by (6-18) in Sec. 6-2. There samples of $v(t)$ were being considered. Here, more generally, the $\mathbf{v}$ represents the result of correlating $v(t)$ against the N orthonormal $\phi_j(t)$'s, as provided by the N correlation outputs in Fig. 6-31.

To demonstrate the validity of these statements consider the quadratic distance measure of (6-82). Expanding this quadratic form term by term, we get

$$(\mathbf{v} - \mathbf{s}_i) \cdot (\mathbf{v} - \mathbf{s}_i) = \|\mathbf{v}\|^2 - 2 \left(\mathbf{v} \cdot \mathbf{s}_i - \frac{\|\mathbf{s}_i\|^2}{2}\right) \qquad (6\text{-}82a)$$

Since $\|\mathbf{v}\|^2$ is independent of the transmitted signal, it is apparent that minimizing the distance corresponds precisely to *maximizing*

$$\left(\mathbf{v} \cdot \mathbf{s}_i - \frac{\|\mathbf{s}_i\|^2}{2}\right)$$

Consider now the term $\mathbf{s}_i \cdot \mathbf{s}_i = \|\mathbf{s}_i\|^2 = \sum_{j=1}^{N} s_{ij}^2$. It is easily shown that this is just the expression $\int_0^T s_i^2(t)\, dt = E_i$, the *energy* in signal $s_i(t)$. For, using the orthogonal expansion of $s_i(t)$, we have

$$\int_0^T s_i^2(t) \, dt = \int_0^T \left[\sum_{j=1}^N s_{ij}\phi_j(t) \right] \left[\sum_{k=1}^N s_{ik}\phi_k(t) \right] dt$$

Interchanging summation and integration, and recalling that the $\phi_j(t)$ functions were defined to be orthonormal over the interval $(0, T)$, it is left to the reader to show that

$$\|\mathbf{s}_i\|^2 = \sum_{j=1}^N s_{ij}^2 = \int_0^T s_i^2(t) \, dt = E_i \qquad (6\text{-}83)$$

This is just a special case of the general Parseval theorem discussed earlier (see Sec. 6-3, for example), relating integrals of time functions to the sums of their Fourier (orthogonal function) coefficients. It is thus apparent that the MAP processor of Fig. 6-31 for the AWGN channel can be replaced by the equivalent processor shown in Fig. 6-32. The energy terms shown subtracted out in the figure have a physical justification. They eliminate a bias in favor of higher-energy signals when comparing them on the basis of the decision parameter $\mathbf{v} \cdot \mathbf{s}_i$. As an additional by-product of these manipulations, note from (6-83) that the points in the signal constellation are all located at distances from the origin corresponding to the square root of their respective energies. Signal $\mathbf{s}_i$ is located at a distance $\sqrt{E_i}$, for example (see Fig. 6-25). We shall use this result in the next section in evaluating probabilities of error.

The optimum matched-filter receiver for the AWGN channel mentioned above is also readily obtained by applying Parseval's theorem to the distance

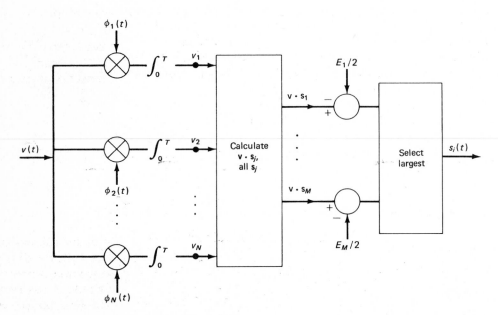

Figure 6-32 Equivalent receiver, AWGN channel.

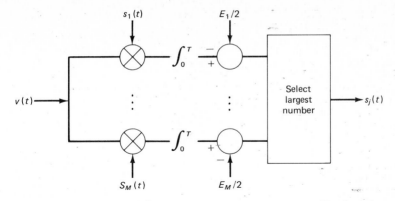

Figure 6-33 Matched-filter receiver, AWGN channel.

decision function of Eq. (6-82). Specifically, it is left to the reader to show, in a manner similar to that used in proving (6-83), that

$$\mathbf{v} \cdot \mathbf{s}_i \equiv \sum_{j=1}^{N} v_j s_{ij} = \int_0^T v(t) s_i(t) \, dt \tag{6-84}$$

Since it is the quantity $\mathbf{v} \cdot \mathbf{s}_i - E_i/2$ that is to be maximized in deciding on the most probable signal transmitted, it is apparent from (6-83) and (6-84) that one could equally well operate *directly* on the received signal waveshape $v(t)$ rather than first correlating against the carriers as in Figs. 6-31 and 6-32. The resultant optimum receiver that maximizes $\mathbf{v} \cdot \mathbf{s}_i - \|\mathbf{s}_i\|^2/2$ directly is shown sketched in Fig. 6-33. This is called a *matched-filter* receiver since the incoming received signal $v(t)$ is correlated with or matched against stored replicas of the M possible signal waveshapes. This extends the binary matched-filter results of Sec. 6-3 to M-ary signals. Rather than carry out the correlation function involving multiplication and integration, one can show as in the binary case that this is identical to passing $v(t)$ through a bank of *filters* and comparing outputs every T seconds.

The choice as to which implementation to use, that of Fig. 6-32 say, or that of Fig. 6-33, depends on the complexity of the signals transmitted, how easily generated, their number, the number N of carriers, and how easily these are generated. Consider as an example the case in which the two carriers are quadrature sine waves $\sqrt{2/T} \cos \omega_0 t$ and $\sqrt{2/T} \sin \omega_0 t$. If there are four possible signals transmitted, as in Fig. 6-25, the correlation receiver of Fig. 6-26 requires just two reference signals at the receiver. The equivalent matched-filter receiver would thus involve correlating against the four possible signals, $\sqrt{2/T} \, a \cos (\omega_0 t + \theta_i)$, with $\theta_i = \pm \pi/4$ and $\pm 3\pi/4$. In practice, all four waveshapes at the receiver would be derived from one sine wave. Note that the correlation receiver requires N correlations or filtering operations, that of the matched filter M such operations. In practice, there is of course always some shaping of the amplitude term multiplying the carrier. (It is not just a fixed number or rectangular shaping, as assumed up to now.) This is particularly true at the transition regions separating the beginning and end of successive T-second intervals. Matched filtering may then be used to account for this shaping. The distinction between the correlation and matched-filter receivers of Figs. 6-32 and 6-33 is in fact often blurred (particularly in the case where the carriers are simple sines and cosines) and both are often called matched-filter devices.

6-5 SIGNAL CONSTELLATIONS AND PROBABILITY OF ERROR CALCULATION

The vector-space representation of signals and the optimum detector solution as that which chooses the signal "closest to" the received signal is particularly useful in signal design as well as in probability-of-error calculations. Geometric concepts and results can very often be brought to bear successfully in handling these questions. The comments made earlier as to the distinction between power-limited and band-limited channels are readily verified in terms of a geometric picture. A power-limited channel (or its equivalent, an energy-limited channel) corresponds to one in which signal points are all constrained to lie on or within a sphere of radians $\sqrt{E}$, with E the maximum energy. ($E = ST$, with S the maximum power allowed for transmission; see Fig. 6-34.)

Consider first the case in which $N = 2$ carriers are used. (This is again the example of either two quadrature sine waves or two orthogonal sine waves of different frequencies.) It is apparent that in order to transmit at higher and higher bit rates, with correspondingly larger numbers of signals used, the signals become packed closer together and the probability of error must increase.

The only way to avoid this dilemma of increasing error probability or, alternatively, the only way to *reduce* the probability of error by purposely choosing larger values of M is by going to higher dimensions. A three-dimensional example with three orthogonal signals appears in Fig. 6-35. These signals are now "farther apart" than they are in two dimensions, so one would

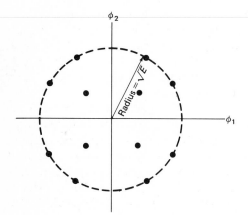

Figure 6-34 Maximum energy (power) constraint, two dimensions.

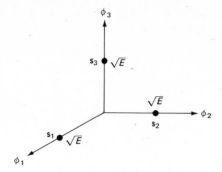

Figure 6-35 Orthogonal signals, three dimensions.

expect the probability of error to decrease. We shall demonstrate this quantitatively in the next section. As we add more *orthogonal signals,* increasing the dimensionality of the vector space, the signals move farther apart and the error probability decreases. However, as noted earlier, this requires correspondingly wider bandwidths for transmission.

Now consider the band-limited channel. In this case we are restricted to the two dimensions of Fig. 6-34. One would also use quadrature carriers rather than orthogonal sine-wave carriers of different frequencies to reduce the bandwidth. This is precisely what was described in Chap. 4, in discussing QAM transmission. As noted above, the probability of error increases if more signals must be packed into a given (energy-limited) region. However, if there is no specific power constraint (at least within limits), the signal points can always be moved farther apart. Equivalently, if the signal-to-noise ratio is "sufficiently high" (as will be shown shortly), signal points may be added without deteriorating the system too much.

An interesting design problem to which approximate answers only can be given is that of determining a set of M signals, constrained in power and in bandwidth, that minimizes the probability of error. The corresponding geometric picture is quite apparent. In the case of two dimensions, for example, how does one choose the optimum location of M points within the circle of Fig. 6-34? Rather than answer this question directly, we shall focus on the calculation of probability of error once a constellation (location of signal points) has been specified. Even here we shall have to restrict ourselves to simple locations of points (most commonly spanning a rectangular region in two dimensions) to keep the problem from becoming computationally too complex. This leads directly to probability-of-error calculations for QAM signals.

The reason for this is readily apparent using the geometric picture. Consider, for example, the three cases of $M = 3$, 4, and 8 signals located in a two-dimensional plane as portrayed in Fig. 6-36. The minimum probability of error decision rule for the AWGN channel with equally probable signals assigns the decision to the signal "closest to" the received signal point [Eq. (6-82)]. This then defines a *decision region* around each signal point. Three sets of decision regions appear in the examples of Fig. 6-36. It is apparent that one

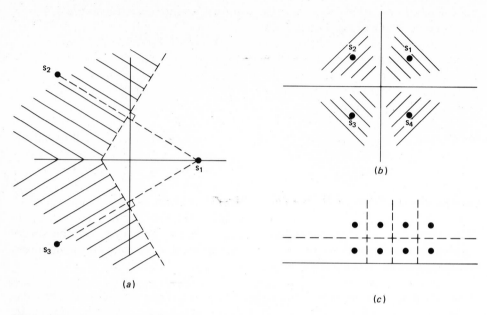

Figure 6-36 Examples of signal constellations and optimum decision regions. (a) $M = 3$. (b) $M = 4$. (c) $M = 8$.

finds these by locating the perpendicular bisector between pairs of signal points, as shown in Fig. 6-36a, for example. Details are left to the reader.

How does one now calculate the probability of error? One can determine the probability that the received signal vector **v** falls *outside* the specified decision region for a particular signal transmitted, and then average appropriately over the entire signal set, after weighting by the probability of occurrence of the given signal. Alternatively (and this is the approach we shall adopt), one can calculate the probability that the received signal vector **v** falls (correctly) within the desired decision region for each signal and again average appropriately. In either case it is apparent that the evaluation of probabilities over irregular regions, as in Fig. 6-36a, can be quite difficult. The rectangular regions of Fig. 6-36b and c are much simpler to handle, however, and we shall focus on these.

Binary Signals

Before proceeding with the probability-of-error calculations for M-ary signaling with rectangular signal constellations (QAM provides a practical example of one such set of signals), we calculate the probability of error for binary signals. The results found agree of course with those already obtained in Sec. 6-3, as well as in Chap. 5. They provide an alternative and instructive way of carrying out the error probability calculations, and serve to introduce the approach used for more complex signal constellations.

As the first example of the calculation of probability of error, consider the polar or antipodal signals of Fig. 6-37 spaced d units apart. With the two signals equally likely to be transmitted (as assumed here) it suffices to determine the probability of error for either one. By symmetry this is the same as for the other, and, averaging over both, is the system probability of error as well. In this case it is apparent that one correlator only in Fig. 6-27 need be used, since the output of the second branch should be zero and could equally well be closed. Alternatively, it could just as well be ignored. This shows up in the decision regions of Fig. 6-37 as well, since it is apparent that v anywhere in the right-half plane should be associated with s_1, while any value in the left-half plane is associated with s_2. It is thus the noise n_1 at the ϕ_1 correlator output of Fig. 6-27 only that can cause an error. Specifically, then, let s_2 be transmitted. An error occurs only if $n_1 > d/2$, for then the vector $v = s_2 + n$ (or the scalar $v_1 = s_{21} + n_1$) moves into the s_1 decision region. The probability of this happening is easily calculated as

$$P_e = \int_{d/2}^{\infty} \frac{e^{-x^2/n_0}}{\sqrt{\pi n_0}}\, dx$$

$$= \frac{1}{2}\, \text{erfc}\, \frac{d}{2\sqrt{n_0}} \tag{6-85}$$

using the known statistics of the gaussian random variable n_1 [see (6-78) and (6-80)]. Here erfc x is again the *complementary* error function defined as

$$\text{erfc}\, x \equiv \frac{2}{\sqrt{\pi}} \int_x^{\infty} e^{-y^2}\, dy \tag{6-86}$$

Obviously, the probability of error depends on the signal separation d. For $d/\sqrt{n_0}$ large enough (approximately >3), P_e decreases exponentially with d^2/n_0. To demonstrate the connection between the error probability P_e and the signal energy, it is instructive to rewrite P_e in (6-85) in terms of the energy E. (We shall do this in fact with all error calculations.) It is

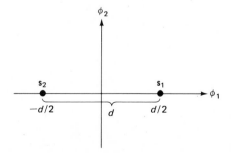

Figure 6-37 Polar (antipodal) signal set.

apparent that $d/2 = \sqrt{E}$ in this case. (Recall that $s_i \cdot s_i = E$.) Hence for the antipodal signal set of Fig. 6-37,

$$\text{PSK:} \quad P_e = \frac{1}{2} \text{ erfc } \sqrt{\frac{E}{n_0}} \qquad (6\text{-}85a)$$

This agrees of course with the results of Sec. 6-3 and Chap. 5.

Now take the two signal points of Fig. 6-37, keep their distance separated at d units, and move them arbitrarily over the (ϕ_1, ϕ_2) plane. An example is shown in Fig. 6-38a. A little thought will indicate that the *probability of error remains the same:* the points are still spaced the same distance apart. An error will occur if noise added to one signal point moves the sum more than halfway to the other point. This is exactly the calculation made in obtaining Eq. (6-85). Alternatively, one may choose two other orthonormal signals, ϕ_1' and ϕ_2', linearly dependent on ϕ_1 and ϕ_2, in terms of whose coordinates the s_1 and s_2 of Fig. 6-38a look just like those of Fig. 6-37. This obviously corresponds to rotating and translating the axes of Fig. 6-38 to line up appropriately with s_1 and s_2. The choice of axes does not affect the probability of error. So rotation and translation of a constellation does not affect P_e as long as the points in the constellation all move together. This is a very useful observation in the calculation of probability of error.

What *is* affected is the energy (and hence power) requirement. For as the points move farther away from the origin, the energy required increases. A little thought will in fact indicate that the minimum energy location of the two signal points of Fig. 6-38a is exactly that of Fig. 6-37 (or the rotated equivalent). In general, the minimum energy location of *any* constellation has its center of gravity located at the origin. As an example, consider the two special orientations of the two-signal set shown in Fig. 6-38. The one in Fig. 6-38b has its two signals orthogonal to one another. This thus corresponds to the binary FSK case, with the two frequencies chosen orthogonal to one another. In Fig. 6-38c one signal is always zero. This thus corresponds to on-off or on-off-keyed (OOK) transmission. It is apparent that the FSK signal set requires twice the

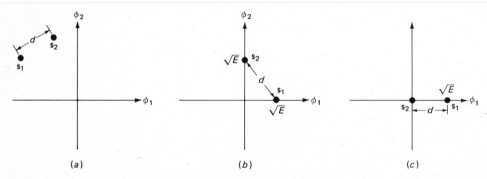

Figure 6-38 Translation and rotation of two-signal set. (*a*) Arbitrary location. (*b*) FSK. (*c*) OOK.

energy of the polar set for the same probability of error, while the nonzero component of the OOK set requires four times the signal energy. (Its *average* energy, assuming equally likely signals, is only twice as much.)

If the energy in these two examples is kept *fixed,* the spacing d must decrease. This is reflected in the following error-probability equations for the FSK and OOK signals, respectively, obtained from (6-85) by substituting in the appropriate values for d from Fig. 6-38b and c.

$$\text{FSK:} \quad P_e = \frac{1}{2} \text{ erfc } \sqrt{\frac{E}{2n_0}} \tag{6-85b}$$

$$\text{OOK:} \quad P_e = \frac{1}{2} \text{ erfc } \sqrt{\frac{E}{4n_0}} \tag{6-85c}$$

{These results agree of course with those found earlier in Chap. 5 and in Sec. 6-3 [see (6-57) to (6-59)].}

Quadrature Amplitude Modulation (QAM)

The probability-of-error calculation is easily generalized to signals in two dimensions that occupy a rectangular region as in Fig. 6-36c. One particular example is the class of QAM or quadrature-amplitude-modulated signals. These signals were discussed in some detail in Sec. 4-3 in connection with the need to transmit at relatively high bit rates over band-limited channels such as the telephone channels. As noted in that chapter, QAM signal sets have been implemented in high-speed modems. The binary signal probability of error calculations are easily extended to cover this class of signals. In terms of the notation of the previous section, this class has in general M signals and $N = 2$ orthogonal carriers. Hence the receiver of Fig. 6-26 applies here.

Consider the $M = 16$ signal set of Fig. 6-39 as an example. The signal points are shown spaced d units apart and symmetrically oriented about the origin. This is thus a minimum energy set.

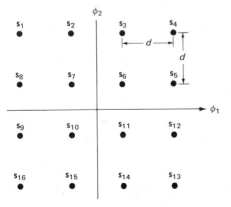

Figure 6-39 Quadrature amplitude modulation: 16-signal set.

The signals are of varying energy (and hence power) in this type of signal set. This contrasts with a pure M-ary PSK signal set in which the signals are all of equal energy, spaced at different angles around a circle in the (ϕ_1, ϕ_2) plane. Since the signals are of varying amplitude, one can only evaluate the transmission energy requirements in a statistical sense. For M signals equally likely to be transmitted, the *average* energy is given by

$$\overline{E} = \frac{1}{M} \sum_{i=1}^{M} E_i \tag{6-87}$$

with E_i the energy of signal s_i. For the specific case of Fig. 6-39 with the signals spaced d units apart,

$$\overline{E} = \frac{1}{16} \left(\frac{4d^2}{2} + \frac{8 \times 10d^2}{4} + \frac{4 \times 18d^2}{4} \right)$$

$$= \frac{5}{2} d^2 \tag{6-88}$$

Hence in terms of the average energy,

$$d^2 = \tfrac{2}{5} \overline{E}$$

This serves to connect the spacing d to the average energy.

Now consider the calculation of probability of error of this set. As noted earlier, it is simpler to first find the probability P_c of *correct* reception. Then $P_e = 1 - P_c$. It is apparent from Fig. 6-39 that the decision region for each signal is of rectangular shape. Three types of regions appear in Fig. 6-39. These are indicated in Fig. 6-40. The signal point in question appears at the center of the region in each case.

Consider region a in Fig. 6-40 as an example. It is apparent from Fig. 6-39 that this corresponds to the decision region for signals s_6, s_7, s_{10}, and s_{11}. A received signal vector v falling inside one of these regions would result in that signal being declared present. In corresponding fashion, *if* one of these signals s_i *is* transmitted, the probability $P(c|s_i)$ that the signal is received correctly corresponds to the probability that the received signal vector v lies within the square

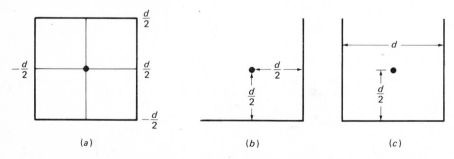

Figure 6-40 Typical decision regions in Fig. 6-39.

of Fig. 6-40a. For the QAM signal in question, with $N = 2$ carriers, the correlation detector of Fig. 6-27 produces two output samples with independent gaussian noise terms n_1 and n_2, respectively. For $\mathbf{v}$ to lie within the square of Fig. 6-40a, it is apparent that both n_1 and n_2 must lie within the range 0 to $\pm d/2$. The probability of correct reception, conditioned on s_i being transmitted, $i = 6, 7, 10,$ or 11, is thus

$$P(c|s_i) = \text{Prob} \left(-\frac{d}{2} < n_1 < \frac{d}{2} \right) \cdot \text{Prob} \left(-\frac{d}{2} < n_2 < \frac{d}{2} \right)$$

$$= p^2 \tag{6-89}$$

with

$$p = 2 \int_0^{d/2} \frac{e^{-x^2/n_0}}{\sqrt{\pi n_0}} \, dx$$

$$= 2 \int_0^{d/2\sqrt{n_0}} \frac{e^{-x^2}}{\sqrt{\pi}} \, dx \tag{6-90}$$

Here the property, previously proved, that each of the random variables n_1 and n_2 has variance $n_0/2$ has been used.

In similar manner, the probability of correct reception of signals s_1, s_4, s_{13}, and s_{16} in Fig. 6-39 with the decision region of Fig. 6-40b is readily shown to be

$$P(c|s_i) = r^2 \tag{6-91}$$

$$r = \int_{-d/2}^{\infty} \frac{e^{-x^2/n_0}}{\sqrt{\pi n_0}} \, dx$$

$$= \frac{1}{2} + \frac{p}{2} \tag{6-92}$$

Finally, the probability of correct reception of signals s_2, s_3, s_5, s_{12}, s_{14}, s_{15}, s_8, and s_9 is found from Fig. 6-40c to be given by

$$P(c|s_i) = pr \tag{6-93}$$

The probability of correct reception of an entire signal set is found in general by multiplying the conditional probabilities by the a priori probabilities of transmission and summing:

$$P_c = \sum_{i=1}^{M} P(s_i) P(c|s_i) \tag{6-94}$$

In the special case of the 16-signal set of Fig. 6-39 with the signals equally likely to be transmitted, the probability of correct reception is given simply by

$$P_c = \frac{1}{16} \left[4p^2 + 4p(1 + p) + 4 \left(\frac{1 + p}{2} \right)^2 \right]$$

$$= \frac{(3p + 1)^2}{16} \tag{6-95}$$

with p defined by Eq. (6-90). As a check, if the signals all move far apart, with $d \rightarrow \infty$, $p \rightarrow 1$ and $P_c \rightarrow 1$.

As in the two-signal case considered previously, it is of interest to write the probability of error directly in terms of energy and noise spectral density. From (6-88), $d^2 = \frac{2}{5}\overline{E}$. Using this in (6-90) for p, letting $P_e = 1 - P_c$, and assuming that $P_e \ll 1$ (the desired case), it is readily shown from (6-95) that

$$P_e \doteq \frac{3}{2}(1 - p) = \frac{3}{2}\,\text{erfc}\,\sqrt{\frac{\overline{E}}{10n_0}} \qquad (6\text{-}96)$$

Comparing with (6-85a), it is apparent that the 16-signal QAM constellation of Fig. 6-39 requires somewhat more than 10 times the signal power of the PSK signals for the same probability of error. On the other hand, the system is effectively transmitting at four times the bit rate possible with PSK alone. This is the price one pays for packing more signals into the same "signal space."

An example serves to put this in perspective. Assume that we have a telephone channel over which data are to be transmitted for which the useful transmission bandwidth is 2,400 Hz (see Fig. 6-41). If PSK or OOK transmission were to be used, the carrier would be centered in the useful band. Depending on the Nyquist signal shaping used, 1,200- to just under 2,400-bits/s signals could then be transmitted. Vestigial sideband techniques, with the carrier located at either the upper or lower part of the band, could be used to increase the bit rate allowable to almost 4,800 bits/s. If FSK transmission were used instead, two carriers with a modulation index of $\beta = 0.7$ could be chosen within the bandwidth. In this case the modulation bandwidth available would be about 700 Hz and the *maximum* bit rate possible 1,400 bits/s.

But assume now that 16-signal QAM is used instead. If noise is no problem on the channel, the data rate allowable can range from 4,800 to just under 9,600 bits/s, depending on the signal shaping used. From Eq. (6-96), however, the signal-to-noise ratio $\overline{E}/n_0$ should be at least 100 or 20 dB to have $P_e \leq 10^{-5}$. This signal-to-noise ratio is quite acceptable for the telephone channel, with 30-dB signal-to-noise ratios commonly measured. In the case of the telephone channel, limitations on data transmission arise not from gaussian noise but from such

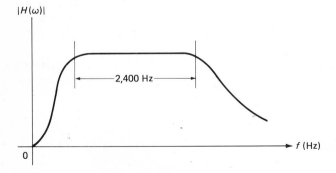

Figure 6-41 Amplitude-frequency characteristic, typical telephone channel.

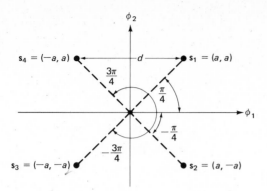

Figure 6-42 Four-signal set.

other factors, such as intersymbol interference, phase jitter, and impulse noise. This has already been noted. As was also noted in Sec. 4-3, data modems allowing 9,600-bits/s transmission over telephone channels are commonly available, but only over specially conditioned channels to widen the bandwidth available. They also incorporate adaptive equalizers to further shape or equalize the channel characteristics to reduce intersymbol interference, and the signal constellation used may be chosen to reduce errors due to phase jitter rather than to improve the error performance due to noise.

Although the AWGN channel model is thus not a very accurate one for the telephone channel, the concepts developed here are nonetheless quite useful in describing data transmission over that channel. Matched filtering is used, in addition to further equalization to reduce intersymbol interference. QAM transmission is commonly used, and the signal constellation description is found extremely useful as well. In addition, as the effects of intersymbol interference and phase jitter are reduced, the ultimate limitations due to thermal noise eventually come into focus. The AWGN channel model thus serves as a bound on the transmission capability of a channel such as the band-limited telephone channel. We shall return to this point shortly in discussing the Shannon capacity bound.

Another example of a signal set whose error performance in additive white gaussian noise is easily evaluated is the $M = 4$ QAM set used to motivate this material, and plotted in Fig. 6-25. The signal constellation plot is repeated here in Fig. 6-42. It may also be considered as an example of a QPSK (four-phase PSK) signal set.

Letting the spacing between pairs of points again be d units, it is apparent that the four decision regions are identically those of Fig. 6-40b. The probability of correct reception is, because of the symmetry, thus given by

$$P_c = r^2 \tag{6-97}$$

as in (6-91) and (6-92). Since $E = d^2/2$ in this case, as may be noted from Fig. 6-42, it is left to the reader to show that

$$P_e = 2 \int_{\sqrt{E/2n_0}}^{\infty} \frac{e^{-x^2}}{\sqrt{\pi}} \, dx = \text{erfc} \, \sqrt{\frac{E}{2n_0}} \tag{6-98}$$

Comparing with (6-85b), it is apparent that this signal set has an error performance similar to that of an FSK signal. It requires just a little more than twice the energy of the PSK signal but allows a doubling of the bit rate.

6-6 M-ary ORTHOGONAL SIGNALS

We now consider the case of M orthogonal signals. As noted earlier in this chapter, this class of signals is most useful for a power-limited channel: as M increases the transmission bandwidth goes up in proportion, but the probability of error is found to decrease exponentially with M. An example of such a signal set for $M = 3$ has already been shown in Fig. 6-35. More generally, we have $N = M$ carriers required, and we can write

$$s_1(t) = \sqrt{E} \, \phi_1(t), \, s_2(t) = \sqrt{E} \, \phi_2(t), \, \ldots \, , \, s_M(t) = \sqrt{E} \, \phi_M(t) \tag{6-99}$$

(The signals are all taken to be of equal energy.) In vector form, we have

$$\begin{aligned} \mathbf{s}_1 &= (\sqrt{E}, 0, \, \ldots \, , 0) \\ \mathbf{s}_2 &= (0, \sqrt{E}, \, \ldots \, , 0) \\ \mathbf{s}_M &= (0, 0, \, \ldots \, , \sqrt{E}) \end{aligned} \tag{6-100}$$

Such signals satisfy the orthogonality condition

$$\int_0^T s_i(t)s_j(t) \, dt = E\delta_{ij} \tag{6-101}$$

with δ_{ij} the Kronecker delta:

$$\begin{aligned} \delta_{ij} &= 1 \quad i \neq j \\ &= 0 \quad i \neq j \end{aligned}$$

In the vector-space formulation,

$$\mathbf{s}_i \cdot \mathbf{s}_j = E\delta_{ij} \tag{6-102}$$

Examples of such signals are M frequencies, displaced roughly $1/T$ hertz to make them orthogonal; M pulses or pulsed sine waves, each lasting T/M seconds, each displaced from one another in time by T/M seconds (see Fig. 6-43); and coded orthogonal signals (Fig. 6-44). The M pulses of Fig. 6-43 are quite inefficient in their use of power, since they only occupy $1/M$ of the entire time interval T seconds long. For the same energy (or average power) the peak power must be M times as large. Figure 6-44 shows an example of three signals lasting the entire T-second interval that are orthogonal to one another. These are more efficient than the pulses of Fig. 6-43. Generally, one divides the T-second interval into M subintervals T/M seconds apart. If binary transmission is used on each one of these subintervals, there are 2^M possible signal

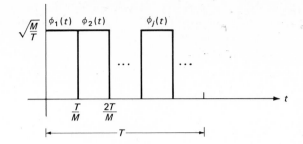

Figure 6-43 Example of orthogonal signals.

variations. One would then select M of these that are orthogonal. The pulse at each subinterval could then be used to modulate a sine-wave carrier or either of two quadrature carriers. An example of two orthogonal signals for the case of $M = 32$ appears in Fig. 6-45.

It is apparent that the bandwidth requirements for a set of M baseband orthogonal signals ranges from $M/2T$ to M/T, depending on the pulse shaping used. For a bit rate of R bits/s, $k = RT$ bits are encoded to form the M-ary signals. We thus have $M = 2^{RT}$, and the bandwidth actually increases exponentially with the encoding time T. This is one major drawback in the use of M-ary orthogonal signals. For high-frequency M-ary transmission the bandwidth doubles. Consider as an example M-ary FSK transmission, with one of $M = 32$ different frequencies to be transmitted every T seconds. A typical sine wave is

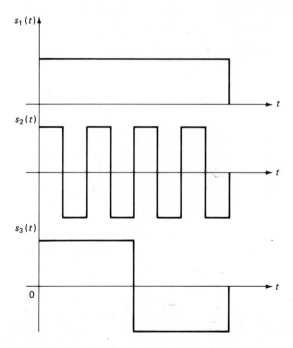

Figure 6-44 Another example of orthogonal signals.

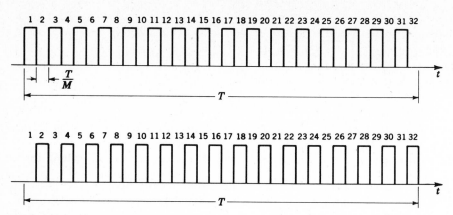

Figure 6-45 Orthogonal signals, $M = 32$.

shown in Fig. 6-46a. Assuming almost optimum Nyquist shaping in this case (the shaping is not shown in Fig. 6-46a), and spacing signals $1/T$ hertz apart to ensure orthogonal signaling, one gets as the overall transmission bandwidth for this M-ary signal set,

$$W = \frac{32}{T}$$

This is shown in Fig. 6-46b. Figure 6-46c shows the original binary FSK signals, each lasting $1/R = T/5$ seconds. (Shaping is again not shown in this case.) Spacing these two signals R hertz apart to ensure orthogonality (in practice one might pick $0.7R$ as the frequency spacing), and assuming almost ideal Nyquist shaping again, the original bandwidth is $W = 10/T$ hertz. The increase in bandwidth is thus about 3 to 1. As RT increases, the exponential increase $M = 2^{RT}$ comes into play more and more.

Despite this increase in bandwidth, these signals do provide a desirable improvement in error performance. To demonstrate this, we again calculate the probability of *correct* reception of a signal.[11]

Specifically, assume, as previously, that the orthogonal signals are equally likely to be transmitted. An optimum receiver then declares $\mathbf{s}_i$ transmitted if

$$\|\mathbf{v} - \mathbf{s}_i\|^2 < \|\mathbf{v} - \mathbf{s}_j\|^2 \qquad \text{all } j \neq i \qquad (6\text{-}103)$$

Equation (6-103) then defines the decision region for signals $\mathbf{s}_i$ if it is transmitted. The decision will thus be correct, when $\mathbf{s}_i$ is transmitted, if the received signal $\mathbf{v}$ is such as to satisfy (6-103). Alternatively, since all signals have the same energy, we must have

$$\mathbf{v} \cdot \mathbf{s}_i > \mathbf{v} \cdot \mathbf{s}_j \qquad \text{all } j \neq i \qquad (6\text{-}104)$$

---

[11] Wozencraft and Jacobs, *op. cit.*, p. 257.

This is exactly the receiver implementation of Fig. 6-32 if all the signals have the same energy. But s_i may be written $s_i = \sqrt{E}\phi_i$ with ϕ_i a unit vector in the ϕ_i direction (see Fig. 6-47). Then $\mathbf{v} \cdot \mathbf{s}_i = v_i\sqrt{E}$, with v_i the projection of $\mathbf{v}$ along the s_i axis. (v_2 is shown, for example, in Fig. 6-47.) The optimum receiver for the orthogonal signal case thus declares signal s_i present if

$$v_i > v_j \qquad \text{all } j \neq i \qquad (6\text{-}105)$$

This also defines the region of correct reception of s_i if that is the signal transmitted.

In words, signal s_i is correctly received if the component of $\mathbf{v}$ along the i axis is greater than the components along all the other axes. But with s_i transmitted,

$$v_i = \sqrt{E} + n_i \qquad v_j = n_j \qquad \text{all } j \neq i \qquad (6\text{-}106)$$

(see Fig. 6-48). As previously, the n_j's are all zero-mean, independent gaussian variables, with variance $= n_0/2$. Note from (6-105) and (6-106) that signal s_i will be correctly received if, *given* v_i, *all* $n_j < v_i$. Thus, *conditioned* on a particular

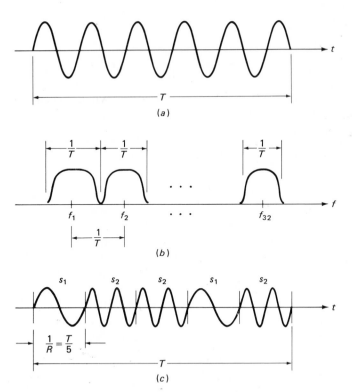

Figure 6-46 M-ary FSK transmission, $M = 32$, minimum bandwidths. (a) 1 of 32 frequencies. (b) Bandwidth $W \doteq 32/T$ Hz. (c) Original binary signals: Bandwidth $\doteq 10/T$ Hz.

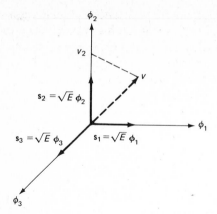

Figure 6-47 Signal reception in orthogonal signaling ($M = 3$).

value of v_i, the probability of correct reception of $\mathbf{s}_i$ is just

$$P(c|v_i, \mathbf{s}_i) = \text{Prob (all } n_j < v_i)$$
$$= [\text{Prob } (n_j < v_i)]^{M-1} \tag{6-107}$$

since all the noise terms are independent with the same (gaussian) distribution. To find the desired probability $P(c|\mathbf{s}_i)$ of the correct reception of $\mathbf{s}_i$ we must multiply (6-107) by the probability of obtaining a particular value of v_i and then sum over all possible values:

$$P(c|\mathbf{s}_i) = \int P(c|v_i, \mathbf{s}_i)f(v_i) \, dv_i \tag{6-108}$$

From (6-106), however, v_i is gaussian with average value $\sqrt{E}$. Using this plus the fact that n_j is zero-mean gaussian it is left to the reader to show that (6-108) becomes

$$P(c \,|\mathbf{s}_i) = \int_{-\infty}^{\infty} \frac{e^{-(y-\sqrt{E})^2/n_0}}{\sqrt{\pi n_0}} [P(n < y)]^{M-1} \, dy \tag{6-109}$$

where
$$P(n < y) = \int_{-\infty}^{y} \frac{1}{\sqrt{\pi n_0}} e^{-x^2/n_0} \, dx \tag{6-110}$$

Since the signals are assumed equally likely, and all have the same probability $P(c|\mathbf{s}_i)$ of being received correctly, (6-109) represents the probability of an error-free transmission for the M orthogonal signals transmitted over the

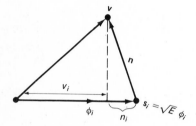

Figure 6-48 Geometry of received signal, M-ary signaling.

AWGN channel. Note that it depends on E/n_0 and M. Equation (6-109) cannot be integrated exactly. Tables and curves for $P_e = 1 - P_c$ have been obtained numerically. One such set of curves appears in Fig. 6-49.[12] Here, in place of E/n_0 and M, use has been made of the relation

$$\frac{E}{n_0} = \frac{ST}{n_0} = \frac{S}{n_0 R} (RT)$$

$$= \frac{S \log_2 M}{n_0 R} \tag{6-111}$$

S again represents the average *power* of the signal, as received at the receiver, and the relation $M = 2^{RT}$ has been introduced. The curves of Fig. 6-49 show P_e versus $S/n_0 R$ for various values of $k = \log_2 M = RT$. Note that they demonstrate the improvement in error performance mentioned earlier as M increases. It is theoretically possible to drive the error probability down to as low a value as desired by increasing M (and hence the dimensionality of the signal space) indefinitely. This requires encoding the original bit stream over correspondingly longer time intervals, since $\log_2 M = RT$. As indicated in Fig. 6-24a, the T-second encoding delay at the transmitter thus increases indefinitely, while a correspondingly long delay at the receiver is incurred in carrying out the required correlation operations (Fig. 6-32 or 6-33).

Figure 6-49 also indicates a remarkable phenomenon: this improvement in error performance by the use of M-ary orthogonal signals (with the corresponding increase in bandwidth already noted) is only possible provided that $S/n_0 R > 0.69$. This thus puts a lower limit on the signal-to-noise ratio. For $S/n_0 R < 0.69$ the probability of error approaches 1, and the system fails! Alternatively, error-free transmission as $M \to \infty$ is only possible if $R < S/0.69 n_0$. For a given power-to-noise ratio S/n_0, this puts an upper bound on the rate R with which one can communicate through the channel. For example, if $S/n_0 = 1,000$, R may not exceed 1450 bits/s. As we shall see shortly, this is a special case of a more general result obtained by Shannon that the probability of error on the AWGN channel may be reduced as much as desired *provided* that the bit rate R does not exceed a characteristic *capacity* of the channel. This capacity depends, in general, on S, n_0, and the channel bandwidth. The capacity concept has been extended to other types of channels as well. For the case of an AWGN channel, with no bandwidth limitation the capacity of the channel is precisely $S/0.69 n_0$. The Shannon capacity expression was introduced in Sec. 5-2. It will be discussed again in the next section.

The asymptotic behavior of error probability with M for the M-ary orthogonal signal case and the AWGN channel is demonstrated, at least approximately, by the use of bounds on the probability of error.[13] Specifically, assume

---

[12] A. J. Viterbi, *Principles of Coherent Communication*, McGraw-Hill, New York, 1966, fig. 8.3.

[13] Wozencraft and Jacobs, *op. cit.*, p. 264.

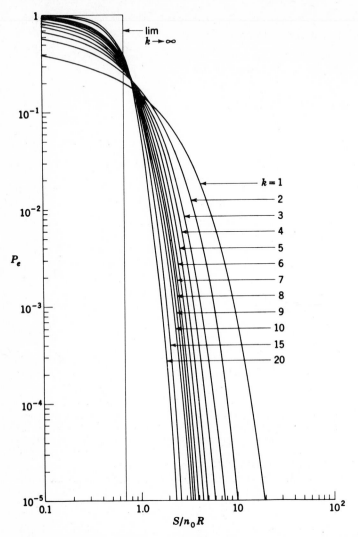

Figure 6-49 Error probability for orthogonal signals ($k = \log_2 M$). (From A. J. Viterbi, *Principles of Coherent Communication*, McGraw-Hill, New York, 1966, fig. 8.3, with permission.)

s_i transmitted, and let $P_e(s_i, s_k)$ denote the probability that the received signal vector v will be received closer to another signal s_k. This is just the probability of error of the two-signal sets of Figs. 6-37 and 6-38 given by Eq. (6-85):

$$P_e(s_i, s_k) = \frac{1}{2} \text{ erfc } \frac{d}{2\sqrt{n_0}} \qquad (6\text{-}112)$$

If we now have one of M signals present, as is the case here, an error will occur if v is closer to s_k than to the transmitted s_i for at least one such signal

s_k, $k \neq i$. As a loose upper bound we can say the probability of this happening is just the sum of $(M - 1)$ terms like that of (6-112). Thus

$$P_e \leq \sum_{k=1}^{M-1} P_e(s_i, s_k)$$

$$= \frac{(M - 1)}{2} \text{ erfc } \frac{d}{2 \sqrt{n_0}} \tag{6-113}$$

But the orthogonal signals under consideration are pairwise $d = \sqrt{2E}$ units apart (see Fig. 6-47). Hence we have

$$P_e < \frac{M}{2} \text{ erfc } \sqrt{\frac{E}{2n_0}} \tag{6-113a}$$

and the right-hand term is a loose upper bound on the exact expression found from (6-109).

A simple bound on erfc x is readily found to be given by

$$\tfrac{1}{2}\text{erfc } x < e^{-x^2} \tag{6-114}$$

Hence $$P_e < Me^{-E/2n_0} \tag{6-113b}$$

from (6-113a). Letting $M = 2^{RT} = e^{0.69RT}$ and $E = ST$, with S the average power of the signals, (6-113b) becomes

$$P_e < e^{-T[(S/2n_0)-0.69R]} \tag{6-113c}$$

This is the desired bound. It indicates that P_e decreases exponentially with T *provided* that the bit rate R is less than a specified number. The number shown here is half that noted earlier and as indicated in Fig. 6-49. This is due to the looseness of the original bound of (6-113). A tighter bound actually provides the desired result that $P_e \to 0$ exponentially with T, provided that $R < S/0.69n_0$.

This exponential decrease in the probability of error with T for orthogonal signals is obviously quite desirable, since it shows that power-limited channels may be used effectively in the presence of noise by proper coding of signals. The improvement in error performance is not as dramatic or practical as might at first be expected, however, because of the extremely large values of M required. Specifically, consider that $P_e = 10^{-5}$. From Fig. 6-49 note that the signal-to-noise ratio required to attain this error probability decreases *very* slowly with M. Thus in going from $M = 2$ ($k = 1$) or binary FSK to $M = 32$ ($k = 5$) orthogonal signals the signal-to-noise ratio required is reduced only 6 dB, from $S/n_0 R = 20$ to 5. A 10-dB reduction to $S/n_0 R = 2$ would require the storage and coding of 20 consecutive bits. This by itself is not too difficult (as we shall see, practical binary codes store far more than 20 bits), but the system then requires the generation of 2^{20} different orthogonal signals, as well as a bandwidth expansion of this order. For this reason error-reduction techniques for use with power-limited channels have moved more in the direction of sophisticated coding procedures, to be discussed in a later section.

The geometric reason for the error improvement with increasing T goes as follows. The signal points are each located at the point $\sqrt{E}$ on one axis of an $N = M$ dimensional space. As T increases, so does $E = ST$. The noise variance remains the same, however, so the possibility of noise moving a given signal point to another decreases. Alternately, if the received signal is divided by $\sqrt{T}$, renormalizing the geometric hyperplane of $M = N$ dimensions, all signal points remain at a fixed distance $\sqrt{S}$ from the origin. The noise variance is reduced by T, however. As the dimensionality of the signal space increases, then, the noise tends to cluster more about the signal point corresponding to the one transmitted. The noise terms tend to fall within spheres more and more tightly packed about the signal points. As long as $S/n_0R > 0.69$, these spheres do not overlap, and the probability of an error becomes rarer as T gets larger.

Space Channel Applications

The discussion of M-ary orthogonal signaling and its comparison with binary signaling can be clarified further by considering a few simple examples drawn from space communication. Recall that in Sec. 5-14 of Chap. 5 we obtained a simple expression (5-227) for the received power intercepted by a receiving antenna of area A_R, with efficiency η_R, located at a distance d meters from the transmitter. Repeating that expression here, we have

$$S_R = \frac{S_T}{4\pi d^2} G_T A_R \eta_R \tag{6-115}$$

G_T represents the transmitter antenna gain over an isotropic radiator and is in turn itself given by

$$G_T = \frac{4\pi\eta_T A_T}{\lambda^2} \tag{6-116}$$

with A_T the transmitting antenna area (cross-sectional aperture), η_T the antenna efficiency, and λ the wavelength of the transmission.

Given the various parameters of the transmitter-receiver system, and the known spacing d between them, we can calculate the received power. For a given bit rate of transmission this enables us to calculate the energy E intercepted by the receiver. Recall that the noise spectral density was given by $n_0 = kT$. Knowing the effective temperature T and the energy we can then calculate the fundamental ratio E/n_0 that determines the performance of the system.

Alternatively, and more properly, we use the approach of Sec. 5-14. We *specify* the performance, in bit-error probability, desired, or tolerable, and from this determine the bit transmission rate R (in bits/s) allowable. In particular, how much improvement does the introduction of M-ary orthogonal signaling provide us? We have already seen that this improvement increases rather slowly with M, but the use of some examples makes this point more dramatic.

1. $T = 100$ K. $S_T = 100$ W. Let the antenna efficiencies be 1, for simplicity. We also neglect system power losses. The frequency

$$f = 2,000 \text{ MHz} \ (\lambda = 0.15 \text{ m}) \qquad d = 10^6 \text{ mi} = 1.6 \times 10^9 \text{ m}$$

The receiving antenna has a 13 by 13 ft aperture so that $A_R \sim 19 \text{ m}^2$ and $G_R \sim 10^4$. Let the transmitting antenna be much smaller, with $G_T \sim 100$. Then from (6-115), $S_R \sim 6 \times 10^{-15}$ W and $S_R/n_0 \sim 3.5 \times 10^6$.

If binary PSK transmission is used, and $P_e \leq 10^{-5}$ desired, we must have $E/n_0 \geq 10$. It is then apparent, since $E = S_R/R$, with R the binary transmission rate in bits/s, that $R \leq 3.5 \times 10^5$ bits/s. So the use of binary transmission, with no additional encoding, limits the rate of transmission to less than 3.5×10^5 bits/s in order to provide a tolerable error probability. Note again that in all examples discussed in this chapter it was E/n_0 that determined the probability of error. The only way to increase the effective signal energy E received at the antenna, and reduce P_e, without resorting to further signal encoding, is by slowing down the signal rate, allowing the signal to be received over a longer interval of time.

The use of M-ary orthogonal signaling can improve this situation. Thus, for $M = 32$ ($RT = 5$ input binary symbols must be stored), and $P_e \leq 10^{-5}$, Figure 6-49 indicates that $S_R/n_0 R \geq 5$. Hence $R \leq 700,000$ bits/s.[14] This is an improvement of $2:1$ over the previous maximum binary rate. Although there is improvement, it is quite slow with M, as already noted previously. The resultant system is also much more complex than the binary system, requiring, in addition to the encoder and decoder, a transmitter capable of producing 32 orthogonal signals and a 32-bank correlation receiver. The bandwidth required is also several times that of the binary signal case. Continuing to increase M by storing additional binary digits (with a corresponding increase in complexity and bandwidth), one finds from Fig. 6-49, that for $M = 2^{10} = 1,024$, $E/n_0 = S_R/n_0 R \geq 3.2$ for $P_e \leq 10^{-5}$. For the example given here, $R \leq 1.09 \times 10^6$ bits/s. In the limit, as M gets very large, $S_R/n_0 R = 0.69$ is required to drive P_e to as low a value as possible. In this example the *maximum* rate of transmission, with M very large, is $R = 5 \times 10^6$ bits/s, an increase of $10/0.69 = 14.5$ in the transmission rate over binary PSK transmission, but clearly impractical using M-ary orthogonal signaling techniques.

2. This is the same as the previous example, but with receiving antennas increased considerably to 55 by 55 ft dishes. Then $A_R \sim 300 \text{ m}^2$, $G_R \sim 1.5 \times 10^5$, $S_R/n_0 \sim 18 \times 10^6$, and $R \leq 5.3 \times 10^6$ bits/s for binary transmission with $P_e \leq 10^{-5}$. The 15-fold increase in antenna size leads to a corresponding 15-fold increase in the binary transmission rate. The $M = 32$ signaling rate can be increased 15-fold as well to $R \leq 10.5 \times 10^6$ bits/s.

3. Let $f = 500$ MHz and use the same receiving antenna as in (2) above. Then G_R is now approximately 10^4. Assume that the transmitter average

---

[14] The curves of Fig. 6-49 actually refer to *character*-error probability, i.e., the probability that one of the M signals will be in error. This may be readily converted to bit-error probability, if so desired. See Viterbi, *op. cit.*, p. 227, fig. 8.6. For $M \gg 1$, there is no essential difference between the two.

power is now 1 W, let the temperature be 200 K, and let $G_T = 10$ at this lower transmitting frequency. Then, with $d = 10^6$ mi again, 2,900 bits/s is the maximum data transmission rate at $P_e = 10^{-5}$ if PSK transmission is used. It can be increased to 5,800 bits/s with $M = 32$ signal orthogonal transmission and in the limit approaches 42 kbits/s for very large M. If the transmission distance increases to $d = 10^7$ mi, it is apparent that the received power drops by 100, and the maximum data rate drops to 29 bits/s using PSK transmission.

4. $f = 2,000$ MHz, $S_T = 8$ W, and $d = 8 \times 10^{11}$ m. (This is the distance to Jupiter.) Assume that the effective temperature of the sky plus noise temperature of the antenna and receiving system result in $T = 50$ K. A 45 by 45 ft ground antenna is used, and its efficiency is $\eta = 0.6$. Then $G_R = 8 \times 10^4$. Let the space antenna gain be $G_T = 8 \times 10^{13}$. (This corresponds to a 15 by 15 ft or 5 by 5 m dish, much larger than in the previous example.) It is left to the reader to show that $R \leq 140$ bits/s for binary transmission and $R \leq 280$ bits/s for 32-orthogonal signal transmission, if $P_e = 10^{-5}$ is desired, while the limiting bit rate using M-ary signaling is 2,000 bits/s.[15]

These numbers indicate the dilemma a designer faces in a power-limited AWGN channel as exemplified by the space channel. As distances between the earth and the space probe increase, it appears that the data transmission rate must decrease correspondingly (going as $1/d^2$), unless the power and antenna size can be increased. These are precisely the trade-offs that were discussed earlier in Sec. 5-14. Alternatively, it is possible to get some further improvement by appropriate encoding of data symbols.

This is of course the procedure adopted in going to M-ary orthogonal signal transmission. We showed in this case that it was in fact possible to improve the bit rate, at a given error probability, by using large values of M, but that the rate of improvement was rather slow. As $M \to \infty$, however, an improvement of 14.5 in the bit rate is possible at $P_e = 10^{-5}$, and the error probability in fact can be reduced as low as desired in this case. The question then raised is: Are there other encoding schemes, perhaps more effective than M-ary orthogonal signaling, that can provide a desired improvement over binary signaling alone with no further encoding? The answer is "yes." There exist many such schemes, and we shall explore some of these in following sections. Interestingly, however, the best possible encoding scheme can do no better than the M-ary orthogonal signal case with $M \to \infty$. To demonstrate this we discuss next the famous Shannon capacity expression that puts a bound on the rate of *error-free* digital signal transmission.

An important point to note, however: we look for encoding schemes because theory indicates that they might provide more than 10 times the bit rate

---

[15] Note that by adjusting the various design parameters appropriately, as discussed in Sec. 5-14, communications with Jupiter at a rate of 115 kbits/s has actually been achieved. The example here is purely hypothetical.

allowed by noncoded binary transmission. (The number, again, is $10/0.69 = 14.5$ at $P_e = 10^{-5}$.) However, just as in the case of M-ary orthogonal signaling, very wide bandwidths are required. This is obviously not acceptable with band-limited channels such as the telephone channel. Complex coders and decoders may thus not be appropriate in that case. It is for power-limited channels such as the space channel, with relatively wide bandwidths allowed, that one thinks in terms of sophisticated coding techniques to improve digital transmission capability.

6-7 THE SHANNON CAPACITY EXPRESSION

Although the Shannon capacity expression has been referred to previously in this book (see Sec. 5-2, for example), we now explain its meaning and significance in detail, in order to motivate the need for encoding techniques for digital communications.

Figure 6-22 indicates the best performance possible using binary signals in the presence of additive white gaussian noise. Figure 6-49 shows that one may improve the performance by going to M-ary orthogonal signaling, although relatively large improvements are obviously impractical, since they require enormous values of M. Is it then possible to improve the performance in some other way, using some other form of signaling?

This question was essentially answered in Chap. 5 in discussing the Shannon capacity expression of Eq. (5-17). We indicated there that Shannon had proven that virtually error-free digital transmission was possible over a channel with additive gaussian noise, providing one did not try to exceed the channel capacity in bits per second. Specifically, assume as in Chap. 5 an available transmission bandwidth of W Hz. Band-limited gaussian noise of spectral density $n_0/2$ is added during transmission. (This is just the assumption made in the optimum binary and M-ary transmission cases). Then Shannon was able to show that by appropriately *coding* a binary message sequence before transmission it should be possible to achieve as low an error rate as possible providing the channel capacity C was not exceeded. The capacity expression in this case of band-limited white noise was found by Shannon to be given by[16]

$$C = W \log_2 \left(1 + \frac{S}{N} \right) \qquad \text{bits/s} \qquad (6\text{-}117)$$

with S the average signal power and $N = n_0 W$ the average noise power. (S/N is then the signal-to-noise ratio at the receiver.)

[16] C. E. Shannon, "Communication in the Presence of Noise," *Proc. IRE,* vol. 37, pp. 10–21, January 1949. See also Wozencraft and Jacobs, *op. cit.,* p. 323, and R. G. Gallager, *Information Theory and Reliable Communication,* Wiley, New York, 1968, pp. 373, 389.

Thus, providing one did not attempt to transmit more than C bits/s over such a channel, one could hope to attain tolerable error rates. Specifically, if the binary transmission rate is R bits/s (the binary interval is then $1/R$ seconds), and if $R < C$, it may be shown that the error probability is bounded by

$$P_e \leq 2^{-E(C,R)T} \qquad R < C \tag{6-118}$$

with $E(C, R)$ a positive function such as the one shown in Fig. 6-50. As the transmission R approaches C the probability of error $\to 1$. This is apparent from Eq. (6-118) and Fig. 6-50.

The parameter T appearing in Eq. (6-118) indicates the time required to transmit the *encoded* signal. With binary transmission rate R and channel capacity C fixed, the probability of error may be reduced by increasing T. Although Shannon proved the capacity expression of Eq. (6-117) quite generally for the *gaussian channel,* he did not provide a formula for actually designing the required encoder. His proof indicates that it is *possible* to transmit at as low an error rate as possible, but does not show how. A great deal of research activity in the years since Shannon developed his capacity expression has been devoted to the investigation of various types of encoders.

Not only does the need to design encoders provide a drawback to the application of Shannon's theorem, but an additional complication enters into the picture. As one tries for lower and lower probabilities of error, the encoding time T becomes longer and longer. A delay of T seconds in transmission is then incurred at the transmitter. A corresponding delay of T seconds at the receiver is incurred in decoding the actual transmitted message, for a total delay of $2T$ seconds. As usual, then, a price must be paid for the required decrease in SNR. The circuitry becomes much more complex and large time delays are incurred. This is precisely what was found with the case of M-ary orthogonal signals. Larger values of M require correspondingly more binary digits to be stored ($\log_2 M = RT$), and the same T-second delay at both the transmitter and receiver is encountered there as well. As will be seen shortly, the M-ary orthogonal signal set, with $M \to \infty$, provides in fact an example of a Shannon-type encoder.

The general form of the encoding scheme suggested by Shannon's work appears in Fig. 6-51. Note that it is similar to the binary-M-ary encoder of Fig. 6-24. The basic difference is that the encoder and decoder of Fig. 6-51 are left completely unspecified. A modulator is not shown in Fig. 6-51.

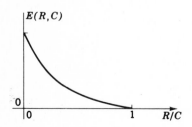

Figure 6-50 Shannon error exponent.

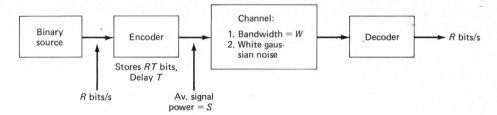

Figure 6-51 Encoding for optimum transmission.

Note that in the encoding-decoding process the same bit rate of R bits/s is always maintained. There is no increase or decrease in rate of transmission. A total delay of $2T$ seconds is entailed, however, because of the encoding at the transmitter (time taken to examine RT bits) and the decoding at the receiver (time taken to examine the encoded message T seconds long).

Aside from providing a stimulus for work in the area of coding, the Shannon formulation does provide another concurrent result: it enables us to compare transmission over real channels with the maximum possible rate of error free transmission. One can thus judge the capability of a particular transmission scheme and determine how close one is to the real ideal scheme. Consider, for example, the band-limited telephone channel already mentioned several times in passing in this chapter. We have noted that the key impairments for data transmission over this channel are intersymbol interference, signal phase jitter, and impulse noise. It has been found experimentally that additive gaussian noise on this channel is such as to provide a signal-to-noise ratio S/N of 10^3 (30 dB). If the other impairments were eliminated, this would then provide the remaining, irreducible, limit on transmission. The Shannon capacity bound applied to this channel, assuming the other impairments not present, thus provides an upper bound on the maximum rate of error-free transmission. If we assume $W \sim 3$ kHz for this channel, a reasonable figure, (6-117) indicates that the capacity is about $C = 30,000$ bits/s. Actually, the ideal band-limited frequency model assumed in Shannon's derivation is not quite appropriate for the telephone channel. Shannon's analysis can be extended to include nonideal frequency characteristics as well, and the application of this work to the telephone channel[17] indicates that its capacity is closer to 23,000 bits/s. Since 9,600-bits/s modems already come within 50 percent of the Shannon limit, it is not clear that sophisticated coding techniques are worthwhile for this channel. Instead of forward error correction, error detection with feedback is used routinely to reduce the error rate to tolerable values.[18]

[17] R. W. Lucky, J. Salz, and E. J. Weldon, Jr., *Principles of Data Communication,* McGraw-Hill, New York, 1968, pp. 35–38.

[18] M. Schwartz, *Computer Communication Network Design and Analysis,* Prentice-Hall, Englewood Cliffs, N.J., 1977.

The Shannon's capacity expression applied to the power-limited channel produces quite a different conclusion. It indicates that coding *can* play a role in improving the transmission rate over such channels, with the error probability maintained at a tolerable value. We have already noted this fact in our discussion of M-ary orthogonal transmission.

Consider, for example, the capacity expression of (6-117) applied to a channel with no limits on bandwidth. It is apparent that if S/N is held fixed, the capacity is directly proportional to bandwidth. So increasing the bandwidth should provide an improvement in transmission capability. (Note that for a fixed bandwidth an increase in power only provides a logarithmic increase in the capacity. For example, in the telephone case cited earlier, if S/N were doubled to 2,000, the capacity would only increase by 10 percent.) Actually, the noise increases with bandwidth as well, since $N = n_0 W$.

In particular, for large W the capacity levels off and approaches a limiting value C_∞ given by

$$C_\infty = \lim_{W \to \infty} C = \frac{S}{n_0 \log_2 e} = \frac{S}{0.69 n_0} \qquad (6\text{-}119)$$

This is shown in Fig. 6-52. Note that this is precisely the bound discussed earlier under M-ary transmission! It appears in Fig. 6-49 as well. [This limiting value of C is obtained from (6-117) by noting that $\log_e (1 + \epsilon) \doteq \epsilon$, $\epsilon \ll 1$.]

There is thus a limit on the rate at which one can transmit error-free over a power-limited channel with the bandwidth allowed to get as large as desired. It is still significantly higher than the data rates attainable with binary PSK and $M = 32$ orthogonal signal transmission, as noted in the examples at the end of the previous section, accounting for the great interest in coding techniques for space channels to which reference has already been made. Although transmission at a rate of C_∞ bits/s is clearly unattainable, just as in the case of M-ary orthogonal transmission with $M \to \infty$, transmission at one-half this rate is a

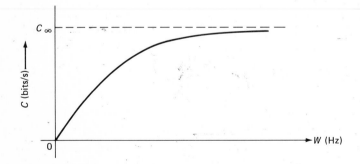

Figure 6-52 Capacity variation with bandwidth.

realistic goal. This is still seven times the uncoded PSK rate and is therefore in many cases a desirable goal. As an example, in the hypothetical space problem of example 4 in the previous section, we found digital transmission with an error probability of 10^{-5} could only be carried out at a rate of 140 bits/s, using uncoded PSK transmission. Since $C_\infty = 2,000$ bits/s for this example, a transmission rate of 1,000 bits/s could be considered a realistic figure for which to strive.

6-8 BLOCK CODING FOR ERROR DETECTION AND CORRECTION

Introduction

In previous sections on the additive white gaussian noise (AWGN) channel, we discussed at length probability-of-error calculations and showed that it is possible, by encoding k binary digits into one of $M = 2^k$ orthogonal signals, to reduce the probability of error P_e as low as we like by increasing k indefinitely. The binary source rate R bits/s had to be less than the channel capacity C, in bits/s. We found the transmission bandwidth increasing with M, or *exponentially* with k, however. We also found the improvement, or reduction of P_e, rather slow with M. In discussing Shannon's capacity theorem in the last section, we indicated that M-ary orthogonal signaling is just one possible way of encoding binary digits to reduce the error rate.

In this section we discuss a class of encoding procedures on which a great deal of work has been carried out since Shannon's pioneering efforts in the area. This is the class of parity-check block codes, in which r parity-check bits, formed by linear operations on the k data bits, are appended to each block of k bits. These codes can be used to both detect and correct errors in transmission. We shall focus here on a particular class of block codes, cyclic codes, for which relatively simple digital encoders and decoders may be found. Such codes are commonly used not only for data transmission but for such applications as magnetic tape encoding, enhanced computer system reliability, etc. As was the case with orthogonal signals, the forward error-correcting properties of such codes have not found a great deal of utilization in band-limited channels, however. Instead, there the prime application has been in the error-*detecting* properties of these codes. If an error is detected anywhere in the received code block the transmitter is notified to repeat the signal.

Space and satellite channels have made extensive use of the forward-error correcting capabilities of codes. In addition to block codes, convolutional codes have been used a great deal. In these codes a sliding sequence of past data bits is used to generate several code bits. Distinct blocks are no longer sent, and successive transmitted bits contain in them the history of a sequence of data bits. Various decoding algorithms have been developed for such codes, and it

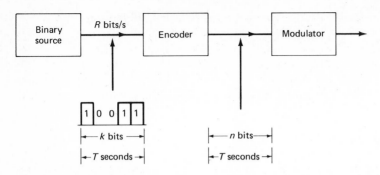

Figure 6-53 Binary encoding operation.

has been found possible to get quite respectable improvements in error performance by using relatively short encoders.[19]

Block Coding[20]

Recall from our previous discussion of the AWGN channel that we store up k successive binary digits and use these to generate a new set of signal waveshapes. In the previous discussion we considered one encoder-modulator device that carried out this operation. We now separate these two functions, as shown in Fig. 6-53. The encoder takes a block of k successive binary digits and converts this to an equivalent block of $n > k$ binary digits. These digits are in turn fed into a modulator which generates the analog waveshapes for transmission. This modulator can group several bits or all of the n-bit block in outputting the transmitted signals, just as in the modulators discussed in the AWGN sections. More commonly, each individual bit of the code block is used to modulate a carrier, resulting in binary PSK, FSK, or other of the common binary transmission signals mentioned in Chap. 4. (For space communication applications PSK transmission is used because of its inherently better error performance.)

In this section we focus on the encoder design. One simple example is that of a single-parity-check device, used to detect an odd number of errors. Here $n = k + 1$, and for every k data bits one bit is added. Most commonly, this bit represents the modulo-2 (mod-2) sum of the data bits (in which case it is called an even-parity bit since the total number of 1's in the code word is even) or the complement of this sum (hence called an odd-parity bit).

[19] A. J. Viterbi, "Convolutional Codes and Their Performance in Communication Systems," *IEEE Trans. Commun. Technol.*, vol. COM-19, no. 5, part II, p. 751, October 1971.

[20] W. W. Peterson and E. J. Weldon, Jr., *Error-correcting Codes,* MIT Press, Cambridge, Mass., 2nd ed., 1972; Lucky, Salz, and Weldon, *op. cit.,* chaps. 10 and 11; S. Lin, *An Introduction to Error-correcting Codes,* Prentice-Hall, Englewood Cliffs, N.J., 1970. The Peterson-Weldon book is an authoritative advanced text on the subject. The Lin book focuses on codes that show promise for practical applications. The Lucky, Salz, and Weldon material is more concisely presented and stresses applications as well.

By mod-2 sum we recall that we mean the following definitions:

$$0 \oplus 0 = 0$$
$$0 \oplus 1 = 1$$
$$1 \oplus 0 = 1 \qquad (6\text{-}120)$$
$$1 \oplus 1 = 0$$

The symbol $\oplus$ will henceforth represent the mod-2 sum operation. This notation has already been introduced in Sec. 3-4 in connection with the Gray code [Eq. (3-25)]. More than two bits may similarly be combined. As an example the mod-2 sum of the data sequence 1 0 0 1 0 1 is 1. An even-parity check code word would thus be

$$\overbrace{\underbrace{1\ 0\ 0\ 1\ 0\ 1}_{k\ \text{data bits}}\ 1}^{n\text{-bit code word}}$$

while an odd-parity check code word would be

$$1\ 0\ 0\ 1\ 0\ 1\ 0$$

(The bit stream will be assumed read in from left to right. The first three bits in order are then, 1, 0, 0.) Either code word could be used to detect an odd number of errors at the receiver by simply repeating the parity-check calculation and comparing with the received parity bit, or by carrying out the mod-2 sum (exclusive-or) operation over *all* the received bits and checking to see if the resultant sum is 0 (even parity) or 1 (odd parity), as the case may be.

More generally with codes, $r = n - k$ check bits are added to every k bits input. One thus speaks of (n, k) codes. Here n represents the total number of bits in a code word, while k is the original block size. An encoder outputs a unique n-bit code word for each of the 2^k possible input k-bit blocks. As an example, a $(15, 11)$ code has $r = 4$ parity-check bits for every 11 data bits. A $(7, 4)$ code would be one generating $r = 3$ parity-check bits for every 4 data bits. A little thought will indicate that as the number of parity-check bits r increases it should be possible to correct more and more errors. (Recall again that with $r = 1$ error correction is not possible. The code will only detect an odd number of errors.) In addition, as k increases, Shannon's theorem indicates the overall probability of error should decrease. Long codes with a relatively large number of parity-check bits should thus provide better performance. Such codes are more difficult and more costly to implement, however. As r increases, the required transmission bandwidth goes up as well ($n = k + r$ bits are now packed into a time slot previously allocated to k bits). The efficiency of transmission, k/n, goes down accordingly, since relatively fewer data bits are using a given channel. An example of a $(7, 4)$ code with 3 parity-check bits and 4 data bits appears in Fig. 6-54. This example demonstrates the time compaction noted above.

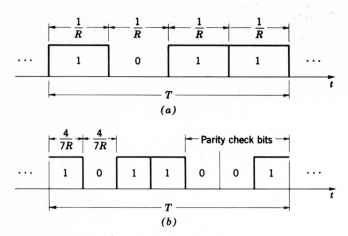

Figure 6-54 (7, 4) error-correcting code. (*a*) Original binary stream. (*b*) Encoded binary word.

A critical objective is therefore to choose the parity-check bits to correct as many errors as possible, but with the efficiency kept as high as possible.[21] Innumerable codes exist, with different properties. There are various types of codes for correcting independently occurring errors, for correcting burst errors, for providing relatively error-free synchronization of binary data, etc. Much of the effort in research on codes and the development of coding theory in recent years has gone into the development of codes with a great deal of structure to them. This limits the possible types of codes but enables a great deal to be said about the coder developed. We shall focus on *linear codes* only in this discussion. In these codes the parity-check bits are given by appropriate mod-2 sums of the data bits. In addition, we shall limit ourselves to *systematic codes,* in which the parity-check bits always appear at the end of the code word, following the original binary data sequence.[22] Much of the material later will concentrate still further on the class of *cyclic codes,* which are particularly simple to generate and decode.

Consider a k-bit data sequence $d_1, d_2, \ldots, d_k$. We shall denote this by the vector **d:**

$$\mathbf{d} = (d_1, d_2, \ldots, d_k)$$

The corresponding code word will be denoted by the n-bit vector **c:**

$$\mathbf{c} = (c_1, c_2, \ldots, c_k, c_{k+1}, \ldots, c_n)$$

---

[21] Although increasing r, and hence n, results in more errors corrected, the resultant required increase in bandwidth introduces more noise and hence increases the probability of error among the uncorrectable error sequences. The net result is to reduce the error probability gain expected. We shall carry out some sample calculations shortly to demonstrate this effect.

[22] It can be shown that this restriction does not affect the performance of the resultant code.

For a systematic code $c_1 = d_1$, $c_2 = d_2$, . . . , $c_k = d_k$. The r parity-check bits, c_{k+1}, c_{k+2}, . . . , c_n, are in turn given by the weighted mod-2 sum of the data bits:

$$c_{k+1} = h_{11}d_1 \oplus h_{12}d_2 \oplus \cdots \oplus h_{1k}d_k$$

$$c_{k+2} = h_{21}d_1 \oplus h_{22}d_2 \oplus \cdots \oplus h_{2k}d_k$$

$$\cdot$$
$$\cdot$$
$$\cdot$$

(6-121)

$$c_n = h_{r1}d_1 \oplus h_{r2}d_2 \oplus \cdots \oplus h_{rk}d_k$$

The h_{ij} coefficients are either 0 or 1. It is the choice of these coefficients that determines the properties of the particular code.

As an example, consider a (15, 11) code with the following parity-check equations:

$$
\begin{aligned}
c_{12} &= d_1 \oplus d_2 \oplus d_3 \oplus d_4 \quad\ \oplus d_6 \quad\ \oplus d_8 \oplus d_9 \\
c_{13} &= \quad\ d_2 \oplus d_3 \oplus d_4 \oplus d_5 \quad\ \oplus d_7 \quad\ \oplus d_9 \oplus d_{10} \\
c_{14} &= \quad\quad\ d_3 \oplus d_4 \oplus d_5 \oplus d_6 \quad\ \oplus d_8 \quad\ \oplus d_{10} \oplus d_{11} \\
c_{15} &= d_1 \oplus d_2 \oplus d_3 \quad\ \oplus d_5 \quad\ \oplus d_7 \oplus d_8 \quad\quad\ \oplus d_{11}
\end{aligned}
$$

(6-122)

It is left for the reader to show that if the data vector is $\mathbf{d}$ = (0 1 0 1 1 1 0 1 0 1 1) the 4 check bits are, respectively, $c_{12} = 0$, $c_{13} = 0$, $c_{14} = 0$, $c_{15} = 0$.

With the parity-check bits given by (6-121), it is apparent that the vector form $\mathbf{c}$ for a systematic code word can be written as a matrix operation on the data word $\mathbf{d}$:

$$\mathbf{c} = \mathbf{d}G \tag{6-123}$$

G must be a $k \times n$ matrix, with the first k columns an identity matrix I_k representing the fact that the first k bits of $\mathbf{c}$ are just the original data bits. The remaining r columns of G represent the transposed array of h_{ij} coefficients of (6-121). Thus we must have

$$G = [I_k \quad P] \tag{6-124}$$

with

$$
P = \begin{bmatrix}
h_{11} & h_{21} & \cdots & h_{r1} \\
h_{12} & h_{22} & \cdots & h_{r2} \\
\cdot & \cdot & & \cdot \\
\cdot & \cdot & & \cdot \\
\cdot & \cdot & & \cdot \\
h_{1k} & h_{2k} & \cdots & h_{rk}
\end{bmatrix}
\tag{6-125}
$$

G is called the *code generator matrix*.

For the (15, 11) code example of (6-122), we have

$$P = \begin{bmatrix} 1 & 0 & 0 & 1 \\ 1 & 1 & 0 & 1 \\ 1 & 1 & 1 & 1 \\ 1 & 1 & 1 & 0 \\ 0 & 1 & 1 & 1 \\ 1 & 0 & 1 & 0 \\ 0 & 1 & 0 & 1 \\ 1 & 0 & 1 & 1 \\ 1 & 1 & 0 & 0 \\ 0 & 1 & 1 & 0 \\ 0 & 0 & 1 & 1 \end{bmatrix} \tag{6-126}$$

The G matrix is similarly given by

$$G = \begin{bmatrix} 1 & 0 & 0 & 0 & 0 & 0 & 0 & 0 & 0 & 0 & 0 & 1 & 0 & 0 & 1 \\ 0 & 1 & 0 & 0 & 0 & 0 & 0 & 0 & 0 & 0 & 0 & 1 & 1 & 0 & 1 \\ 0 & 0 & 1 & 0 & 0 & 0 & 0 & 0 & 0 & 0 & 0 & 1 & 1 & 1 & 1 \\ 0 & 0 & 0 & 1 & 0 & 0 & 0 & 0 & 0 & 0 & 0 & 1 & 1 & 1 & 0 \\ 0 & 0 & 0 & 0 & 1 & 0 & 0 & 0 & 0 & 0 & 0 & 1 & 1 & 1 \\ 0 & 0 & 0 & 0 & 0 & 1 & 0 & 0 & 0 & 0 & 0 & 1 & 0 & 1 & 0 \\ 0 & 0 & 0 & 0 & 0 & 0 & 1 & 0 & 0 & 0 & 0 & 0 & 1 & 0 & 1 \\ 0 & 0 & 0 & 0 & 0 & 0 & 0 & 1 & 0 & 0 & 0 & 1 & 0 & 1 & 1 \\ 0 & 0 & 0 & 0 & 0 & 0 & 0 & 0 & 1 & 0 & 0 & 1 & 1 & 0 & 0 \\ 0 & 0 & 0 & 0 & 0 & 0 & 0 & 0 & 0 & 1 & 0 & 0 & 1 & 1 & 0 \\ 0 & 0 & 0 & 0 & 0 & 0 & 0 & 0 & 0 & 0 & 1 & 0 & 0 & 1 & 1 \end{bmatrix} \tag{6-127}$$

It is apparent from (6-125) and (6-126) that no columns of P can be alike, for then the same parity-check bit would be generated by these columns, defeating the purpose of the encoding operation.

The encoding operation, represented in equation form by (6-121) and in matrix form by (6-123), may be schematized by the block diagram of Fig. 6-55. Other encoder representations, using mod-2 operations on shift register outputs, will be considered later. The commutator in Fig. 6-55 must rest on the register k time slots while the k data bits are being read out. Note that the k data bits plus r parity-check bits must all be read out during the time the k bits are being read in.

Another example of a code is a (7, 3) code with the P matrix

$$P = \begin{bmatrix} 1 & 1 & 0 & 0 \\ 0 & 1 & 1 & 0 \\ 1 & 1 & 1 & 1 \end{bmatrix} \tag{6-128}$$

Note that there are 4 check bits. The code efficiency is thus $\frac{3}{7}$. There are eight possible code words:

$\mathbf{d}$	$\mathbf{c} = \mathbf{d}G = \mathbf{d}[I_k \quad P]$	
0 0 0	0 0 0	0 0 0 0
0 0 1	0 0 1	1 1 1 1
0 1 0	0 1 0	0 1 1 0
0 1 1	0 1 1	1 0 0 1
1 0 0	1 0 0	1 1 0 0
1 0 1	1 0 1	0 0 1 1
1 1 0	1 1 0	1 0 1 0
1 1 1	1 1 1	0 1 0 1

A close look at these indicates that they differ in at least three positions. Any *one* error should then be correctible since the resultant code word will still be *closer* to the correct (transmitted) one, in the sense of the number of bit positions in which they agree, than to any other. This is thus an example of a *single-error-correcting code*. The difference in the number of positions between any two code words is called the *Hamming distance*. The Hamming distance plays a key role in assessing the error-correcting capability of codes. For two errors to be correctible (other error patterns may be detectable as well) the Hamming distance d should be at least 5. In general, for t errors correctible, $d \geq 2t + 1$, or $t = [(d-1)/2]$, where the $[x]$ notation refers to the integer less than or equal to x.

What decoding operation is now required at the receiver? One simple procedure is to repeat the parity-check calculation of the encoder and compare the resultant $r = (n - k)$ parity-check pattern with that actually received. If the

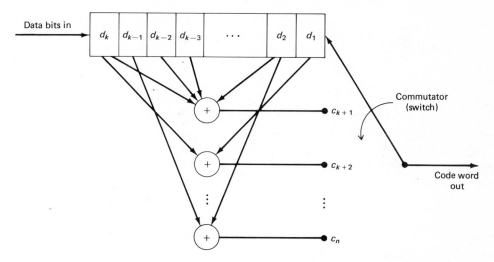

Figure 6-55 One possible encoder (commutator rests on register k time slots, then moves to other positions for one time slot each).

calculated and received patterns do not agree, one or more errors are indicated. Appropriate error-correction procedures can then be initiated. (The simplest error-correction procedure conceptually is to compare the received code word against a stored table of code words, selecting the one most likely to have been transmitted. It is apparent that this table grows exponentially with k, however, so that for long codes this method proves self-defeating.)

More precisely, consider a systematic code word $c = dG$, with d the k-bit data sequence. It is apparent that this may be written

$$c = [d \quad dP] = [d \quad c_P] \tag{6-129}$$

from (6-123) and (6-124). The parity-check bit sequence c_P is simply given by

$$c_P = dP \tag{6-130}$$

Assume now that c represents a received sequence of n digits. The first k digits representing some vector d *must* be part of a transmitted code word, since they *always* represent one of 2^k possible sequences of k digits. To check to see whether the full n-bit sequence c represents a possible code word the decoder can carry out the operation dP and compare with the received parity-check sequence c_P. Since mod-2 subtraction is the same as addition, we must have

$$\underset{\substack{\text{calculated} \\ \text{at} \\ \text{the decoder}}}{dP} \quad \oplus \quad \underset{\substack{\text{received} \\ \text{parity-check} \\ \text{sequence}}}{c_P} \quad = 0 \tag{6-131}$$

if the received sequence is a proper code word. The vector summation in (6-131) implies a mod-2 comparison bit by bit. This conceptual way of checking for possible errors is sketched in Fig. 6-56. [Note that the mere fact that (6-131) is found valid does not guarantee error-free reception. Any code word can obviously have error patterns occur which are not detectable.]

Equation (6-131) may be rewritten in matrix form as follows:

$$dP + c_P = [d \quad c_P] \begin{bmatrix} P \\ I_{n-k} \end{bmatrix} = 0 \tag{6-132}$$

Here I_{n-k} is an identity matrix of order $n - k$. Very generally, then, for any code word c, we must have

$$cH^T = 0 \tag{6-133}$$

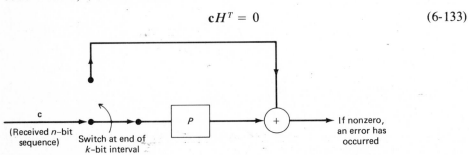

Figure 6-56 Conceptualized decoder.

The $n \times r$ matrix H^T is given by

$$H^T = \begin{bmatrix} P \\ I_{n-k} \end{bmatrix} \tag{6-134}$$

Its transpose

$$H = [P^T \quad I_{n-k}] \tag{6-135}$$

is called the *parity-check* matrix. This matrix plays an important role in the theory of error-correcting codes.[23] Note from (6-125) and (6-135) that H is in general given by

$$H = \begin{bmatrix} h_{11} & h_{12} & \cdots & h_{1k} & 1 & 0 & \cdots & 0 \\ h_{21} & h_{22} & \cdots & h_{2k} & 0 & 1 & \cdots & 0 \\ \cdot & \cdot & & \cdot & & \cdot & & \cdot \\ \cdot & \cdot & & \cdot & & \cdot & & \cdot \\ \cdot & \cdot & & \cdot & & \cdot & & \cdot \\ h_{r1} & h_{r2} & \cdots & h_{rk} & 0 & 0 & \cdots & 1 \end{bmatrix} \tag{6-136}$$

Consider as an example the (7, 3) code mentioned earlier, whose P matrix was given by Eq. (6-128). For this code we have as the H^T and H matrices, respectively,

$$H^T = \begin{bmatrix} 1 & 1 & 0 & 0 \\ 0 & 1 & 1 & 0 \\ 1 & 1 & 1 & 1 \\ 1 & 0 & 0 & 0 \\ 0 & 1 & 0 & 0 \\ 0 & 0 & 1 & 0 \\ 0 & 0 & 0 & 1 \end{bmatrix} \tag{6-137}$$

and

$$H = \begin{bmatrix} 1 & 0 & 1 & 1 & 0 & 0 & 0 \\ 1 & 1 & 1 & 0 & 1 & 0 & 0 \\ 0 & 1 & 1 & 0 & 0 & 1 & 0 \\ 0 & 0 & 1 & 0 & 0 & 0 & 1 \end{bmatrix} \tag{6-138}$$

Summarizing what has been said thus far we can generate a code word by carrying out the operation

$$\mathbf{c} = \mathbf{d}G \tag{6-123}$$

A code word must in turn satisfy the condition

$$\mathbf{c}H^T = 0 \tag{6-133}$$

Now say that an error occurs in one or more of the digits $\mathbf{c}$. The received vector $\mathbf{r}$ may then be written in the form

$$\mathbf{r} = \mathbf{c} \oplus \mathbf{e} \tag{6-139}$$

[23] Peterson and Weldon, *op. cit.*

with **e** an n-bit vector representing the error pattern. [If errors occur in the second and third bits, for example, $\mathbf{e} = (0\ 1\ 1\ 0\ 0\ \cdot\ \cdot\ \cdot\ 0)$.] Operating on **r** with the matrix H^T, we have

$$\mathbf{r}H^T = (\mathbf{c} \oplus \mathbf{e})H^T = \mathbf{e}H^T = \mathbf{s} \tag{6-140}$$

The r-element vector **s**, called the *syndrome,* will be nonzero if an error has occurred. (Conversely, since not all error patterns are detectable by any code, the condition $\mathbf{s} = 0$ is no guarantee of error-free transmission.)

Assume now, as an example, that a *single* error has occurred in the ith digit of the n-bit code word transmitted. Then

$$\mathbf{e} = (0\ 0\ 0\ \cdot\ \cdot\ \cdot\ \underset{\substack{i\text{th} \\ \text{digit}}}{1}\ \cdot\ \cdot\ \cdot\ 0) \tag{6-141}$$

From the defining relation for the syndrome it is then easy to show that

$$\mathbf{s} = (h_{1i}\ h_{2i}\ \cdot\ \cdot\ h_{ri}) \tag{6-142}$$

Comparing with (6-136), this is just the ith column of the parity-check matrix H! So the error is not only detectable, but correctable as well, provided that all columns of H (or corresponding rows of H^T) are uniquely defined. Any code satisfying this condition is a single-error-correcting code.

The (7, 3) code again provides a simple example. Say that 1101010 is the code word that was transmitted, while the received code word is 1111010 (there is an error in the third position). The **e** vector then has the elements 0010000 and the syndrome vector is $\mathbf{s} = (1111)$, after post-multiplying $\mathbf{r} = (1111010)$ by the matrix H^T of (6-137). It is apparent that **s** is just the third row of H^T or the third column of the parity-check matrix H [eq. (6-138)] for this code.

What kind of constraint is put on codes to satisfy the single-error-correcting condition that all columns of H are uniquely defined? This says that the first k columns of (6-136) must be unique. They must also differ from the last r columns containing a single 1 in each column, and cannot include an all-0 column. (Why is this so?) It is thus apparent that to have each of the k r-bit columns uniquely defined, we must have

$$2^r - (r + 1) \geq k \tag{6-143}$$

(r bits can be arranged in 2^r possible ways, $r + 1$ of which must be ruled out because of the stipulation on the all-zero column and the r single-one columns.) This inequality that must be satisfied by a single-error-correcting code is a special case of the Hamming inequality for a more general t-error correcting code, to be discussed briefly below. As an example of the use of this inequality, note that the (7, 3) code has $r = 4$. Hence $2^4 - 5 = 11 > k = 3$. The code should therefore be single-error-correcting, as we have already seen. A (7, 4) code, with $r = 3$, is also single-error-correcting. The (15, 11) code of Eq. (6-126) is obviously single-error-correcting since all 11 rows of P (corresponding to the 11 columns of H^T that must be uniquely defined) *are* unique, nonzero, and with no single 1's. Note that this code has $r = 4$ check bits. Hence it satisfies (6-143) as

Table 6-1 Examples of error-correcting codes

Single-error-correcting codes: $t = 1$ ($d = 3$)			
n	k_{max}	Code	Efficiency
4	1		
5	2	(5, 2)	0.4
6	3	(6, 3)	0.5
7	4	(7, 4)	0.57
15	11	(15, 11)	0.73
Double-error-correcting codes: $t = 2$ ($d = 5$)			
n	k_{max}	Code	Efficiency
10	4	(10, 4)	0.4
11	4	(11, 4)	0.36
15	8	(15, 8)	0.53
Triple-error-correcting codes: $t = 3$ ($d = 7$)			
n	k_{max}	Code	Efficiency
10	2	(10, 2)	0.2
15	5	(15, 5)	0.33
23	12	(23, 12)	0.52
24	12	(24, 12)	0.5

an equality. It is left for the reader to show that a (6, 3) code is single-error-correcting, while a (6, 4) code cannot be designed for this purpose.

More generally, what can we say about the parity-check requirements for higher-order error-correcting codes? This is obviously quite useful in beginning the selection process for appropriate codes. More precisely, for a code to correct *at least* t errors, what values of k and n are required, and what efficiencies are possible? Various bounds are available to help in this selection process,[24] but we focus only on the Hamming bound, a special case of which appeared in (6-143). This bound says simply that the number of possible check-bit patterns must at least equal the number of ways in which up to t errors can occur. For an (n, k) code, then, we have

$$2^{n-k} \geq \sum_{i=0}^{t} \binom{n}{i} \qquad (6\text{-}144)$$

[Note that with $t = 1$, and $n = r + k$, we get just Eq. (6-143).] Table 6-1 shows some typical (n, k) representations and their efficiency, found by solving (6-144) and selecting the codes with the largest k value only. The $t = 1$, or single-error-correcting cases, are those already discussed. The (n, k) doublets shown represent possible candidates for error correction. Systematic ways of finding

[24] Lucky, Salz, and Weldon, *op. cit.*, pp. 313–316.

(n, k) codes will be described later. These codes must then be investigated further to see whether they in fact have the desired error-correction capability, as well as other desirable properties. Note that the efficiency decreases as the requirement on the error-correcting capability increases. Longer codes are then needed to recoup this efficiency.

It was pointed out earlier that it is not sufficient to characterize a code by its error-correcting capability only. By introducing r parity-check bits the bit rate must of necessity increase by a factor of n/k, or fractionally by r/k, to maintain real-time transmission. There is thus an increased bandwidth requirement that must be charged against the code. Additional noise is also let into the system. Alternatively, because of the shorter time interval over which a bit is transmitted, there is correspondingly less energy in the received signal bit. The critical detection parameter E/n_0, with E the signal energy in each received bit, and n_0 the noise spectral density for white noise, is thus reduced. This tends to *increase* the probability of error, partially reducing the effectiveness of the code in correcting errors. A complete study of a code must take this reduced E/n_0 into account.

To demonstrate this procedure, we consider a simple example. Assume that a (7, 4) single-error-correcting block code is used. Binary PSK is used for transmitting the successive bits in each code word. The AWGN channel is assumed for simplicity. It is desired to compare the probability of error in the coded case with that in the uncoded case. We do this by comparing the error probability of a *block* in the two cases, coded and uncoded.

Specifically, let the probability of error of a bit in the uncoded case be P_{e1}. For the AWGN channel with PSK transmission this is just Eq. (6-59) found earlier in this chapter, as well as in Chap. 5.

$$P_{e1} = \frac{1}{2} \operatorname{erfc} \sqrt{\frac{E}{n_0}} \tag{6-145}$$

The probability of error of the uncoded block in the (7, 4) case is just the probability that at least one bit in four will be in error, and is thus given by

$$P_{e,\text{uncoded}} = 1 - (1 - P_{e1})^4 \doteq 4 P_{e1} \qquad P_{e1} \ll 1 \tag{6-146}$$

Now consider the coded case. Let the corresponding bit-error probability be p. This must be of the same form as P_{e1}, but with the signal-to-noise ratio E/n_0 reduced by $\frac{4}{7}$. Hence

$$p = \frac{1}{2} \operatorname{erfc} \sqrt{\frac{4E}{7n_0}} \tag{6-147}$$

The probability of an error in this coded case is now the probability that at least two independent errors will occur in a pattern of 7 bits. This is given by the cumulative binomial probability

$$P_{e,\text{coded}} = \sum_{j=2}^{7} \binom{7}{j} p^j (1 - p)^{7-j}$$

$$\doteq 21 p^2 \qquad p \ll 1 \tag{6-148}$$

Table 6-2 Comparison of block error probability, coded and uncoded, (7, 4) code

E/n_0, dB	P_{e1}	$P_{e,\text{uncoded}}$	$P_{e,\text{coded}}$
-1	10^{-1}	0.344	0.34
4.2	10^{-2}	0.0394	0.026
6.8	10^{-3}	4×10^{-3}	1.9×10^{-3}
8.3	10^{-4}	4×10^{-4}	1.6×10^{-4}
9.6	10^{-5}	4×10^{-5}	8.6×10^{-6}

Using (6-145) and (6-147), (6-146) and (6-148) may be compared. A comparison for various values of E/n_0 appears in Table 6-2. The (7, 4) block code thus does not provide significant improvement in the error probability until $P_{e1} = 10^{-5}$ or less. One would have to go to much longer codes to demonstrate appreciable improvement.

Hamming Distance and the Binary Symmetric Channel

Implicit in the discussion of errors, and their detection and correction thus far, has been the assumption that signals in each binary interval at the receiver, prior to decoding, were individually detected and decisions made on each as to whether it was a 1 or a 0. The incoming signal, even though analog in nature because of distortion and additive noise encountered during transmission, is thus assumed converted to a sequence of binary digits before entering the decoder. This process of bit-by-bit quantization into either of two levels is called *hard limiting*. It is not at all obvious that this is a "good" procedure, particularly if there is memory in the channel. One can in fact show, using convolutional decoders, for example, that it is possible to improve system performance in that case by retaining a sequence of analog received signals, and making binary decisions on the composite set.[25] This procedure of basing decisions on the analog signals (or at least a finely quantized version of these) is called *soft limiting*.

We noted in passing, earlier, that the Hamming distance d, the minimum number of bit positions by which code words for a given code differ, appears to play a critical role in the performance of block codes. In fact, we indicated that the error-correcting capability t was simply $[(d - 1)/2]$. Here too we were implicitly assuming bit-by-bit hard limiting. If the 2^k possible code words differ in at least d bit positions a hard-limited received block should be correctable if fewer than $(d - 1)/2$ bits have been changed during transmission. This procedure of assigning a hard-limited received block to the code word to which it is closest in the Hamming distance sense can in fact be shown to be optimum for a

[25] A. J. Heller and I. M. Jacobs, "Viterbi Decoding for Satellite and Space Communications," *IEEE Trans. Commun. Technol.*, vol. COM-19, no. 5, part II, p. 835, October 1971.

particular class of *memoryless* channels called the *binary symmetric channel* (BSC). The AWGN channel with hard limiting is one example of this class.

The binary symmetric channel is, as its name indicates, a channel for which only two digits (0 and 1) appear at the transmitter and receiver (hence the emphasis on hard limiting). Either digit is assumed converted to the other during transmission with the same probability $p \leq \frac{1}{2}$. (This is the reason for the word "symmetry" in the title.) Since the channel is assumed memoryless, *each* digit in a sequence has the same probability p of being received in error. A schematic representation of the BSC appears in Fig. 6-57. To demonstrate the optimality of the Hamming distance rule for this set of channels say a specific n-bit code word c_i of the 2^k possible is transmitted. An n-bit binary sequence r is received. Then for the minimum probability of error we must use the maximum a posteriori (MAP) rule discussed earlier in this chapter. Specifically, we select code word c_i if

$$P(c_i|r) > P(c_j|r) \qquad \text{all } j, \quad j \neq i \tag{6-149}$$

Now assume all 2^k code words equally likely. Rule (6-149) is then converted to the equivalent maximum-likelihood rule that selected c_i if

$$P(r|c_i) > P(r|c_j) \qquad \text{all } j, \quad j \neq i \tag{6-150}$$

Consider a particular code word c_j and compare it now to r. Say that the two differ in n' positions. For the binary symmetric channel, then, we must have

$$p(r|c_j) = p^{n'}(1 - p)^{n-n'} = \left(\frac{p}{1-p}\right)^{n'} (1 - p)^n \tag{6-151}$$

From (6-150) we are to calculate this probability for *all j*, and select the one that is largest. Since $p/(1 - p) < 1$, this is the same as choosing the probability that has the smallest n'. Hence the best rule, in the sense of minimizing probability of error, is to select as the appropriate code word of the 2^k available the one that is closest to r in the Hamming distance sense. This type of decoder is optimum for the BSC. It is *not* optimum for burst-type channels (in which a series of successive bit changes occur), or for channels using convolutional coding, in which memory is introduced into the transmission, and other examples of channels with memory.

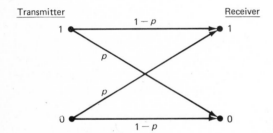

Figure 6-57 Binary symmetric channel.

Cyclic Codes

The Hamming bound discussed earlier is one example of a technique used to help us determine the size of a code needed to check for a specified error-correction capability. We pointed out earlier as well that any (n, k) code satisfying the Hamming inequality will automatically provide single-error-correcting capability provided that the columns of the parity-check matrix H (or, equivalently, the rows of the G matrix) are uniquely defined. To go further with higher-order error-correction capability in a systematic way, it is necessary to introduce additional structure into the code formulation. Although this of necessity limits the choice of codes, as pointed out previously, it at least provides a systematic way to select codes. We therefore focus on the class of codes called *cyclic codes*. [26] These are codes such that code vectors are simple lateral shifts of one another.

For example, if $\mathbf{c} = (c_1, c_2, \ldots , c_{n-1}, c_n)$ is a possible code vector, then so are $(c_2, c_3, \ldots , c_n, c_1)$ and $(c_3, c_4, \ldots , c_n, c_1, c_2)$, etc. Consider $\mathbf{c} = (1\ 0\ 1\ 1\ 0\ 1)$. Then it is apparent that $(0\ 1\ 1\ 0\ 1\ 1)$ and $(1\ 1\ 0\ 1\ 1\ 0)$ are cyclic shifts and must serve as code words as well. These cyclic codes have a great deal of structure and specific rules of generation may be set up. They are commonly used, not only for error correction, but in error detection as well, since they are easily implementable.

An interesting property of the generator matrix of a cyclic code is that its last element, in the kth row and nth column, must always be a 1. For consider the data-bit sequence $\mathbf{d} = (0\ 0\ 0\ \cdots\ 0\ 1)$ operated on by a generator matrix G whose last element is 0. Then the code word is $\mathbf{c} = (\underbrace{0\ 0\ 0\ \cdots\ 0\ 1}_{\substack{\text{information} \\ \text{bits}}}\ \underbrace{\cdots\ 0}_{\substack{\text{parity} \\ \text{bits}}})$.

Shifting this once to the right, we obtain $\mathbf{c}' = (\underbrace{0\ 0\ 0\ \cdots\ 0\ 0}_{\substack{\text{information} \\ \text{bits}}}\ \underbrace{1\ \cdots\ }_{\substack{\text{parity} \\ \text{bits}}})$. This is obviously impossible, for our codes are such that the all-zero data sequence must produce an all-zero parity sequence. For a systematic cyclic code we must thus have the generator matrix of the form

$$
G = \underbrace{\begin{bmatrix}
1 & 0 & 0 & \cdots & 0 & \vdots & \cdots & \\
0 & 1 & 0 & \cdots & 0 & \vdots & & \\
0 & 0 & 1 & \cdots & 0 & \vdots & & \\
& & \cdot & & & \vdots & \cdot & \\
& & \cdot & & & \vdots & \cdot & \\
& & \cdot & & & \vdots & \cdot & \\
0 & 0 & 0 & \cdots & 1 & \vdots & \cdots & 1
\end{bmatrix}}_{I_k}
\tag{6-152}
$$

We shall see shortly that it is the last row of the matrix that determines the properties of the code.

[26] Peterson and Weldon, *op. cit.;* Lucky, Salz, and Weldon, *op. cit.*

Cyclic codes are describable in polynomial form, a property that is extremely useful in their analysis and implementation. The code word $\mathbf{c} = (c_1, c_2, \cdots, c_n)$ may be expressed as the $(n-1)$-degree polynomial

$$c(x) = c_1 x^{n-1} + c_2 x^{n-2} + \cdots + c_{n-1} x + c_n \qquad (6\text{-}153)$$

Each power of x represents a one-bit shift in time. The highest-order coefficient c_1 in the polynomial represents the first bit of the code word, the last coefficient c_n, the last bit of the code word. Successive shifts to generate other code words are then repeated by the operation $xc(x) \bmod (x^n + 1)$. Thus, shifting once, we have

$$xc(x) \bmod (x^n + 1) = c_2 x^{n-1} + c_3 x^{n-2} + \cdots + c_{n-1} x^2 + c_n x + c_1 \quad (6\text{-}154)$$

Shifting a second time, we have

$$x^2 c(x) \bmod (x^n + 1) = c_3 x^{n-1} + c_4 x^{n-2} + \cdots + c_n x^2 + c_1 x + c_2 \quad (6\text{-}155)$$

Using this polynomial representation, the G matrix may be represented with polynomials of x as well. Specifically, we insert the appropriate power of x in any element with a 1, and leave blank the elements containing a zero. Thus, as an example, consider the following matrix for a (7, 3) code:

$$G = \begin{bmatrix} 1 & 0 & 0 & 1 & 1 & 1 & 0 \\ 0 & 1 & 0 & 0 & 1 & 1 & 1 \\ 0 & 0 & 1 & 1 & 1 & 0 & 1 \end{bmatrix} \qquad (6\text{-}156)$$

Its polynomial representation is then

$$G = \begin{bmatrix} x^6 & - & - & x^3 & x^2 & x & - \\ - & x^5 & - & - & x^2 & x & 1 \\ - & - & x^4 & x^3 & x^2 & - & 1 \end{bmatrix} \qquad (6\text{-}156a)$$

More generally, for an (n, k) cyclic code, we must have

$$\leftarrow g(x) \qquad G = \begin{bmatrix} x^{n-1} & - & - & - & - & \cdots \\ - & x^{n-2} & - & - & - & \cdots \\ - & - & \ddots & - & - \\ - & - & & \ddots & - \\ - & - & - & & \ddots & - \\ - & - & - & - & x^{n-k} & \cdots & 1 \end{bmatrix} \leftarrow g(x) \qquad (6\text{-}157)$$

Note that the last row must always be representable by a polynomial of the form

$$g(x) = x^{n-k} + \cdots + 1 \qquad (6\text{-}158)$$

by our observation earlier that for cyclic codes the last element must always be a 1.

This polynomial, called the *generator polynomial* of the code, determines the characteristics of the code. For consider a matrix G' made up of k rows generated by successive multiplications of $g(x)$ by x:

$$G' = \begin{bmatrix} x^{k-1}g(x) \\ \cdot \\ \cdot \\ \cdot \\ x^2g(x) \\ xg(x) \\ g(x) \end{bmatrix} \qquad (6\text{-}159)$$

Operating on G' by the vector $\mathbf{d} = (d_1, \ldots, d_{k-1}, d_k)$, with d_1 the first bit in the k-bit data sequence and d_k the last, we get as a code polynomial

$$\begin{aligned} c(x) &= d_1 x^{k-1} g(x) + d_2 x^{k-2} g(x) + \cdots + d_k g(x) \\ &= d(x) g(x) \end{aligned} \qquad (6\text{-}160)$$

with $d(x)$ a $(k - 1)$-degree (or lower) polynomial whose coefficients are the components of $\mathbf{d}$. Note that $c(x)$ is a cyclic polynomial of degree $(n - 1)$ or less, and there are just 2^k possible code words, corresponding to the 2^k possible k-bit data sequences. The code words are not systematic, however, in the sense that the first k bits represent the information bits and the remaining $r = (n - k)$ bits the parity bits. To generate the systematic form matrix, G' must be transformed into a new matrix G which has the identity matrix I_k in the first k columns. This is called the *standard form* of G and turns out to be exactly the desired matrix of the type of (6-157).

The recipe for doing this goes as follows:

1. Use $g(x)$ as the kth row.
2. To generate the $(k - 1)$ row, cyclically shift the kth row one column to the left. This corresponds of course to the operation $xg(x)$. But the kth column entry must be zero to have the standard form. If this entry is 1, add the kth row to it. Thus the $(k - 1)$ row is $xg(x)$ if the coefficient of x^{n-k-1} in $g(x)$ is 0, or $xg(x) + g(x)$ (assuming mod-2 addition again) if the coefficient is 1.
3. To generate the $(k - 2)$ row repeat the same process: shift the $(k - 1)$ row entries one column to the left. Add $g(x)$ if the kth column entry is not zero. Repeat this for all the rows until the topmost one is reached. Note that this corresponds to successive row additions on G' until the appropriate standard form of G is reached.

As an example, say that $g(x) = x^4 + x^3 + x^2 + 1$, and let $n = 7$. [This is thus a (7, 3) code.] Then

$$G' = \begin{bmatrix} x^6 & x^5 & x^4 & - & x^2 & - & - \\ - & x^5 & x^4 & x^3 & - & x & - \\ - & - & x^4 & x^3 & x^2 & - & 1 \end{bmatrix} \qquad (6\text{-}161)$$

This of course does not have the appropriate identity form in the first $k = 3$ columns. To generate the standard form of G, we follow the rules above and find precisely the matrix shown previously in (6-156a):

$$G = \begin{bmatrix} x[xg(x) + g(x)] \\ xg(x) + g(x) \\ g(x) \end{bmatrix} = \begin{bmatrix} x^6 & — & — & x^3 & x^2 & x & — \\ — & x^5 & — & — & x^2 & x & 1 \\ — & — & x^4 & x^3 & x^2 & — & 1 \end{bmatrix} \quad (6\text{-}156a)$$

The generator matrix for this code, replacing x's by 1's and blanks by 0's in (6-156a), is of course the one given earlier by (6-156).

$$G = \begin{bmatrix} 1 & 0 & 0 & 1 & 1 & 1 & 0 \\ 0 & 1 & 0 & 0 & 1 & 1 & 1 \\ 0 & 0 & 1 & 1 & 1 & 0 & 1 \end{bmatrix} \quad (6\text{-}156)$$

It is apparent that the standard form of G is obtained from G' by successive addition of rows. That this does not change the code words but corresponds simply to their reordering is readily demonstrated. Consider an arbitrary matrix G_1 with n-element row vectors $\mathbf{r}_1, \mathbf{r}_2, \ldots, \mathbf{r}_k$. Specifically, let

$$G_1 = \begin{bmatrix} \mathbf{r}_1 \\ \mathbf{r}_2 \\ . \\ . \\ . \\ \mathbf{r}_k \end{bmatrix} \quad (6\text{-}162)$$

These row vectors are said to serve as the basic vectors for the code words. For consider a data vector

$$\mathbf{d}_1 = (d_1, d_2, \cdots d_{k-1}, d_k) \quad (6\text{-}163)$$

Then the code word corresponding to this vector is given by

$$\mathbf{c}_1 = \mathbf{d}_1 G_1 = d_1 \mathbf{r}_1 + d_2 \mathbf{r}_2 + \cdots + d_k \mathbf{r}_k \quad (6\text{-}164)$$

Another matrix G_2 is now formed by adding the jth row of G_1 to the ith row to form a new ith row. Hence

$$G_2 = \begin{bmatrix} \mathbf{r}_1 \\ \mathbf{r}_2 \\ . \\ . \\ . \\ \mathbf{r}_i + \mathbf{r}_j \\ . \\ . \\ . \\ \mathbf{r}_k \end{bmatrix} \quad (6\text{-}165)$$

The code word for the same k-bit data word $\mathbf{d}_1$ is now

$$\begin{aligned} \mathbf{c}_2 = \mathbf{d}_1 G_2 &= d_1 \mathbf{r}_1 + \cdots + d_i (\mathbf{r}_i + \mathbf{r}_j) + \cdots + d_k \mathbf{r}_k \\ &= d_1 \mathbf{r}_1 + \cdots + d_i \mathbf{r}_i + \cdots + (d_j + d_i)\mathbf{r}_j + \cdots \\ &= \mathbf{d}' G_1 \end{aligned} \quad (6\text{-}166)$$

Since every one of the 2^k possible sequences k bits long must be a data word, the sum of two words must give rise to a new code word:

$$\mathbf{d}_1 + \mathbf{d}_2 = \mathbf{d}_3 \tag{6-167}$$

It is apparent that $\mathbf{d}'$ in (6-166) must be a data word, and hence $\mathbf{c}_2$ is one of the 2^k code words. Both matrices G_1 and G_2, the latter obtained by linear transformations on the former, give rise to the same code-word set. Hence both G and G' discussed earlier produce the same set of code words. The only difference is that G produces a *systematic* set, with information bits always corresponding to the first k bits.

Each cyclic code is thus derivable from a generator matrix $g(x)$. A code word $c(x)$, in polynomial form, may always be written in the form

$$c(x) = a(x)g(x) \tag{6-168}$$

since, from the rules for finding G, each row of the generator matrix must be a polynomial times $g(x)$. As a check, the polynomial $a(x)$ must be of the $(k - 1)$ order to have $c(x)$ an $(n - 1)$ polynomial. There are thus k coefficients of this polynomial with a total of 2^k possible code words.

Since all cyclic codes are generated by an appropriate generator polynomial $g(x)$, all that remains to determine them is to indicate how one finds $g(x)$. This turns out to be very straightforward. We state without proof the following theorem[27]: *the generator polynomial $g(x)$ for an (n, k) cyclic code is a divisor of* $x^n + 1$.

As an example, consider the class of $(7, k)$ codes. We have already indicated that the $(7, 4)$ and $(7, 3)$ codes are single-error-correcting. To find the cyclic codes of this group we need the appropriate divisors of $x^7 + 1$. It is left for the reader to show that

$$x^7 + 1 = (x + 1)(x^3 + x + 1)(x^3 + x^2 + 1) \tag{6-169}$$

Products of divisors are divisors as well. This has two polynomials of the fourth order that can serve as generator polynomials of a $(7, 3)$ code. [Recall that $g(x)$ is of $(n - k)$ order.] Consider, in particular, the generator polynomial

$$g(x) = (x + 1)(x^3 + x + 1) = x^4 + x^3 + x^2 + 1 \tag{6-170}$$

This is precisely the example used in (6-161) and (6-156a). The generator matrix for this code is given by (6-156a). Its eight possible code words, found using (6-123), are

0	0	0	0	0	0	0
0	0	1	1	1	0	1
0	1	0	0	1	1	1
0	1	1	1	0	1	0
1	0	0	1	1	1	0
1	0	1	0	0	1	1
1	1	0	1	0	0	1
1	1	1	0	1	0	0

[27] Lucky, Salz, and Weldon, *op. cit.*, p. 294; Peterson and Weldon, *op. cit.*, p. 208.

Note that except for the all-zero code word these are all cyclic versions of one another. Note also that the minimum Hamming distance is 4, corresponding to a single-error-correction capability. This code has a limited burst-error correction capability as well.[28]

To summarize our results thus far for cyclic codes, we have shown that the Hamming inequality of (6-144) can be used to find an appropriate (n, k) combination as a possible candidate for specified error-correction capability. A divisor of $x^n + 1$ of the $(n - k)$ order will then serve as the generator polynomial $g(x)$. Using the rules outlined earlier, we can then find the generator matrix G. The resultant codes found must then be tested to see if they in fact possess the desired error-correction capability. We shall demonstrate some simple procedures shortly for generating the code words directly from $g(x)$. This is in fact one of the reasons for focusing on cyclic codes: they are often easily implemented. In addition to the random-error-correction capability stressed thus far, one can get other properties as well by choosing $g(x)$ appropriately. These include, among others, burst-error-correction capability, synchronization capability, ability to detect (but not correct) various error patterns, etc. We shall discuss the error-detection property of cyclic codes later in this section.

As another example of a set of cyclic codes, consider the class of $(15, k)$ codes.[29] These must be generated by divisors of $x^{15} + 1$. Carrying out the division, one finds

$$x^{15} + 1 =$$
$$(x + 1)(x^2 + x + 1)(x^4 + x + 1)(x^4 + x^3 + 1)(x^4 + x^3 + x^2 + x + 1) \quad (6\text{-}171)$$

These five divisors may in turn be multiplied together to form new divisors. In all, there are

$$\binom{5}{1} + \binom{5}{2} + \binom{5}{3} + \binom{5}{4} = 30$$

possible polynomials from which generator polynomials may be obtained. It turns out 26 of these are nontrivial. Some of these 26 codes and their generator polynomials are given in Table 6-3.[30]

Cyclic-Code Generation: Polynomial Encoding

It has already been noted above that cyclic codes lend themselves readily to generation directly from the generator polynomial $g(x)$. Shift register implementation can be used to carry out the code generation serially, if desired. To demonstrate this, recall from (6-168) that a code-word polynomial $c(x)$ must be divisible by the generator polynomial $g(x)$. It is also of degree $(n - 1)$ or less.

---

[28] Lucky, Salz, and Weldon, *op. cit.*, p. 373.

[29] *Ibid.*, p. 295.

[30] *Ibid.*, p. 295, table 10-1. To simplify the notation, $(2, 1, 0)$ is used to represent $x^2 + x + 1$; $(1, 0) (4, 1, 0)$ is $(x + 1) (x^4 + x + 1) = x^5 + x^4 + x^2 + 1$.

Table 6-3 Some $(15, k)$ cyclic codes

k	t (error-correction capability)	d (Hamming distance)	$g(x)$
14	0	2	$(1, 0) \equiv x + 1$
11	1	3	$(4, 1, 0) \equiv x^4 + x + 1$
10	1	4	$(1, 0)(4, 1, 0)$
7	2	5	$(4, 1, 0)(4, 3, 2, 1, 0)$
6	2	6	$(1, 0)(4, 1, 0)(4, 3, 2, 1, 0)$
5	3	7	$(4, 1, 0)(4, 3, 2, 1, 0)(2, 1, 0)$
4	3	8	$(1, 0)(4, 1, 0)(4, 3, 2, 1, 0)(2, 1, 0)$
2	4	10	$(1, 0)(4, 1, 0)(4, 3, 2, 1, 0)(4, 3, 0)$
1	7	15	$(4, 1, 0)(4, 3, 2, 1, 0)(4, 3, 0)(2, 1, 0)$

Consider now a data sequence $d_1, d_2, \ldots, d_{k-1}, d_k$ and write this as the polynomial

$$d(x) = d_1 x^{k-1} + \cdots + d_{k-1}x + d_k \qquad (6\text{-}172)$$

of degree $(k - 1)$ or less. The operation $x^{n-k}d(x)$ then generates a polynomial of degree $(n - 1)$ or less. We now take $x^{n-k}d(x)$ and divide this by the $g(x)$ polynomial of degree $(n - k)$:

$$\frac{x^{n-k}d(x)}{g(x)} = q(x) + \frac{r(x)}{g(x)} \qquad (6\text{-}173)$$

The division results in a polynomial $q(x)$ of degree $(k - 1)$ or less and a remainder polynomial $r(x)$. Since $r(x) + r(x) = 0$ under mod-2 addition, it is apparent that the $(n - 1)$-degree polynomial $x^{n-k}d(x) + r(x)$ is divisible by $g(x)$ and must therefore be a code word, from (6-168). Thus

$$c(x) = a(x)g(x) = x^{n-k}d(x) + r(x) \qquad (6\text{-}174)$$

But $x^{n-k}d(x)$ corresponds to a simple left shift by $(n - k)$ units of the data bits. Hence the remainder $r(x)$ must represent the parity check bits. Specifically, then,

$$r(x) = \text{rem} \; \frac{x^{n-k}d(x)}{g(x)} \qquad (6\text{-}175)$$

with "rem" denoting remainder.

As an example, consider the $(7, 3)$ code with generator polynomial $g(x) = x^4 + x^3 + x^2 + 1$ discussed earlier [see Eqs. (6-156), (6-156a), and (6-170)]. Say that the data word is $\mathbf{d} = (0\ 0\ 1)$, or $d(x) = 1$. Then it is left for the reader to show that

$$r(x) = \text{rem} \; \frac{x^4}{x^4 + x^3 + x^2 + 1} = x^3 + x^2 + 1 \qquad (6\text{-}176)$$

The code word is thus

$$\mathbf{c} = (0\ 0\ 1\ 1\ 1\ 0\ 1) \tag{6-177}$$

agreeing with the second code word in the set of eight tabulated earlier. Similarly, say that $\mathbf{d} = (1\ 1\ 1)$ or $d(x) = x^2 + x + 1$. Then

$$r(x) = \operatorname{rem} \frac{x^6 + x^5 + x^4}{x^4 + x^3 + x^2 + 1} = x^2 \tag{6-178}$$

and $\mathbf{c} = (1\ 1\ 1\ 0\ 1\ 0\ 0)$, just the last code word listed in the set of eight.

The polynomial representation of cyclic codes and the calculation of the parity-check bit remainder polynomial $r(x)$ by dividing the left-shifted $d(x)$ by the generator polynomial $g(x)$ suggest various ways of implementing the parity-check bit calculation. These give rise to simple shift-register encoders. One such scheme, using $r = n - k$ shift register stages, is shown in Fig. 6-58. With the switch at the right held in the O position, as shown, the k data bits are shifted in, one at a time. The shift register elements are designated by the 1-bit delay symbol D. As these k bits are moving through the encoder, they are being shifted out onto the output line as well, since they form the first k bits of the n-bit code word. The data bits continue moving through the shift registers until the last (kth) data bit clears the last ($n - k$) register. The mod-2 addition units shown are exclusive-or devices. The gain controls $g_{n-k-1}, g_{n-k-2}, \cdot\ \cdot\ \cdot\ , g_1$ are either present (a 1) or absent (a 0), depending upon whether the corresponding coefficients in the $g(x)$ polynomial given by $g(x) = x^{n-k} + g_1 x^{n-k-1} + \cdot\ \cdot\ \cdot\ + g_{n-k-1}x + 1$ are 1 or 0. At the time the last data bit clears the last register, the $r = n - k$ registers contain the parity-check bits. The switch is now thrown to position P, and the r check bits are shifted out, one at a time, onto the line. In effect, multiplication through the feedback elements shown in Fig. 6-58 provides the division called for by (6-175) in the calculation of the remainder polynomial.

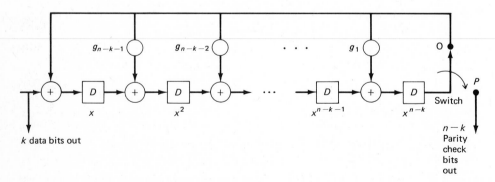

Figure 6-58 Cyclic-code encoder, $r = n - k$ registers.

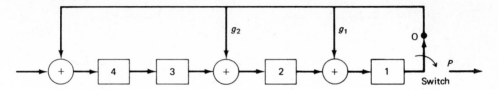

Figure 6-59 Encoder, (7, 3) code, $g(x) = x^4 + x^3 + x^2 + 1$.

The encoder for the (7, 3) code discussed previously as an example appears in Fig. 6-59. Since

$$g(x) = x^4 + x^3 + x^2 + 1 = x^4 + g_1 x^3 + g_2 x^2 + 1 \qquad g_1 = g_2 = 1, g_{n-k-1} = g_3 = 0$$

in this case. To check the operation of the encoder in this case, consider the data sequence 0 0 1, applied in that order to the input of the encoder. We trace the operation of the device by indicating the contents of each of the four registers as the bits shift through (the $*$ represents the last bit being shifted through the system).

contents of →	4	3	2	1	bit interval
	0	—	—	—	↓
	0	0	—	—	
last data bit enters →	1*	0	0	—	
	—	1*	0	0	
	0	0	1*	0	
	0	0	0	1*	
last data bit clears →	1	0	1	1	→ parity bits, shift out

Note that the output bit sequence, 1 1 0 1, in that order, is in fact the desired parity bit sequence for the data sequence 0 0 1.

An alternative encoder implementation is found by noting that since $g(x)$ is a divisor of x^{n+1}, from the theorem quoted earlier, we can always write

$$g(x)h(x) = x^{n+1} \tag{6-179}$$

The $h(x)$ polynomial as defined is called the *parity-check polynomial*. $h(x)$ must be a polynomial of order k, and hence can always be written in the form

$$h(x) = x^k + h_1 x^{k-1} + h_2 x^{k-2} + \cdots + h_{k-1} + 1 \tag{6-180}$$

The coefficients h_j, $1 \le j \le k - 1$, are either 1 or 0, depending on whether the corresponding term in the polynomial appears or not. As an example, for the (7, 3) code discussed above, it is readily shown that $h(x) = (x^7 + 1)/g(x) = x^3 + x^2 + 1$.

It turns out that a cyclic-code encoder may be implemented using a k-stage shift register with mod-2 operations involving the $(k - 1)$ coefficients in the

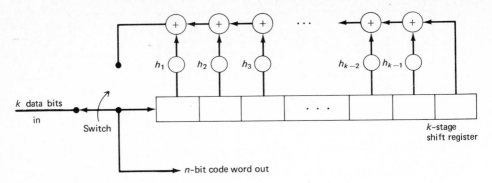

Figure 6-60 Alternate encoder, cyclic codes.

expansion of (6-180). The specific implementation appears in Fig. 6-60. We state this without proof, relying on the (7, 3) code to provide a specific example.

For that code, with $g(x) = x^4 + x^3 + x^2 + 1$ and $h(x) = x^3 + x^2 + 1$, as noted above, the encoder takes the form of Fig. 6-61. The device operates as follows. With the switch in the horizontal position, as shown, the k data bits are shifted in, one at a time, until all k registers are filled. They are simultaneously fed out onto the output line, as shown, to provide the first k bits of the n-bit code word. As the k bits shift through the various steps of the register, the mod-2 (exclusive-or) operations defined in Figs. 6-60 and 6-61 are carried out.

At time $k+$, just after the last register is filled, the switch is thrown to the vertical position. The $r = n - k$ parity bits are then shifted out, onto the output line. The bits fed out represent the successive outputs of the leftmost exclusive-or device in Figs. 6-60 and 6-61.

To demonstrate the operation of this encoder, consider the same example used previously in demonstrating the operation of the encoder of Fig. 6-59. Let the input data sequence again be 0 0 1, in that order. Call the bit outputted at time j, c_j (the output of the mod-2 adder in Fig. 6-61 after the switch is thrown up). The contents of the three registers in Fig. 6-61, as well as c_j, at successive intervals following the throwing of the switch appear as follows:

time j	contents of →	1	2	3	c_j
3+		1	0	0	← switch up
4		1	1	0	1
5		1	1	1	1
6		0	1	1	0
7		1	0	1	1

parity-check bits, in sequence

Note that the parity-check sequence agrees with that found previously using the encoder of Fig. 6-59.

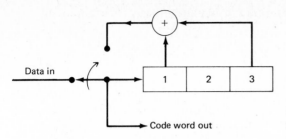

Figure 6-61 Encoder, (7, 3) code, $h(x) = x^3 + x^2 + 1$.

As another example, consider the (15, 11) code appearing in Table 6-3, with generator polynomial $g(x) = x^4 + x + 1$. It is left to the reader to show that

$$h(x) = x^{11} + x^8 + x^7 + x^5 + x^3 + x^2 + x + 1$$

for this code. The two encoders, one corresponding to operations with $g(x)$, the other to operations with $h(x)$, appear in Fig. 6-62. It is left as an exercise to the reader to trace through the operation of these two encoders and to show that they do in fact provide the same parity-check outputs.

Note that the two encoder implementations discussed here operate on the serial data to provide serial output. The implementation of Fig. 6-55, with $r = n - k$ output leads, each connected in a different manner to the k registers, could provide either serial or parallel output. Figure 6-55 may also be looked on as an implementation in which the k information (data) bits are read in sequentially,

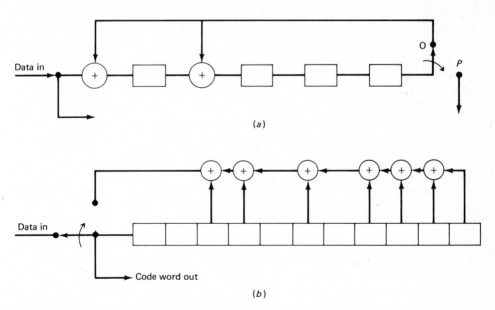

Figure 6-62 Encoders, (15, 11) code. (a) $g(x) = x^4 + x + 1$. (b) $h(x) = x^{11} + x^8 + x^7 + x^5 + x^3 + x^2 + x + 1$.

stored, and then read out, either serially or in parallel, after combining through an appropriate matrixing switch.

Error Detection Using Cyclic Codes

Although most of the effort in coding has gone into the study and development of forward error-correcting codes, error detection is used extensively in modern data communications as well. Error detection normally requires a feedback path over which to signal the transmitting station that its data message has been received in error. The transmitter can then repeat the message. Various versions of such schemes use negative acknowledgments (NAKs) to signal incorrect reception of a message; positive acknowledgments only (ACKs) to signal correct reception of a message, with a message automatically repeated if no ACK has been received at the transmitter within a specified timeout interval; and combinations of both ACKs and NAKs.[31] Such schemes are variously called ARQ (automatic repeat-request), or acknowledgment with retransmission techniques.

All packet-switching data networks and computer networks use error detection with retransmission in the event of an error detected. Examples of such networks include time-shared networks, public data networks, private corporate networks (banking, airline reservation, manufacturing retail, etc.) and newly developing networks that tie a multiplicity of computers together.[32] These networks are made up of a multiplicity of communication lines (or links), connected in various topologies, with error detection carried out commonly on each of the links. Cyclic codes are used most frequently in performing the error-detection function. The blocks of data transmitted in these networks are frequently called *packets*. Packets can range from 500 to 1,000 bits in length, or even longer. It is apparent that error correction for code words of this length can be a formidable task. Error detection, with a fixed number of parity-check bits appended, is easily carried out, however.

We have already noted a simple form of error detection—the use of a single added bit to determine whether an odd number of bits have been received in error. More generally, r parity-check bits enable any burst of errors r bits or less in length to be detected. This is independent of the length of the packet, accounting for the utility of the technique. Since $r \ll k$, most commonly, the $g(x)$ or remainder-type cyclic encoder is generally used in the error-detection application. (Recall that this requires an r-stage shift register.)

Burst errors are, as the name implies, those errors that wipe out some or all of a sequential set of bits. A burst of length b by definition consists of a sequence of bits in which the first and the bth bits are in error, with the $(b - 2)$ bits in between either in error or received correctly. A theorem then states that

[31] M. Schwartz, *Computer Communication Network Design and Analysis*, Prentice-Hall, Englewood Cliffs, N.J., 1977, chap. 14.

[32] *Ibid.*, chaps. 1 and 2.

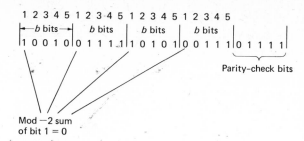

Figure 6-63 Exclusive or parity-check calculation for 5-bit burst error detection.

b parity-check symbols as part of a linear block code are necessary and suffi-cient for detecting all burst errors of length b or less in a block of length n.[33] Note, as stated earlier, that this detection capability is independent of n, which is what makes it so useful in the error detection of data packets, where the number of information bits is generally quite large and usually varies from packet to packet. All that is necessary in this case is to signal to the receiver the beginning and end of the packet so that it can carry out the error checking. Rather than prove the theorem, we simply show by construction how b parity-check bits can be used to detect any error burst of bits or less.

The construction consists of grouping the k information (data) bits into segments b bits long. A simple even-parity check is then applied to all equal-numbered bits in each of the segments. (This is equivalent to taking the mod-2 sum of equal-numbered bits and then adding a parity check bit that equals the mod-2 sum.) The b parity-check bits so formed are then placed at the end of the entire code word. This thus generates a systematic linear code. An example with $b = 5$ appears in Fig. 6-63. As shown in the figure, the mod-2 sum of the number 1 bits is 0. The mod-2 sum of the number 2 bits is similarly found to be 1, etc. The full 5-bit parity-check vector is then 0 1 1 1 1, as shown.

Now note from this construction that only one bit in any burst of b bits or less will affect any parity-check bit, and will thus be detected. This is true whether the burst appears in one of the b-bit segments into which the data sequence has been grouped, or overlaps two such segments. This simple code thus detects any burst of b bits or less.

In addition to detecting a burst of b bits or less, a linear code with $r = b$ parity-check bits will detect a high percentage of longer bursts as well. This is what makes the error-detection property so useful. In particular, a theorem we prove below states that the fraction of bursts of length $b > r$ that remain *undetected* by a *cyclic* (n, k) *code* $(n - k = r)$ is 2^{-r}, if $b > r + 1$, or is $2^{-(r-1)}$, if $b = r + 1$.[34] If the number of parity bits, r, is large enough, almost all errors are detected. As an example, if $r = 16$ bits, all bursts of length 16 bits or less will be detected and the fraction of bursts of length $b > 17$ remaining undetected is $2^{-16} \sim 4 \times 10^{-6}$, an extremely small number.

[33] Peterson and Weldon, *op. cit.*, p. 109 et seq.
[34] *Ibid.*, pp. 228–230.

To prove this theorem, say that a burst of length b starts with the ith bit and ends with the $(i + b - 1)$ bit. Again using powers of x to represent shifts in time, we can define a burst polynomial $b(x)$, such that

$$b(x) = x^i b_1(x) \tag{6-181}$$

and $b_1(x)$ is a polynomial of degree $(b - 1)$:

$$b_1(x) = x^{b-1} + \cdots + 1 \tag{6-182}$$

This is apparent from Fig. 6-64, which shows an error, represented by a 1, appearing at bit i, and an error appearing at bit $(i + b - 1)$, at the end of the burst. The error burst b bits long can have 2^{b-2} possible symbols (1 or 0) between its beginning and end, each constrained, by definition, to be a 1. This corresponds to 2^{b-2} possible burst patterns each b bits long, or 2^{b-2} possible forms of the polynomial $b_1(x)$.

Since the parity-check calculation at the receiver will be carried out by dividing by the generator polynomial $g(x)$, an error will remain undetected if and only if $b_1(x)$ is divisible by $g(x)$. (Why is this so?) Hence the condition for a burst-error pattern to remain undetected is that $b_1(x)$ have $g(x)$ as a factor. In this case $b_1(x)$ must take on the form

$$b_1(x) = g(x)Q(x) \tag{6-183}$$

$Q(x)$ is a polynomial of degree $(b - 1) - r$, since $g(x)$ is of degree r and $b_1(x)$ is of degree $(b - 1)$. There are two possible cases to consider. In the first case the burst length is such that $b - 1 = r$. Then $Q(x) = 1$, and there is only *one* undetected burst pattern. The fraction of undetected bursts is then simply given by

$$\text{fraction of undetected bursts} = \frac{1}{2^{b-2}}$$

$$= 2^{-(r-1)} \qquad b - 1 = r \tag{6-184}$$

In the second case, $b - 1 > r$. Since the polynomial $Q(x)$ is of degree $(b - 1) - r$ and must end with a 1, it has $(b - 1) - r - 1$ terms whose coefficients can be 1 or 0. The number of undetected burst patterns in this case is therefore $2^{(b-1)-r-1}$, and the

$$\text{fraction of undetected bursts} = \frac{2^{b-1-r-1}}{2^{b-2}}$$

$$= 2^{-r} \qquad b - 1 > r \tag{6-185}$$

These two results prove the theorem cited above. They verify the statement made earlier that an error-detecting code with a moderate number of parity-

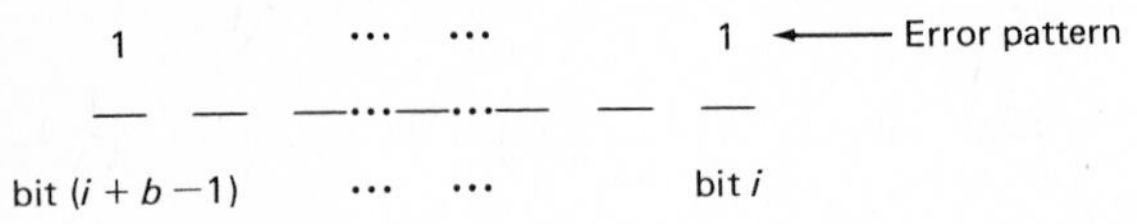

1     $\cdots$ $\cdots$ 1 $\longleftarrow$ Error pattern

bit $(i + b - 1)$ $\cdots$ $\cdots$ bit i

Figure 6-64 Error pattern, burst of length b.

check bits will detect the occurrence of a large majority of error burst patterns. As noted previously, all modern data and computer networks use some form of error detection with requests for repeats of incorrectly received packets. The use of an error-detection scheme reduces the raw probability of error of a packet by the factor 2^{-r}. Typical error-detection procedures have the number of parity-check bits ranging from 8 to 32 bits.[35] The corresponding reduction in error probability is thus 2^{-8} to 2^{-32}. An international standard for link data control, the HDLC (High-Level Data Link Control) protocol, uses a 16-bit cyclic check sequence.[36] The generator polynomial for this standard is prescribed to be

$$g(x) = x^{16} + x^{12} + x^5 + 1 \tag{6-186}$$

The actual calculation of probability of error of a packet or block of data protected by a cyclic check scheme is difficult to carry out because of a lack of detailed knowledge of the mechanisms producing typical bursts. If the underlying mechanism is the ever-present thermal noise and its effect may be modeled as additive white gaussian noise, successive bit errors in a burst are independent of one another. The probability that a packet or block n bits long is then received in error is $[1 - (1 - p)]^n \doteq np$, $np \ll 1$, with p the bit-error probability. It has already been noted, however, that the band-limited telephone channel, which is used most frequently as the backbone communication link for data networks, is not modeled accurately as an AWGN channel. Error bursts on this channel do have memory and do introduce error dependence into successive bits in a data stream. This makes the calculation of error probabilities quite difficult. Nevertheless, tests have shown that the effect of error on blocks of data is to make a block error more likely as the block length n increases and that a reasonable model for block error probability has it proportional to block length.[37] Thus

$$P_b \doteq np \tag{6-187}$$

with p a parameter to be determined from experiment. This simple result agrees with the intuitive feeling that the chance of an error should go up as the block length increases.

If cyclic error checking is now carried out, our simple result says that 2^{-r} of the error events will be undetected as such. (This assumes the burst length $b > r$, or that $r \gg 1$.) The overall block error probability can thus be written, approximately, as

$$P_e \doteq np2^{-r} \tag{6-188}$$

This equation is useful in assessing the performance of various block checking schemes. Some examples follow.

[35] Schwartz, *op. cit.*, chaps. 2, 14.
[36] *Ibid.*, chap. 14.
[37] H. O. Burton and D. D. Sullivan, "Errors and Error Control," *Proc. IEEE*, vol. 60, no. 11, pp. 1293–1301, September 1972.

1. $p = 10^{-5}, n = 500$ bits, $r = 8$ bits. Then

$$P_e \doteq 2 \times 10^{-5}$$

2. $p = 10^{-5}, n = 500$ bits, $r = 16$ bits

$$P_e \doteq 10^{-7}$$

The effect of adding another 8 bits to the parity check scheme is to reduce the error probability by $2^{-8} = 1/256$.

3. $p = 10^{-5}, n = 1,000$ bits, $r = 16$ bits

$$P_e \doteq 2 \times 10^{-7}$$

4. $p = 10^{-5}, n = 1,000$ bits, $r = 32$ bits

$$P_e \doteq 4 \times 10^{-12}$$

It is apparent from these simple calculations that effective error protection is obtained for relatively long data blocks using comparatively few parity-check bits. (Note that the ratio of r/n remains small in all these cases.)

6-9 SUMMARY

In this chapter we attempted to unify discussions in previous chapters on digital communications in the presence of noise.

Starting first with binary signals we asked the question: Are there optimum binary waveshapes and optimum receiver mechanizations or processing techniques to minimize the error probability? To answer this question we applied known techniques of statistical decision theory. Starting first with single-received-signal samples and then generalizing to multiple independent samples drawn from known probability distributions we found that the optimum processing procedure consisted of setting up a likelihood ratio and determining whether this ratio was greater than or smaller than a known constant. Alternatively, the optimum procedure consisted of subdividing the m-dimensional space of the m received signal samples into two disjoint decision regions, one corresponding to one binary signal transmitted, the other to the other signal. In most cases considered, the likelihood ratio could be simplified considerably to provide simple processing procedures for the m samples.

Specializing to the important case of additive white gaussian noise as the disturbance on the channel, we found that the optimum processor consisted of a pair of matched filters, one for each signal transmitted for binary transmission. In the digital version of these filters, each received signal sample is weighted by the corresponding stored transmitted sample, all m weighted samples then being added together. The optimum signal shapes then turned out to be any pair of equal and opposite signals. As shown first in Chap. 5, the error probability then depends solely on the signal-energy-to-noise spectral density.

The analysis of binary signal transmission in the presence of noise was then generalized to M-ary symbol transmission in the presence of additive white gaussian noise. This analysis of the Additive White Gaussian Noise (AWGN) channel led naturally to the use of matched filters, or, equivalently, correlation detection at the receiver. As an application of this material, we showed how one calculates the probability of error for the QAM signal constellations first introduced in Chap. 4. (It must again be noted, however, that these error calculations provide the effect of gaussian noise only. In telephone practice, where such signaling schemes are commonly used for higher-speed data transmission, errors are more often due to other sources.) As expected, the packing of more signal points into a two-dimensional space results in a deteriorating performance.

By going to multidimensional signaling, using M-ary orthogonal signals, however, we found that the error probability could be reduced. The price paid is increased bandwidth and complexity. This reduction in error probability using M-ary orthogonal signaling was then known to be a special case of the Shannon channel-capacity theorem. This theorem demonstrates that it is theoretically possible to drive the probability of error in the presence of additive gaussian noise to as low a value as desired by appropriate encoding and decoding operations at the transmitter and receiver, respectively. This is possible provided the binary transmission rate R in bits per second does not exceed the channel capacity, a number determined by the channel bandwidth, average signal power, and noise spectral density.

We concluded this chapter by examining some methods of detecting and correcting binary errors as a means of further improving the performance of binary systems. The most common schemes for error detection and correction consist of inserting check bits in the binary stream to enable a specified number of errors to be detected and/or corrected. The trick here is to choose coding schemes that approach the error performance predicted by Shannon.

By focusing on codes with a great deal of structure, the cyclic codes studied in this chapter being the most common example, one can develop systematic ways of carrying out the encoding process. Although cyclic codes are used commonly for their forward error-correcting properties, they are also used quite frequently for coding where error detection only is desired. In this latter case it was shown that the error probability is reduced by a factor of 2^{-r}, where r is the number of parity-check bits added.

PROBLEMS

6-1 Consider the received-signal conditional-density functions $f(v|1)$ and $f(v|2)$ shown in Fig. P6-1.

(*a*) Indicate the two decision regions V_1 and V_2 for the following values of P_1, the a priori probability of transmitting a 1: 0.3, 0.5, 0.7.

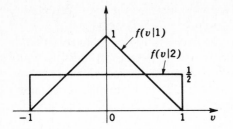

Figure P6-1

(*b*) Calculate the probability of error in each of the three cases of (*a*).

6-2 The probability-density function of a sample of received signal v corresponding to $s_1(t)$ transmitted is given by

$$f(v|1) = k_1 e^{-|v|} \qquad -\infty < v < \infty$$

while the corresponding density function corresponding to signal $s_2(t)$ is

$$f(v|2) = k_2 e^{-2|v|} \qquad -\infty < v < \infty$$

(*a*) Find the appropriate values of k_1 and k_2.

(*b*) The a priori probabilities are $P_1 = \frac{3}{4}$, $P_2 = \frac{1}{4}$. Find the values of v for which we choose s_1, and the values for which we choose s_2.

6-3 An OOK signal is transmitted over a fading medium, and gaussian noise of mean-squared value N added at the receiver. The composite signal plus noise is envelope detected before binary decisions are made. It may then be shown[38] that at the decision point the sampled envelope r has either one of the two density functions

$$f(r|1) = \frac{re^{-r^2/2N_T}}{N_T} \qquad \text{or} \qquad f(r|2) = \frac{re^{-r^2/2N}}{N}$$

corresponding, respectively, to signal plus noise received, and to noise alone (zero signal). Here $N_T = N + S$, with S the mean signal power averaged over the fading ($0 < r < \infty$).

(*a*) Show that in the case of equally likely binary signal transmission the optimum decision test consists of deciding on a 1 ("on") signal transmitted if the envelope r exceeds a threshold $b = \sqrt{2N(1 + N/S) \log_e (1 + S/N)}$.

(*b*) $S/N = 10$. Calculate b and evaluate the overall probability of error. Repeat for $S/N = 1$.

(*c*) m independent samples r_j, $j = 1, \ldots, m$, of r are taken before a decision is made. Show that the optimum test consists of determining whether $\sum_{j=1}^{m} r_j^2$ is greater or less than $2mN(1 + N/S) \log_e (1 + S/N)$.

6-4 A polar binary signal $\pm A$ is received in the presence of additive gaussian noise of variance N. Find the appropriate decision levels if one sample of signal plus noise is taken, for $P_1 = 0.3$, 0.5, and 0.7.

6-5 A polar binary signal of amplitude ± 1 has added to it noise $n(t)$ with density function $f(n) = \frac{3}{32}(4 - n^2)$. Find the minimum probability of error if the a priori probabilities are $P_1 = \frac{2}{5}$ and $P_{-1} = \frac{3}{5}$.

6-6 The received voltage for binary transmission has the two conditional-density functions $f(v|1)$ and $f(v|2)$ shown in Fig. P6-6. Find the optimum decision rule and minimum probability of error in the three cases $P_1 = \frac{1}{2}, \frac{2}{3}, \frac{1}{3}$.

[38] M. Schwartz, W. R. Bennett, and S. Stein, *Communication Systems and Techniques,* McGraw-Hill, New York, 1966.

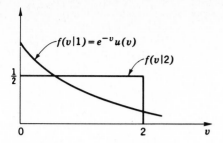

Figure P6-6

6-7 A received waveform $v(t)$ is of the form

$$v(t) = \begin{matrix} +2E \\ \text{or} \\ -E \end{matrix} + n(t)$$

with $n(t)$ zero-mean gaussian noise, of variance N. The a priori signal probabilities are $P(+2E) = \frac{1}{3}$, $P(-E) = \frac{2}{3}$. A single sample of $v(t)$ is taken.

(a) For what values of v should we choose $+2E$ in order to minimize the overall probability of error?

(b) Give an expression for the overall probability of error.

6-8 One of two signals is transmitted over a noisy channel. The conditional-density functions of the received random variable v are

$$f(v|1) = \frac{1}{2\pi} \qquad |v| \le \pi$$

$$= 0 \qquad \text{otherwise}$$

$$f(v|2) = \frac{1}{2\pi}(1 + \cos v) \qquad |v| \le \pi$$

$$= 0 \qquad \text{otherwise}$$

(a) With $P_1 = P_2 = \frac{1}{2}$, find the decision region of v corresponding to minimum probability of error.

(b) Find the minimum probability of error in (a).

(c) Find the values of P_1 and P_2 such that the optimum decision rule says *always* decide on signal 1. What is the probability of error in this case?

6-9 Either one of two noiselike signals is transmitted. The density functions of the received signal are

$$f(v|1) = \frac{e^{-v^2/2\sigma_1^2}}{\sqrt{2\pi\sigma_1^2}} \qquad f(v|2) = \frac{e^{-v^2/2\sigma_2^2}}{\sqrt{2\pi\sigma_2^2}}$$

$P_1 = P_2 = \frac{1}{2}$. Find the optimum receiver processing in the case of one, and then two, independent samples. Show in this latter case that signal 1 is declared present if $v_1^2 + v_2^2 > d$, d a prescribed decision level. What are the two-dimensional regions V_1 and V_2 in this latter case?

6-10 Binary signals with $P_1 = P_2 = \frac{1}{2}$ are received in additive gaussian noise of rms value of 0.5 V. Two independent samples are used at the receiver to decide on signal s_1 or s_2. At the (known) sampling times the two-dimensional signal vectors are, respectively, $s_1 = (+15 \text{ V}, +15 \text{ V})$, and $s_2 = (-7 \text{ V}, -7 \text{ V})$.

(a) Find the decision rule that minimizes the probability of error.

(b) Sketch regions V_1 and V_2 in the two-dimensional plane of the received vector $\mathbf{v}$. Indicate s_1 and s_2 in the same sketch.

(c) Find the minimum probability of error in terms of the complementary error function erfc x defined as $1 - (2/\sqrt{\pi}) \int_0^x e^{-x^2} \, dx$.

6-11 Devise a detection scheme for *three* signals $s_1(t) = +a$, $s_2(t) = 0$, and $s_3(t) = -a$, received in additive gaussian noise of variance N. Assume that the signals are equiprobable. Find the optimum thresholds and the minimum error probability. *Hint:* Symmetry may be used in locating the optimum thresholds.

6-12 *Diversity transmission* (use of more than one channel to improve performance). Consider the system shown in Fig. P6-12. Polar signals $\pm a$ are sent out, in parallel over two channels as shown. Because of differing attenuation (or fading) along the two paths, the signals arrive as $\pm a_1$ and $\pm a_2$, respectively, at each receiver. Gaussian noise of variance N_1 and N_2, respectively, is added at each receiver as shown.

(a) Show that the summed output v is a gaussian variable of expected value $\pm(A_1 a_1 + A_2 a_2)$, and variance $A_1^2 N_1 + A_2^2 N_2$.

(b) Show that the probability of error depends on the effective SNR $(a_1 + Ka_2)^2/(N_1 + K^2 N_2)$, with $K = A_2/A_1$. $K = 0$ ($A_2 = 0$) and $K = \infty$ ($A_1 = 0$) correspond to the single-receiver case. Show that the diversity system provides SNR, and hence error probability, improvement over the single-receiver case.

(c) Show that the optimum choice of the gain ratio K is given by

$$K_{\text{opt}} = (a_2/a_1)(N_1/N_2)$$

This is equivalent to setting $A_1 = (a_1/N_1)g$, $A_2 = (a_2/N_2)g$, g some arbitrary gain constant. The optimum diversity system hence weights each receiver input by the ratio of signal to noise (a_i/N_i) measured at that input. This type of combining is called *maximal-ratio combining*.[39] Show that the effective SNR for this case is given by $a_1^2/N_1 + a_2^2/N_2$, the sum of the two SNR's.

(d) As a special case of diversity combining, assume that the two signal terms a_1 and a_2 represent samples in *time* of a transmitted signal. Let $N_1 = N_2 = N$ be the variance of the noise added, the two noise samples assumed independent. Show that the optimal processing of the two signal-plus-noise samples, in the sense of minimum probability of error, consists of adding them after weighting the first by a_1, the second by a_2. A little thought indicates this is the same as *matched filtering*. Compare this *matched filtering* for time diversity with *maximal-ratio combining* for diversity techniques in general.

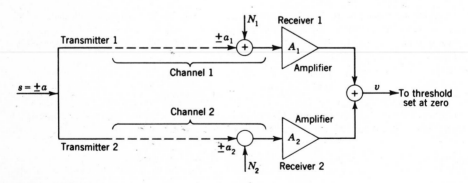

Figure P6-12

[39] *Ibid.*

6-13 A binary message (0 or 1) is to be transmitted in the following manner. The transmitter has two coins labeled C_0 and C_1. Coin C_0 has a probability of a head p_0 and coin C_1 has a probability of a head p_1. If message i ($i = 0$ or 1) is to be transmitted, coin C_i is flipped n times (independent tosses) and the sequence of heads and tails is observed by the receiver.

(*a*) Assuming the two messages are equally likely, find the optimum (minimum probability of error) decision rule for deciding between the two messages. Indicate one simple method for implementing this rule.

(*b*) Set up an expression for the resultant probability of error.

6-14 Consider the binary detection problem where we receive (after processing) the random variable v given by

$$v = s + n$$

where s, the signal, is either equal to 0 or 1 with equal prior probabilities. n is an exponential random variable with density

$$f(n) = \tfrac{1}{2}e^{-|n|} \qquad -\infty < n < \infty$$

Find:

(*a*) The decision rule which leads to the minimum probability of error.

(*b*) The resultant minimum probability of error.

6-15 One of two equally likely signals is transmitted and received in additive gaussian noise of variance 2 V^2. The signals are $s_1(t) = 4$ V $= -s_2(t)$, the binary interval being 1 ms long. Eight equispaced independent samples of the received signal

$$v(t) = \begin{matrix} s_1(t) \\ \text{or} \\ s_2(t) \end{matrix} + n(t)$$

are taken, and comprise an 8-vector **v**.

(*a*) For the following sets of observed data **v** which signal would you decide was sent?

(1) **v** $= (4.5, 0, -1.5, 2, -6, 10, 1, -4)$.

(2) **v** $= (-5, -3, -4, -5, -3, 20, 15, 5)$.

(*b*) What is the probability of error?

6-16 $s_1(t)$ is a triangle 1 msec long with peak voltage of 4 V. $s_2(t) = 0$. Repeat Prob. 6-15 for the following sets of observed data samples:

(*a*) **v** $= (-2, -6, +1, +5, +6, +2, +1, -8)$.

(*b*) **v** $= (0, -2, -4, +5, +6, +2, -4, -2)$.

6-17 One of two equally likely signals $s_1(t)$ and $s_2(t)$ is transmitted, and gaussian noise added. m independent samples of signal plus noise are taken and comprise the vector **v**. Show the decision is to choose signal s_1 if the distance between **v** and s_1 is less than the distance between **v** and s_2. [This verifies Eq. (6-16).]

6-18 Show that the optimum processing in Prob. 6-17 is equally well given by Eq. (6-18) or (6-19).

6-19 Two independent samples for the case of Prob. 6-17 are taken. Draw the decision line dividing region V_1 from V_2. What is the effect of having unequal a priori probabilities?

6-20 Consider the integral

$$\int_{-\infty}^{\infty} \frac{\sin \pi(2Bt - k)}{\pi(2Bt - k)} \cdot \frac{\sin \pi(2Bt - m)}{\pi(2Bt - m)} \, dt$$

[See Eq. (6-41).] Show by a simple change of variables that this may be written as the convolution integral

$$\frac{1}{2B\pi} \int_{-\infty}^{\infty} \frac{\sin (\tau - x)}{(\tau - x)} \frac{\sin x}{x} \, dx$$

with $\tau \equiv (k - m)\pi$. Recalling that the Fourier transform of $(\sin ax)/\pi x$ is 1, $|\omega| \leq a$; 0, $|\omega| > 0$, take Fourier transforms and show that the integral is $(1/2B)\delta_{km}$, where δ_{km} is the Kronecker delta. The $(\sin x)/x$ functions are thus examples of *orthogonal* functions.

6-21 (*a*) As a generalization of Prob. 6-20 above, prove that

$$\int_{-\infty}^{\infty} \frac{\sin \omega_1(t - x)}{\pi(t - x)} \frac{\sin \omega_2 x}{\pi x} \, dx = \frac{\sin \omega_1 t}{\pi t}$$

assuming $\omega_1 \leq \omega_2$. *Hint:* This is already in the form of a convolution integral. Use the approach suggested in Prob. 6-20.

(*b*) As a special case let $\omega_1 = \omega_2 = 2\pi B$; $t = (k - m)/2B$. Show that this gives the same result as in Prob. 6-20.

6-22 *Orthogonal functions.* Consider a set of functions $\phi_i(t)$ with the property

$$\int_a^b \phi_i(t)\phi_j(t) \, dt = \delta_{ij}$$

The $\phi_i(t)$'s then constitute a normalized orthogonal or *orthonormal* set of functions, over the integration range (a,b).

(*a*) We desire to approximate an arbitrary function $f(t)$ by a linear sum of orthonormal functions:

$$f(t) \sim \sum_{j=1}^n b_j\phi_j(t) \equiv f_n(t)$$

Show that the mean-squared error between $f(t)$ and $f_n(t)$,

$$\epsilon^2 \equiv \int_a^b [f(t) - f_n(t)]^2 \, dt$$

is minimized by choosing

$$b_j = \int_a^b f(t)\phi_j(t) \, dt$$

Use the symbol a_j to denote this special case of b_j. $\sum_{j=1}^n a_j\phi_j(t)$ then approximates $f(t)$ best in a least mean-squared sense. The a_j's are sometimes called the generalized Fourier coefficients.

(*b*) The orthogonal set $\phi_j(t)$ is said to be *complete* if, using the Fourier coefficients a_j, that is,

$$f_n(t) = \sum_{j=1}^n a_j\phi_j(t)$$

$\epsilon^2 \to 0$, as $n \to \infty$. Show that for this case

$$\int_a^b f^2(t) \, dt = \sum_{j=1}^\infty a_j^2$$

This is a generalized form of Parseval's theorem, first met in Chap. 2. That is, the energy in the signal equals the energy in the orthogonal functions. We then write

$$f(t) = \sum_{j=1}^\infty a_j\phi_j(t)$$

where the equality is meant in this sense of equal energy.

(*c*) Let the interval (a,b) be $(-T/2, +T/2)$. Find the normalized set of sines and cosines that are orthogonal over this interval.

(*d*) According to Prob. 6-20 [and Eq. (6-41)], the $(\sin x)/x$ functions are orthogonal over the interval $(-\infty, \infty)$. Normalize these functions and show how the a_j coefficients are related to the sampled values $f(j/2B)$ of a function $f(t)$ expanded in terms of the $(\sin x)/x$ functions [see Eq. (6-40)]. Show that the Parseval theorem in this case is given by

$$\int_{-\infty}^{\infty} f^2(t)\, dt = \frac{1}{2B} \sum_{j=-\infty}^{\infty} f^2\left(\frac{j}{2B}\right)$$

(*e*) Let

$$f_1(t) = \sum_{j=1}^{\infty} a_j \phi_j(t)$$

$$a_j = \int_{a}^{b} f_1(t)\phi_j(t)\, dt$$

$$f_2(t) = \sum_{j=1}^{\infty} b_j \phi_j(t)$$

$$b_j = \int_{a}^{b} f_2(t)\phi_j(t)\, dt$$

Show that

$$\int_{a}^{b} f_1(t)f_2(t)\, dt = \sum_{j=1}^{\infty} a_j b_j$$

Use this to verify Eq. (6-43) as a special case.

6-23 Equally likely polar signals of amplitude $\pm A$ are received in the presence of additive gaussian noise of variance σ^2. $A/\sigma = 1$.

(*a*) One sample of the received signal plus noise is taken and optimumly processed. Show that the error probability $P_e = 0.159$. Show that taking three independent samples and optimumly processing these reduces P_e to 0.0418.

(*b*) A suboptimum processor is used for the three samples of (*a*): each sample is independently checked for polarity. If two or three samples are positive, $+A$ is declared present. Otherwise, $-A$ is declared present. Show $P_e = 0.068$. Compare this procedure and its performance with that of the optimum processor in (*a*).

Note: In Probs. 6-24 to 6-32, appropriate signal shaping factors normally included are not shown explicitly.

6-24 A binary source outputs 7,200 bits/s. A modem is used to convert the binary symbols to a format capable of being transmitted over a telephone channel of 2,400 Hz bandwidth.

(*a*) Determine which of the following signal sets is suitable for this purpose.

(1) The QAM set of Fig. P6-24.

$$\phi_1 = \sqrt{\frac{2}{T}} \cos \omega_0 t \qquad \phi_2 = \sqrt{\frac{2}{T}} \sin \omega_0 t$$

(2) The set is the same as that of (1), but

$$\phi_1 = \sqrt{\frac{2}{T}} \cos \omega_1 t \qquad \phi_2 = \sqrt{\frac{2}{T}} \cos \omega_2 t$$

The spacing $\Delta f = f_2 - f_1 = 0.7/T$. (This is the minimum spacing possible to keep ϕ_1 and ϕ_2 orthogonal.)

(3) The *M*-ary AM set of Fig. P6-24. $\phi_1 = \sqrt{2/T} \cos \omega_0 t$. All points are spaced d apart.

(4) The set of Fig. P6-24. Here $\phi_1 = \sqrt{2/T} \cos \omega_1 t$, $\phi_2 = \sqrt{2/T}\cos \omega_2 t$, $\phi_3 = \sqrt{2/T}\cos \omega_3 t$, and the eight signal points appear at the vertices of a cube: $(d/2, d/2, -d/2)$, $(d/2, -d/2, -d/2)$, etc.

(*b*) Write specific expressions, as a function of time, for each of the signals in those signal sets deemed suitable.

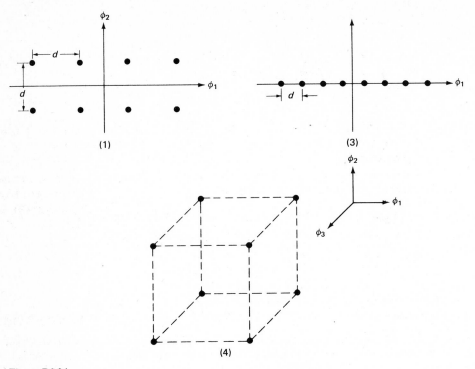

(1)

(3)

(4)

Figure P6-24

6-25 Two equally likely polar signal vectors s_1 and s_2 appear as in Fig. P6-25.

$$v = s_i + n \qquad i = 1 \text{ or } 2$$

n is zero-mean white gaussian. We find $P_e = 0.01$. *Note:* Why are v and n written as scalars here?

(*a*) Find P_e for each of the six cases shown in the figure and compare with that for the polar signals. Compare powers as well. All signals are equally likely. The noise is the same in all cases. *Hint:* In (5) and (6), find P_c first. Show that $P_e = 0.013$ in (5) and 0.015 in (6).

(*b*) What is the effect on P_e and the power in all cases if all signals are shifted *up* by 2 units? *Down* by 2 units?

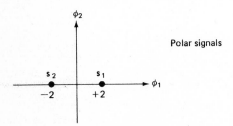

Polar signals

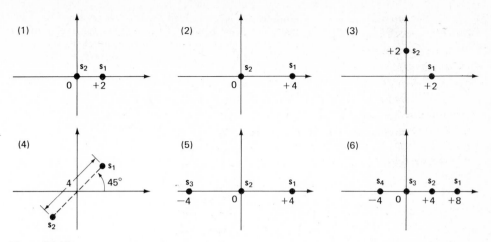

Figure P6-25

6-26 One of four equally likely signals is transmitted. The vector representation of each is on a circle of radius $\sqrt{E}$, as shown in Fig. P6-26.

(*a*) Indicate the optimum decision regions when white gaussian noise is added.

(*b*) The noise components each have variance $n_0/2$. Show that P_e is given approximately by (6-98).

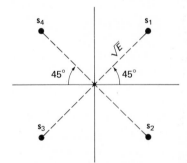

Figure P6-26

6-27 A data terminal outputs information at the rate of 1,200 bits/s. Three successive bits are stored and used to generate one of eight possible waveshapes.

(*a*) Find the minimum bandwidth required to transmit each of the following signal sets:

(1) The signal vectors are at the corners of the cube shown in Fig. P6-27. $\phi_1 = \sqrt{2/T}\cos\omega_1 t$, $\phi_2 = \sqrt{2/T}\cos\omega_2 t$, $\phi_3 = \sqrt{2/T}\sin\omega_2 t$. Both signal shaping and the appropriate choice of frequencies are used to minimize the bandwidth.

(2) Same as (1), but $\phi_3 = \sqrt{2/T}\cos\omega_3 t$, with all three frequencies chosen to minimize the bandwidth.

(3) See the figure. $\phi_1 = \sqrt{2/T}\cos\omega_0 t$, $\phi_2 = \sqrt{2/T}\sin\omega_0 t$.

(4) A cosinusoidal pulse $T/3$ seconds long of amplitude $+a$ or $-a$ is transmitted in each of three adjacent time slots. Then $s_i(t) = \pm a\phi_1(t) \pm a\phi_2(t) \pm a\phi_3(t)$. Each orthogonal function is given by $\sqrt{6/T}\cos\omega_0 t$, but defined in the ranges $0 \le t < T/3$, $T/3 \le t < 2T/3$, and $2T/3 \le t < T$, respectively.

(b) White gaussian noise of spectral density $n_0/2$ is added during transmission. The eight signals are equally likely to be transmitted. Find the probability of error of each signal set in terms of E/n_0. *Note:* For signal set (3), assume that a point at the origin is present in calculating P_e.

(c) Can you find a set of signals equivalent to set (1) that yields the same P_e but with less average energy?

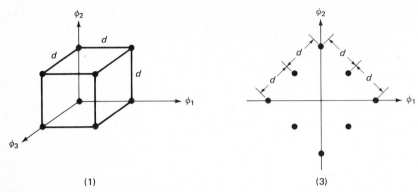

(1)   (3)

Figure P6-27

6-28 A data source outputs digits at a rate of R bits/s. These are to be transmitted over a telephone line using a modem. The modem outputs one of the following eight signals every T_0 seconds:

$$s_i(t) = \begin{array}{ll} a\phi_1(t) + a\phi_2(t) & -a\phi_1(t) + a\phi_2(t) \\ 3a\phi_1(t) + a\phi_2(t) & -3a\phi_1(t) + a\phi_2(t) \\ a\phi_1(t) - a\phi_2(t) & -a\phi_1(t) - a\phi_2(t) \\ 3a\phi_1(t) - a\phi_2(t) & -3a\phi_1(t) - a\phi_2(t) \end{array}$$

Here $$\phi_1(t) = \sqrt{\frac{2}{T_0}} \cos \omega_c t \qquad \phi_2(t) = \sqrt{\frac{2}{T_0}} \sin \omega_c t \qquad 0 \leq t \leq T.$$

Shaping parameters, not shown here, are also used to pack $s_i(t)$ into the minimum possible bandwidth. The eight signals are equally likely to be transmitted.

(a) What are the telephone channel bandwidth B, T_0, and R if the modem output rate is 7,200 bits/s?

(b) White gaussian noise with spectral density $n_0/2$ is added during transmission. Indicate the optimum receiver structure schematically. Show the optimum decision regions. Find an expression for the probability of error. Show this may be written in terms of E/n_0, with E the average signal energy in the T_0-second interval.

6-29 A transmitter sends one of four possible waveforms (with equal probability) over an additive white gaussian noise channel with spectral density $n_0/2$. Each waveform is constructed as a combination of three possible "tones":

$$\phi_i(t) = E \cos \omega_i t \qquad \omega_i = \frac{2\pi i}{T} \qquad i = 1, 2, 3$$

The signal duration is T seconds. The four signals are given by

$$s_1(t) = \phi_1 \qquad s_2(t) = \phi_2 \qquad s_3(t) = \phi_3 \qquad s_4(t) = \phi_1 + \phi_2 + \phi_3$$

(a) Do the ϕ_i's constitute an orthogonal set?

(b) Sketch the form of an optimal receiver for this system.

(c) What geometric figure do the s_i's form in the signal space?

(d) Find an equivalent set of signals which would yield the same probability of error, but with less average signal energy.

6-30 (a) Find P_e for each of the signal sets in Fig. P6-30. Each of the four signals is equally likely to be transmitted and white gaussian noise of spectral density $n_0/2$ is added during transmission. Compare the average energy required for transmission as well.

(b) For the three signal sets above, $\phi_1 = \sqrt{2/T} \cos \omega_0 t$, $\phi_2 = \sqrt{2/T} \sin \omega_0 t$. Express each of the signal sets in the form $s_i(t) = a_i \sqrt{2/T} \cos \omega_0 t + b_i \sqrt{2/T} \sin \omega_0 t$. For sets 2 and 3, express each in the form $s_i(t) = c_i \sqrt{2/T} \cos (\omega_0 t + \theta)$.

(c) The signals in (b) are to be transmitted over a telephone channel of 2,400 Hz bandwidth. For this purpose $T = 1/1,200$ s is used as the transmission interval. What is the bit rate over the channel? With appropriate signal shaping $T = 2/3(1,200)$ s might be used as the interval. What bit rate would then be transmitted?

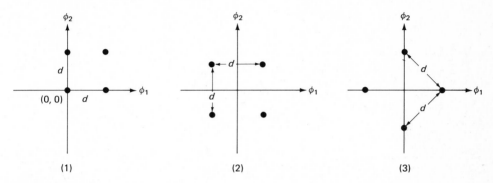

(1)   (2)   (3)

Figure P6-30

6-31 $N = 3$ orthogonal waveshapes are used to transmit $M = 2^3 = 8$ equally likely signals. Two possible sets of such signals appear in Figs. P6-24(4) and P6-27(1).

(a) Find P_e for those two sets in white gaussian noise of spectral density $n_0/2$. *Hint:* Show that P_e is of the form r^3, while in Prob. 6-30 it is given by r^2. Can you extend this to $M = 2^N$ signals located at the vertices of a hypercube in N dimensions?

(b) Which of the two signals requires less energy to transmit? If the spacing d is the same in this problem and Prob. 6-30, which signal sets in the two problems have a smaller probability of error?

(c) $T = 1/1,200$ s in this problem. Let $\phi_1 = \sqrt{2/T} \cos \omega_1 t$, $\phi_2 = \sqrt{2/T} \sin \omega_1 t$, $\phi_3 = \sqrt{2/T} \cos \omega_2 t$, $f_2 - f_1 = 1/T$. What bandwidth is required for transmission? What is the bit rate?

(d) $P_e = 10^{-5}$ is desired. Find E/n_0 required for the eight signals in Fig. P6-24(4). How does this compare to E/n_0 for PSK transmission, for the same $P_e = 10^{-5}$? *Hint:* With $P_e \ll 1$, as here, $r \doteq 1 - \epsilon$. Then $P_e \doteq 3\epsilon$. (Why?) Use the approximation

$$\int_a^\infty \frac{e^{-x^2} \, dx}{\sqrt{\pi}} \doteq \frac{e^{-a^2}}{2a \sqrt{\pi}} \qquad a > 1$$

and show that

$$P_e \doteq \frac{3}{\sqrt{2}} \frac{e^{-2E/3n_0}}{\sqrt{2E/3n_0}}$$

Solve for E/n_0.

6-32 A binary stream is to be encoded into one of four signals. Various signal sets are available for this purpose. Compare the following signal sets in terms of probability of error and bandwidth required. The signal energy is the same in all cases. The encoded signal is transmitted over a white gaussian noise channel. T is the transmission interval.

(1) $\pm \sqrt{\dfrac{2}{T}} \, a \cos \omega_0 t \pm \sqrt{\dfrac{2}{T}} \, a \sin \omega_0 t$

(2) $\pm \sqrt{\dfrac{2}{T}} \, a \sin \omega_0 t$

$\sqrt{\dfrac{2}{T}} \, a \cos \omega_0 t \pm \sqrt{\dfrac{2}{T}} \, a \sin \omega_0 t$

(3) $\sqrt{\dfrac{2}{T}} \, \sqrt{2} \, a \cos (\omega_0 t + \theta) \qquad \theta = \pm \dfrac{\pi}{4}, \pm \dfrac{3\pi}{4}$

(4) $\pm \sqrt{\dfrac{2}{T}} \, a \cos \omega_0 t \pm \sqrt{\dfrac{2}{T}} \, a \cos \left(\omega_0 + \dfrac{2\pi}{T}\right) t$

6-33 Refer to Prob. 6-24. Find and compare P_e for sets (1), (2), and (3), if $E/n_0 = 24$.

6-34 S/n_0 at the receiver of a digital communications system is 10^3.

(a) Show the maximum bit rate is 50 bits/s if binary FSK is used and $P_e = 10^{-5}$ is desired.

(b) It is desired to transmit at a rate of 250 bits/s with $P_e = 10^{-5}$. Indicate by block diagram and with numbers how this may be done. Approximately what bandwidth is required?

(c) What is the *maximum* channel capacity in this case?

6-35 (a) Communications between a space vehicle and the earth is under investigation. A probability of error $P_e \le 10^{-5}$ is to be maintained. Compare the binary rates R (bits/s) and the approximate transmission bandwidths required for the following modes of transmission: (1) PSK; (2) binary FSK; (3) one of 32 orthogonal signals. $S/n_0 = 800$, with S the received signal power and $n_0/2$ the gaussian noise spectral density.

(b) Repeat if the distance between space vehicle and earth doubles.

(c) It is desired to transmit at a rate of 320 bits/s, with $P_e = 10^{-5}$. $S/n_0 = 800$. Indicate if this is possible and, if so, how this may be done. What is the *maximum* possible rate, with the probability of error reduced as low as desired?

6-36 A digital communications channel has $S/n_0 = 10$ at the receiver. With PSK transmission and $P_e \le 10^{-5}$ required, at most 1-bits/s transmission rate is possible.

(a) What is the *maximum* bit rate allowed over this channel if arbitrarily wide bandwidths are allowed and $P_e \to 0$?

(b) A bandwidth of 1 kHz is available. $P_e = 10^{-5}$ is required. Indicate the maximum bit rate attainable using M orthogonal signals. What value of M is required? How many bits are encoded per M-ary signal transmitted?

6-37 (a) Plot the capacity in bits per second versus bandwidth W of a channel with additive band-limited gaussian noise of spectral density $n_0/2$ and average power S.

(b) $S/n_0 = 100$. Find the maximum rate of transmission of binary information if PSK is used and a maximum probability of error of 10^{-5} is to be maintained. Repeat for $P_e = 10^{-4}$. What is the maximum rate in both cases if FSK transmission is used?

(c) $S/n_0 = 100$ again. The channel bandwidth is $W = 10$ Hz. If the binary digits of (b) may be encoded using as complicated a digital scheme as desired, what is the maximum rate of transmission in bits per second with a probability of error as small as desired? Compare with (b). What is the SNR in this case?

(d) $S/n_0 = 100$. The channel bandwidth may be made as large as necessary. Repeat (c) and again compare with (b).

6-38 Digital communications for a deep-space probe (10^8 miles from earth) is to be investigated. Assume 500-MHz transmission with space vehicle and earth antenna gains of 10 dB and 40 dB, respectively. The transmitter power is limited to 10 W. $T = 100$ K. Bit-error probability is to be less than 10^{-5}.

(a) Find the maximum rate of binary transmission if a PSK system with synchronous detection is considered.

(b) The binary data are to be encoded into one of 64 orthogonal signals. Find the maximum binary rate in this case. What is the encoding or storage time required at both transmitter and receiver?

(*c*) What is the maximum possible transmission rate if an arbitrarily large bandwidth and complex encoding are allowed, and the probability of error is to be made as small as desired? Compare (*a*), (*b*), (*c*).

(*d*) The antenna sizes are fixed. Repeat the problem for two different frequencies: 2,000 MHz and 1,000 MHz.

(*e*) Repeat if a maser receiver providing an overall temperature of 30 K is used on earth.

6-39 Refer to the discussion of the Mariner 10 deep-space mission in Sec. 5-14. Find the allowable bit transmission rate if $M = 32$ orthogonal signals were to be used. What is the Shannon capacity ($W \to \infty$) for this channel?

6-40 Repeat Prob. 6-39 for the Voyager mission (see Prob. 5-78 for the necessary parameters).

6-41 Consider the binary code with the P matrix

$$P = \begin{bmatrix} 1 & 0 & 1 \\ 0 & 1 & 1 \\ 1 & 1 & 1 \end{bmatrix}$$

(*a*) Is the word (101010) a code word?

(*b*) A code word is of the form (X1100). Is X a 0 or a 1?

(*c*) Suppose the code word (001111) is transmitted, but (001101) is received. What is the resultant syndrome? Where would this syndrome indicate an error had occurred?

(*d*) How many code words are in this code? List them.

(*e*) What is the smallest number of errors that could change one code word into another code word? Why?

6-42 Consider a binary communication system consisting of two links as shown in Fig. P6-42. The noise in *each* channel is such that (1) errors occur independently; and (2) the probability that a 1 is received when a 0 is transmitted is p. The probability that a 0 is received when a 1 is transmitted is also p.

(*a*) Find the following four probabilities for the entire system:

A 0 is received when a 0 is transmitted
A 1 is received when a 0 is transmitted
A 0 is received when a 1 is transmitted
A 1 is received when a 1 is transmitted

(*b*) Assume that a simple coding scheme is used such that a 0 is transmitted as three successive 0's and a 1 as three successive 1's. At the detector, the following (majority) decision rule is used:

$$\begin{cases} \text{Decide 0 if 000, 001, 010, or 100 is received} \\ \text{Decide 1 if 111, 110, 101, or 011 is received} \end{cases}$$

If a 0 and 1 are equally likely, what is the probability of deciding incorrectly? Evaluate for $p = \tfrac{1}{3}$.

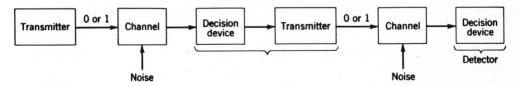

Figure P6-42

6-43 A binary message consists of words which are 5 bits long. The message words are to be encoded using a single-error-correcting code. The first 5 bits of each code word must be the message bits d_1, d_2, d_3, d_4, d_5, while the remaining bits are check bits.

(a) What is the minimum number of check bits? What are the P and G matrices?
(b) Construct an appropriate H matrix for this code.
(c) Find the syndrome at the receiver if there is an error in d_5.
(d) How does this code respond to double errors?

6-44 Show, by construction of an appropriate H matrix, that a (6, 3) code is single-error-correcting. Show that this agrees as well with the single-error-correcting inequality of Eq. (6-143). Show that a (6, 4) code cannot be designed to provide a single-error-correcting capability.

6-45 Consider a single-error-correcting code for 11 message bits.
(a) How many check bits are required?
(b) Find a suitable G matrix.
(c) Find the syndrome if the single error occurs in the seventh position.

6-46 A code consists of three message digits d_1, d_2, d_3 and three check digits c_4, c_5, c_6. The transmitted sequence is $d_1d_2d_3c_4c_5c_6$. At the transmitter the check digits are formed from the following equations:

$$c_4 = d_1 \oplus d_3$$
$$c_5 = d_1 \oplus d_2$$
$$c_6 = d_1 \oplus d_2 + d_3$$

(a) For the message $d_1 = 0$, $d_2 = 1$, $d_3 = 1$, find the transmitted sequence.
(b) Write down the G matrix.
(c) Will this code correct single errors? Why?
(d) Assume that the sequence 011100 is received and that no more than one error has occurred. Decode this sequence: find the location of the error and the transmitted message d_1, d_2, d_3.

6-47 Consider a binary code with three message digits and three check digits in each code word. The code word is of the form

$$d_1d_2d_3c_4c_5c_6$$

where the d_i's are the message digits and the c_i's are the check digits. Assume that the check digits are computed from the set of equations

$$c_4 = d_1 \oplus d_2 \oplus d_3$$
$$c_5 = d_1 \oplus d_3$$
$$c_6 = d_2 \oplus d_2$$

(a) How many code words are there in the code?
(b) Find the code word that begins 110. . . . .
(c) Suppose that the received word is 010111. Decode to the closest code word (i.e., the code word that differs from the received word in the fewest positions).

6-48 Calculate the probability of error for the following binary codes and compare:

(n, k)	t
(7, 4)	1
(15, 11)	1
(15, 7)	2
(15, 5)	3
(31, 26)	1
(31, 21)	2
(31, 16)	3
(31, 11)	5
(31, 6)	7

where n is the total number of digits (message digits plus check digits), k is the number of check digits, and t is the error-correcting capability of the code; that is, t or fewer errors can be corrected.

Assume that the probability of bit error when no coding is used is 10^{-5}. Adjust the duration of the binary digits in each code so that the transmission rate (message digits per second) is constant. *Hint:*

$$P_e = \binom{n}{t+1} p^{t+1}(1-p)^{n-(t+1)} + \binom{n}{t+2} p^{t+2}(1-p)^{n-(t+2)} + \cdots + p^n$$

Why? For very small p, as true here, $P_e \doteq \binom{n}{t+1} p^{t+1}$.

6-49 A (7, 4) code is to be used.

(a) Using the Hamming bound, find the potential random-error-correcting capability of such a code.

(b) The 16 code words are 0000000, 111111, and cyclic variations of 0001011 and 0011101. List the 16 code words and show that the minimum distance d agrees with (a).

(c) Show that these are generated by the G matrix

$$G = \begin{bmatrix} 1 & 0 & 0 & 0 & 1 & 0 & 1 \\ 0 & 1 & 0 & 0 & 1 & 1 & 1 \\ 0 & 0 & 1 & 0 & 1 & 1 & 0 \\ 0 & 0 & 0 & 1 & 0 & 1 & 1 \end{bmatrix}$$

(d) Find the H^T matrix and show, for a few of the code words of (b), that $cH^T = 0$.

6-50 Use the Hamming bound of Eq. (6-144) to

(a) Show that a (15, 11) code may correct one error.

(b) Show that (10, 4) and (11, 4) codes may correct two errors.

(c) Find the error-correcting capability of the following $(15, k)$ codes: $k = 14, 11, 10, 7, 6, 5, 4,$ 2, 1.

6-51 Verify the potential error-correcting capabilities of the codes listed in Table 6-1.

6-52 (a) Show that $x^3 + x + 1$, $x + 1$ and $x^3 + x^2 + 1$ are divisors of $x^7 + 1$. *Note:* Additions are all mod-2. $x^i \oplus x^i = 0$.

(b) Show that $x^2 + 1$ and $x^4 + x^2 + 1$ are divisors of $x^6 + 1$.

(c) Consider the polynomial $g(x) = x^4 + x + 1$. Divide this into $x^{15} + 1$ to show that

$$h(x) = \frac{x^{15} + 1}{g(x)} = x^{11} + x^8 + x^7 + x^5 + x^3 + x^2 + x + 1$$

6-53 In Prob. 6-52 it is shown $x^3 + x + 1$ is a divisor of $x^7 + 1$. Use this as a generating polynomial $g(x)$.

(a) What (n, k) code will this give rise to?

(b) Using $g(x) = x^3 + x + 1$, find the G matrix. Compare this matrix with that given in Prob. 6-49c.

6-54 Repeat Prob. 6-53 for a (6, 2) code. *Hint:* Use Prob. 6-52b.

(a) Find the G matrix.

(b) From the G matrix generate all 2^k code words. What is the minimum (Hamming) distance d? What is the error-correcting capability of this code?

6-55 Refer to Prob. 6-53. Generate a new matrix G' by writing $g(x)$, $xg(x)$, $x^2g(x)$, etc., for successive rows. Use this matrix to generate the 16 code words. Compare them with the ones found in Prob. 6-49. Show that G' found here can be put into the form of G by adding rows appropriately.

6-56 Take any three 4-bit information vectors $\mathbf{d}$. Write these as $d(x)$. These are to be encoded into the appropriate (7, 4) code words of Prob. 6-53. For this purpose use $g(x) = x^3 + x + 1$. Calculate

$$r(x) = \text{rem } \frac{x^{n-k}d(x)}{g(x)} = \text{rem } \frac{x^3 d(x)}{g(x)}$$

in this case. Show that $x^{n-k}d(x) + r(x) = c(x)$, the appropriate code words for this case. The $r(x)$ polynomial thus provides the parity-check bits for these three information vectors. The three code words should of course agree with those found using the G matrix in Prob. 6-49.

6-57 Consider the $(7, 3)$ code with generator polynomial $g(x) = x^4 + x^3 + x^2 + 1$. Find the parity-check bits for each of the seven nonzero data sequences by calculating the remainder polynomial $r(x)$ for each.

6-58 Consider the $(15, 11)$ cyclic code listed in Table 6-3.

(a) Show that the generator polynomial $g(x)$ generates the G matrix of Eq. (6-127).

(b) Using the G matrix, find the 15-bit code word for the information vector

$$\mathbf{d} = [10001001010]$$

(c) Check the result of (b) first finding the data polynomial $d(x)$, and then the remainder polynomial $r(x)$.

6-59 Consider the $(7, 4)$ code with $g(x) = x^3 + x + 1$, of Prob. 6-53. Find $h(x)$ for this code. Implement the two encoders discussed in the text. Pick any 4-bit data sequence. Calculate the parity-check bits using both encoders. Compare with the results of the G matrix in Prob. 6-49 or Prob. 6-56.

6-60 Consider the $(15, 11)$ cyclic code with generator polynomial $g(x) = x^4 + x + 1$ discussed in the text.

(a) Find the parity-check polynomial $h(x)$.

(b) Verify that the shift register devices of Fig. 6-62 do represent encoders for this code. Take any 11-bit data word, find the parity-check bits obtained with the two encoders, and verify that they are the same.

(c) Find the G matrix for this code and show that it is the G matrix given by (6-127). Use the G (or P) matrix to calculate the parity-check bits for the same data word used in part (b). It should agree with the parity bits found there.

6-61 Refer to the two encoder implementations for the $(15, 11)$ cyclic code shown in Fig. 6-62. Apply the 11-bit data sequence $\mathbf{d} = [10001001010]$ to the input of each encoder, trace through the calculation of the 4 parity-check bits for each, and show that they agree. Show that they are the same as would be calculated using the G matrix and polynomial remainder calculations.

6-62 Consider a code word made up of ten 8-bit characters. The last (parity) character has as each of its bits the exclusive-or of all 9 previous bits in the same time slot. Using the block error probability model described in the text, calculate P_e if $p = 10^{-5}$. Show, by example, that bursts of 8 bits or less will be detected. Take any burst pattern at random, of more than 8 bits, and test to see if it is detected. Pick any code word at random for this purpose.

6-63 Refer to Prob. 6-62. The performance of the exclusive-or error detection scheme is to be determined through simulation. For this purpose write a computer program that provides the exclusive-or parity check of all information bits in the same time slot. Randomly generate information sequences nine or more characters long, perturb these with random burst patterns of varying length, including patterns greater than 8 bits in length. Run your program enough times to verify the error-detecting capability of the code.

6-64 A code word consists of eight 8-bit information characters and parity-check bits generated by the polynomial

$$g(x) = x^{16} + x^{12} + x^5 + 1$$

(a) What burst lengths is this code guaranteed to detect? For longer burst lengths, what is the fraction remaining undetected?

(b) Using the error model described in the text, with block error probability proportional to block length, determine the block error probability if $p = 10^{-4}$.

(c) Repeat (a) and (b) if the information sequence is 1,000 bits long. How could one improve the detection capability of this system?

6-65 Refer to the generator polynomial of Eq. (6-186) used in data link control. (See also Prob. 6-64.) Pick any information sequence two characters long. Use the remainder theorem to calculate the parity-check bits. Repeat for an information sequence four characters long. In both cases show the complete code word to be transmitted.

6-66 A cyclic (7, 3) code has as its generator polynomial $g(x) = x^4 + x^2 + x + 1$.

(a) Find the G matrix for this code.

(b) Find all possible code words. What is the minimum distance of this code? Show that it is single-error-correcting.

(c) Select any 3-bit data sequence. Use the remainder theorem to find the parity-check bits, and compare with the corresponding code word in (b).

(d) Data bits are transmitted over a satellite link using PSK transmission. Matched filtering is used at the receiver. $E/n_0 = 9.6$ *without* encoding. The corresponding *bit* error is then

$$p = \frac{1}{2} \text{ erfc } \sqrt{\frac{E}{n_0}} \doteq \frac{e^{-E/n_0}}{2\sqrt{\pi E/n_0}} = 10^{-5}$$

Calculate P_e for an uncoded 3-bit sequence and compare with P_e for the coded 7-bit sequence. Make all reasonable approximations. *Hint:* You showed in (b) that the code was single-error-correcting.

FUNDAMENTALS OF PROBABILITY[1]

A-1 INTRODUCTION

The need for statistical analysis arises in many branches of science. In many cases, measurements of various parameters may be deterministic in nature. They follow classical laws, and results may presumably be predicted exactly if all pertinent information is known. If a great many variables are involved, however, it frequently becomes extremely difficult to analyze the problem exactly. Instead, various average properties may be defined.

For example, if we are interested in investigating various properties of a gas trapped in a container, we could conceivably do this analytically by following the path of each molecule as it moves along, colliding with its neighbors, with the walls of the container, etc. Application of the simple laws of mechanics would presumably tell us all we want to know about each gas molecule, and therefore about the gas as a whole. But with millions of molecules to treat simultaneously, each moving with a possibly different velocity from the next, this theoretically possible calculation becomes practically impossible. Instead, because we deal with large numbers of molecules, we can determine average values for the velocity, force, momentum of the molecules, etc., and from these determine such average properties of the gas as temperature and pressure. (We shall define the concept of average more precisely later on, but we rely on our intuition now.)

[1] A. Papoulis, *Probability, Random Variables, and Stochastic Processes,* McGraw-Hill, New York, 1965; H. Cramer, *Mathematical Methods of Statistics,* Princeton, Princeton, N.J., 1946; E. Parzen, *Modern Probability Theory and Its Applications,* Wiley, New York, 1960; W. Feller, *An Introduction to Probability Theory and Its Applications,* 3rd ed., Wiley, New York, 1968.

The crucial point here is the phrase "large numbers." Whenever we deal with large numbers of variables (whether the millions of gas molecules in a small container, the millions of possible messages in a particular language to be transmitted, or the millions of different pictures that could possibly be seen on a small TV screen), we can talk about the average properties, obtained by applying statistical concepts to the variables in question.

These large numbers we deal with and the statistics based on them can be generated in different ways. For example, we could conceivably set up 1,000 blackboards and ask 1,000 persons to start simultaneously writing anything that came into their minds. If at a given instant we scanned the particular letter being written by each, we could determine the relative frequency of occurrence of letters in the language (say, English) being used. (This assumes that the persons were selected randomly, have no connection with one another, were not briefed beforehand, etc.) Alternatively, we could watch one typical person writing and perhaps pick out the first letter in each new paragraph as he writes. This, too, could be used to determine the relative frequency of occurrence of letters in the language. But we would need many paragraphs of writing here (say, 1,000 or more) to get a good approximation to our desired frequencies. The same type of experiments could be used to determine the average properties of gas; one could either measure simultaneously the velocity of each molecule in a container or follow one molecule about over a long period of time, and from this determine its average velocity and other statistical properties.

The distinction between these two methods of determining average or statistical properties of a particular quantity will be considered later. Now we simply emphasize that they both involve dealing with large numbers of events.

How do we now make use of large numbers of measurements to determine the statistical properties of a particular occurrence? Assume, as the simplest type of experiment, that we are engaged in repeatedly tossing a coin. Can we determine whether heads or tails will appear in any particular throw? Of course, given enough information about the way in which the coin is dropped, initial velocity, etc., we could predict heads or tails. But this is extremely difficult. Instead, we say ordinarily that we *do not know exactly* which side will appear on any one throw but that the odds are 50–50 either way; i.e., in the *long run* (large number of tries), as many heads will appear as tails. Either event (heads or tails) is equally likely.

How do we arrive at this conclusion? We may perhaps say intuitively that by virtue of complete symmetry, assuming an unloaded coin, there can be no preference as to heads or tails. Or perhaps we have come to the same conclusion after long experience with coin tossing (a valuable and instructive use of time!).

Thus, although in only a few throws the frequency of occurrence of heads or tails may not be the same, we feel sure that over many repetitions (tosses) of the experiment, either event will occur very nearly the same number of times. If we plotted, for example, the ratio of the number of heads H to the total number of tosses N, we might obtain the curve of Fig. A-1.

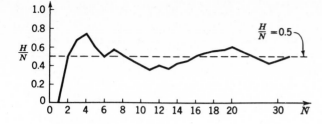

Figure A-1 Coin-tossing experiment. Fraction of heads thrown.

Initially, (N small) H/N takes on greatly differing values. As N increases, however, H/N approaches 0.5 more and more closely. For large N, then, we are presumably justified in saying that heads will occur as often as tails. We then generalize by saying that in the next throw there is a 50 percent chance of heads occurring.

The same conclusions would of course be drawn from an experiment with signals consisting just of on-off pulses. If a long series of tests with all types of messages show that pulses appear as frequently as spaces (no pulses),we would presumably say there is a 50 percent chance of getting a pulse (or a space) in any one interval. The pulses and spaces are then equally likely.

We can consider similarly the tossing of a die: we feel intuitively, or might perhaps demonstrate by actual tosses of a die, that in many tosses each face should come up very nearly as often as any other. We say that, with six faces, each face has a $\frac{1}{6}$ chance of coming up on any one toss. Any two faces have a $\frac{1}{3}$ chance; any three faces, a $\frac{1}{2}$ chance; etc.

These intuitive ideas of chance as associated with many repetitions of an experiment may be formalized somewhat, as in Sec. 1-4, by defining the probability of one event out of several possible occurring as follows: Say there are n possible outcomes of an experiment given by $A_1, A_2, \ldots, A_n$. (In the case of the coin, A_1 is H, A_2 is T, n is 2. In the case of the die, n is 6, and the A's represent different faces of the die.) If the experiment is repeated N times and the relative frequency of occurrence of event A_k is found to approach a limiting value for N much larger than n, this limiting value is defined to be the probability of occurrence, $P(A_k)$, of event A_k. Thus

$$P(A_k) = \frac{\text{number of times } A_k \text{ occurs in } N \text{ tries}}{N} \qquad N \gg n \qquad (A-1)$$

If the occurrence of any one event excludes the occurrence of any others (a head excludes the occurrence of a tail), the events are said to be *mutually exclusive*. Then, if all possible events A_1 and A_n are included,

$$P(A_1) + P(A_2) + \cdots + P(A_n) = 1 \qquad (A-2)$$

Equation (A-2) is an obvious statement of the fact that some one of the n events must occur in the N tries, and the sum of all the events must be equal to N. In probability terms, then, the probability of *any* event occurring is 1. (The probability of a head or a tail is 1; the probability of any one of the six faces of a die coming up is 1; etc.)

We can now generalize our definition of probability a bit. If the n events are mutually exclusive, the probability of one of k events of the n occurring ($k \leq n$) must be the sum of the probabilities of each event occurring. Thus

$$P(A_{j+1} \text{ or } A_{j+2} \text{ or } \cdots \text{ or } A_{j+k})$$
$$= P(A_{j+1}) + P(A_{j+2}) + \cdots + P(A_{j+k}) \leq 1 \quad \text{(A-3)}$$

For the relative frequency of the k events is simply

$$\frac{N_{j+1} + N_{j+2} + \cdots + N_{j+k}}{N} = \frac{N_{j+1}}{N} + \frac{N_{j+2}}{N} + \cdots + \frac{N_{j+k}}{N}$$

or the sum of the probabilities, as above.

The simplest examples of the calculation of probabilities relate to games of chance. We shall refer to some of these as we proceed, for the very same ideas involved appear in the calculation of noise and signal statistics.

As an example, assume that a box contains three white and seven black balls. What is the probability of drawing a white ball? If we were repeatedly to draw one ball at a time from the box, replacing each ball after it was drawn, we would again expect intuitively to find the white balls appearing 30 percent of the time, the black balls 70 percent of the time. There are thus two possible events here: A_1 is a white ball, A_2 is a black ball, and $P(A_1) = 0.3$, $P(A_2) = 0.7$.

Alternatively, we could tag each white and black ball separately. Any one ball would thus appear very nearly $\frac{1}{10}$ of the time in many ($N \gg 10$) drawings of a ball. But we can lump together the chances of drawing any one of the three white balls and get a probability of 0.3 of drawing a white ball. (Here we could also say that each ball is equally likely to be drawn. Any one ball must thus have a $\frac{1}{10}$ chance of being drawn.) This latter approach is simply one example of adding the probabilities of each of a group of mutually exclusive events to determine the probability of occurrence of the overall group.

For another example, consider two dice to be thrown. What is the probability of getting a 6? Again two approaches are possible:

1. Since each die has six faces, a total of $6 \times 6 = 36$ possible outcomes exists. Each possibility is equally likely to occur (as determined by experiment, or, more probably, by intuition) so that the probability of drawing any combination of two faces is $\frac{1}{36}$. But there are five different face combinations that give a 6 (5,1; 4,2; 3,3; 2,4; 1,5). The probability of the overall event is $\frac{1}{36} + \cdots + \frac{1}{36} = \frac{5}{36}$.

2. More directly, of 36 possible outcomes, 5 are the favorable ones corresponding to a 6 occurring. We can call the drawing of a 6 the desired event rather than the drawing of any one of the combinations giving 6 as above. Then 6 will occur $\frac{5}{36}$ of the time in many repeated tosses. [As a check, $P(2) = \frac{1}{36}$,

$P(3) = \frac{2}{36}$, $P(4) = \frac{3}{36}$, $P(5) = \frac{4}{36}$, etc. Then

$$P(2) + P(3) + \cdot \ \cdot \ \cdot + P(11) + P(12) = 1$$

as expected.]

For the third example, say two coins are thrown. What is the probability of one head and one tail? Here there are four possible outcomes: H,H; H,T; T,H; T,T. Each outcome is equally likely and has a probability of $\frac{1}{4}$ of appearing. Head-tail can occur in two ways if no distinction is made as to which coin turns up head. So

$$P(H,T) = \tfrac{1}{4} + \tfrac{1}{4} = \tfrac{1}{2}$$

Alternatively, we can say that the head-tail combination is one of three possible: *HT, TT, HH*. But in many throws we would expect *HT* to come up 50 percent of the time.

Note that from our definition of probability as the relative frequency of occurrence of a specified event we have gradually and almost unconsciously moved to another interpretation of probability:

We enumerate the total number r of possible outcomes of an experiment that are *equally likely* and *mutually exclusive* (i.e., in a long series of repetitions of the experiment each outcome would occur on the average as often as any other). Then the probability of one of these events occurring is simply $1/r$. Our specified event could be just one of these equally likely events, in which case its probability of occurrence would be $1/r$. Or it might comprise a group, say, k ($k < r$), of the r events. Its probability would then be k/r by summing the probabilities of the mutually exclusive events.

For example, in the case of the two dice thrown, all face combinations are equally likely. Then $r = 36$, and the probability of any one face combination occurring is $\frac{1}{36}$. But of these 36 combinations $k = 5$ constitutes our desired event (a 6 appearing); so $\frac{5}{36}$ is the probability of throwing a 6.

By reducing the calculation of probability to the determination of the equally likely events first we can often avoid the necessity of actually performing an experiment many times to determine relative frequencies of occurrence. Thus, in calculating the probabilities of getting various head-tail combinations in the tossing of a coin, there is usually no need actually to carry out a coin-tossing experiment. We know intuitively that tossing a head or a tail on an unbiased coin is equally likely and begin our calculations from that point.

This is not possible in many cases, however. The calculation of the probability of occurrence of the different letters in the English alphabet cannot be based on any "equally likely" argument. We must actually calculate the relative frequency of occurrence of a particular letter in a long series of trials.

Where the "equally likely" approach is fruitful, however, use can sometimes be made of some of the fundamental notions of permutations and combinations. As an example, say a set of 2 cards is to be drawn from a deck of 52. What is the probability of drawing one spade and one heart?

The total number of ways in which two successive cards may be drawn is obviously 52×51. (For each card drawn, 51 remain. But 52 different cards may be drawn on the first try.) This number is just the permutations of 52 elements taken 2 at a time. Each one of these 52×51 possibilities or permutations is equally likely, so that the probability of drawing any one pair of cards is $1/(52 \times 51)$. But 13 different spades could have been drawn as the first card and for each such spade 13 different hearts as the second card. 13×13 is thus the number of spade-heart possibilities. But a heart followed by a spade would have done just as well; so $2(13 \times 13)$ is the number of favorable possibilities of the 52×51 possible outcomes. The probability of drawing one spade and one heart is thus $2(13 \times 13)/(52 \times 51)$.

In general we may have n items (say, n books, n symbols of an alphabet, n voltage levels, etc.) which can be selected. m of these items are to be selected successively, with no duplication allowed. Thus, if a particular book or symbol is selected as the first of m selections, it can no longer be used in further selections. (If the 2-volt level is picked in the first interval, it can no longer be chosen the second or succeeding intervals.) What is the probability of selecting a particular symbol combination, book combination, or voltage-level combination? This probability will depend on the number of permutations or combinations of n items, m at a time.

If we pay attention to the *order* of the selection, the number of ways of picking m of n events is the number of *permutations* of n items taken m at a time. Thus, if two of the letters a, b, c, d, e are to be selected and *order* is important, ab differs from ba.

The number of ways of selecting m of n events is called the number of *combinations* if the order is unimportant. Thus, if an a and a b are all that are desired, ab is the same as ba.

If the m elements selected are *ordered,* the first may be selected in n ways, the second in $(n - 1)$ ways, etc. (Remember that once a particular item is chosen it cannot be selected again.) Then the number of possible selections is $P_{n,m}$ (number of permutations of n items taken m at a time).

$$P_{n,m} = n(n - 1)(n - 2) \cdots [n - (m - 1)] = \frac{n!}{(n - m)!} \qquad \text{(A-4)}$$

As an example, if two of the five letters a, b, c, d, e are to be selected, there are $5 \times 4 = 20$ possibilities, with order important. (The reader may wish to tabulate these to convince himself of this result.) In particular, if n selections are made,

$$P_{n,n} = n! \qquad \text{(A-5)}$$

For example, there are $3! = 6$ permutations of the three letters a, b, c selected in three successive tries: $abc, acb, bac, bca, cab, cba$.

How do we now find the number of unordered selections or combinations? Call this number $C_{n,m}$. Then, for example, $abcd$ and $abce$ are different combinations, but $abcd$ and $abdc$ are not to be distinguished. There are $m!$ ways of obtaining each particular combination of m units, or, alternatively, $m!$ ways of rearranging a combination of m items, once obtained. (For example, abc can be rearranged in $3! = 6$ ways.) Then $C_{n,m}m!$ must equal the total number of possible selections of $P_{n,m}$.

$$C_{n,m} = \frac{P_{n,m}}{m!} = \frac{n!}{m!(n-m)!} \tag{A-6}$$

Very commonly, the symbol $\begin{pmatrix} n \\ m \end{pmatrix}$ is used to denote the number of combinations of n objects taken m at a time. We shall follow this practice in writing equations where required. By definition, we then have

$$\begin{pmatrix} n \\ m \end{pmatrix} \equiv \frac{n!}{m!(n-m)} \tag{A-6a}$$

The card problem we discussed previously is a good example of using permutations and combinations to calculate probabilities. We draw 2 cards from a 52-card pack. The number of combinations possible is $C_{52,2} = (52 \times 51)/2$. Each is equally likely. But of these the possible number of ways for a spade-heart to appear is 13×13. Then $P(\text{heart, spade}) = [2(13 \times 13)]/(52 \times 51)$, as before.

Alternatively, there are $P_{52,2} = 52 \times 51$ permutations of the cards taken two at a time. Of these, $2(13 \times 13)$ represent possible ways of getting a spade and a heart, and we get the same result for the probability of getting a spade and a heart.

A-2 CONDITIONAL PROBABILITY AND STATISTICAL INDEPENDENCE

Up to this point we have discussed the probability of one particular event occurring: a head or a tail in coin tossing, a particular number in the tossing of dice, a particular letter (say, e or g, for example) in the English alphabet, etc. It is appropriate at this point to extend our definition of probability as the relative frequency of occurrence to include the *joint probability* of two or more events occurring.

Random signals and noise offer many examples of the joint occurrence of two random events. Consider, for example, a binary pulse train (Fig. A-2a). What is the probability that two pulses spaced T seconds apart will both be 0's (or 1's)? Figure A-2b represents the voltage at the output of an amplifier. What is the joint probability that the noise voltage at time t_1 exceeds 4 V (or some other specified voltage), and the voltage at time t_2 exceeds 5 V?

In Sec. 1-4 we indirectly hinted at the need for determining such probabilities. There we pointed out that in determining the statistical structure of a language such as English, a count of the relative frequency of occurrence of different letters of the alphabet does not provide full information about the

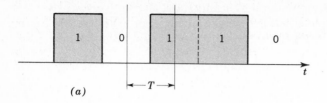

(a)

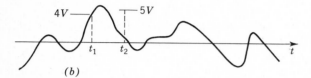

(b)

Figure A-2 Joint occurrence of two random events. (*a*) Binary signals. (*b*) Noise voltage at output of an amplifier.

language statistics. Letters commonly occur in pairs (for example, *qu, th, tr, st*) or even in groups of three or more. Some word combinations occur more frequently than others. All these joint occurrences affect the overall statistics and must be included in an accurate determination, for example, of the language information content.

The fact that some events occur *jointly* indicates that there may be some dependence of one event on another: the letter *u* obviously follows *q* much more frequently than does the letter *e*, for example. So we should be able to determine the independence of several events by measuring the frequency of joint occurrence. We shall restrict ourselves here to two events, *A* and *B*, only. The reader is referred to the literature on probability previously cited for an extension to more than two events of interest.

A simple example of the calculation of a joint probability will clarify these ideas and enable us to generalize quite readily. We determined previously, as an example in the calculation of probability, the chance of drawing a heart and a spade from a deck of cards. Using permutations and combinations, we were able to show that

$$P(\text{hearts, spades}) = \frac{2(13 \times 13)}{52 \times 51} = 2\frac{13}{4 \times 51}$$

Although we considered the heart-spade drawing at that time to be a single event of interest, we could just as well have treated it as a problem in joint probability: What is the probability that in the drawing of two cards one will be a heart, the other a spade?

To answer this question, consider first the case in which we are interested in drawing a heart as the first card, a spade as the second. Let *A* represent the drawing of a heart, *B* the drawing of a spade. Then $P(AB)$ represents the desired probability of drawing first a heart, then a spade.

The chance of drawing a heart as the first card is $P(A) = \frac{1}{4}$, since 13 of the 52 cards are hearts. The probability of drawing a spade as the second card is not $\frac{1}{4}$, however, but $\frac{13}{51}$, since 51 cards are left. The second drawing is thus dependent upon, or *conditioned* by, the first drawing. We designate this possible dependence of the second of the two events on the first by the symbol $P(B|A)$: the probability of event B occurring, it being known that A has occurred. This is called a *conditional probability*. Here $P(B|A) = \frac{13}{51}$. [Had the first card been replaced, we would have had $P(B) = \frac{1}{4}$, independent of the first card.]

The chance of drawing the desired sequence heart-spade (AB) is then

$$P(AB) = \tfrac{1}{4} \times \tfrac{13}{51} = P(A)P(B|A)$$

since, for every heart we draw ($\frac{1}{4}$ of the time on the average), there is a $\frac{13}{51}$ chance of drawing a spade.

But we could just as well have asked for a spade first, then a heart in this particular problem. (The original problem specified, not a particular sequence of cards, but merely a heart and a spade.) This sequence obviously gives $P(BA) = \frac{1}{4} \times \frac{13}{51} = P(B)P(A|B)$ again. Since the probabilities of the two sequences are mutually exclusive (*either* a heart *or* a spade is drawn as the first card), the chance of a heart-spade combination being drawn is the sum of the two probabilities, or

$$\tfrac{1}{4} \times \tfrac{13}{51} + \tfrac{1}{4} \times \tfrac{13}{51} = 2 \times \tfrac{1}{4} \times \tfrac{13}{51}$$

This agrees of course with the result using permutations and combinations. To recapitulate, there we considered the heart-spade drawing as the one desired event and calculated the corresponding probability of this event occurring from a consideration of its relative frequency of occurrence. Here we have chosen to treat it as two events (one dependent on the other in this case).

We can now generalize the concepts of joint and conditional probability and their relation to the dependence of two events on one another. Assume that we perform an experiment and look for the occurrence of two events A and B in that order. We repeat this experiment many times and measure the relative frequency of occurrence of each event separately and as a pair (AB). (We shall actually give an example of such an experiment, and calculations based on it, shortly.)

Our problem is to determine the relative dependence (or independence) of B on A. The conditional probability $P(B|A)$, or the probability of B occurring, given A having occurred, will serve as a measure of this dependence. How do we then determine $P(B|A)$ from our measurements?

Let n_{AB} represent the number of times in n repetitions of the experiment that the combination AB appears. For n a "large number," the joint probability of first A and then B occurring is

$$P(AB) = \frac{n_{AB}}{n} \tag{A-7}$$

by using our previous definition of probability. In the same n trials the outcome A is found to appear n_A times, the outcome B n_B times. The term n_A must include n_{AB}, since some of the times that A appears it is followed by a B. The ratio

$n_{AB}/n_A \leq 1$ represents the relative frequency of occurrence of B preceded by the event A and is just the desired conditional probability $P(B|A)$ (n_A "large" enough). Thus

$$P(B|A) = \frac{n_{AB}}{n_A} \leq 1 \tag{A-8}$$

As an example, n might represent a large number of drawings of two successive cards, n_A the number of times a heart appears, n_B the number of times a spade appears, n_{AB} the number of times a heart is followed by a spade. Dividing the numerator and denominator of Eq. (A-8) by n, we get

$$P(B|A) = \frac{n_{AB}/n}{n_A/n} = \frac{P(AB)}{P(A)} \tag{A-9}$$

from Eq. (A-7) and the definition of $P(A)$. Equation (A-9) is the defining equation relating conditional and joint probabilities. Multiplying through by $P(A)$ we get

$$P(AB) = P(A)P(B|A) \tag{A-10}$$

This is the relation used in the two-card problem given as an example.

Now assume that $P(B|A) = P(B)$. This implies that the probability of event B happening is independent of A. Such a situation would be true in the two-card problem if the first card were immediately replaced after having been drawn. In this case Eq. (A-10) gives

$$P(AB) = P(A)P(B) \tag{A-11}$$

We thus multiply the two separate probabilities together to find the probability of the event AB, if B is independent of A.

Equation (A-11) can also be written

$$P(AB) = P(B)P(A)$$

This implies that $P(AB)$ is independent of the order of occurrence of A and B in this case. It must thus be the same as $P(BA)$, the probability of A following B. But we also write

$$P(BA) = P(B)P(A|B)$$
$$= P(B)P(A) \tag{A-12}$$

in this case. Then $P(A|B) = P(A)$, and A is independent of the occurrence of B.

Two events A and B are said to be *statistically independent* if their probabilities satisfy the equations

$$P(AB) = P(BA) = P(A)P(B) \tag{A-13}$$

and $$P(B|A) = P(B) \qquad P(A|B) = P(A) \tag{A-14}$$

Some examples of conditional probability are in order at this point.

1. *An urn containing two white balls and three black balls.* Two balls are drawn in succession, the first one not being replaced. What is the chance of

picking two white balls in succession? (Note that this is similar to the two-card problem.)

Letting event A represent a white ball on draw 1, event B a white ball on draw 2,

$$P(AB) = P(A)P(B|A) = \tfrac{2}{5} \times \tfrac{1}{4} = \tfrac{1}{10}$$

Here two of the five balls are white so that the chance of drawing a white is $\tfrac{2}{5}$. Once a white is drawn, however, only one white ball remains among the four balls left. The chance of drawing a white ball now, assuming a white drawn on the first try, is $\tfrac{1}{4}$.

Alternatively, we can say that there are $C_{5,2} = 5!/3!2! = 10$ possible combinations of five balls arranged in groups of two. Of these only one combination, two white balls, is of interest. The chance of drawing this combination is thus $P(AB) = 1/C_{5,2} = \tfrac{1}{10}$ again. (Combinations are used here instead of permutations because the two white balls drawn cannot be distinguished from one another.)

If the first ball were replaced before drawing the second, the two events would be *independent*:

$$P(AB) = P(A)P(B) = (\tfrac{2}{5})^2 = \tfrac{4}{25}$$

2. *Two urns contain white and black balls*. Urn A contains two black balls and one white ball; urn B contains three black balls and two white balls (Fig. A-3). One of the urns is selected at random, and one of the balls in it chosen. What is the probability $P(W)$ of drawing a white ball?

There are two ways of satisfying the desired event W—the drawing of one white ball:

(*a*) Pick urn A; draw W.

(*b*) Pick urn B; draw W.

These two subevents are mutually exclusive, so that

$$P(W) = P(AW) + P(BW)$$

But $\qquad\qquad\qquad P(AW) = P(A)P(W|A) = \tfrac{1}{2} \times \tfrac{1}{3} = \tfrac{1}{6}$

Thus the probability of drawing a white ball from urn A is the probability $P(A) = \tfrac{1}{2}$, of first picking A, times the probability $P(W|A) = \tfrac{1}{3}$, that a white ball will be selected once A is chosen. [$P(A) = P(B) = \tfrac{1}{2}$, since the urns are selected at random.] Similarly,

$$P(BW) = P(B)P(W|B) = \tfrac{1}{2} \times \tfrac{2}{5} = \tfrac{1}{5}$$

Therefore, $\qquad\qquad\qquad P(W) = \tfrac{1}{6} + \tfrac{1}{5} = \tfrac{11}{30}$

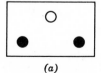 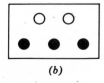

(*a*) (*b*)

Figure A-3 Two-urn problem. (*a*) Two black, one white. (*b*) Three black, two white.

[Note that if the balls were in one box $P(W)$ would be $\frac{3}{8}$. Because of the two urns the balls are not equally likely to be drawn.]

3. *Color blindness.* [2] Assume that 5 men out of 100 and 25 women out of 10,000 are color-blind. A color-blind person is chosen at random from a representative sample of 10,000 men and 10,000 women. What is the probability that he will be male?

It may be assumed that 500 men and 25 women of this sample are color-blind. This gives a total of 525 color-blind persons. The probability that the person chosen will be male is thus $\frac{500}{525} = \frac{20}{21}$.

Alternatively, let $N = 20,000$, $N_M = N_W = 10,000$ be the number of men or women in the sample, and $N_C = 525$ be the total number of color-blind persons. $N_{MC} = 500$ represents the number of color-blind men. Then the conditional probability that a man is selected if a color-blind person is chosen is

$$P(M|C) = \frac{N_{MC}}{N_C} = \frac{500}{525}$$

As a check, let $P(MC)$ be the probability of selecting a color-blind man, $P(C)$ the probability of selecting a color-blind person.

$$P(MC) = P(C)P(M|C) = \tfrac{525}{20000} \times \tfrac{500}{525} = \tfrac{500}{20000}$$

as expected. We could also determine the probability of selecting a color-blind man by first picking a man at random. The chance of doing this is $P(M) = \frac{1}{2}$. Once a man is selected, the probability that he is color-blind is $P(C|M) = \frac{5}{100}$. The probability of picking a color-blind man is again $\frac{5}{200}$.

4. *Statistics of three-letter alphabets.* Suppose that we have an alphabet containing three letters A, B, C. We wish to determine the statistics of messages using this alphabet. In particular what is the relative frequency of occurrence, or probability, of each letter, and the probability of two-letter groups such as AA, AB, BC, CA, etc., occurring? From this we can determine the statistical dependence (or independence) of successive letters.

We take a typical example of a message using this three-letter alphabet and proceed to count the frequency of occurrence of the individual letters and groups of two successive letters. This is then an example of the repeated experiment mentioned previously as a method of determining the different probabilities of interest. A typical message of 50 letters appears as follows:

CBACABCABABCCBBABBCABAAACACCBBCCACCBBBBACCAACBABC

Letting n = the total number of letters, n_A the number of A's, n_{AA} the number of pairs (diads) of AA, n_{AB} the number of pairs of AB, n_{BA} the pairs of BA, etc., we

---

[2] W. Feller, *An Introduction to Probability Theory and Its Applications*, 2nd ed., Wiley, New York, 1957.

get $n = 50$: $n_A = 16$, $n_B = 17$, $n_C = 17$, and

$$
\begin{array}{ccc}
n_{AA} = 4 & n_{AB} = 6 & n_{AC} = 6 \\
n_{BA} = 6 & n_{BB} = 6 & n_{BC} = 5 \\
n_{CA} = 6 & n_{CB} = 5 & n_{CC} = 5
\end{array}
$$

From the almost equal number of times that the different letters appear we can conclude that the letter probabilities are equal. The different letters are thus equally likely in this particular three-letter language.

$$
P(A) = \frac{n_A}{n} = 0.32 \qquad P(B) = 0.34 = P(C)
$$

are the calculated probabilities, and 0.33 would be the actual probabilities if the letters were equally likely to occur.

The conditional probabilities can be calculated quite easily from the number of times the different pairs appear. For example,

$$
P(B|A) = \frac{P(AB)}{P_A} = \frac{n_{AB}}{n_A} = \frac{6}{16}
$$

Thus $P(B|A)$ represents the number of times that an A is followed by a B. Repeating this calculation for the different pair combinations—nine in number—we get the following conditional-probability table.[3]

Calculated probabilities

$P(j\|i)$		A	B	C
			j	
i	A	$\frac{4}{16}$	$\frac{6}{16}$	$\frac{6}{16}$
	B	$\frac{6}{17}$	$\frac{6}{17}$	$\frac{5}{17}$
	C	$\frac{6}{17}$	$\frac{5}{17}$	$\frac{5}{17}$

Theoretical probabilities, assuming statistical independence

$P(j\|i)$		A	B	C
			j	
i	A	$\frac{1}{3}$	$\frac{1}{3}$	$\frac{1}{3}$
	B	$\frac{1}{3}$	$\frac{1}{3}$	$\frac{1}{3}$
	C	$\frac{1}{3}$	$\frac{1}{3}$	$\frac{1}{3}$

Both the table and the list of the number of times each pair appears lead us to conclude that the letters in the pair combinations are *statistically independent*. For statistical independence $P(j|i) = P(j)$, or the probability of occurrence of any letter is independent of the letter preceding. Here, theoretically, we would expect all the $P(j|i)$'s to be $\frac{1}{3}$. Because of the relatively short length of the message used (small sample, small number of experiment repetitions) the actual frequencies calculated differ somewhat.

---

[3] C. E. Shannon, "A Mathematical Theory of Communication," *Bell System Tech. J.*, vol. 27, pp. 379–423, July 1948.

C. E. Shannon has published an example of a three-letter alphabet in which any letter in a particular sequence is dependent on the letter immediately preceding.[4] (There is no dependence on letters before that one.) This alphabet consists of the letters A, B, C with the following probability tables:

i	$P(i)$
A	$\frac{9}{27}$
B	$\frac{16}{27}$
C	$\frac{2}{27}$

$P(j\mid i)$		A	B	C
	A	0	$\frac{4}{5}$	$\frac{1}{5}$
i	B	$\frac{1}{2}$	$\frac{1}{2}$	0
	C	$\frac{1}{2}$	$\frac{2}{5}$	$\frac{1}{10}$

The letter B should thus occur most frequently, the letter C only occasionally. The letter A has 0 probability of being followed by another A, as is true also for C following a B. Each time A appears there is a $\frac{4}{5}$ probability that a B will follow, a $\frac{1}{5}$ probability that a C will follow. A C is followed by a B $\frac{2}{5}$ of the time, by an A $\frac{1}{2}$ of the time, by another C $\frac{1}{10}$ of the time.

A typical message given by Shannon for this three-letter language is

ABBABABABABABABBBABBBBBAB
ABABABABBBACACABBABBBBABB
ABACBBBABA

Calculating the number of times each letter and each pair of letters occurs, just as we did previously, we get $n = 60$:

$$n_A = 22 \qquad n_B = 35 \qquad n_C = 3$$
$$n_{AA} = 0 \qquad n_{AB} = 18 \qquad n_{AC} = 3$$
$$n_{BA} = 19 \qquad n_{BB} = 16 \qquad n_{BC} = 0$$
$$n_{CA} = 2 \qquad n_{CB} = 1 \qquad n_{CC} = 0$$

Again calculating a conditional probability table based on the actual relative frequencies, we get

Calculated

$P(j\mid i)$		A	B	C
	A	0	$\frac{18}{22}$	$\frac{2}{22}$
i	B	$\frac{19}{35}$	$\frac{16}{35}$	0
	C	$\frac{2}{3}$	$\frac{1}{3}$	0

Theoretical

$P(j\mid i)$		A	B	C
	A	0	$\frac{4}{5}$	$\frac{1}{5}$
i	B	$\frac{1}{2}$	$\frac{1}{2}$	0
	C	$\frac{1}{2}$	$\frac{2}{5}$	$\frac{1}{10}$

[4] *Ibid.*

Note again that the calculated probabilities, based on relative frequencies of occurrence, and the theoretical probabilities agree reasonably well. The reader should check these results for himself. He should also calculate the joint probabilities $P(ij)$ and compare for both the theoretical and actual cases.

A-3 AXIOMATIC APPROACH TO PROBABILITY[5]

The relative-frequency approach to probability discussed in the last two sections lends itself to experimental measurements of probability. It also helps one develop a "physical feel" for some of the concepts introduced. The theory of probability has achieved its greatest impetus, however, by being developed on an axiomatic basis. Although this approach using set theory is more abstract, it is at the same time much more general than the relative-frequency approach, and hence more readily extendable to more complex situations.

We shall outline this approach very briefly, contenting ourselves with paralleling the equations developed using the relative-frequency approach. The reader is referred to the references cited on probability at the beginning of Sec. A-1 for much more comprehensive discussions.

We begin with some simple definitions from set theory, and then follow by setting up the axioms in terms of these. Assume that a typical experiment among those discussed earlier (coin and die tossing, cards, balls in urns, three-letter alphabets, etc.) has S possible outcomes. We call S the set and each outcome an element in it. For example, assume that three balls, $a, b, c,$ are to be placed into two urns. There are then $2^3 = 8$ possible elements in the space S: a may be in either urn, b may be in either urn, and c may be in either urn. The eight elements are tabulated in the following table:

Element	Urn I	Urn II
1	abc	—
2	—	abc
3	ab	c
4	ac	b
5	bc	a
6	a	bc
7	b	ac
8	c	ab

This total set may in turn be split into possibly overlapping subsets. Call these $A, B, C, \ldots$. As examples, four such subsets are defined below and the

[5] A. Papoulis, *Probability, Random Variables, and Stochastic Processes,* McGraw-Hill, New York, 1965, chap. 2.

elements contained within them shown bracketed.

A : all the balls appear in 1 urn	$\{1, 2\}$
B : ball a is in urn I	$\{1, 3, 4, 6\}$
C : at least two balls in urn I	$\{1, 3, 4, 5\}$
D : no balls in urn II	$\{1\}$

Many more subsets may obviously be defined.

Note that subset D is contained within A (all elements in D are in A), that A and B overlap by having element 1 in common, and that B and C overlap by having 1, 3, 4 in common.

We define the set $A + B$ (or *union* of A and B) to be another set containing all elements in either A or B or both. In the example above $A + B$ contains the elements $\{1, 2, 3, 4, 6\}$; $B + C$ contains $\{1, 3, 4, 5, 6\}$. The set AB (or *intersection* of A and B) contains elements common to A and B. Again in this example AB contains the element $\{1\}$, while BC contains $\{1, 3, 4\}$. $\bar{A}$ is defined to be the *complement* of A : this is the set containing all elements *not* in A. Again in this example $\bar{A}$ has the elements 3, 4, 5, 6. It is apparent that $A + \bar{A} = S$. The zero or null set 0 contains no elements. These set-theory definitions and relations are easily visualized by using two-dimensional diagrams called *Venn diagrams*. S, A, B, $A + B$, $\bar{A}$, and AB are thus diagrammed in Fig. A-4. The set S has been chosen rectangular here solely for simplicity's sake.

As a second example, indicating that the elements of a set do not necessarily have to be related to *discrete* outcomes, let x represent the rectified (positive) voltage at the output of an amplifier and y the rectified (positive) voltage at the output of another amplifier. The set S is given by all possible combinations of x and y, or by the elements $\{x,y > 0\}$. Let A be the subset $\{x > 10\}$, that is, all x voltages greater than 10 V; B the subset $\{y > 10\}$ (Fig. A-5). AB is then the set "both voltages greater than 10," as shown. It is apparent that $S = AB + \overline{AB}$. Define the set C as $\{x > y\}$. This is the set for which all x voltages are greater than y voltages. On the Venn diagram in Fig. A-5 this is shown bounded by the 45° line (Fig. A-5*c*). It is left for the reader to show that AC and ABC are the subsets indicated in Fig. A-5*d*.

The Venn diagrams may be used to prove the following set identities:

$$AS = A$$

$$A\bar{B} + AB = A(\bar{B} + B) = A$$

$$A(B + C) = AB + AC$$

$$\overline{A + B} = \bar{A}\bar{B}$$

$$\overline{AB} = \bar{A} + \bar{B}$$

We now introduce probability by assigning to each subset or *event* a number we *define* to be the probability associated with event A. This number must have the property that it is positive. Since the set S contains all possible outcomes, $P(S) = 1$ is an obvious condition. This simply states the truism that some event

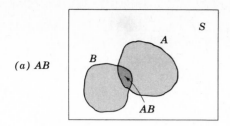

(a) AB

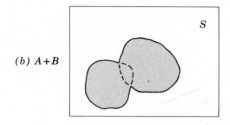

(b) A+B

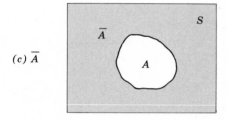

(c) $\bar{A}$

Figure A-4 Venn diagrams.

must occur if an experiment is performed. To complete the basic definitions and make them consistent with the relative-frequency approach we have already discussed, we also require that if two sets A and B have no elements in common ($AB = 0$), their probabilities add. (Such events are said to be *mutually exclusive*.) We summarize by tabulating three basic axioms:

1. $$P(S) = 1 \tag{A-15}$$

2. $$P(A) \geq 0 \tag{A-16}$$

3. If $AB = 0$ $$P(A + B) = P(A) + P(B) \tag{A-17}$$

From these three axioms other relations among the probabilities follow quite simply and often almost mechanically.

For example, since $A + \bar{A} = S$, and $A\bar{A} = 0$, by definition,

$$P(A + \bar{A}) = P(A) + P(\bar{A}) = P(S) = 1$$

But $P(A)$ and $P(\bar{A})$ are both positive by hypothesis. Then

$$P(A) \leq 1 \tag{A-18}$$

Our probabilities as defined thus range between zero and one, as expected.

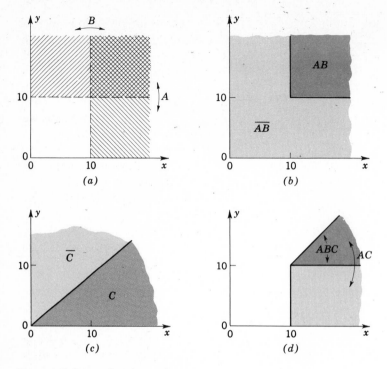

Figure A-5 Sets and various subsets.

Similarly, we can calculate the probability of overlapping events by using the three axioms and referring to an appropriate Venn diagram. Thus, consider A and B overlapping as in Figs. A-4 and A-5. What is the probability of event $(A + B)$? It is apparent that we cannot simply add the two probabilities, since $AB \neq 0$. A and B are *not* mutually exclusive in this case. As a matter of fact, from the Venn diagrams we note that by adding $P(A)$ and $P(B)$ we have included AB twice. It is thus apparent that we must have

$$P(A + B) = P(A) + P(B) - P(AB) \tag{A-19}$$

Here $P(AB)$ is exactly the kind of joint probability discussed earlier. As a check, note that $A + B = A + \bar{A}B$. But A and $\bar{A}B$ have no elements in common and are thus mutually exclusive events. $[A(\bar{A}B) = 0.]$ Then

$$P(A + B) = P(A) + P(\bar{A}B)$$

But $B = \bar{A}B + AB$, with $(\bar{A}B)(AB) = 0$. Hence

$$P(B) = P(\bar{A}B) + P(AB)$$

Finally, then, with $P(\bar{A}B) = P(B) - P(AB)$, we get Eq. [A-19].

Although the manipulations to determine probabilities of various events follow readily from the basic axioms, the actual *numbers* used must still be provided from other sources. Thus one either guesses at the probabilities of various outcomes of experiments, assumes them, or calculates them, as in the previous sections. The numerical part of probability does *not* come automatically.

As an example assume in Fig. A-5 that $P(A) = 0.5$ and $P(B) = 0.3$. Thus we have good reason to believe that the rectified voltage of amplifier x has a 50 percent chance of exceeding 10 V, while that of amplifier B has a 30 percent chance. If we *know* the joint probability $P(AB)$ is 0.4 (this is the probability that x and y will both exceed 10 V when measured), the total probability $P(A + B) = 0.5 + 0.3 - 0.4 = 0.4$. In other words, the probability that either amplifier A output or amplifier B output exceeds 10 V is 0.4.

Conditional probabilities are introduced by axiom as well. Thus we define the conditional probability of event A occurring, given B having occurred, as given by

$$P(A|B) = \frac{P(AB)}{P(B)} \tag{A-20}$$

This is, of course, in agreement with Eq. (A-9), obtained from relative-frequency considerations. Here we do not derive it, we simply take it as an axiom. The motivation is apparent from the Venn diagram of Fig. A-4. With B having first occurred, it is apparent that the compound region of occurrence is immediately restricted to AB. Thus $P(A|B)$ must be proportional to $P(AB)$. Division by $P(B)$ serves to *normalize* $P(A|B)$ so that it is always ≤ 1. As a check let A be wholly in B. Then $AB = A$, and $P(A|B) = P(A)/P(B) \geq P(A)$. If we *know* we are in region B, our chance of further being in A (a subset of B) is greater than if we did not possess this additional information.

As an example of the use of these conditional probabilities, in addition to the examples discussed in Sec. A-2, assume that one out of four symbols is transmitted in a communication system. The four symbols could be four levels in a four-level digital system, or simply four letters in a four-letter alphabet. Call these symbols A_1, A_2, A_3, and A_4. Call the four corresponding symbols received at the receiver B_1, B_2, B_3, B_4. Ideally A_1 is received as B_1, A_2 as B_2, etc. Because of noise or other distortion during transmission, however, A_1 as received may be mistaken for B_2, or B_3, or B_4. The same is true for the other transmitted symbols. Alternately, at the receiver, B_1 as received may actually have come from any one of the four transmitted symbols. We would like to determine the probability that B_1, as received, actually came from A_1. Some of the transitions that may occur are shown in Fig. A-6.

Specifically, we would like to find the conditional probability $P(A_1|B_1)$, that is, the probability that A_1 was transmitted, given B_1 received. (This is often called the a posteriori probability, the probability *after the fact*. We have occasion to refer to such probabilities in more detail in the book.) To calculate this we first have to specify some probabilities. Let $P(A_1) = P(A_2) = \frac{1}{8}$, $P(A_3) = \frac{1}{4}$,

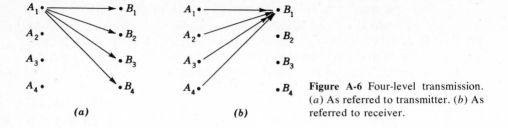

Figure A-6 Four-level transmission. (*a*) As referred to transmitter. (*b*) As referred to receiver.

$P(A_4) = \frac{1}{2}$. (These are called the *a priori* probabilities.) Then level A_4 is the most probable one to be transmitted, followed by A_3, A_1, and A_2 in that order. Note that A_1, A_2, A_3, and A_4 are all mutually exclusive events and incorporate the complete transmission set. Thus

$$\sum_{j=1}^{4} P(A_j) = P(S) = 1$$

We must also specify the various probabilities that transitions will take place. Specifically, let

$$P(B_1|A_1) = \tfrac{1}{2} \qquad P(B_1|A_2) = \tfrac{1}{4} \qquad P(B_1|A_3) = P(B_1|A_4) = \tfrac{1}{8}$$

From Eq. (A-20) the desired a posteriori probability $P(A_1|B_1)$ is given by

$$P(A_1|B_1) = \frac{P(A_1 B_1)}{P(B_1)} \tag{A-21}$$

This is easily written, but how do we find the two terms $P(A_1 B_1)$ and $P(B_1)$? The first term, the joint probability of transmitting A_1 and receiving B_1, is easily calculated from the known conditional probabilities:

$$P(A_1 B_1) = P(A_1)P(B_1|A_1) = (\tfrac{1}{8})(\tfrac{1}{2}) = \tfrac{1}{16} \tag{A-22}$$

The second term, the probability $P(B_1)$ of receiving B_1, is obviously related to the transitions shown in Fig. A-6. To calculate it consider the Venn diagram shown in Fig. A-7. The set S contains the mutually exclusive events A_1, A_2, A_3, A_4, as noted above. It must also contain the four mutually exclusive events B_1, B_2, B_3, B_4. In particular, event B_1 is shown overlapping with all four transmitted events. It is apparent from the figure that B_1 is the sum of four mutually exclusive events:

$$B_1 = A_1 B_1 + A_2 B_1 + A_3 B_1 + A_3 B_1 \tag{A-23}$$

The probabilities then add, and we have

$$P(B_1) = \sum_{j=1}^{4} P(A_j B_1)$$

$$= \sum_{j=1}^{4} P(A_j)P(B_1|A_j) \tag{A-24}$$

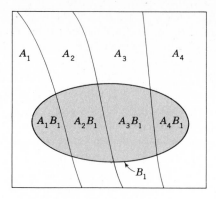

Figure A-7 Venn diagram for four-level transmission.

using the conditional probability axiom to find each term $P(A_j B_1)$. In words: The probability of jointly transmitting A_j and receiving B_1 is the probability of first transmitting A_j, and then receiving B_1 with A_j transmitted. Summing over the mutually exclusive events $A_j, j = 1, 2, 3, 4$, we find the probability of receiving B_1. In particular, it is readily found that $P(B_1) = \frac{3}{16}$. The desired a posteriori probability $P(A_1 | B_1)$ is then found from Eq. (A-21) to be $\frac{1}{3}$.

Note that nowhere was it necessary to discuss the relative frequency of occurrence of the various events. The form of the expression $P(A_1 | B_1)$ was obtained solely from the use of axioms of probability. In actually evaluating this probability numerically, however, we had to assume a knowledge of the various a priori and transition probabilities. These would normally be estimated from repeated experiments, guessed, or calculated. (Generally, the a priori probabilities are estimated from many repeats of the experiment. One then falls back on the relative-frequency approach to actually evaluate probabilities numerically. The transition probabilities, however, are generally calculated from models of noise on the transmission channel. Examples of this type of calculation appear in Chaps. 5 and 6.)

A-4 DISCRETE AND CONTINUOUS PROBABILITY DISTRIBUTIONS

In this introductory discussion of probability we have thus far been primarily concerned with calculating the probability of occurrence of a finite number of discrete events. For instance, in the coin-tossing example there were two possible events: head or tail. In the case of a die we might be interested in the probability of one of the six possible faces coming up. In the case of the English alphabet there are 26 letters, and we may ask for the probability that one of them, say f, will occur. (Only in the two-amplifier example of Fig. A-5 did we discuss continuous outcomes.) In the discrete examples we can represent the different possible outcomes by a discrete variable x. If x has n distinct values x_1,

$x_2, \ldots, x_n$, each of which has probability $P_1, P_2, P_3, \ldots, P_n$, x is called a discrete chance variable, or discrete stochastic variable, or *discrete random variable*. As an example, say that we toss two dice. Let x represent the number coming up on any throw. Then x takes on the discrete values 2, 3, 4, . . . , 12, each one with its associated probability.

If the variable x is now allowed to take *any* value in a whole interval, however, and to each subinterval in the overall interval there is associated a probability of occurrence, x is called a *continuous random variable*. For example, if we start a pointer on a wheel spinning, the pointer will stop at any position on the wheel's circumference. It is not limited to discrete positions only.

We are primarily concerned with continuous chance variables because of our interest in the statistics of noise and signals in noise. If we were to plot the noise voltage at the output of an amplifier, for example, it might have the continuous appearance of the curve of Fig. A-2b. All values of voltage are possible, not just discrete values.

How do we now compute probabilities in the case of continuous chance variables?

We shall start by first using the relative-frequency approach. In dealing with continuous variables, we shall find it convenient to introduce a *probability-density function*. We shall compare this probability-density function with the analogous mass density with which we are presumably quite familiar. Charge-density functions occur of course in electric-field problems also and are used to compute the field due to a continuous, or smoothed-out, array of charge, rather than discrete charges.

As an example of the determination of probability for continuous chance variables, say that we are interested in determining the distribution of height among American males. The height of a man may take on *any* value within a specified interval and thus represents a *continuous* random variable. Once we have such a height distribution, we may use this to calculate the probability that the height of a given male will lie between 5 ft 7 in and 5 ft 9 in. We may use this to determine an average height, etc.

Assume that we select for this determination a representative sample of 1,000 men. We measure their heights (this is an example of the repeated experiment of the previous sections) and group them according to the nearest even inch (5 ft 0 in, 5 ft 2 in, 5 ft 4 in, etc.). All heights from 4 ft 11 in to 5 ft 1 in will, for example, be grouped in the 5-ft 0-in category. The relative frequency of men found in each height interval is then the number grouped in that interval divided by the total number (1,000). A typical plot of such a height distribution is shown in Fig. A-8. x represents the height, and n_x/n the relative number of men grouped in the interval Δx inches about x. n is the 1,000-man sample here. Since all heights in a 2-in interval about the even heights are grouped together, they are shown as horizontal lines covering the 2-in grouping. For a large enough sample we can say that n_x/n represents the probability $P(x_j)$ that the height of an American male will lie between $x_j - \Delta x/2$ and $x_j + \Delta x/2$.

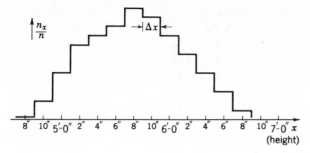

Figure A-8 Relative distribution of heights.

This probability $P(x_j)$ obviously depends on the choice of interval size Δx (2 in here). For if we decrease Δx, say, to 1 in, the number of men in this height interval decreases also. In particular, if we begin using smaller and smaller intervals, the ordinate will become smaller and smaller also. Eventually, if the height intervals are made very small (0.1 in, for example), very few men will be found in any one interval. The plot of relative distribution versus height approaches zero and is of little value to us.

To do away with the dependence on the size of the interval Δx chosen for the height distribution, we represent the height distribution, at a particular height and of interval Δx, by a rectangle of area n_x/n. The height of this rectangle will be $(1/\Delta x)(n_x/n)$, its width Δx. Such a relative-frequency curve is called a *histogram*. Two histograms for the height example used here are shown in Fig. A-9*a* and *b*, one for $\Delta x = 2$ in, the other for $\Delta x = 1$ in. For a large enough sample size the probability that a man's height will lie between two given heights, 5 ft 7 in and 5 ft 9 in, for example, will now be the area of the histogram between these two limits.

If the histogram approaches a smooth curve as $\Delta x \rightarrow 0$, the ordinate takes on the form of a *probability-density function*, with the *area* under the curve between two points giving the probability that the height will be found between these two points. The probability-density function as the limiting case of the two histograms of Fig. A-9 is shown in Fig. A-10. The area between the two

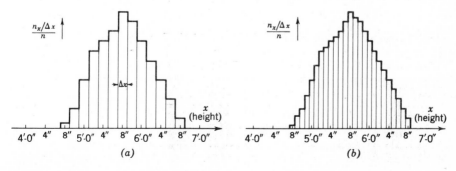

Figure A-9 Histograms of height distribution. (*a*) $\Delta x = 2$ in. (*b*) $\Delta x = 1$ in.

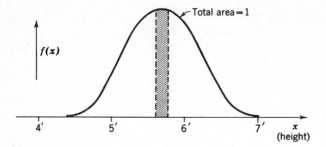

Figure A-10 Probability-density function corresponding to height histogram.

points, 5 ft 6 in and 5 ft 8 in, shown crosshatched in the figure, represents the probability that a man's height will be found in that range.

The assumption of a smooth curve corresponds mathematically to assuming that the probability density is continuous. Under this assumption we can formally define the probability density $f(x)$ by the limit

$$f(x) = \lim_{\substack{\Delta x \to 0 \\ n \to \infty}} \frac{n_x/\Delta x}{n} \tag{A-25}$$

where n_x represents the number of samples of the total n found in the range $x - (\Delta x/2)$ to $x + (\Delta x/2)$. The probability that the variable x will lie in the range x_1 to x_2 is then the area under the $f(x)$ curve, or

$$\text{Prob } (x_1 < x < x_2) = \int_{x_1}^{x_2} f(x)\, dx \tag{A-26}$$

In particular, the probability that x will lie somewhere in its allowable range of variation must be 1. In general, x can range from $-\infty$ to ∞ (the voltage at the output of an amplifier is an example). In special cases it may range only between 0 and ∞ (the variation of height, for example) or between 0 and 2π (for example, the rotation of a pointer on a wheel). For the general case of x ranging between $-\infty$ and ∞

$$\int_{-\infty}^{\infty} f(x)\, dx = 1 \tag{A-27}$$

The histogram, or probability-density curves, must be normalized to have unity area.

From the definition of probability density $f(x)$ must be a function which is always positive. Thus

$$f(x) \geq 0 \tag{A-28}$$

Equations (A-27) and (A-28) represent two conditions that must be satisfied by any probability-density function, as, for example, the one of Fig. A-10.

To emphasize the fact that $f(x)$ represents only a probability-*density* function, with $f(x)\, dx$ the probability that x will be found in the range $x \pm dx/2$, we define another important function, the *cumulative-distribution function* $F(x)$. (For conciseness this will often be referred to as the distribution function.) This is defined to be the probability that the variable will be less than or equal to some value x. Since all values of x are mutually exclusive (a noise voltage can have only one value at any instant; a pointer may stop only at one point), $F(x)$ must be the sum of all the probabilities from $-\infty$ to x. This is just the area under the $f(x)$ curve from $-\infty$ to x. $F(x)$ is thus given by

$$F(x) = \int_{-\infty}^{x} f(x)\, dx \tag{A-29}$$

Equation (A-29) is the indefinite integral of $f(x)$. If $F(x)$ possesses a first derivative, we have

$$f(x) = \frac{dF(x)}{dx} \tag{A-30}$$

Equations (A-29) and (A-30) relate the probability-density function and the distribution function.

Since $f(x) \geq 0$ and $\int_{-\infty}^{\infty} f(x)\, dx = 1$, $F(x)$ must satisfy the inequality

$$0 \leq F(x) \leq 1 \tag{A-31}$$

$F(x)$ is a continuously or monotonically increasing function, going from 0 to 1. A typical density function and its corresponding distribution function are shown in Fig. A-11. $F(x_1)$ corresponds to the crosshatched area of Fig. A-11a. It is left to the reader to show very simply that the probability that x will lie somewhere between x_1 and x_2 can be found from the distribution function by the relation

$$\text{Prob}\,(x_1 < x \leq x_2) = \int_{x_1}^{x_2} f(x)\, dx = F(x_2) - F(x_1) \tag{A-32}$$

The rotating pointer on a wheel serves as a good example of these different relations. Intuitively, we feel that a freely rotating pointer has an equally

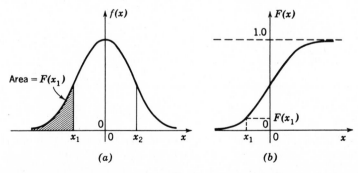

(a)   *(b)*

Figure A-11 Probability-density and probability-distribution functions. $(a)\, f(x)$ = density function. $(b)\, F(x)$ = distribution function.

likely chance of stopping anywhere on the wheel. The probability that it will stop at any one angle θ is zero since there is an infinity of points, but the probability of its stopping within some angular range $d\theta$ is proportional to $d\theta$ and independent of the particular value of θ (Fig. A-12). Since this probability must be $f(\theta)\,d\theta$, with $f(\theta)$ the density function, $f(\theta)$ must be a constant K for this example. The constant is found by invoking the specification that the area under the $f(\theta)$ curve must be 1. (The pointer will obviously stop somewhere on the wheel rim.) Thus

$$\int_0^{2\pi} f(\theta)\,d\theta = 1 = K\int_0^{2\pi} d\theta = 2\pi K \tag{A-33}$$

and
$$f(\theta) = 1/2\pi \tag{A-34}$$

This serves to normalize $f(\theta)$ properly. $f(\theta)$ in this case is one example of the rectangular density function to which we shall have occasion to refer in more detail later.

The cumulative-distribution function $F(\theta)$ is the probability that the variable will be less than or equal to θ. In this case the lower limit of θ is 0, and

$$F(\theta) = \int_0^{\theta} f(\theta)\,d\theta = \frac{\theta}{2\pi} \tag{A-35}$$

This agrees with our intuitive feeling that all values of θ are equally likely. Both $f(\theta)$ and $F(\theta)$ are shown sketched in Fig. A-12. The probability that θ will be less than $\pi/4$, for example, is $\frac{1}{8}$. The probability that it will be less than π is $\frac{1}{2}$.

The probability that θ will lie between $\pi/2$ and π is

$$F(\pi) - F\left(\frac{\pi}{2}\right) = \frac{\pi}{4\pi} = \frac{1}{4}$$

as is to be expected.

We have been discussing the probability-density function $f(x)$ for continuous variables. Can we also define such a function for discrete variables? This will be of particular interest in problems where a variable may have both discrete and continuous ranges. We shall now show that the impulse, or delta, function serves to connect the discrete and continuous cases.

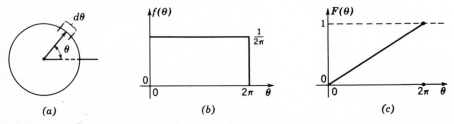

Figure A-12 Rotating pointer and probability functions. (*a*) Rotating pointer. (*b*) Probability-density function. (*c*) Distribution function.

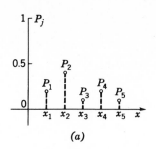

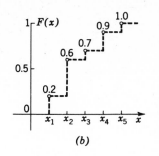

Figure A-13 Probability-distribution function for discrete variable. (*a*) Individual probabilities. (*b*) Cumulative-distribution function.

We can develop the required relationship quite simply from the cumulative-distribution function. Assume a discrete random variable x with values $x_1, x_2, \ldots, x_n$. Corresponding to each value of x is a probability $P_1, P_2, \ldots, P_n$. An example is depicted in Fig. A-13a. Since the values of x are mutually exclusive and some one value of x must occur,

$$P_1 + P_2 + \cdots + P_n = 1 \tag{A-36}$$

The probability that x will be less than or equal to some value x_j can again be defined as a cumulative-distribution function $F(x_j)$. Since all values of x are mutually exclusive, $F(x_j)$ will be equal to the sum of the probabilities $P_1 + P_2 + \cdots + P_j$. Thus

$$F(x_j) = P_1 + P_2 + \cdots + P_j \tag{A-37}$$

For example, assume that x represents the number given by the toss of two dice. The numbers possible are $x = 2, 3, \ldots, 8, \ldots, 12$, and the corresponding probabilities are $\frac{1}{36}, \frac{2}{36}, \ldots, \frac{5}{36}, \ldots, \frac{1}{36}$, respectively. (Readers can check this for themselves.) The probability that a number less than 5 will come up is then the sum of the probabilities of the numbers 2, 3, 4, or $\frac{1}{36} + \frac{2}{36} + \frac{3}{36} = \frac{1}{6}$.

Since the individual probabilities P_1, P_2, etc., are all greater than or equal to 0, the distribution function $F(x)$ must monotonically increase with x, just as in the case of the continuous random variable. Although it is defined only at the points $x_1, x_2, \ldots, x_n$, we may arbitrarily draw it as a series of ascending steps, as in Fig. A-13b. The jump at step x_2 is just $P_2 = 0.4$, for example.

Although the derivative of $F(x)$ for the discrete variable does not exist in the usual sense (limit of ratio $\Delta F / \Delta x$ as $\Delta x \to 0$), we may define the derivative in terms of impulse functions. Thus, at x_j the derivative of $F(x)$ is $P_j \delta(x - x_j)$. This we *define* to be the probability-density function $f(x)$ at $x = x_j$.

$$f(x_j) \equiv P_j \delta(x - x_j) \tag{A-38}$$

The area under the impulse function is thus the actual probability at the point in question. In terms of this impulse-type density function the sum representation of the distribution function $F(x)$ given by Eq. (A-37) becomes the integral

$$F(x) = \int_{-\infty}^{x} f(x)\, dx \tag{A-39}$$

where
$$f(x) = \sum_{j=1}^{n} P_j \delta(x - x_j) \tag{A-40}$$

We thus generalize our concept of probability density to include discrete random variables as well as continuous random variables by utilizing the delta function.

Mass Analogy

It is instructive to relate the concepts of density and distribution functions to the analogous and familiar relations involving the distribution of mass. (Another analogy is that involving charge distributions.)

Consider first a weightless bar as shown in Fig. A-14 with masses M_1, M_2, . . . , M_n suspended at distances x_1, x_2, . . . , x_n from the left end. The weights are normalized to a total weight of 1 lb, so that

$$M_1 + M_2 + \cdots + M_n = 1$$

The masses are then analogous to the probabilities at different values of the discrete variable and could be plotted as in Fig. A-13a.

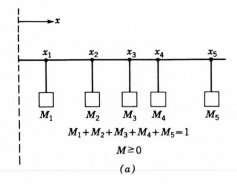

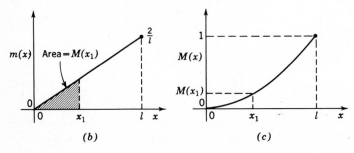

Figure A-14 Mass distributions. (a) Discrete mass distribution. (b) Density function. (c) Distribution function.

The cumulative mass distribution $M(x_j)$ can be defined to be the total mass to the left of and including x_j and is obviously

$$M(x_j) = M_1 + M_2 + \cdot \cdot \cdot + M_j \le 1$$

$M(x_j)$ would plot in the manner of $F(x_j)$ in Fig. A-13.

Now consider a second bar with no hanging weights, but with a mass density $m(x)$, or (mass)/(unit length), varying in some manner along the bar. The *total* mass is still normalized to be 1. If the total length of the bar is l ft, $m(x)$ must satisfy the relation

$$\int_0^l m(x) \, dx = 1$$

Between any two infinitesimally separated points the mass included is $m(x) \, dx$ pounds. For example, if the mass is uniformly distributed, $m(x)$ is a constant equal to $1/l$ lb/ft. A sketch of $m(x)$ would be similar to the $f(\theta)$ function shown in Fig. A-12b.

The cumulative-distribution function $M(x)$, or the mass of the bar included at all points less than x, is

$$M(x) = \int_0^x m(x) \, dx = \frac{x}{l} \qquad \text{pounds}$$

for the uniformly distributed case. This corresponds to the sketch of $F(\theta)$ in Fig. A-12c.

In all cases we must have $m(x) \ge 0$ and $0 \le M(x) \le 1$ is a monotonically increasing function. The mass included between two points x_1 and x_2 is given by

$$\int_{x_1}^{x_2} m(x) \, dx = M(x_2) - M(x_1)$$

All these relations are obviously identical with those set down for the probability-density and distribution functions. Just as in the probability case we must remember that $m(x)$ is a mass *density,* that only $m(x) \, dx$ represents mass (in particular, the mass between $x - dx/2$ and $x + dx/2$), and that the mass at any *one* point is zero if the mass is distributed smoothly throughout the bar.

Another example of a mass-density function might be one in which the density progressively increases along the bar. For example, let $m(x) = Kx$ lb/ft. With the total weight maintained at 1 lb the constant K is given by

$$\int_0^l m(x) \, dx = 1 = K \int_0^l x \, dx = \frac{Kl^2}{2}$$

Then $K = 2/l^2$, and $m(x) = 2x/l^2$ lb/ft. The mass-distribution function is now

$$M(x) = \int_0^x m(x) \, dx = \frac{x^2}{l^2}$$

Both $m(x)$ and $M(x)$ for this example are shown sketched in Fig. A-14b and c.

We can of course extend this analogy, just as in the probability case, to include a bar with both discrete weight and distributed mass. The mass-density function would now include delta functions corresponding to the discrete masses.

Generalized Approach to Distribution Functions

The concepts of probability-distribution function and probability-density function may be developed in a more formal way, without resource to the relative frequency of occurrence of events and histograms, by expanding on the set-theoretic approach of the last section.[6] This again is a much more powerful approach because of its generality. Here one first defines a random variable $\underline{x}$ by associating every possible outcome in the set S with a real number x. The real number x can, in general, range between $-\infty$ and $+\infty$. For any real number x the subset $\{\underline{x} \leq x\}$ of S then corresponds to all outcomes with numbers less than or equal to x. We may then define the probability-distribution function $F(x)$ as the probability that the random variable $\underline{x} \leq x$:

$$F(x) \equiv P(\underline{x} \leq x) \tag{A-41}$$

Two conditions must be appended here:

$$P(\underline{x} = +\infty) = P(\underline{x} = -\infty) = 0$$

It is then easy to demonstrate that $F(x)$ has the following properties:

1. $F(+\infty) = P(\underline{x} \leq \infty) = 1$. [This is the probability $P(S)$ and corresponds to all possible outcomes in the set S.]
2. $F(-\infty) = P(\underline{x} = -\infty) = 0$.
3. If $x_1 < x_2$, $F(x_1) \leq F(x_2)$. (The proof here consists of associating $\{\underline{x} \leq x_1\}$ with the event A, $\{\underline{x} \leq x_2\}$ with B, and using the definitions of probability of the last section.) The cumulative-distribution function is thus a monotonically increasing function, starting at 0 and increasing to 1 as x increases.

Note that both continuous and discrete random variables are contained within this development of the distribution function. If the random variable $\underline{x}$ is continuous, $F(x)$ may be defined as the integral of a density function $f(x)$:

$$F(x) = \int_{-\infty}^{x} f(x)\, dx \qquad f(x) = \frac{dF(x)}{dx} \tag{A-42}$$

It is then apparent that the density function $f(x) \geq 0$, and that $\int f(x)\, dx = F(+\infty) = 1$. [If $\underline{x}$ has discrete values only, $F(x)$ will be stepwise continuous and $f(x)$ is definable in terms of impulse functions, as previously.]

[6] A. Papoulis, *Probability, Random Variables, and Stochastic Processes*, McGraw-Hill, New York, 1965, chap. 4.

The basic difference between this approach and that beginning with the histogram and the density function is that the emphasis here is placed on the *distribution function F(x)*. Since this function represents the probability that the random variable $\underline{x} \leq x$, it may always be found once the possible outcomes of an experiment are associated with the real numbers x. The density function $f(x)$ is then defined in terms of this fundamental function, rather than the other way around.

As an example of this approach in first defining a random variable, and then its distribution function, consider a digital system emitting a train of binary symbols. In any one binary interval we have either one of two binary symbols. This is then a particularly simple set with only two possible (nonoverlapping) events. Associate the real number 0 with one event, 1 with the other. The random variable $\underline{x}$ then has only two values here, 0 and 1. (Although in some cases a 0 and 1 may actually be transmitted, more often the binary symbols will be two different and distinguishable pulse shapes. The random variable $\underline{x}$ does not necessarily have to coincide with the numerical values of the events, if such exist. We could equally well let the real numbers be -10 and -5, -1 and 0, or any other pair of numbers. Obviously, 0 and 1 serve as a particularly good representation for binary numbers.) Assume we know beforehand (a priori) that the symbol represented by the 1 has a probability p of being transmitted, while that represented by the 0 has probability $q = 1 - p$. (As noted in the last section, one finds p by a judicious guess, by making measurements on the system, or by knowledge of the signal source and its statistics.) It is apparent that the distribution function here is particularly simple:

$$F(0^-) = P(\underline{x} < 0) = 0$$

$$F(0) = P(\underline{x} \leq 0) = q$$

$$F(1) = P(\underline{x} \leq 1) = q + p = 1$$

This is sketched in Fig. A-15a. Note that $F(x)$ has the monotonic property indicated previously, rising from 0 to a maximum of 1. In this binary case it has discrete jumps of q and p at the values of $x = 0$ and 1, respectively. The density function is shown in Fig. A-15b.

Now consider n successive bits to be transmitted. Each bit is assumed independent of all others. We then ask for the distribution of 0's and 1's among the bits. In particular, what is the probability of obtaining no 1's (and hence n 0's), one 1 (and $n - 1$ 0's), etc.? In this case let the space S include all possible combinations of 1's and 0's. For simplicity's sake let the random variable $\underline{x}$ here take on the numbers corresponding to the possible numbers of 1's transmitted. Then $\underline{x} = k, k = 0, 1, \ldots, n$. Since the successive bits are assumed independent, the probability of transmitting a particular string of k 1's and $(n - k)$ 0's must be $p^k q^{n-k}$. But there are $\binom{n}{k} = n!/(n - k)!k!$ possible ways in which such a string could be transmitted. Since these different occur-

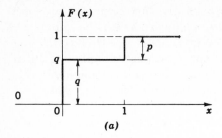

(a)

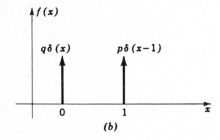

(b)

Figure A-15 Distribution function, binary random variable. (*a*) Distribution function. (*b*) Density function.

rences are mutually exclusive, the probability of transmitting exactly k 1's and $(n - k)$ 0's is just the sum of the probabilities for each possible occurrence, or

$$P(\underline{x} = k) = \binom{n}{k} p^k q^{n-k} \tag{A-43}$$

The cumulative distribution $F(m)$, or the probability that $\underline{x} \leq m$, is then just the sum of the probabilities:

$$F(m) = \sum_{k=0}^{m} \binom{n}{k} p^k q^{n-k} \tag{A-44}$$

Note that this too is a monotonically increasing function, changing with steps of $P(\underline{x} = k)$ at $\underline{x} = k$ (Fig. A-16). As a check, let $m = n$. Then

$$F(n) = \sum_{k=0}^{n} \binom{n}{k} p^k q^{n-k} = (p + q)^n = 1 \tag{A-45}$$

by the binomial expansion, and invoking the identity $p + q = 1$.

This distribution corresponding to the probability of k occurrences in n tries of a binary variable is called the *binomial distribution*. It is probably the most common discrete distribution and occurs quite often in science and engineering. The probability of k heads in n tosses of a coin is of course given by the binomial distribution. (If $p = q = \frac{1}{2}$, the coin is unbiased.) If in a particular electronic system successive voltage samples are divided into two groups—those above a certain level, those below—the distribution of the number exceeding the specified level is, of course, binomial.

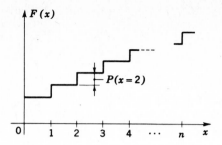

Figure A-16 Binomial distribution.

Note that the density function for the binomial distribution is again made up of a sequence of delta functions. In particular, we have, from Eq. (A-44),

$$f(x) = \sum_{k=0}^{n} P(\underline{x} = k)\delta(x - k) \tag{A-46}$$

A-5 STATISTICAL AVERAGES AND EXAMPLES OF DENSITY FUNCTIONS

The probability-density functions discussed in the previous section (including impulse functions for discrete variables) provide us with information as to the probability or chance that a random variable will occupy a specified portion of its range. As such, they may also be interpreted as weighting functions. Because of this weighting property, the density function also provides us with information as to the average value of a given random variable.

To demonstrate this, assume first that we have a discrete variable x, whose average value we would like to calculate. x can take on the values $x_1, x_2, \ldots , x_n$. We again perform repetitions of a hypothetical experiment, measuring the number of times N_1 that x_1 appears, the number of times N_2 that x_2 appears, etc. The total number of trials N is

$$N_1 + N_2 + \cdots + N_n = N \tag{A-47}$$

The arithmetic or sample average of x is then found in the usual way by *weighting* each value of x by the number of times it appears, summing the weighted values of x, and dividing by the total number of trials N. Thus, av x, the average value of x, is simply

$$\text{av } x = \frac{x_1 N_1 + x_2 N_2 + \cdots + x_n N_n}{N} \tag{A-48}$$

For a large enough number of trials, however, N_1/N is just the relative frequency of occurrence of x, or its probability P_1. (The expression "large enough number of trials" is taken to mean in an intuitive sense the number of trials beyond which we can detect no significant difference in the calculation of N_1/N and of av x.) We can thus write

$$\text{av } x = \sum_{j=1}^{n} P_j x_j \qquad \sum_{j=1}^{n} P_j = 1 \qquad \text{(A-49)}$$

The average value of x in terms of its probability of occurrence, as given by Eq. (A-49), is taken as the *definition* of statistical average. Although it is apparent that the sample average of Eq. (A-48) should presumably approach the statistical average in some sense as the number of samples N becomes very large,[7] we shall henceforth distinguish between the two by using the symbol E to represent statistical average. This symbol stands for the phrase *expected value*, which is commonly used as a synonym for average value. Thus we define the statistical average or expected value of x to be

$$E(x) = \sum_{j=1}^{n} P_j x_j \qquad \sum_{j=1}^{n} P_j = 1 \qquad \text{(A-49a)}$$

The term *mean* value is also commonly used to denote the statistical average.

In a similar way we could find the average of the square of x, the average of the square root of x, etc. In general, if we are interested in the average value of some function $g(x)$, we determine it in the same manner and write

$$E[g(x)] = \sum_{j=1}^{n} P_j g(x_j) \qquad \text{(A-50)}$$

Note that the individual probabilities P_j behave like weighting parameters; those values of x most likely to occur (i.e., those with the highest probabilities) are weighted most heavily in determining the average value. As a special case, if x_5 has a probability $P_5 = 1$ of occurring and all other values of x have zero probability, $E[g(x)] = g(x_5)$, as expected.

As an example, consider the binary variable introduced in the previous section. The average value of x in any one binary interval is just

$$E(x) = q \cdot 0 + p \cdot 1 = p$$

as might be expected. The expected value of the binomial distribution is more interesting. Recall that we asked for the probability of k 1's occurring in an n-bit binary sequence. If n becomes very large, we might expect from relative-frequency considerations to have roughly np 1's and nq 0's occurring in the sequence. (See Chap. 1 for a discussion of this point.) We now show in fact that the *average* number of 1's, $E(k)$, is just np!

---

[7] *Ibid.*, pp. 245, 246. See also M. Schwartz and L. Shaw, *Signal Processing*, McGraw-Hill, New York, 1975, pp. 92–94.

To demonstrate this, we use Eqs. (A-49a) and (A-43) and write

$$E(k) = \sum_{k=0}^{n} kP(\underline{x} = k)$$

$$= \sum_{k=0}^{n} k \binom{n}{k} p^k q^{n-k} \tag{A-51}$$

To evaluate this sum we use a simple trick. Note that

$$k \binom{n}{k} = \frac{n!}{(n-k)!(k-1)!} = n \binom{n-1}{k-1} = n \binom{n-1}{j}$$

with $j = k - 1$, a newly defined index. Rewriting Eq. [A-51] in terms of j, we get

$$E(k) = np \sum_{j=0}^{n-1} \binom{n-1}{j} p^j q^{(n-1)-j} = np \tag{A-51a}$$

(The sum over j is just 1 from the binomial theorem.) On the average, then, one expects to have np 1's and nq 0's appearing in a sequence of n binary symbols. The qualification *average* is highly important here. For the actual number of 1's in n bits is random, varying statistically (and unpredictably) from one n sequence to another. Thus k may generally take on any value from 0 to n. But over many repetitions of this experiment one would expect to find the average becomes np. Alternatively, as n gets very large (effectively equivalent to many independent repeats of the experiment), the number of 1's approaches np. We shall demonstrate this somewhat more quantitatively later on after introducing the concept of variance.

The calculation of the average value of x or of some function $g(x)$ is carried out in the same way for continuous variables. Assume that we perform a hypothetical series of measurements on x. We choose x at Δx intervals, just as in the previous section, in determining the histogram of x. From these measurements we calculate $E(x)$ or $E[g(x)]$ as a straight arithmetic average. Alternatively, we can utilize the information that $f(x) \Delta x$ is the probability that x will be found in the range $x \pm \Delta x/2$. From Eq. (A-50), then,

$$E[g(x)] = \sum_{j} g(x_j) f(x_j) \Delta x$$

In the limit, as $\Delta x \to 0$,

$$E[g(x)] = \int_{-\infty}^{\infty} g(x) f(x) \, dx \tag{A-52}$$

As a special case the average value of x is

$$E(x) = \int_{-\infty}^{\infty} x f(x) \, dx \tag{A-53}$$

in agreement also with Eq. (A-49a). Equations (A-52) and (A-53) are taken as *definitions* in the set-theoretic approach to probability.

The average value of x, $E(x)$, is frequently also called the first moment m_1 of x by analogy with the concept of moments in mechanics. In mechanics, the first moment of a group of masses is just the average location of the masses, or their center of gravity. For example, the center of gravity of the masses M_1, M_2, . . . in Fig. A-14a can be found simply by taking moments about $x = 0$. The average value of x, or the center of gravity, is

$$\text{av } x = \frac{M_1 x_1 + M_2 x_2 + \cdots}{M_1 + M_2 + \cdots}$$

identical with the form of Eq. (A-48). If the mass is not concentrated at discrete points but is smeared out over the bar, we can calculate the center of gravity by considering a differential mass $m(x)\, dx$ [with $m(x)$ the linear mass density] located x units from the origin. The center of gravity is then found by summing all the mass contributions. This sum again becomes an integral, and we have

$$\text{av } x = \int_0^l x m(x)\, dx$$

just as in Eq. (A-53).

The second moment in mechanics is just the moment of inertia of a mass or the turning moment of a torque about a specified point. By analogy with mechanics the second moment of a random variable is the average value of the square of the variable. For discrete variables this is given by

$$m_2 = \sum_{j=1}^{n} P_j x_j^2 \tag{A-54}$$

as a special case of Eq. (A-50). For a continuous variable we have

$$m_2 = E(x^2) = \int_{-\infty}^{\infty} x^2 f(x)\, dx \tag{A-55}$$

as a special case of Eq. (A-52).

We can continue defining higher moments if we wish. In general, the nth moment of x, or av x^n, is given by

$$m_n = E(x^n) = \int_{-\infty}^{\infty} x^n f(x)\, dx \tag{A-56}$$

The density function $f(x)$ plays the role of a weighting function throughout. If some particular range of x, say, in the vicinity of x_j, occurs most frequently, the probability-density function will be highly peaked about that point. The nth moment will then be very nearly x_j^n. In particular, if

$$f(x) = \delta(x - x_j) \tag{A-57}$$

$$m_n = \int_{-\infty}^{\infty} x^n\, \delta(x - x_j)\, dx = x_j^n \tag{A-58}$$

Here x_j is the only point weighted, to the exclusion of all other points.

The delta function is one example of a probability-density function. It enables us to include the case of a variable defined to have one value only. If x is a discrete variable, the density function may be written as a sum of weighted impulse functions as in Eq. (A-40). Thus

$$f(x) = \sum_{j=1}^{m} P_j \, \delta(x - x_j) \tag{A-59}$$

and

$$m_n = E(x^n) = \int_{-\infty}^{\infty} x^n \left[\sum_{j=1}^{m} P_j \delta(x - x_j) \right] dx$$

$$= \sum_{j=1}^{m} P_j x_j^{\,n} \tag{A-60}$$

checking Eq. (A-50).

What is the significance of these moments?

In Chap. 5, in discussing signal and noise problems, we note that m_1 gives just the dc voltage or current. m_2 is found to give the mean-squared voltage (current) or the mean power. These quantities can easily be measured with meters. Since m_1 and m_2 can be determined from the probability-density function $f(x)$, it should be possible in turn to say something about $f(x)$, given the measured values of m_1 and m_2.

Consider, for example, the two possible probability-density functions shown in Fig. A-17. What are the distinguishing characteristics of these two density functions? Obviously $f_1(x)$ is more highly peaked about point a. $f_2(x)$ is more squat and spread out over a larger range of x. A wider range of values of x will thus appear in the case of $f_2(x)$, while in the case of $f_1(x)$ values close to a will appear much more frequently than those somewhat removed. We would expect that the average value of x would be close to, or equal to, a in both cases.

In this particular case the curves happen to have two distinguishing parameters: the location of the peak (a in this case), and the width, or spread, of the curves. We would probably agree intuitively that the first moment m_1 serves as some measure of the location of the peak in this case. (If the curve is symmetri-

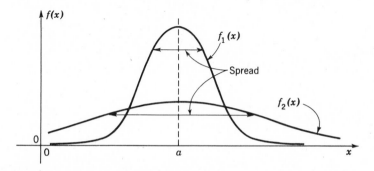

Figure A-17 Two possible density functions.

cal and unimodal, i.e., one peak, the peak will occur at m_1.) In general m_1 is one possible measure of the location of the range of most probable values of x. Other measures used and, in general, providing results not too different from m_1 are the median, or point at which the cumulative distribution $F(x)$ is 0.5, and the mode, or actual value at which the peak occurs if a single peak exists. Figure A-18 shows one example of an unsymmetrical density function encountered in Chap. 5. Here $m_1 >$ modal point.

How do we set up a measure of the width of the curves? We could use the 0.707 or 3-dB points as in Chap. 2 for the amplitude-frequency-response curves. We could also use a specified percentage of the total area about point a as a measure of this spread. In general, the measure of the width or criterion used to define a spread parameter is completely arbitrary. One possible measure of the width of the curve about a is the mean-squared variation about a:

$$s = \text{spread} \equiv \int_{-\infty}^{\infty} (x - a)^2 f(x) \, dx \tag{A-61}$$

Because of the squared term, values of x to either side of a are equally significant in measuring variations away from a. If $f(x)$ happens to be a broad function such as $f_2(x)$ in Fig. A-17 values of x far removed from a will still have sufficiently large values of $f_2(x)$ to be weighted strongly and provide a large value of spread, as desired. If $f(x)$ is a narrow function, such as $f_1(x)$, only those values of x close to a will be weighted significantly, and the spread will be correspondingly smaller.

In general this measure of the width of the curve $f(x)$ will vary with the value of a chosen. Although a was picked as the peak location in Fig. A-17 it could have been chosen anywhere else, with Eq. (A-61) defined to be the spread about that point. The value of a for which the spread is a minimum is of interest. To find this point, differentiate Eq. (A-61) with respect to a. This gives

$$\frac{ds}{da} = 0 = -2 \int_{-\infty}^{\infty} (x - a) f(x) \, dx$$

$$= 2a \int_{-\infty}^{\infty} f(x) \, dx - 2 \int_{-\infty}^{\infty} x f(x) \, dx \tag{A-62}$$

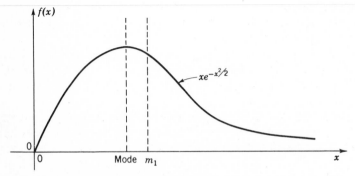

Figure A-18 Unsymmetrical density function.

But $\int_{-\infty}^{\infty} f(x)\, dx = 1$ and $m_1 = \int_{-\infty}^{\infty} xf(x)\, dx$ by definition. Then

$$a = m_1 = \int_{-\infty}^{\infty} xf(x)\, dx$$

or the point about which this mean-squared measure of the curve width is a minimum is just the average value, or first moment. In the symmetrical curves of Fig. A-17 this happens to coincide with the peak. In the unsymmetrical curve of Fig. A-18 it is to the right of the peak. The fact that $a = m_1$ gives a minimum and not a maximum is shown very simply by taking the second derivative of Eq. (A-61).

The spread about m_1, or the mean-squared variation about m_1, is called the *variance* or second central moment μ_2.

$$\mu_2 \equiv \int_{-\infty}^{\infty} (x - m_1)^2 f(x)\, dx = E[(x - m_1)^2] \qquad \text{(A-63)}$$

The square root of this term is called the *standard deviation* and is frequently given the symbol σ. We shall use σ as the measure of the spread of the density function about m_1. Thus

$$\mu_2 = \sigma^2$$

Multiplying out the squared term in the integral of Eq. (A-63) and integrating term by term, we get

$$\mu_2 = \sigma^2 = \int_{-\infty}^{\infty} (x - m_1)^2 f(x)\, dx$$

$$= \int_{-\infty}^{\infty} x^2 f(x)\, dx - 2m_1 \int_{-\infty}^{\infty} xf(x)\, dx + m_1^2 \int_{-\infty}^{\infty} f(x)\, dx$$

$$= m_2 - m_1^2 \qquad \text{(A-64)}$$

from Eq. (A-53).

The variance is thus the second moment less the square of the first moment. m_2 represents the spread of the curve about $x = 0$, μ_2 about $x = m_1$. Equation (A-64) is analogous to the parallel-axis theorem of mechanics, with m_2 the moment of inertia about the origin, μ_2 the moment of inertia about the center of gravity. In terms of mean-squared voltages and mean power, σ^2 is considered to be the mean ac power and σ the rms ac voltage.

We now consider three typical distributions as illustrative examples of the above ideas. The first is the binomial distribution again, as an example of discrete random variables; the other two refer to continuous random variables.

Binomial distribution. We found earlier that the expected value or first moment in this case was $E(k) = np$ [Eq. (A-51a)]. We also indicated that as n increased, one would expect the number of 1's in a sequence of n binary symbols to approach the expected value np. In fact, from relative-frequency considerations one would expect p, the a priori probability of a 1 occurring, to be given "very closely'' by k/n, as n becomes very large. The number k here represents the actual number of 1's counted in a sequence n bits long. All this intuitive reasoning implies that the variance or spread about p becomes small as n gets large. We now show by calculating the variance that these qualitative arguments are in fact true. This will then also serve as a quantitative justification of the relative frequency argument.

From Eq. (A-64), $\sigma^2 = m_2 - m_1^2$. (Although derived for continuous variables, this is easily proved as well for discrete variables. The proof is left as an exercise for the reader.) Since we already know

$$E(k) = m_1 = np$$

for the binomial distribution, we must now find m_2 to determine the variance σ^2. Specifically, then,

$$m_2 = E(k^2) = \sum_{k=0}^{n} k^2 \binom{n}{k} p^k q^{n-k} \tag{A-65}$$

We use a trick similar to the one used to find the first moment $E(k)$. Subtracting and adding $E(k)$ to the right-hand side of Eq. (A-65), we have

$$m_2 = \sum_{k=2}^{n} k(k-1) \binom{n}{k} p^k q^{n-k} + np$$

$$= n(n-1)p^2 \sum_{j=0}^{n-2} \frac{(n-2)!}{j!(n-2-j)!} p^j q^{(n-2)-j} + np$$

(Here we have introduced the dummy index $j = k - 2$.) But the sum is again just 1, by the binomial theorem. Then

$$m_2 = n(n-1)p^2 + np \tag{A-65a}$$

and the variance σ^2 is

$$\sigma^2 = m_2 - m_1^2 = np - np^2 = npq \tag{A-66}$$

with $q = 1 - p$. The standard deviation or spread about $E(k) = np$ is thus

$$\sigma = \sqrt{npq} \tag{A-67}$$

As an example, let $n = 9$, $p = q = \frac{1}{2}$. Then $E(k) = 4.5$, and $\sigma = 1.5$. The sketch of the density function in Fig. A-19 shows the clustering in the region $E(k) \pm \sigma$ or 3 to 6. Note that as n increases, the *relative* spread about $E(k)$, $\sigma/E(k)$, approaches zero:

$$\frac{\sigma}{E(k)} = \frac{1}{\sqrt{n}} \sqrt{\frac{q}{p}} \tag{A-68}$$

This indicates that for n ''large enough,'' the number of 1's in an n-bit sequence is highly concentrated about the average value np, the concentration

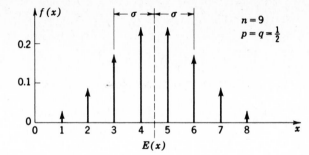

Figure A-19 Binomial distribution: density function. (After A. Papoulis, *Probability, Random Variables, and Stochastic Processes,* McGraw-Hill, New York, 1965, fig. 4.9, p. 102.)

packing in more and more as n increases. Alternatively, assume we take a long string of 1's and 0's and measure the number k of 1's occurring. We say this is a "good" measure or *estimate* of the (unknown) a priori probability. Call this estimate $\hat{p}$. Thus

$$\hat{p} = \frac{k}{n} \tag{A-69}$$

Since k is a random number, $\hat{p}$ is a random variable as well. What are its properties? We have as its average value

$$E(\hat{p}) = \frac{E(k)}{n} = p \tag{A-70}$$

so that *on the average* our estimate gives us p. (Such an estimate is said to be *unbiased.*) The variance of this estimate is then seen to be

$$\hat{\sigma}^2 = E(\hat{p} - p)^2 = \frac{\sigma^2}{n} = \frac{pq}{n} \tag{A-71}$$

and the relative spread about the expected value p of the estimate is

$$\frac{\hat{\sigma}}{p} = \frac{\sigma}{E(k)} = \frac{1}{\sqrt{n}} \sqrt{\frac{q}{p}} \tag{A-68a}$$

identical to the relative spread about $E(k)$. (This is obvious, since $\hat{p} = k/n$, and we are simply normalizing, dividing through by n.) The density function $f(\hat{p})$ is shown sketched in Fig. A-20 for large n. As an example, say that $p = 10^{-2}, q = 0.99 \approx 1, n = 10^5$. Then one would expect the number of 1's in a binary string 10^5 bits long to cluster about an average value of 10^3. The spread about 10^3 is $\sigma = \sqrt{npq} \doteq \sqrt{1,000} \doteq 33$. The relative spread is $\sigma/E(k) = 0.033$. If n is now increased to $10^7, E(k) = 10^5, \sigma \doteq \sqrt{10^5} \doteq 330$, and $\sigma/E(k) = 0.0033$, indicating the relative reduction in the spread by a factor of 10, as n increases by 100.

Rectangular distribution. Assume that the random variable x is continuous and uniformly distributed with density function $f(x) = K$ between $x = a$ and

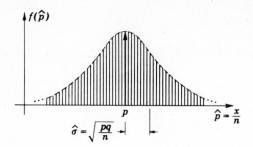

Figure A-20 Estimation of p, binomial distribution.

$x = b$ (Fig. A-21). The pointer problem of the previous section is one example of this density function. (There we had $a = 0$, $b = 2\pi$.) The PCM quantization error discussed in Chap. 3 is assumed to be uniformly distributed and hence has this form of density function. The variable x has no values less than a and greater than b and is equally likely to be found anywhere from a to b.

Judging from the curve of Fig. A-21, we should expect the average value to be halfway between a and b, or at $m_1 = (a + b)/2$. The width of the curve should obviously be related to $b - a$.

Since $f(x)$ must be normalized to have unity area [the probability that x lies somewhere between a and b is $\int_a^b f(x)\ dx = 1$], the constant K must be $1/(b - a)$. m_1 is given by

$$m_1 = \int_a^b xf(x)\ dx = \frac{1}{b - a} \int_a^b x\ dx = \frac{a + b}{2} \tag{A-72}$$

as expected. The second moment m_2 is given by

$$m_2 = \int_a^b x^2 f(x)\ dx = \frac{1}{b - a} \int_a^b x^2\ dx = \frac{b^3 - a^3}{3(b - a)}$$

$$= \frac{b^2 + ab + a^2}{3} \tag{A-73}$$

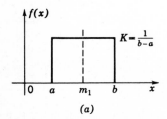
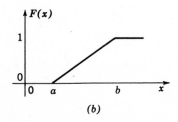

(*a*)   (*b*)

Figure A-21 Rectangular distribution. (*a*) Probability density. (*b*) Distribution function.

The variance is then

$$\sigma^2 = m_2 - m_1{}^2 = \frac{(b-a)^2}{12} \tag{A-74}$$

This result could have been obtained just as well by finding σ^2 directly from Eq. (A-63). The standard deviation σ is then

$$\sigma = \frac{b-a}{2\sqrt{3}} \tag{A-75}$$

It thus appears in terms of $b - a$, as expected.

Note that if the width of the curve, $b - a$, is made smaller, the height $1/(b - a)$ increases correspondingly. The variable x is constrained to lie within a narrower range of variation and has a greater probability of doing so. Ultimately, as $b - a \to 0$, by letting b approach a, the height $\to \infty$, but with the area under the curve always equal to 1. In the limit $f(x) \to \delta(x - a)$, and the variable has only one possible value, $x = a$, with a probability of 1.

As an example of the use of the rectangular distribution, we can let the variable x be the PCM quantization error ϵ of Chap. 3. This error is the error due to quantization of continuous voltages measured within a specified range about a particular voltage level. If all voltages about this level are assumed equally likely to occur, ϵ will be uniformly distributed with zero average value. The voltage separation between adjacent levels is just the $b - a$ variation here. If this is set equal to a volts, as in Chap. 3, the rms quantization error is the standard deviation σ, or $a/(2\sqrt{3})$ volts. This is exactly the result obtained in Chap. 3.

As pointed out by Bennett,[8] these results are applicable in general to quantizing noise arising when analog signals are converted to their quantized digital form and are not limited to the PCM problem alone.

The cumulative-distribution function, or the probability that x will be less than some specified value, is found to be

$$F(x) = 0 \qquad\qquad x < a$$

$$F(x) = \frac{x - a}{b - a} \qquad a < x < b \tag{A-76}$$

$$F(x) = 1 \qquad\qquad x \geq b$$

Gaussian or normal distribution. The gaussian density function is one of the most important density functions in probability theory and statistics. We find it recurring over and over again, for example, in discussing noise statistics in

---

[8] W. R. Bennett, "Methods of Solving Noise Problems," *Proc. IRE,* vol. 44, no. 5, pp. 609–638, May 1956.

this book. The gaussian function for one variable is given by the equation

$$f(x) = \frac{e^{-(x-a)^2/2\sigma^2}}{\sqrt{2\pi\sigma^2}} \tag{A-77}$$

When plotted, it has the characteristic bell-shaped curve of Fig. A-22. The curve is symmetrical about the point $x = a$ and has a width proportional to σ. This is apparent from Eq. (A-77), for if we pick the point $x - a = \sqrt{2}\sigma$ at which the exponential is unity and let σ increase, $x - a$ increases correspondingly. Intuitively, we would expect a to be just the first moment, or average value of x, for this distribution. We can check this by actually performing the required integration. Thus

$$m_1 = \int_{-\infty}^{\infty} xf(x)\, dx = \int_{-\infty}^{\infty} \frac{xe^{-(x-a)^2/2\sigma^2}\, dx}{\sqrt{2\pi\sigma^2}} \tag{A-78}$$

Upon changing variables by letting $y = (x - a)/\sqrt{2\sigma^2}$, $dx = \sqrt{2}\,\sigma\, dy$. The limits of integration remain $-\infty$ and ∞, so that we get

$$m_1 = \int_{-\infty}^{\infty} \frac{(\sqrt{2}\sigma y + a)e^{-y^2}}{\sqrt{\pi}}\, dy$$

But

$$\int_{-\infty}^{\infty} ye^{-y^2}\, dy = 0$$

(either by direct integration, or by noting that the function is odd), and

$$\int_{-\infty}^{\infty} e^{-y^2}\, dy = \sqrt{\pi}$$

from any table of definite integrals. This gives us immediately

$$m_1 = a \tag{A-79}$$

for this particular function.

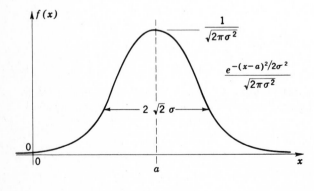

Figure A-22 Density function, gaussian distribution.

In a similar manner we can show by integration that σ^2 in Eq. [A-77] is the second central moment or variance μ_2 of this distribution.

$$\mu_2 = \int_{-\infty}^{\infty} \frac{(x - a)^2 e^{-(x-a)^2/2\sigma^2} \, dx}{\sqrt{2\pi\sigma^2}} = \sigma^2 \tag{A-80}$$

The actual calculation is left as an exercise for the reader.

Since σ in Eq. (A-77) has been shown to be a measure of the width of the gaussian curve, this result agrees with our interpretation of the standard deviation as a measure of the spread or width of a probability-density curve.

That $f(x)$ as given by Eq. (A-77) is properly normalized can be shown by direct integration. Thus

$$\int_{-\infty}^{\infty} f(x) \, dx = \int_{-\infty}^{\infty} \frac{e^{-(a-x)^2/2\sigma^2}}{\sqrt{2\pi\sigma^2}} \, dx = 1 \tag{A-81}$$

This integration is also left as a simple exercise for the reader.

This gaussian curve weights values of x near a most heavily. The value of $f(x)$ at the peak is $1/\sqrt{2\pi\sigma^2}$, so that, as the width σ decreases, the height of the curve in the vicinity of $x = a$ increases. Ultimately, for $\sigma \to 0$, this curve approaches the delta function $\delta(x - a)$, and the variable x becomes a constant a with a probability of 1. Note that the sketch of Fig. A-22 is very similar to that of Fig. A-20 for the binomial distribution. It may in fact be shown under some simple conditions that the binomial distribution approaches the gaussian for n large enough.[9]

The cumulative-distribution function, or the probability that the variable will be less than some value x, is

$$F(x) = \int_{-\infty}^{x} \frac{e^{-(x-a)^2/2\sigma^2}}{\sqrt{2\pi\sigma^2}} \, dx \tag{A-82}$$

Since the $f(x)$ curve is symmetrical about $x = a$, half the area is included from $-\infty$ to a. The probability that $x \le a$ is thus 0.5, or

$$F(a) = 0.5 \tag{A-83}$$

As mentioned before, the 0.5 probability point is called the median of a statistical distribution. For the gaussian function the median, the average value, and the modal point [peak of $f(x)$] all coincide.

$F(x)$ is shown plotted in Fig. A-23. The curve is symmetrical about the point $x = a$.

The probability distribution of fluctuation noise discussed in Chap. 5 is of the gaussian form. The average value of the noise is 0 volts, however, so that the curve is symmetrical about the origin. σ then represents the rms value of the noise voltage. A question frequently asked is: What is the probability that the noise voltage will be less than some prescribed $K\sigma$

---

[9] Papoulis, *op. cit.*, p. 269.

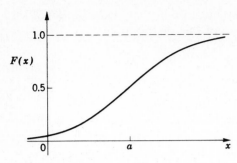

Figure A-23 Cumulative-distribution function, gaussian distribution.

(K a constant)? (Positive and negative voltages are included.) Letting x represent the instantaneous noise voltage, we can answer the question by writing

$$\text{Prob}\,(-K\sigma < x < K\sigma) = \int_{-K\sigma}^{K\sigma} f(x)\,dx$$

$$= \int_{-K\sigma}^{K\sigma} \frac{e^{-x^2/2\sigma^2}}{\sqrt{2\pi\sigma^2}}\,dx \qquad \text{(A-84)}$$

This integral cannot be evaluated in closed form. Instead, the integrand must be expanded in a power series and the resultant integral evaluated term by term. It can be put into a form more convenient for tabulation by letting $y = x/\sqrt{2}\sigma$. Doing this, and utilizing the symmetry of $f(x)$, we get

$$\text{Prob}\,(-K\sigma < x < K\sigma) = \frac{2}{\sqrt{\pi}} \int_{0}^{K/\sqrt{2}} e^{-y^2}\,dy \qquad \text{(A-85)}$$

This integral is frequently called the *error function* and is abbreviated erf $(K/\sqrt{2})$. In general,

$$\text{erf}\, x \equiv \frac{2}{\sqrt{\pi}} \int_{0}^{x} e^{-y^2}\,dy \qquad \text{(A-86)}$$

and

$$\text{Prob}\,(|x| < K\sigma) = \text{erf}\, \frac{K}{\sqrt{2}} \qquad \text{(A-87)}$$

The error function is tabulated in various books of mathematical tables[10] and in books on probability and statistics. Using these tables, we find that, for $K = 1$,

$$\text{erf}\, \frac{1}{\sqrt{2}} = 0.683$$

[10] See, e.g., B. O. Peirce, *A Short Table of Integrals*, Ginn, Boston, 1929.

and, for $K = 2$,

$$\text{erf } \frac{2}{\sqrt{2}} = 0.955$$

The probability that the noise voltage will be less than σ volts in magnitude is thus 0.68. The probability that the voltage will be less than twice the rms noise voltage (2σ) is 0.95.

Although we have used fluctuation noise as an example here, these results are more general. Thus it is easy to show that for any variable which has a gaussian probability-density function, the probability that the variable will deviate from the average value by less than σ is 0.68, while the chance of a deviation greater than 2σ is $1 - 0.95 = 0.05$. For example, assume that 100,000 resistors are to be manufactured with 100 kΩ nominal resistance. Owing to variations in the raw material and in the manufacturing process used the resistors will actually vary about the 100-kΩ nominal value. Assume that the variations away from this value follow a gaussian curve with a standard deviation of 10 percent of the average value. σ is then 10 kΩ, and 68 percent of the group of 100,000, or 68,000 resistors, should on the average have resistances within ± 10 kΩ of the 100-kΩ rated value. On the average 95 percent, or 95,000, should lie within ± 20 kΩ of the rated value.

The cumulative probability distribution of Eq. (A-82) can be related quite simply to the error function defined by Eq. (A-86). It is left as an exercise to the reader to show that

$$F(x) = \frac{1}{2} \left(1 + \text{erf } \frac{x - a}{\sqrt{2}\sigma} \right) \tag{A-88}$$

A-6 FUNCTIONAL TRANSFORMATIONS OF RANDOM VARIABLES

Up to this point in our discussion of random variables we have been assuming the probability density known. In the case of random noise this has been assumed to be the gaussian distribution.

We point out in discussing modulation systems in Chaps. 3 and 4 that many operations, both linear and nonlinear, may be performed on a signal. Noise passing through a system is also operated on by modulators, rectifiers, amplifiers, filters, etc. It is of interest to investigate the effect of the system operation on the probability distribution. We shall restrict ourselves in this section primarily to nonlinear operations, with the output voltage y of a device related to the input v by the relation

$$y = G(v)$$

If v is a random variable with known probability density $f_v(v)$ (fluctuation noise of zero dc level, for example), what is the probability density $f_y(y)$ at the output?

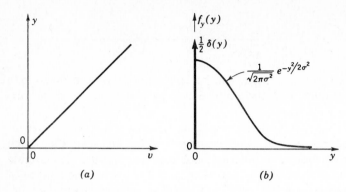

Figure A-24 Piecewise-linear rectification of noise. (*a*) Piecewise-linear rectifier. (*b*) Probability density, output noise.

As an example, assume that random noise is passed through a piecewise-linear rectifier of the type analyzed as a modulator and demodulator in Chap. 4. We assume that the characteristics are given by

$$y = y \qquad v \geq 0$$
$$y = 0 \qquad v < 0 \qquad \text{(A-89)}$$

What is the probability density $f_y(y)$ of the noise at the output (see Fig. A-24*a*)?

If the input noise happens to be of positive amplitude, it will be passed undisturbed through the rectifier. The probability that $y > 0$ must be the same as the probability that $v > 0$. For $v > 0$, then,

$$f_y(y) = f_v(v) = \frac{1}{\sqrt{2\pi\sigma^2}} \, e^{-v^2/2\sigma^2} \qquad v > 0 \qquad \text{(A-90)}$$

But for $v < 0$ the rectifier produces zero output ($y = 0$). The probability that $v < 0$ is 0.5, and this must correspond to the probability that $y = 0$. Thus

$$\text{Prob } (v < 0) = 0.5 = \text{Prob } (y = 0) \qquad \text{(A-91)}$$

The probability-density function corresponding to this must be a delta function of area $\frac{1}{2}$ centered at $y = 0$. The total probability density for y is then

$$f_y(y) = \tfrac{1}{2}\delta(y) + \frac{1}{\sqrt{2\pi\sigma^2}} \, e^{-y^2/2\sigma^2} \qquad y \geq 0 \qquad \text{(A-92)}$$

Note that y has both a discrete and a continuous part. $f_y(y)$ is properly normalized, as can be seen by integrating $f_y(y)$ over all values of y (Fig. A-24*b*).

We can use $f_y(y)$ to calculate the expected value or dc component of the noise at the output (now nonzero), the rms voltage, etc.[11] The dc voltage,

---

[11] As indicated earlier, we show in Chap. 5 that the first moment and the dc voltage are the same under some simple conditions. Similarly, the standard deviation and the rms voltage are shown to be the same.

for example, is given by

$$m_1 = E(y) = \int_0^\infty \frac{ye^{-y^2/2\sigma^2}}{\sqrt{2\pi\sigma^2}} \, dy + \int_{0-}^\infty y \, \delta(y) \, dy$$

$$= \frac{\sigma}{\sqrt{2\pi}} \tag{A-93}$$

where σ is the rms voltage of the input noise.

If the transformation $y = G(v)$ can be expressed algebraically and the inverse $v = g(y)$ readily found, a simple relation between $f_y(y)$ and $f_v(v)$ can be developed.

Let v vary between v_1 and v_2. y varies correspondingly between $y_1 = G(v_1)$ and $y_2 = G(v_2)$. The probability that v will range between v_1 and v_2 is $\int_{v_1}^{v_2} f_v(v) \, dv$, and this must be just the probability that y will vary between y_1 and y_2. Thus

$$\int_{v_1}^{v_2} f_v(v) \, dv = \int_{y_1}^{y_2} f_y(y) \, dy \tag{A-94}$$

Now let $v = g(y)$ in the left-hand integral. Assuming that $g(y)$ is a *single-valued function* of y, we get, after transforming the integrand and changing the limits of integration,

$$\int_{v_1}^{v_2} f_v(v) \, dv = \int_{y_1}^{y_2} f_v[g(y)]g'(y) \, dy = \int_{y_1}^{y_2} f_y(y) \, dy \tag{A-95}$$

Comparing the two integrals on the right, we get the relation

$$f_y(y) = f_v[g(y)]g'(y) \tag{A-96}$$

Geometrically, Eq. (A-94) corresponds to equating the two areas shown in Fig. A-25. In particular, as $v_1 \rightarrow v_2$, we must have

$$f_y(y) \, dy = f_v(v) \, dv$$

which is just the relation given by Eq. (A-96).

As an example, assume that v is uniformly distributed in the range $0 \leq v \leq 1$, with $f_v(v) = 1$. What is the distribution at the output of a square-law detector with the characteristic $y = v^2$ (Fig. A-26)?

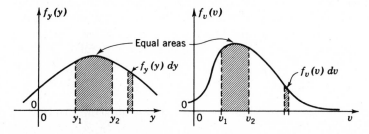

Figure A-25 Transformation of a random variable.

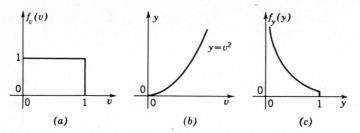

Figure A-26 Example of a transformation. (*a*) Input variable. (*b*) Characteristic. (*c*) Output variable.

Here $g(y) = \sqrt{y}$ (since v is positive, the positive square root must be used); $g'(y) = \frac{1}{2}y^{-1/2}$, and the density function at the output is

$$f_y(y) = \frac{y^{-\frac{1}{2}}}{2} \qquad 0 \le y \le 1 \tag{A-97}$$

This is shown sketched in Fig. A-26c. We can check to see whether or not $f_y(y)$ is properly normalized by calculating the area under the curve.

$$\int_0^1 f_y(y)\ dy = \frac{1}{2}\int_0^1 y^{-1/2}\ dy = 1$$

as expected.

The cumulative-distribution function is given by

$$F_y(y) = \int_0^y f_y(y)\ dy = y^{1/2} \tag{A-98}$$

In particular, the probability that the output will be less than $\frac{1}{2}$ is $F_y(\frac{1}{2}) = 0.707$.

The square-law transformation is also interesting because it gives us the density function of the instantaneous *power* if the input function v happens to be a voltage or a current applied to a 1-Ω resistor. For example, in the example just given, the probability that the voltage will be less than 0.707 is just 0.707 (uniform or rectangular distribution). This must correspond to the probability that the power will be less than $\frac{1}{2}$, checking the answer obtained.

As another simple example of the transformation, assume $f_v(v) = e^{-v}$, $0 \le v < \infty$. (The properties of this density function are left to the reader as an exercise.) What is the density function of $y = v^2$? (This again corresponds to passing the variable v through a square-law detector, or to finding the probability distribution of the power.)

Here $v = g(y) = y^{1/2}$ again, $f_v[g(y)] = e^{-\sqrt{y}}$, $g'(y) = \frac{1}{2}y^{-1/2}$, and

$$f_y(y) = \frac{e^{-\sqrt{y}}}{2\sqrt{y}} \qquad 0 \le y < \infty \tag{A-99}$$

As a third example, assume that an angular variable θ is uniformly distributed between the values $-\pi/2 \le \theta \le \pi/2$. Then $f_\theta(\theta) = 1/\pi$ in that range. We would like to find the distribution of $y = a \sin \theta$. (This corresponds to finding the density function of a sine wave whose angle is uniformly distributed.)

Here $g(y) = \sin^{-1}(y/a)$, $g'(y) = 1/\sqrt{a^2 - y^2}$, and

$$f_y(y) = \frac{1}{\pi \sqrt{a^2 - y^2}} \qquad -a \le y \le a \qquad \text{(A-100)}$$

$f_y(y)$ is shown sketched in Fig. A-27b. This result is as might be intuitively expected, for the sine function spends most of its time in the region of $\theta = \pi/2$ or $-\pi/2$ (where its derivative or rate of change has the smallest value). y would thus be expected to have the largest probability density in the vicinity of a and $-a$.

In all three examples cited thus far $v = g(y)$ has been single-valued because of the range of v chosen.

As a final example, consider the case of gaussian-distributed noise passed through a square-law detector. Alternatively, we may ask for the density function of the instantaneous power corresponding to a gaussian-distributed noise voltage. Thus with

$$f_v(v) = \frac{e^{-v^2/2\sigma^2}}{\sqrt{2\pi\sigma^2}} \qquad -\infty < v < \infty$$

and

$$y = v^2$$

find $f_y(y)$.

Here we have to be very careful. Both positive and negative v contribute to y, and $v = g(y) = \pm\sqrt{y}$ is no longer single-valued. To treat this case, assume that we are interested in the probability that y lies between two values y_1 and y_2. Since *two* sets of values of v give rise to these values of y and the values of v are *mutually exclusive,* we must sum the probabilities that v lies in the two different regions, to find

$$\text{Prob } (y_1 \le y \le y_2)$$

Calling the two sets of values of v, v_1, v_2 and v_1', v_2',

$$\text{Prob } (y_2 \le y \le y_2) = \text{Prob } (v_1 \le v \le v_2) + \text{Prob } (v_1' \le v \le v_2')$$

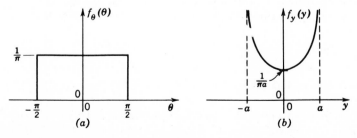

Figure A-27 Phase uniformly distributed. (a) $f_\theta(\theta)$. (b) $f_y(y)$: $y = a \sin \theta$.

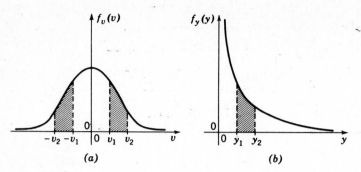

Figure A-28 Gaussian noise at output of square-law device. (*a*) Input distribution. (*b*) Output distribution.

In terms of the density functions we get

$$\int_{y_1}^{y_2} f_y(y) \, dy = \int_{v_1}^{v_2} f_r(v) \, dv + \int_{v_1'}^{v_2'} f_r(v) \, dv \tag{A-101}$$

for a double-valued function.

The gaussian function happens to be symmetrical about $v = 0$, so that $f_r(v)$ is the same for the positive and negative values of v that produce the same y. This is shown in Fig. A-28a. In this case we get

$$f_y(y) = 2f_r[g(y)]g'(y) \tag{A-102}$$

from Eq. (A-101).

In particular, for $y = v^2$, $g'(y) = 1/2\sqrt{y}$, and

$$f_y(y) = \frac{e^{-y/2\sigma^2}}{\sqrt{2\pi\sigma^2 y}} \qquad 0 \le y < \infty \tag{A-103}$$

is the distribution of the noise power, or the distribution at the output of a square-law device.

As a check, it is left to the reader to show that the average value of y is $m_1 = \sigma^2$. Since the input noise voltage has zero average value, its second moment, or total mean power, is just σ^2. (For the voltage, $m_2 = m_1 + \sigma^2 = \sigma^2$ in this case.) The second moment of the voltage is thus just the first moment of the power, as might be expected: the average value of the instantaneous power corresponds to the mean-squared voltage across a 1-ohm resistor.

A-7 DISTRIBUTION OF n INDEPENDENT VARIABLES

Up to this point we have been considering probability-density functions and cumulative distributions for one variable [for example, $f_x(x)$ and $F_x(x)$]. Just as in the case of discrete variables it is of interest to extend the notions of continuous probability functions to include the probability of joint occurrence of two or more continuous independent random variables.

This will enable us to then discuss the probability distributions of the sum of many independent random variables. Summation occurs commonly in many communication systems so that this problem is of substantial interest in its own right. As an example, signal pulses may be repeated several times to enable signals to be better detected in the presence of noise. (Multiple signal samples are discussed in Chap. 6.) The summation of many noise samples is encountered in discussing PCM repeaters in Chap. 5.

In addition we shall show by example that the sum of many independent random variables approaches a gaussian-distributed variable under some rather simple conditions. This is the so-called *central-limit theorem* of probability theory. It provides the motivation for spending so much time on gaussian statistics, as well as the justification for assuming noise statistics to be gaussian.

Two Independent Variables

We recall that in Sec. A-2 we introduced the concept of joint probability for discrete variables by defining a probability $P(AB)$ that both event A *and* event B occur. For example, considering playing cards, if A represents a heart, B a king, $P(AB)$ represents the probability of the king of hearts being drawn. If A represents a spade on the first of two card draws and B a spade on the second draw, $P(AB)$ is the probability of drawing two spades in succession.

If the event B is dependent on the event A, as is true in the two-spade case,

$$P(AB) = P(A)P(B|A)$$

where $P(B|A)$ is the conditional probability that event B will occur, it being known that A has occurred.

In the two-spade case, for example,

$$P(A) = \tfrac{13}{52} = \tfrac{1}{4} \qquad P(B|A) = \tfrac{12}{51} \qquad P(AB) = \tfrac{1}{17}$$

We also showed, in Sec. A-2, that if B is independent of event A, and vice versa,

$$P(B|A) = P(B)$$

$$P(A|B) = P(A)$$

$$P(AB) = P(A)P(B)$$

In the example given, this would correspond to replacing the first card once drawn.

By extension to the continuous-probability case we can express the probability that two variables x and y will jointly take on values between x_1 and x_2, y_1 and y_2, respectively, by the expression

$$\text{Prob } (x_1 \le x \le x_2, \ y_1 \le y \le y_2) = \int_{x_1}^{x_2} \int_{y_1}^{y_2} f_{xy}(x,y) \ dx \ dy \qquad \text{(A-104)}$$

$f_{xy}(x,y)$ is a *two-dimensional* probability density, and the joint probability is the *volume* enclosed under the $f_{xy}(x,y)$ curve in the region bounded by x_1, x_2, y_1, y_2. This compares with the one-dimensional density function $f_x(x)$ for a single continuous variable, with the probability given by an area under the curve. $f_{xy}(x,y)\, dx\, dy$ represents the probability that x and y will lie jointly in the region

$$x \leq x \leq x + dx$$

$$y \leq y \leq y + dy$$

As an example of the need for the joint-probability formulation occurring in our work, we may ask for the probability that a noise voltage $n(t)$ will at time t appear between n_1 and $n_1 + dn$ volts and, τ sec later, between n_2 and $n_2 + dn$ volts. $n(t)$ then corresponds to x, $n(t + \tau)$ to y (Fig. A-29). We discuss this specific problem in Chap. 5.

The two-dimensional cumulative distribution also becomes a volume integral and is given by

$$F_{xy}(x_1,y_1) = \int_{-\infty}^{x_1} \int_{-\infty}^{y_1} f_{xy}(x,y)\, dx\, dy \qquad \text{(A-105)}$$

This represents the probability that x will be less than x_1, y less than y_1.

f_{xy} can be found from the limiting case of a relative-frequency expression, just as in the case of a single variable. For example, we could cut up the noise-voltage curve of Fig. A-29 into many strips T sec long ($T \gg 1/\text{band-width}$), divide the voltage scale into many levels Δn volts wide, and sample the voltage at fixed times t sec and $(t + \tau)$ sec from the beginning of each strip. The fraction of times that the voltages $n_1 \pm \Delta n/2$ at t and $n_2 \pm \Delta n/2$ at $t + \tau$ appeared would give us an expression approximating $f_{n_1 n_2}\, dn_1\, dn_2$. This procedure is of course identical with the method used in Sec. A-2 to find joint-probability expressions from relative-frequency counts and is just an extension of the histogram idea to two variables.

Consider now the case where the two variables x and y are independent. (As shown in Chap. 5, in the case of Fig. A-29 this corresponds to $\tau \gg 1/\text{band-width}$.) By direct extension of our results for the discrete case we define

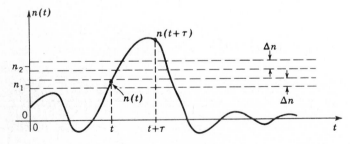

Figure A-29 Joint probability, noise voltage.

the condition of independence to be

$$f_{xy}(x,y) = f_x(x)f_y(x) \tag{A-106}$$

The joint-probability density of two independent variables is then just the product of the individual probability-density functions, and the individual probabilities multiply. The probability that x will range between x_1 and x_2 and y between y_1 and y_2 is

$$\text{Prob } (x_1 \le x \le x_2, \ y_1 \le y \le y_2) = \int_{x_1}^{x_2} \int_{y_1}^{y_2} f_x(x)f_y(y) \ dx \ dy$$

$$= \int_{x_1}^{x_2} f_x(x) \ dx \int_{y_1}^{y_2} f_y(y) \ dy \tag{A-107}$$

For example, assume that the instantaneous noise voltage across each of two resistors R_A and R_B is measured. Assuming the voltage across each resistor is gaussian-distributed and the two voltages are independent of each other (see Chap. 5),

$$f_{A,B}(n_A,n_B) = f_A(n_A)f_B(n_B) = \frac{e^{-(n_A{}^2/2\sigma_A{}^2 + n_B{}^2/2\sigma_B{}^2)}}{\sqrt{2\pi\sigma_A{}^2 \times 2\pi\sigma_B{}^2}} \tag{A-108}$$

where σ_A and σ_B are the rms noise voltages of resistors A and B, respectively. The probability that voltmeter A across R_A will read between 2 and 3 V and voltmeter B across R_B between 1 and 3 V, for example, is found by integrating Eq. (A-108) between these sets of limits or by finding the individual probabilities and multiplying them together.

Sum of Independent Variables: Characteristic Functions

As a special case of the handling of many independent random variables we now consider the situation where two or more variables are summed to find a new random variable. This, as noted earlier, has applicability to many communication problems. The techniques to be discussed also lead to the central-limit theorem mentioned earlier that the distribution of the sum of n independent variables approaches the gaussian (normal) distribution.

We start first by considering two random variables as in the previous paragraph. We then generalize to n independent variables. Consider then a variable $z = x + y$. With $f_x(x)$ and $f_y(y)$ given and x and y independent how do we find $f_z(z)$?

The most direct approach is that of performing a transformation of variables just as in the previous section. We therefore define two new variables given by

$$z = x + y \tag{A-109}$$

and

$$\zeta = x \tag{A-110}$$

The probability that x and y lie within a specified region must equal the probability that z and ζ lie within the corresponding region found from Eqs.

(A-109) and (A-110). Thus

$$\int_z \int_\zeta f_{z\zeta}(z,\zeta) \, dz \, d\zeta = \int_x \int_y f_x(x)f_y(y) \, dx \, dy \tag{A-111}$$

Substituting Eqs. (A-109) and (A-110) into the right-hand side of Eq. (A-111), we obtain

$$\int_z \int_\zeta f_{z\zeta}(z,\zeta) \, dz \, d\zeta = \int_z \int_\zeta f_x(\zeta)f_y(z-\zeta) \, dz \, d\zeta \tag{A-112}$$

and

$$f_{z\zeta}(z,\zeta) = f_x(\zeta)f_y(z-\zeta) \tag{A-113}$$

The distribution for $f_z(z)$ is then found by integrating $f_{z\zeta}(z,\zeta)$ over all values of ζ. Since ζ is exactly x,

$$f_z(z) = \int_{-\infty}^{\infty} f_x(x)f_y(z-x) \, dx \tag{A-114}$$

Note that Eq. (A-114) appears in the form of the convolution integral. (See Chap. 2.) This implies that if we arbitrarily take the Fourier transforms of $f_x(x)$ and $f_y(y)$, the Fourier transform of $f_z(z)$ must be given by the product of the two transforms. This then gives us another (and frequently simpler) method of finding $f_z(z)$. Calling $G_x(t)$ the Fourier transform of $f_x(x)$ (t a new parameter), and $G_y(t)$ the corresponding transform of $f_y(y)$, we have

$$G_x(t) = \int_{-\infty}^{\infty} e^{jtx}f_x(x) \, dx \tag{A-115}$$

$$G_y(t) = \int_{-\infty}^{\infty} e^{jty}f_y(y) \, dy \tag{A-116}$$

(We use e^{+jtx} rather than e^{-jtx}, as in Chap. 2, to conform with the notation of probability theory.) These new functions, Fourier transforms of the probability-density functions, are called *characteristic functions*. The characteristic function $G_z(t)$ of the variable z is then

$$G_z(t) = G_x(t)G_y(t) \tag{A-117}$$

and

$$f_z(z) = \frac{1}{2\pi} \int_{-\infty}^{\infty} e^{-jtz}G_z(t) \, dt \tag{A-118}$$

from Fourier-integral theory. (Note that the $1/2\pi$ now appears in front of the e^{-jtz} term rather than the e^{jtz} term as in Chap. 2.)

Recalling the definition of the average of a function $g(y)$ of a random variable y,

$$E[g(y)] = \int_{-\infty}^{\infty} g(y)f_y(y) \, dy \tag{A-119}$$

we see that the characteristic function is also the average of e^{jty}.

$$G_y(t) = E(e^{jty}) \tag{A-120}$$

We can now extend this procedure to the sum of n independent variables. For $z = x_1 + x_2 + \cdots + x_n$, we have

$$G_z(t) = E(e^{jtz}) = E(e^{jt(x_1+x_2+\cdots+x_n)})$$
$$= E(e^{jtx_1})E(e^{jtx_2}) \cdot \cdot \cdot E(e^{jtx_n})$$
$$= G_{x_1}(t)G_{x_2}(t) \cdot \cdot \cdot G_{x_n}(t) \tag{A-121}$$

(The average of the product of n independent quantities is the product of the averages.) The probability density of z is then

$$f_z(z) = \frac{1}{2\pi} \int_{-\infty}^{\infty} e^{-jzt}G_z(t) \, dt \tag{A-122}$$

with $G_z(t)$ given by Eq. (A-121).

Some examples will demonstrate the utility of the characteristic-function (c-f) method. They will also serve to illustrate some special cases of the central limit theorem.

Sum of n gaussian-distributed variables. Here $z = \sum_{i=1}^{n} x_i$,

$$f_{x_i}(x_i) = \frac{1}{\sqrt{2\pi\sigma_i^2}} e^{-(x_i-a_i)^2/2\sigma_i^2} \tag{A-123}$$

The Fourier transform of this gaussian function is itself a gaussian function. In particular it is left for the reader to show that

$$G_i(t) = e^{ja_i t}e^{-\sigma_i^2 t^2/2} \tag{A-124}$$

Then
$$G_z(t) = \prod_{i=1}^{n} G_i(t)$$

$$= e^{jat}e^{-\sigma^2 t^2/2} \tag{A-125}$$

where
$$a = \sum_{i=1}^{n} a_i$$

and
$$\sigma^2 = \sum_{i=1}^{n} \sigma_i^2$$

Comparing Eqs. (A-125) and (A-124),

$$f_z(z) = \frac{e^{-(z-a)^2/2\sigma^2}}{\sqrt{2\pi\sigma^2}} \tag{A-126}$$

The distribution of the sum of n gaussian-distributed variables is thus also gaussian, with an average value given by the sum of the individual average values and a variance given by the sum of the variances.

As a matter of fact it can be shown that the average value and variance of a random variable

$$z = \sum_{i=1}^{n} x_i$$

with the x_i's independent and of any distribution whatsoever, are given by the sum of the individual average values and variances, respectively.[12] The average fluctuation powers of n independent variables thus add directly. (Here fluctuation refers to variations about the average values.)

Uniform distribution (Fig. A-21).[13] For simplicity's sake we take the special case of a symmetrical uniform distribution. Letting $a = -b = -x_0/2$ in Fig. A-21,

$$f_{x_i}(x_i) = \frac{1}{x_0} \qquad \frac{-x_0}{2} < x_i < \frac{x_0}{2}$$

$$f_{x_i}(x_i) = 0 \qquad \text{elsewhere}$$

(A-127)

This is just the rectangular pulse of Chap. 2, and its Fourier transform is the familiar $(\sin x)/x$ function.

$$G_i(t) = \frac{\sin (tx_0/2)}{tx_0/2}$$

(A-128)

For the sum of n such variables, each assumed uniformly distributed over the *same range*,

$$f_z(z) = \frac{1}{2\pi} \int_{-\infty}^{\infty} \left[\frac{\sin (tx_0/2)}{tx_0/2} \right]^n e^{-jtz} \, dt$$

(A-129)

Consider now the characteristic function

$$G_z(t) = \left[\frac{\sin (tx_0/2)}{tx_0/2} \right]^n$$

This has the value unity at $t = 0$ and damps out rapidly away from the origin for large n. Most of the contribution to the integral thus comes from small values of t. We can expand this function in a power series about $t = 0$ and get

$$G_z(t) = \left[\frac{\sin (tx_0/2)}{tx_0/2} \right]^n = \frac{[tx_0/2 - (tx_0)^3/48 + \cdots]^n}{(tx_0/2)^n}$$

$$= \left[1 - \frac{(tx_0)^2}{24} + \cdots \right]^n$$

$$= 1 - \frac{n}{24} (tx_0)^2 + \cdots$$

(A-130)

---

[12] This is readily shown by taking the expectation of both left- and right-hand sides of $z = \sum_{i=1}^{n} x_i$. Then $E(z) = \sum_i E(x_i)$. Similarly, writing $[z - E(z)]^2 = \left\{ \sum_i [x_i - E(x_i)] \right\}^2$, taking the expectation, and recalling that the x_i's are independent, one finds $\sigma_z^2 = \sum_i \sigma_{x_i}^2$. The details are left for the reader to work out.

[13] W. R. Bennett, "Methods of Solving Noise Problems," *Proc. IRE,* vol. 44, no. 5, pp. 609–638, May 1956.

The first two terms in this power series are the same as those in the series for the exponential

$$e^{-(n/24)(tx_0)^2} = 1 - \frac{n}{24}(tx_0)^2 + \cdots$$

Using the exponential as an approximation to the $[(\sin x)/x]^n$ function in the vicinity of the origin, we have

$$G_z(t) = e^{-(n/24)(tx_0)^2} \qquad n \text{ large, } t \text{ small} \tag{A-131}$$

But this is just a gaussian-type function, and its Fourier transform must also be gaussian. In particular, comparing with Eqs. (A-125) and (A-126), we must have

$$f_z(z) \doteq \frac{1}{\sqrt{2\pi\sigma^2}} e^{-z^2/2\sigma^2} \tag{A-132}$$

where

$$\sigma^2 = \frac{n(x_0)^2}{12} \tag{A-133}$$

We showed previously that the standard deviation of the rectangular distribution was $(b - a)/\sqrt{12} = x_0/\sqrt{12}$ in the present notation. σ^2 here is then just the *sum* of the individual variances as noted in the previous example.

The gaussian approximation is plainly incorrect for large values of z. z can have no values greater than $nx_0/2$; yet the gaussian distribution predicts a finite probability of such values being attained. For large n, however, these values move out to the far tail of the gaussian curve, and the gaussian approximation becomes valid over most of the range of z. In general, just as in the case of the distribution of the envelope of signal plus noise in Chap. 5, the gaussian approximation is valid only in the vicinity of the peak ($z = 0$ here) and definitely not valid for values of z close to $nx_0/2$.

As noted by Bennett[14] this result for the sum of n uniformly distributed variables may be applied to the case of determining the quantizing noise resulting when a number of alternate analog-to-digital and digital-to-analog signal conversions are performed.

Both this example and the previous one are special cases of the central limit theorem.

Random-phase distributions. The final illustrative example concerns the sum of n independent sine waves of random phase. Thus, given

$$x_i = a_i \sin \theta_i$$

with θ_i uniformly distributed over 2π radians and a_i an arbitrary constant, we would like to find $f_z(z)$ for

$$z = \sum_{i=1}^{n} x_i$$

(This is sometimes used as a model for random noise.)

[14] *Ibid.*

We recall that as an example of the calculation of the probability density of a transformed variable, we showed that the density function of x_i was

$$f_{x_i}(x_i) = \frac{1}{\pi} \frac{1}{\sqrt{a_i - x_i^2}} \qquad |x_i| < a_i \qquad \text{(A-134)}$$

The characteristic function $G_i(t)$ is just

$$G_i(t) = \int_{-a_i}^{a_i} \frac{e^{jtx_i}}{\sqrt{a_i^2 - x_i^2}} \frac{dx_i}{\pi}$$

$$= J_0(a_i t) \qquad \text{(A-135)}$$

with $J_0(a_i t)$ the Bessel function of the first kind and zeroth order. This may be shown by the simple change of variables $x_i = a_i \sin \theta$, from which we obtain the integral definition of the Bessel function discussed in Chap. 4.

Alternatively, and much more simply in this case, we can use the fact that $G_i(t) = E(e^{jtx_i})$ directly. Instead of averaging over x_i we can also write $G_i(t) = E(e^{jta_i \sin \theta_i})$ and average over all values of θ_i. Since θ_i is uniformly distributed, this gives

$$G_i(t) = \frac{1}{2\pi} \int_{-\pi}^{\pi} e^{jta_i \sin \theta_i} d\theta_i = J_0(a_i t) \qquad \text{(A-136)}$$

The characteristic function $G_z(t)$, for the sum of the n variables, is given by

$$G_z(t) = \prod_{i=1}^{n} J_0(a_i t) \qquad \text{(A-137)}$$

Now assume, as in the previous example, that the amplitudes a_i are all very small and t not too large. For $a_i t \ll 1$ we note, in Chap. 4,

$$J_0(a_i t) \doteq 1 - \tfrac{1}{4}(a_i t)^2 = e^{-(1/4)a_i^2 t^2} \qquad \text{(A-138)}$$

using the exponential approximation again. $G_z(t)$ now becomes, from Eq. (A-137),

$$G_z(t) \doteq e^{-(1/2)\sigma^2 t^2} \qquad \text{(A-139)}$$

where
$$\sigma^2 = \tfrac{1}{2}(a_1^2 + a_2^2 + \cdots + a_n^2) \qquad \text{(A-140)}$$

Equation (A-140) represents the characteristic function of a normal distribution with zero average value and variance σ^2.

$$f_z(z) = \frac{1}{\sqrt{2\pi\sigma^2}} e^{-z^2/2\sigma^2} \qquad \text{(A-141)}$$

where
$$\sigma^2 = \tfrac{1}{2}(a_1^2 + a_2^2 + \cdots + a_n^2)$$

Note that the term $a_i^2/2$ represents the average power in a sine wave of amplitude a_i. σ^2 is thus the total mean power in the sum of n sine waves of incommensurable frequencies.

Equation (A-141) is again an example of the central limit theorem. Again

care must be exercised in applying Eq. (A-141) to the calculation of probabilities. It is valid only for large n and in the vicinity of $z = 0$. It is again
obviously incorrect for

$$z > \sum_{i=1}^{n} a_i$$

since the probability of this happening should be zero, while Eq. (A-141) predicts a finite probability. However, for n large, $f_z(z)$ falls off extremely rapidly at
large values of z, and any errors become negligible if we stay away from the tail
of the $f_z(z)$ curve.

A-8 DEPENDENT RANDOM VARIABLES

We have devoted considerable space in the past few sections to a discussion of
independent random variables. It is of interest to extend our discussion of
probability theory to include the case of dependent variables.

Dependent discrete variables were considered in Sec. A-2. It was shown
there, by example, that the study of dependent variables was important in
studying the statistical properties of a language. (This is also alluded to in
Chap. 1.) In general, as noted in Chap. 1, there is usually some dependence
between successive signals of a message being transmitted. This reduces the
rate of transmission of information.

In any study of communication in the presence of noise this dependence of
successive voltages on one another must be taken into account. Consider, for
example, the noise-voltage plot of Fig. A-30. We have previously discussed the
chance of noise at any one instant of time exceeding an arbitrary voltage level.

We are now interested in describing the joint statistics of two random
variables, such as n_1 and n_2 in Fig. A-30, separated by a specified time interval.
In the previous section we commented briefly on two such variables [see $n(t)$
and $n(t + \tau)$ in Fig. A-29], but focused only on the case in which these two
variables are spaced "far enough apart" so that they may be considered unre-
lated or independent.

Yet the band-limiting effect of any system through which noise or signal
plus noise are passed will ensure that there is some connection or correlation
between two closely spaced voltages. This in turn must be taken into considera-

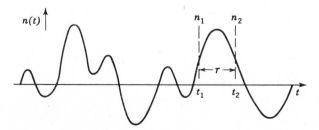

Figure A-30 Noise voltage.

tion in discussing in any thorough manner the statistics of signals in noise. This we do in Chap. 5. We therefore discuss to some extent here the concept of joint statistics of two continuous dependent variables and a particular measure—the correlation function—of this dependency. These concepts can be extended to include dependency between many random variables, as was done with the sum of n independent variables in the previous section. We shall restrict ourselves to two dependent variables, however, in this book.

Consider two random variables x and y. By performing repeated measurements on these two variables jointly, as in the case of the discrete variables, we can develop a set of histogram curves relating the frequency of occurrence of different pairs of x and y. (In the example of Fig. A-30, with $x = n_1$ and $y = n_2$ this would consist of measurements of the different combinations of pairs of voltages occurring.) From such a set of measurements, taken with x broken up into levels Δx units apart and y into levels Δy apart, a joint probability-density function $f(x,y)$ can be constructed. This is the same function introduced in the previous section for the discussion of two independent variables.

$f(x,y)\ dx\ dy$ represents the probability that x and y will jointly be found in the ranges $x \pm dx/2$ and $y \pm dy/2$, respectively. Since some range of x and y must always occur jointly, we must have

$$\int\!\!\int_{-\infty}^{\infty} f(x,y)\ dx\ dy = 1 \qquad (A\text{-}142)$$

The joint density function $f(x,y)$ is thus normalized so that the volume under the curve is 1. As noted in the previous section, $f(x,y)$ may be written as the product $f_x(x)f_x(y)$ if x and y are independent.

We can define a cumulative-probability function as in the case of one random variable, if we wish, and averages can also be computed. For example, the average of $G(x,y)$ is

$$E[G(x,y)] = \int\!\!\int_{-\infty}^{\infty} G(x,y)f(x,y)\ dx\ dy \qquad (A\text{-}143)$$

In particular the averages of x and y raised to integer powers are again called the moments of x and y and are given by

$$m_{ij} = E(x^i y^j) = \int\!\!\int_{-\infty}^{\infty} x^i y^j f(x,y)\ dx\ dy \qquad (A\text{-}144)$$

These moments play an important role in two-variable statistical theory, just as do the moments in the one-variable case. As special cases, we have

$$m_{10} = x_0 = E(x) = \int\!\!\int_{-\infty}^{\infty} xf(x,y)\ dx\ dy$$

$$= \int_{-\infty}^{\infty} xf_x(x)\ dx \qquad f_x(x) = \int_{-\infty}^{\infty} f(x,y)\ dy \quad (A\text{-}145)$$

$$m_{01} = y_0 = E(y) = \int\limits_{-\infty}^{\infty}\!\!\int yf(x,y)\ dx\ dy$$

$$= \int_{-\infty}^{\infty} yf_y(y)\ dy \qquad f_y(y) = \int_{-\infty}^{\infty} f(x,y)\ dx \quad \text{(A-146)}$$

$$m_{11} = E(x,y) = \int\limits_{-\infty}^{\infty}\!\!\int xyf(x,y)\ dx\ dy \qquad\qquad \text{(A-147)}$$

$$m_{20} = E(x^2) = \int_{-\infty}^{\infty} x^2 f_x(x)\ dx \qquad\qquad\qquad \text{(A-148)}$$

$$m_{02} = E(y^2) = \int_{-\infty}^{\infty} y^2 f_y(y)\ dy \qquad\qquad\qquad \text{(A-149)}$$

Just as in the one-variable case we can also define central moments or moments about the average values. For example, the central moments of the second order, comparable with μ_2, are

$$\mu_{20} = E[(x - x_0)^2] = \int_{-\infty}^{\infty} (x - x_0)^2 f_x(x)\ dx.$$

$$= m_{20} - m_{10}^2 = \sigma_x^2 \qquad\qquad \text{(A-150)}$$

$$\mu_{02} = E[(y - y_0)^2] = \int_{-\infty}^{\infty} (y - y_0)^2 f_y(y)\ dy$$

$$= m_{02} - m_{01}^2 = \sigma_y^2 \qquad\qquad \text{(A-151)}$$

and
$$\mu_{11} = E[(x - x_0)(y - y_0)] = E(xy) - E(x)E(y)$$

$$= m_{11} - m_{01}m_{10} \qquad\qquad \text{(A-152)}$$

as may be verified by the reader.

m_{11} and μ_{11} are the two second-order moments which serve as a measure of the dependence of two variables. For, as noted previously, two *independent* variables will have as their joint distribution function the *product* of the individual distribution functions, or

$$f(x,y) = f_x(x)f_y(y) \qquad\qquad\qquad \text{(A-153)}$$

For this case m_{11} and μ_{11} become, respectively,

$$m_{11} = E(x)E(y) = m_{10}m_{01} \qquad\qquad \text{(A-154)}$$

and
$$\mu_{11} = 0 \qquad\qquad\qquad\qquad\qquad \text{(A-155)}$$

μ_{11} is called the *covariance* of the two variables. σ_1^2 and σ_2^2 are again called *variances*. If the two variables are independent, the covariance is zero, and m_{11}, the average of the product, becomes the product of the individual averages. (The converse of this statement is not true in general but does hold for the joint gaussian distribution.)

We may define a normalized quantity called the *normalized correlation coefficient* which serves as a numerical measure of the dependence between two variables. Using the symbol ρ for this quantity, we define

$$\rho \equiv \frac{\mu_{11}}{\sqrt{\mu_{20}\mu_{02}}} = \frac{\mu_{11}}{\sigma_1\sigma_2} \tag{A-156}$$

The correlation coefficient can be shown to be less than 1 in magnitude: $-1 \le \rho \le 1$. In particular assume first that y is completely determined by x and that a linear relation of the form $y = \alpha x + b$ exists between the two. Then $E(y) = \alpha E(x) + b$, or $y_0 = \alpha x_0 + b$. μ_{11} becomes

$$E[(x - x_0)(y - y_0)] = \alpha E[(x - x_0)]^2 = \alpha\mu_{20}$$

Also, $\qquad \mu_{02} = E[(y - y_0)^2] = \alpha^2 E[(x - x_0)^2] = \alpha^2\mu_{20}$

Then $\rho = \pm 1$, depending on the sign of α. On the other hand, if x and y are independent, $\mu_{11} = 0$ and $\rho = 0$. So values of ρ close to 1 indicate high correlation; values close to 0 indicate low correlation.

The covariance function μ_{11} defined by Eq. (A-152), or its normalized parameter ρ [Eq. (A-156)], thus serves as a measure of the dependence of two random variables on one another. This measure of statistical dependence is elaborated on in detail in Chap. 5 in discussing random processes (time functions) and exploring the dependence of samples in time (e.g., n_1 and n_2 in Fig. A-30) on one another.

The two-variable gaussian distribution serves as an example of the joint distribution function. It is given by the expression[15]

$$f(x,y) = \frac{1}{2\pi M} \exp\left[-\frac{1}{2M^2}(\mu_{02}x^2 - 2\mu_{11}xy + \mu_{20}y^2)\right]$$

$$= \frac{1}{2\pi M} \exp\left[\frac{-1}{2(1 - \rho^2)}\left(\frac{x^2}{\sigma_x^2} - \frac{2\rho xy}{\sigma_x\sigma_y} + \frac{y^2}{\sigma_y^2}\right)\right] \tag{A-157}$$

for zero average values ($x_0 = y_0 = 0$). Here

$$M^2 = \mu_{20}\mu_{02} - \mu_{11}^2 = \sigma_x^2\sigma_y^2(1 - \rho^2)$$

Note that if $\rho = 0$ ($\mu_{11} = 0$), $f(x,y)$ becomes the product of two single-variable distribution functions, indicating in this case that x and y are independent. For nonzero x_0 and y_0, x and y are replaced by $x - x_0$ and $y - y_0$, respectively. The two-dimensional gaussian density function is shown sketched in Fig. A-31.[16]

The probability that x and y will be found in the ranges x_1 to x_2, y_1 to y_2, respectively, is just the volume under the curve of Fig. A-31, enclosed by these points. For such a calculation the variances σ_1^2, σ_2^2 and the correlation coefficient ρ must obviously be known from previous information.

[15] A. Papoulis, *Probability, Random Variables, and Stochastic Processes*, McGraw-Hill, New York, 1965.

[16] Bennett, *op. cit.*, fig. 3.

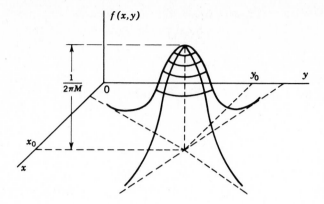

That the parameters in Eq. (A-157) satisfy the defining relations for the moments expressed by Eqs. (A-145) to (A-152) may be verified by evaluating the appropriate integral. For example, assume that $E(x)$ or m_{10} and μ_{20} or σ_x^2 are to be found for the gaussian distribution of Eq. (A-157). Since we are interested only in average values of x, independent of variations in y, we first integrate over all values of y to find the distribution in x alone. This is given by

$$f_x(x) = \int_{-\infty}^{\infty} f(x,y) \, dy = \frac{\exp\{-x^2/[2(1-\rho^2)\sigma_x^2]\}}{2\pi M}$$

$$\times \int_{-\infty}^{\infty} \exp\left[\frac{-1}{2(1-\rho^2)}\left(\frac{y^2}{\sigma_y^2} - \frac{2\rho xy}{\sigma_x \sigma_y}\right)\right] dy \quad \text{(A-158)}$$

The integral may be evaluated by first completing the square in the exponent. Thus

$$\frac{y^2}{\sigma_y^2} - \frac{2\rho xy}{\sigma_x \sigma_y} = \left(\frac{y}{\sigma_y} - \frac{\rho x}{\sigma_x}\right)^2 - \frac{\rho^2 x^2}{\sigma_x^2}$$

$f_x(x)$ now becomes

$$f_x(x) = \frac{\exp(-x^2/2\sigma_x^2)}{2\pi M} \int_{-\infty}^{\infty} \exp\left[-\frac{(y/\sigma_y - \rho x/\sigma_x)^2}{2(1-\rho^2)}\right] dy$$

Using the transformation of variables,

$$z^2 = \frac{(y/\sigma_y - \rho x/\sigma_x)^2}{2(1-\rho^2)}$$

and recalling that $\int_{-\infty}^{\infty} e^{-z^2} \, dz = \sqrt{\pi}$, we get finally

$$f_x(x) = \frac{e^{-x^2/2\sigma_x^2}}{\sqrt{2\pi\sigma_x^2}} \quad \text{(A-159)}$$

as expected from our previous discussion of the one-variable or one-dimensional gaussian function. (Use has also been made here of the defining relation for M.)

From our knowledge of the gaussian function we recognize that $E(x) = 0$ and $\mu_{20} = \sigma_x^2$ for this example.

In a similar manner, $f_y(y)$, μ_{02}, $E(y)$, etc., may be found. These are left as exercises for the reader.

PROBLEMS

A-1 An urn contains two white balls and six red ones. What is the probability that one ball drawn at random will be white?

A-2 An analysis of a long message transmitted in binary digits shows 3,000 zeros and 7,000 ones. What is the probability that any one digit in the message is a zero?

A-3 A selected list of words in the English language is to be transmitted by means of a binary code, with each word represented by 12 binary digits or fewer. How many words can there be on the list?

A-4 The numbers 1 through 10 are selected at random.
   (a) What is the probability that the numbers are selected in the order 1, 2, . . . , 10?
   (b) What is the probability of selecting the number 2 right after the number 1?

A-5 (a) Find the probability of getting a 7 in the toss of two dice.
   (b) Find the probability of throwing a 6, 7, or 8 with two dice.

A-6 What is the probability that 4 cards drawn in succession from a deck of 52 will be aces?

A-7 A box contains five white balls, three red balls, and two black ones. What is the probability that two balls drawn from the box will both be red?

A-8 What is the probability of obtaining four tails if four coins are tossed? What is the probability that at least three heads will appear?

A-9 Calculate the conditional and joint probabilities for the typical message of Shannon's three-letter alphabet at the end of Sec. A-2. Check with the probability tables shown there.

A-10 Refer to Fig. A-5. Express the subsets AC and ABC in words. Show they are in fact given by the regions shown in Fig. A-5d.

A-11 Using a Venn diagram prove the following set identities:

$$A\bar{B} + AB = A$$

$$A(B + C) = AB + AC$$

$$\overline{A + B} = \bar{A}\bar{B}$$

$$\overline{AB} = \bar{A} + \bar{B}$$

A-12 Let x represent the age of husbands, y the age of wives. The set S consists of all possible elements corresponding to $\{15 < x,y < 80\}$. Let A represent the subset "husbands between the ages of 20 and 40," B, the subset "wives between 20 and 40," and C, the subset "husbands older than wives."
   (a) Indicate S, A, B, C on a rectangular xy plot.
   (b) Describe the following subsets in words, sketching them on the xy plot as well: AB, $\overline{AB}$, $\bar{A}$, $\bar{B}$, $\bar{A} + \bar{B}$, AC, $A + C$, ABC.

A-13 Two coins are tossed with the four possible outcomes HT, HH, TT, TH (H stands for head, T for tails). Each outcome is assumed equally likely. Let set S consist of these four elements. Indicate the probability of occurrence of, and elements contained within, the following subsets:

> A : coin 1 comes up head
> B : coin 2 comes up head
> C : coin 1 is a tail
> $A + C, AB, A + B$

Express the latter three subsets in words.

A-14 Consider a binary PCM system transmitting 1's with a probability $P_1 = 0.6$ and 0's with a probability $P_0 = 0.4$. The receiver recognizes 0's, 1's, and a third symbol E, called an erasure symbol. There is a probability $P(0|1) = 0.1$ that the 1's will be received (mistakenly) as 0's, $P(E|1) = 0.1$ that they will be received as E's, and $P(1|1) = 0.8$ that they will be received (correctly) as 1's. Similarly, with 0 assumed transmitted, the appropriate probabilities of events at the receiver are given by $P(1|0) = 0.1$, $P(E|0) = 0.1$, $P(0|0) = 0.8$.

(a) Sketch a diagram indicating two transmit and two receive levels, show the appropriate transitions between them, and indicate the appropriate probabilities. *Note:* In the symbolism used above all conditioning refers to the *transmitter*.

(b) Calculate the probability of receiving a 0, a 1, and an E, respectively. Show these sum to 1, as required.

(c) Show the probability of an error is 0.1, the probability of a correct decision at the receiver is 0.8, and the probability of an erasure is 0.1.

(d) Repeat (b) and (c) if the transition probabilities, with a 0 transmitted, are changed to $P(1|0) = 0.05$, $P(E|0) = 0.05$, $P(0|0) = 0.9$.

A-15 Refer to Prob. A-14. The symbol 1 is received. What is the probability it came from a 0? From a 1? Repeat for the symbols 0, and E, as received. (It may pay to adopt new symbols such as T_0, T_1, and R_0, R_1, R_E, or A_1, A_2, and B_1, B_2, B_3 to keep the appropriate conditional probabilities straight.) Check your results by summing appropriate probabilities.

A-16 Show that the probability of finding a continuous random variable x somewhere in the range x_1 to x_2 is given by $F(x_2) - F(x_1)$, with $F(x)$ the cumulative-distribution function [see Eq. (A-32)].

A-17 Tabulate the probabilities of getting the numbers 2 through 12 on the toss of two dice. Plot these to scale.

A-18 Consider a discrete variable with the probabilities $P_1, P_2, \ldots, P_n$ corresponding to its n possible values. Show that the relation between the variance, or second central moment μ_2, and the first and second moments m_1 and m_2, respectively, is given by $\mu_2 = m_2 - m_1^2$.

A-19 A given time interval T seconds long is divided into $n = T/\Delta T$ time slots each ΔT seconds long. Assume the chance of a phone call occurring in any one ΔT seconds subinterval is $p \ll 1$, and that calls in adjacent subintervals occur independently.

(a) What is the probability of k phone calls occurring in T seconds? At least k calls in T seconds? Relate this problem to the binomial distribution discussed at the end of Sec. A-4 in connection with binary transmission.

(b) What is the average number of phone calls per second, and the standard deviation about this? (See Sec. A-5.)

A-20 *Electron emission.* A heated cathode has a probability $p \ll 1$ of emitting one electron in ΔT seconds.

(a) What is the probability of emitting k electrons in n such ΔT-second time intervals if electron emissions are independent of one another? Compare with Prob. A-19.

(b) What is the average number of electrons emitted per second? If each electron carries e coulombs, what is the average (dc) emission current?

A-21 (a) Consider a binomial distribution with $n = 4$ and $p = 0.4$. Plot the probability $P(k)$ versus k, as well as the cumulative distribution $F(m) = \text{Prob } [k \le m]$.

(b) Repeat for $n = 10$ and $p = 0.4$.

A-22 *Poisson distribution.* A discrete random variable x takes on positive integer values 0, 1, 2, . . . , only. The probability that $x = k$ is then $P(x = k) = a^k e^{-a}/k!$, $k = 0, 1, 2, \ldots$.

(a) $a = 0.5$. Plot $P(x = k)$ and the cumulative-distribution function $F(m) \equiv P(x \leq m)$.

(b) Repeat for $a = 2$.

(c) Show that $F(\infty) = 1$, independent of a.

(d) Show that $E(k) = a$, $\sigma^2 = a$, $\sigma/E(k) = 1/\sqrt{a}$.

A-23 Photons in a laser communication system are emitted with a Poisson distribution. This means that if on the average a is emitted in an arbitrary interval, the probability of k being emitted in the same interval is $a^k e^{-a}/k!$ As noted in the book, the ratio $\sigma/E(k)$ is a measure of the relative fluctuation or spread about the average number emitted. Calculate this ratio for $a = 10^6$, 10^{10}, 10^{14}. (See Prob. A-22d.)

A-24 (a) Show that the distribution of the number of heads appearing in the tossing of a fair coin n times is given by the binomial distribution with $p = q = \frac{1}{2}$.

(b) What is the average number of heads expected in 1,000 tosses, in 10^6 tosses? What is the relative spread, $\sigma/E(k)$, about the average number expected in the two cases? What does this imply about the relative occurrence of heads as the number of tosses gets larger and larger?

A-25 n fair coins are tossed simultaneously.

(a) What is the probability of k heads appearing?

(b) What is the average number of heads expected? Compare with Prob. A-24.

A-26 Consider the function $f(x) = kxe^{-x^2/2N}$, with k and N positive constants.

(a) To what range of values must x be restricted to ensure that the function represents a possible probability-density function?

(b) Determine k such that $f(x)$ is properly normalized. The resultant function is called the Rayleigh distribution and is discussed in Chap. 5. Plot $f(x)$ versus $x/\sqrt{N}$.

(c) What is the probability that a variable x obeying the Rayleigh distribution will have values between $x = \sqrt{2N}$ and $x = 2\sqrt{N}$?

(d) Sketch the cumulative distribution $F(x)$.

A-27 Calculate the mean value m_1 and the standard deviation σ of the Rayleigh distribution of the previous problem.

A-28 Consider the triangular probability-density function of Fig. PA-28.

(a) Find b in terms of a so that the function is properly normalized.

(b) Calculate the mean value and standard deviation of the distribution.

(c) Plot the cumulative-distribution function $F(x)$.

(d) What is the probability that x will be greater than $a/2$?

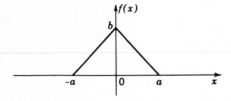

Figure PA-28

A-29 The exponential distribution is given by

$$f(x) = \mu e^{-\mu x} \qquad x \geq 0 \text{ only}$$

(a) Show that $f(x)$ is properly normalized.

(b) Find $E(x)$ and σ for this distribution.

(c) Sketch both $f(x)$ and the cumulative distribution $F(x)$.

A-30 Find the variance of the gaussian, or normal, distribution by evaluating the integral of Eq. (A-80).

A-31 Show that the gaussian function of Eq. (A-77) is properly normalized.

A-32 Show that the cumulative probability distribution for the gaussian function is given in terms of the error function by Eq. (A-88). Plot $F(x)$, using any available table of the error function.

A-33 In the manufacture of resistors nominally rated at 100 Ω it is found that the probability distribution of actual resistance values is very closely given by a normal distribution with standard deviation $\sigma = 5$ Ω. What percentage of the resistors manufactured lie within the range 90 to 110 Ω?

A-34 Two voltmeters are available for a given experiment, and either one may be chosen at random. Depending upon temperature, humidity, etc., the readings of the voltmeters tend to deviate from the true voltage values. (The additional effect of fluctuation noise is assumed negligible here.) The distribution of the readings about each scale value follows a normal error curve. For one voltmeter the standard deviation is 0.1 V, for the other 0.2 V. What is the probability that any voltage read will be within 0.2 V of its true value?

A-35 A density function is given by $f(x) = 1, 0 \leq x \leq 1$. Show that the density function of $y = -\ln x$ is given by the function of Prob. A-29, with $\mu = 1$.

A-36 Show that the density function of Eq. (A-99) is properly normalized.

A-37 Show that the average value of the distribution of Eq. (A-103) is σ^2.

A-38 The random variable x follows the Rayleigh distribution of Prob. A-26. Find the distribution of the square of x. Compare with Prob. A-29.

A-39 Show that the Fourier transform of a gaussian function is itself a gaussian function. In particular, show that if a probability-density function is the gaussian function of Eq. (A-123), its characteristic function is given by Eq. (A-124).

A-40 Find the characteristic function for the probability-density function of Prob. A-29. Using this result, find the probability density for the sum of n such independent random variables. *Hint:* Find the characteristic function for the sum. The probability density is then the inverse transform. To find this note that $\dfrac{1}{n!} \displaystyle\int_0^\infty x^n e^{-ax} \, dx = 1/a^{n+1}$, n a positive integer and $a > 0$.

A-41 Let

$$z = \sum_{i=1}^{n} a_i x_i$$

with the a_i's random variables.

    (*a*) Show

$$E(z) = \sum_{i=1}^{n} a_i \, E(x_i)$$

    (*b*) The x_i's are independent random variables. Show

$$\sigma_z^2 = \sum_{i=1}^{n} a_i^2 \sigma_{x_i}^2$$

Note: Actually, as will be shown in Prob. A-43, the condition that the x_i's are *uncorrelated* suffices.

A-42 A random variable x represents the input voltage, at a specified time, to each of the four devices shown in Fig. PA-42. y is the corresponding random output voltage. Find and sketch the probability-density function of y for each of the following input-density functions:

    (*a*) $f(x) = 1, 0 < x < 1$.
    (*b*) $f(x) = e^{-x}, x > 0$.
    (*c*) $f(x) = (1/\sqrt{2\pi})e^{-x^2/2}$.

Check by calculating the area under each of the output-density functions.

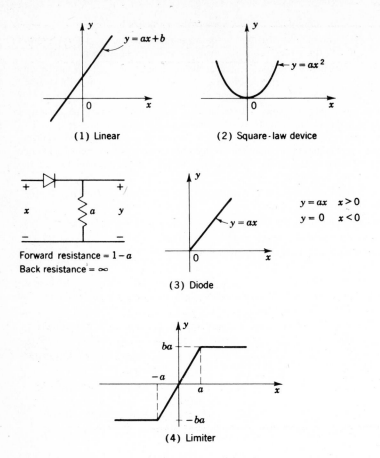

Figure PA-42

A-43 n random variables, x_i, $i = 1, \ldots, n$, are pairwise uncorrelated so that $E[x_i x_j] = E(x_i)E(x_j)$, $i \neq j$. Show that the variance of the sum

$$z = \sum_{i=1}^{n} a_i x_i$$

is given by

$$\sigma_z^2 = \sum_{i=1}^{n} a_i^2 \sigma_{x_i}^2$$

A-44 The two-variable (or two-dimensional) gaussian density function is given by Eq. (A-157). Find the one-dimensional density function $f_y(y)$, $E(y)$, and $\mu_{02} \equiv E[y - E(y)]^2$, and show they agree with previous (one-dimensional) results. Using the definition of the covariance μ_{11} given in Eq. (A-152), verify by actual integration that μ_{11} in Eq. (A-157) is appropriately labeled as such.

INDEX